Rubber - A Story of Glory and Greed

Editors:
Howard Wolf
Ralph Wolf

RubberWorld

SMITHERS
*i*Smithers

*i*Smithers – A Smithers Group Company

Shawbury, Shrewsbury, Shropshire, SY4 4NR, United Kingdom
Telephone: +44 (0)1939 250383 Fax: +44 (0)1939 251118
http://www.rapra.net

First Published in 2009 by
J.J. Little and Ives Company, New York

*i*Smithers
Shawbury, Shrewsbury, Shropshire, SY4 4NR, UK

Reprinted, with revisions, 2009

A catalogue record for this book is available from the British Library.

ISBN (Hard-backed): 978-1-84735-375-7
e-book: 978-1-84735-376-4

Typeset by *Argil Services*
Printed and bound by Lightning Source

Contents

Introduction

This is the first attempt at a history of rubber, the raw material; the first attempt at a history of rubber invention and research; the first attempt at a history of rubber as big business - manufacturing, distribution, finance, labor. Neither in popular, scholarly nor technical fashion has anyone of the three stories making the story of rubber been handled in the large before.

That there is a need for a general history of rubber is a point that needs no laboring in a modern world dependent upon *caoutchouc* in the matter of its transportation, communications and electric power. Whether the present work is well or badly done, then, we like to think that it will be of some value until others come along.

Our target has been the general reader; our hope has been that we might be both entertaining and accurate. If we have failed in the first it is entirely our own fault, for rubber's record is so packed with the odd, the melodramatic and the tragic that writer neglect of the subject is inexplicable. If we have failed in the second it has not been because of failure to consult the authorities - where authorities existed. One venture into conjecture there is in this book, a fanciful account of rubber's discovery, and that is so labeled.

On rubber there is, of course, an immense technical library. This is especially true of its chemistry, but there is also considerable literature dealing with physics, botany, practical planting, processes of manufacture and suggested uses. Some of these works are encyclopedic in scope and offer a brief chapter of history. The majority are specialised studies of some one property or technical application of the tree milk. We have found a number of them extremely valuable.

Forming a little classification of their own are the books which escape the technical listing but make their appeal to those specifically interested in rubber rather than to the average reader. This catalogue includes the first-hand reports on a couple of the old wild rubber lands by J. F. Woodroffe and several others but is dominated by the personal chronicles of the founding fathers: Charles Goodyear, the Connecticut Yankee on whose discovery of vulcanisation the whole structure of the rubber industry is reared; Thomas Hancock, greatest of the English inventors; Henry A. Wickham, chief physical agent in the transferring of the rubber crop from one hemisphere to another;

John Boyd Dunlop, who re-invented the pneumatic tire as the need for it arose. Not commonly met with are any of these source books despite the fact that Hancock's little volume was reprinted in England as an advertising device some years ago. And one of them - Goodyear's - is a museum item of which but a dozen copies are known. Our perusal of this *caoutchouc* Old Testament was under the stern auspices of the rare book room in the Library of Congress at Washington, but such formal approach will shortly be no longer necessary. In April of 1936 the weekly *India-Rubber Journal (London)* began the installment publication of the entire work in photographic reproduction. And it will issue this facsimile in book form some time next year.

Books on one or more aspects of rubber aimed at the general reader form but a meager list - half a dozen in stiff covers, chiefly corporation-subsidised, and a variety of pamphlets mostly from the typewriters of the publicity men. Two of the books rate mention. They are the non-subsidised *The Reign of Rubber* (1922) by W. C. Geer and the frankly sponsored *The House of Goodyear* (1936) by Hugh Allen. Dr. Geer, rubber scientist and former rubber company executive, deals in popular fashion with the variety of rubber uses and with processes of manufacture. Allen, long time rubber factory press agent, argues his employer company's case in a detailed corporation history that makes for instructive reading even when you disagree with his every premise and conclusion.

The pamphlets are mostly exercised over the Romance of Rubber, a catch phrase apparently tracing to the belief that since *caoutchouc* comes from the globe's far corners, the business must be romantic as all hell. On that basis, of course, the salted nut industry is Frederic March. We ourselves have been increasingly impressed with the Tragedy of Rubber.

So much for *caoutchouc's* formal library. A wealth of material we found in the periodical files, not previously mined, and especially in the older trade journals, *India Rubber World* (New York) and *India Rubber Journal*. And indispensable were the assorted publications of the United States Department of Commerce ranging from book-thick surveys of the rubber lands to its semi-monthly *Rubber News Letter*.

A goodly part of our bibliography is composed of books not specifically concerned with rubber, books referring to rubber but briefly or incidentally yet containing information not available in caoutchouc's own library.

Our Amazon chapters, for instance, represent a synthesis of material gleaned from the writings of non-elastic explorers, travelers, naturalists and historians as well as from works primarily dealing with *caoutchouc* in one locality or another. It is, we believe, the first connected story presenting the whole panorama of rubber in South America. So with the chapters on the plantation lands which represent a similar fitting together of jig-saw pieces.

The Congo land atrocities, again, have given rise to a literature running into many thousands of items but publication of the present volume marks the first appearance of that exploitation story in any book on rubber. Nor has the Putumayo story before been fitted into place as part of a *caoutchouc* chronicle. A stack of government publications it has all to itself, however, and several books, dealing exclusively with the atrocities, bowed in when it was a hot story in those forgotten days before the war. The present version is the first complete and compact one, incidentally, since it includes all of the important information developed at the Parliamentary inquiry in London. These angles we have quarried from the British Government Blue Book containing minutes of evidence running to six hundred thousand words and obtainable in that form only.

Continuing the three-sheeting, this is the first book of any sort to deal with: Father Julius A. Nieuwland and his discovery of synthetic rubber; the history of the Rubber Trust; the Federal Trade Commission case against the Goodyear Tire - Sears, Roebuck contract; the story of organised labor in the rubber industry and its 'sitdown' strikes.

So far as the story of raw rubber is concerned this book is entirely a product of research into the written word. We have no first hand information on the old wild rubber lands and the new plantation world and we pretend to none. In dealing with the chemistry of rubber and synthetic rubber, the processes of manufacture, Akron boom and deflation, the tire distribution wars and the rubber labor situation, however, we have drawn on personal knowledge and observation as well as book, pamphlet and periodical. On some of these subjects, indeed, the bibliography is limited to newspaper and occasional magazine pieces, making it essential that other sources of information be tapped.

To the Akron rubber corporations — Goodyear, Goodrich, Firestone — we are indebted for specific data as requested and, in the case of Firestone and Goodyear, for free access to their libraries. Considering our past criticisms of them and their certainty that there was more to come, this expresses an attitude unusual, we believe, in industry.

For information and assistance thanks are due also to: Mr. E. G. Holt, Assistant Chief of the Leather and Rubber Division, Bureau of Foreign and Domestic Commerce, United States Department of Commerce; Mr. Estrada Lubin, Commissioner of Labor Statistics, United States Department of Labor; Mr. M. A. Roberts, Superintendent of Reading Room, Library of Congress; Mr. B. J. Widick, associate editor the *United Rubber Worker*; Mr. Jerry Shaw, editor *Tires*; Mr. John M. Pittenger, associate editor *Tire Review*; Mr. Charles Watney, London agent The Incorporated Society of Planters (Malaya); Mr. Frank G. Smith, secretary The Rubber Growers' Association, London; Mrs. Jean McClintock, Eastbourne, England; Mr. William A. Reid, the Pan American

Union; Mr. Tom May, Catholic University of America; Mr. V. A. Cosler, Wilmington, Delaware; Mrs. Hazel Fetzer, Mr. Eberhard Faber and Mr. Edward Thomas, New York; Mr. W. D. Hines, Miss Verna Neal, Mr. Clyde Schetter, Mrs. Ada Kennedy, Mr. Herburt W. Maxson, Mr. R. Emmett Powers, Mr. A. H. Englebeck, Mr. M. W. Conant, Akron; Mr. A. M. Smith, *the Detroit News*; Mr. Roy L. Adams, *the Buffalo Times*; Mrs. Leona Aiken and Miss Ruth Gibson, *the Akron Beacon Journal*; Miss Helen L. Purdum, Akron Public Library; Mr. L. H. Gerge1y, Cleveland Public Library; Mrs. David Ireland, Municipal University of Akron.

Akron, July 1936.

1 Genesis

In the beginning was the tree. And darkness was upon the face of the jungle. And man, created in God's own image, moved furtively upon the forest floor.

The narrow path he followed was a green and dripping tunnel paved with a mold that was the daily leaf drift of thousands of years, roofed with the foliage of the taller trees and the connecting parasite vines, walled with palm trunks, root buttresses, ferns, bushes, orchids, climbing and trailing plants. Underfoot were pincer-tipped scorpions, deadly spiders, ants whose sting is like the touch of fire, grasses that cut like knife-blades. Above might be the coiling boa constrictor; ahead, perhaps, padded the jaguar. There was good reason, then, for the stealthy progress of this Indian of the Amazon.

Man, it has been written, was formed of the dust of the ground. Of his ancestry on that side, the skulker in the labyrinth carried proof — his earth-brown color. He was naked and not ashamed. He should have been. Of the mean height, for instance, the horsetail hair, the low brow, the ridiculous paunch. This belly it was that was leading the citizen of Amazonia that day as every day. Behind was the temporary leaf-hut that was his nearest approach to home; ahead was the next meal to be found. Meanwhile, he was due for the pause that refreshes. The dull eyes looked about for a milk tree or a water-bearing vine to wound. They fell, instead, on a trunk oozing a creamy fluid drop by drop from some chance cut. The thirsty one applied tongue to gash and tested a drop. The taste was oddly bitter. The experimenter spat disgustedly and moved on. Not, however, until he had noted details of the pale trunk bulging at the top into a dense foliage of triple, long-stemmed, dark green leaves.

This South American aborigine had discovered the tree today called *Hevea brasiliensis*, the tree that today and for decades past has been the chief source of the world's supply of *caoutchouc*—rubber in the raw. And he had learned that all trunks yielding a milk-like fluid are not to be trusted. Nor was it difficult to remember this. The rubber tree insisted on calling itself to his attention. So peremptorily, in fact, that he was genuinely frightened the first time it happened. On the trail of another meal he was when a series of pistol-shot explosions halted him in his tracks and a projectile the size of a bird's egg pinged against his brown hide. He did not stay to investigate.

The next time he was less startled. Ultimately came familiarity with the explosions and discovery that it was the tree with the evil-tasting milk that had been waylaying him. There was curiosity in this dark, squat man despite the fractional forehead, the eyes where no light gleamed. And whether as tree or spirit, his fellow wood's denizen interested him. So from respectful spying he came to know of the annual shading of old leaves into new; of the arrival of the fruit, a three-seeded brown capsule, smooth, mottled and pulpy on the outside, hard within; of the drying of the fruit and its tensing as it dried until the day when the capsules exploded in a musketry rattle of shooting seeds landing sixty to a hundred feet away.

There came, then, the day when he picked up a fist-sized mass of the leaking fluid or latex that had spontaneously solidified at the foot of the trunk. Certainly, he was watched by the serpent, more subtle than any beast of the field, for the base of the *Hevea* is headquarters of the venomous Surucucu during seeding time. The naturalist's explanation is that the snake is wise enough to know that the rodent Agouti and many another walking meal presently will be along in response to the lure of the oily nuts that are irresistible to almost every four-footed jungle dweller. The godly reader will know better. He will know exactly why the serpent took up that station for the *Hevea* surely was the Amazon's tree of the knowledge of good and evil. Being an Indian, the Amazon Adam could not be expected to realise this, however. Monkey-inquisitive, he inspected the rubber apple, took a good sniff of its offensive odor and tossed the object from him. Up from the forest floor it bounced, a thing alive. From the vicinity of that spirit, small but patently potent and most probably malignant, rubber's discoverer removed himself with record celerity.

On the second occasion, he didn't run as far. The day he purposely bounced a rude sphere, fear of this devil died. And rubber went back with him to that day's shelters of his nomad tribe. There it was examined, marveled over, made much of for an hour. Something newer diverted general attention then and the bored botanist flung his find aside. A newcomer just emerging into the near-clearing saw the lump whizzing towards him. Instinctively, he threw hands before face and made an unintentional catch. All he wanted was to get rid of this unwelcome present. Promptly, he hurled it from him in the direction whence it had come. The ball struck its original owner, rebounded to the newcomer. And a great yelp of senseless mirth went up from the people of the leaves as they scrambled to clutch, to throw, to interpose body before ball. That was a great day in the history of the Western hemisphere - the discovery of play.

Now the rubber tree was a marked one. Jungle man had discovered that the liquid fruit of this tree was good, despite its foul odor and its unspeakable taste, and he came rapping with stone axes at its base.

Out of the tree there dripped not only bouncing balls but almost everything that man needs beyond bread alone. That these things were present in the milk of the tree is not so remarkable as that the primitive American extracted them. Plants containing some amount of rubber are found over almost all the globe; plants rich in rubber have always grown wild in tropical and sub-tropical Africa, on the Asiatic continent, in Madagascar, the East Indies, more islands of the *estrada* hemisphere than we have space to name. But no use was made of it by India's intellects, by Malays, Africans or the whites who went among them. Finished rubber products of the New World Indians were seen by Europeans possibly within the first two or three years after the discovery of America, certainly within the first thirty years. Yet it was nearly three centuries after the first voyage of Columbus that Madagascar's wild rubber was discovered; still later that the substance came to light in Asia and on the African mainland. Bearing that in mind, it is apparent that the American Indian's discovery and development of rubber was a considerable feat. The Amazon tree's predilection for popping off made it impossible for him to overlook the source itself, but that he should have gone on from this to investigate the latex, to devise methods of gathering, coagulating, molding it, and to invent a score or so of uses, is something else again. Lucky accidents figured in that growth of a primitive industry, but mostly it had to be a deliberate, determined progress.

To begin with, it was far from being a case of raw material dropping into the lap. The familiar *Hevea* tree, seventy-five to a hundred feet high and a dozen feet in circumference, is the survivor of hundreds that died as seed or shoot in the relentless competition of the tropical vegetable world. The Amazon had, and has, the most diverse tree and plant life anywhere on the globe and it is seldom that more than a few trees of anyone species can be counted in an acre of jungle. In the case of the *Hevea* it is estimated that the average is three or four to the acre. That means that rubber trees had to be picked out of the matted tangle one by one and at considerable distances from each other. That means that finding them and traveling about to collect a usable quantity of the rubber grudgingly yielded drop by drop was real work. Gashing the *Hevea* with stone axes, fabricating receptacles in which to accumulate the slowly dripping fluid - these were primary steps the evolution of which is easy to imagine. But the step up from spontaneous coagulation to purposeful, artificial coagulation and simultaneous shaping into manufactured articles is something that is much more difficult to account for. It is the more difficult when one reflects that the greatest native intelligences of these continents never stumbled upon such primary inventions as the wheel, the plow, the smelting of iron ore; that iron and plow and wheel were as little known to the 'civilised' Aztec, Mayan or Inca as to the aboriginal tapir, toucan or three-toed sloth. It approaches the amazing when it is further recalled that the white races were more than three hundred years in evolving any improvement over the system of manufacture they found in operation among some of the Amazon Indians.

Comparatively few tribes advanced this far in working rubber, of course. Yet from the Southernmost boundary of rubber territory in South America to its Northern limit in Mexico, some use was made of the substance wherever man lived, existed or passed through. On conditions in all of these territories we shall touch briefly, because there was a difference in tree-sources, a difference in collecting methods, as well as a difference in the articles of manufacture. The emphasis, however, will be on the rubber of the Amazon. One reason for this is that the otherwise backward Amazonians were the master rubber workers of all the American aborigines, as we have indicated. The other is that *Hevea brasiliensis* and *Hevea benthamiana*, the leading rubber trees, are native only to the lands drained by the great South American river and its tributaries. So long as wild rubber dominated the world market, the best and the most came from the Amazon territory. And from out of those jungles were smuggled the seeds that became the millions of acres of Malayan, Sumatran and other *estrada* hemisphere plantations supplying the world's rubber today.

Discovery of the bouncing ball was, of course, but the feeblest of first steps towards putting rubber to work. It is the nature of the *Hevea* fluid that, when it is exposed to the air, more or less complete coagulation results within twenty-four hours. So it was possible for the Indian to pick up a lump that had accumulated over the months and have a plaything completed by Nature, save perhaps for a bit of rough shaping. This was a substance which he found impossible to work, however. For that he needed the liquid latex, and in the Amazon he soon learned to come by it. It was a matter of purposefully gashing a series of trees, of arranging a leaf or clay cup to catch the drops and of returning in a few hours with a calabash into which to pour the tablespoonful of liquid found in each receptacle. This milky juice he learned to fashion into any desired form by applying successive coats to clay or wooden molds and by exposing them to a dense smoke, constantly turning them until, layer by layer, the shoe or bottle or jar was built up. Later he had only to wash out the clay or remove the wooden lasts and the operation was complete. For his smudge-fire, the Amazon Indian came to rely exclusively upon burning the nuts of certain palms and, especially, the urucuri nut. This thick, oily smoke he learned to concentrate by the use of a clay cone. The smoking of rubber must have originated as the result of a mere casting about for ways with which to work the gummy substance. In the smoked product he found a strength and a durability never encountered in a spontaneously solidified substance, which quickly went dead in those equatorial regions. And so the Amazon Indian became the first rubber chemist. So great was the difference between smoked and naturally coagulated substance, in fact, that there grew up a belief in the uniquely valuable properties of urucuri smoke, a belief that was universally held but short years ago and has not completely died out today. The exploding of that myth does not detract from the Amazon Indian's achievement. Write off the notion of his having achieved a smoke with unique and mystic qualities and there remains the fact that, consciously or unconsciously, he worked out the best possible method of treating rubber with the means at his command. It is essential that rubber be properly dried or there sets up a disastrous fermentation and mold growth. *Hevea* latex, as it comes from the tree, is about sixty per cent water and twenty-eight per cent rubber, the remainder being accounted for by resins, mineral substances and proteins present in the liquid and in the rubber particles, and by sugars dissolved in the water. Let nature take its course and bacterial action occurs, acids are developed and the imperfectly coagulated substance becomes moldy and of little worth. Prompt and proper drying of rubber is, thus, a necessity and this antiseptic function the smoke performed. By adding chemicals to the latex and tearing and pressing the coagulum between heavy rollers, the modern plantation owner can make the smoking unnecessary. Manifestly, this is something that was quite beyond the Indian's reach. He had to resort to fire for his drying, and it had to be just the sort he evolved, for it was necessary both that the smoking be uniform and that the latex not be subjected to too great a heat.

To tap the trees methodically, collect the latex before spontaneous solidification could set in, shape the sticky stuff over appropriate molds and artificially harden it in just the right amount of heat sounds easy when here set down. Long years of slow, discouraging experimentation by the trial and error system must have gone into the perfecting of the routine, however. In return for his labors the Amazon Indian received his rubber shoes, his unbreakable jars and a variety of other articles. In the field a rubber quiver-cover protected his poison darts. At play he now had hollow as well as solid balls. Faces of animals were molded in the stuff, and syringes - pear shaped balloons with wooden or quill tubes inserted - were made extensively. Finding these last in use among the aborigines, the first Portuguese in Brazil applied the name to rubber and rubber tree, a name later adapted in coining words to signify rubber collector and rubber estate. Whether injections then as now had the leading role in Brazil's lay catalogue of cures is not entirely clear, but certain it is that the instrument was in high esteem as a practical joker's squirt gun. On this point we have the testimony of early explorers, who found that when the Omagua Indians met together for private merrymaking, the master of the hut never failed to pass them around for that purpose — one to each guest. Which establishes, perhaps, the origin of the Great American Sense of Humor.

Dullest of the jungle tribes found rubber useful without resorting to any manufacturing process whatsoever. For one thing, it was in repute as a substance with curative powers, perhaps because, smeared on the feet, it would protect against further bruises or cuts. As an aid to beautification it entered heavily into the social life of savages who found the gummy substance admirably adapted to the attaching of handsome feathers to the human form. A step up in the scale were the head-hunting Antipas, who practiced a tattooing which involved pricking a pattern in the skin with a thorn, holding a lighted piece of rubber under the prepared area until it became well sooted from the black smoke and rubbing the soot into the wounds to leave blue marks similar to powder burns. Deep in the continent it served to cover the heads of the clubs with which the drums were beaten. And widespread was its use as torches.

In South America most of the rubber came from the *Hevea*. Numerous other trees yielded it, too, however, and chief among them were the *Castilloa*, *Hancornia*, *Ceara* (Manihot) and *Sapium*. In Central America and Mexico, the *Castilloa* was almost the sole source. Second in importance to *Hevea* among all rubber plants, this tree gave a poorer quality of latex and inspired an entirely different manner of collecting. Never relinquishing more than a little latex at a single cutting, the *Hevea* can yet be tapped daily, while the *Castilloa*, giving up an immensely greater amount of rubber at one operation, can be milked not more than a few times a year. Often the solution was to cut down the tree; ring the trunk, roots, limbs with incisions and stamp depressions in the earth to hold the flowing milk which was allowed to solidify spontaneously into pancake-shaped masses.

In Mexico the *Castilloa* was tapped after the fashion of the *Hevea* in the Amazon, the milk being collected in calabashes. A shortage of these vessels was no uncommon thing and in such event the Indian would smear his body with the milk as it came from the tree, peeling it off in sheets after it dried. So the early Spanish chroniclers report. They record also that an oil extracted from the gum by the application of heat had figured in native medical practice from time immemorial. The chroniclers ascribed to it great virtues in removing any tightness of the chest and in stopping hemorrhages, when taken internally.

A novel Mexican use of rubber involved the fashioning of shoes quite different from the Amazon variety. The construction was such, in fact, that no man could walk in them without wobbling and tumbling grotesquely. With these the palace dwarfs and hunchbacks were shod by kings and nobles so that the jesting masters might have matter for mirth. Serious uses among the Aztecs included the making of breastplates or shields reputedly proof against the sharpest arrows. So far back as the beginning of the seventeenth century we come upon reports of the waterproofing of garments by Indians, but whether this was an indigenous art or one evolved after contact with the white man remains a question.

These, then, were the varieties of use to which rubber was put by the original American. Interesting, indeed, is the fact that the one universal application involved the fashioning of rubber play balls. Of the games played by the city-builders much is on record. Of the jungle versions we know less, but that they go on today as centuries ago is constant matter of surprise to explorers. So recently as 1914, for instance, the ubiquitous Theodore Roosevelt paused on his progress towards the River of Doubt to exclaim over the strenuous life of Parecis Indians, who were addicted to a primitive soccer in which the ball could be hit only with the head. Rules, stakes, purposes of the games varied in most places, of course, but the same sort of sphere bounded from player to player in Aztec Tenochtitlan, at Chichen-Itza in Yucatan, in the Mayan cities of Central America, in the heart of the darkest Amazon forests, perhaps even in the islands of the West Indies and in the South-Western United States.

Almost, then, rubber was the only thing known in common to all the multitudinous groups and tribes and nations of South America, Central America and Southern North America with their varying languages, their diversified employments in temple-building, head-hunting or nomad wandering. We know from the Spanish conquerors that Mexico never had heard of Peru, for instance, nor Peru of Mexico; that the potato was as unknown in one country as was the use of tobacco in the other. Yet here was a rubber network tying together the Aztec aristocrat and the stoneage man of the deepest American jungles, the Mayan of the written language and the monkey-like cousin far to the South who hardly had a spoken word.

So far had the rubber industry advanced in its first thousands of years. And there it stood still for several centuries more. It was in these centuries of marking time that rubber entered history, however. Found out, experimented with, subjected to manufacturing processes by the Indians of the Western hemisphere, this substance was not, of course, officially 'discovered' until the white man had looked upon it. Supposedly, this necessary inspecting was done by Christopher Columbus.

The island of Hispaniola (today's Haiti), to which rubber is not indigenous, was the scene of the alleged finding and the date would seem to have been sometime between 1493 and 1496. Those were the years of The Admiral's second voyage and, while he spent considerable time at Hispaniola during every expedition, it was immediately after his return to Spain from trip two that he is understood to have turned in his rubber report.

The surviving writings of Columbus himself are of no help in establishing the truth of this tale. Of personally seeing 'three mermaids standing high out of the water' does the great man make mention, of allegedly being told by the Indians that at a distance there were men with one eye only and that in the middle of the forehead, of tapping towering mastic trees in order to extract the resin. There is no single reference which can be interpreted as having to do with rubber, however.

Actually, the oft-repeated story rests wholly upon the authority of Antonio de Herrera y Tordesillas, grand historiographer of the Indies to Philip II of Spain. In his 1601-1615 *General History of the Vast Continent and Islands of America*, this chronicler concludes an account of The Admiral's second voyage with a description of Hispaniola and its natives. One of their pastimes, he records, was a game played with balls 'made of the Gum of a Tree' which 'tho' heavy would fly and bound better than those filled with Wind in Spain.'

Herrera did his reporting a century after the event, but all public archives were thrown open to him and his chronicle was strung together mainly from the manuscripts of contemporaries of Columbus and other discoverers figuring in his work. It is possible, however, to raise so many objections, real and hair-splitting, against the acceptance of Columbus as rubber discoverer that the matter remains decidedly open to question.

Because of that, Pietro Martyre d'Anghiera, unquestionably the first writer on rubber, acquires additional importance as the chronicler establishing the earliest date to which we can trace rubber's history without venturing into the realms of the debatable. His *Decades of the New World* (the Latin original was published in parts from 1511 onward) contains a contemporary recording of the testimony of a reputable witness direct from the courts where Aztecs played with balls which were 'made of the juice of a certain herbe' hardened by heat and which 'beeing stricken upon the ground but softly' would rebound 'incredibly into the ayer.'

Martyre, a Lombard who had become a chaplain in the royal household of Ferdinand and Isabella, had this description from the very lips of Juan Ribera, trusted secretary to Hernando Cortez and bearer of spoils from Montezuma's land for his captain. How many centuries of rubber history were written by the Indians can only be guessed. How many Europeans inspected the gum playthings of the Indians before Cortez and followers moved into Mexico, we do not know. But the fact that one, Pietro Martyre, had a long nose for the marvelous and an obsession for daily recording his wonders in writing makes it possible to state without reservation that in 1519, in the city of Tenochtitlan, men of Spain became acquainted with rubber and with one of the first uses to which it ever was put. Apparently, however, his bouncing ball bulletin aroused no interest in readers of those pages swarming with tales of new-world harpies, of tritons, of giants, of salamanders, of a beast with the snout and hair of a fox, 'wonderful for the gesture and behavior, for in steede of dounge it voydeth snakes of a cubite long.' Nor did the substance seem to have aroused much curiosity in those Spaniards who were on the American scene. The five lengthy letters comprising the known literary output of Hernando Cortez contain no more sports news out of the new world than do the writings of Columbus, and the companions of Cortez are similarly silent. Several of the Spanish chroniclers thought it worth at least a mention, however, and so it is that we find Salahan and Oviedo bridging the gap between Martyre and Herrera. This last-mentioned historiographer had more than the Columbus story to offer. His account of Aztec tennis (or could it be basketball, supposedly invented by a YMCA instructor of the Gay Nineties?) is the most detailed, the most spirited and the most interesting of the somewhat repetitious ball game reports by the early rubber writers and so we have picked it to represent them all. We quote from the 1725 translation of Capt. John Stevens:

> The king took much delight in seeing the sport at ball, which the Spaniards have since prohibited, because of the mischief that often happened at it; and was by them called Tlachtli, being like our tennis. The ball was made of the gum of a tree that grows in hot countries, which having holes made in it distils great white drops, that soon harden, and being worked and moulded together turn as black at pitch. The balls made thereof though hard and heavy to the hand, did bound and fly as well as our footballs, there being no need to blow them, nor did they use chaces but vied to drive the adverse party, that is to hit the wall the others were to make good, or to strike it over. They struck it with any part of their body, as it happened, or they could most conveniently; and sometimes he lost that touched it with any other part but his buttock, or hip, which was looked upon among them as the greatest dexterity; and to this effect that the ball might rebound the better, they fastened a piece of stiff leather on their buttocks. They might strike it every time it rebounded, which it would do several times one after another, in so much that it looked as if it had been alive. They played in parties, so many on a side, for a load of mantles, or what the gamesters could afford at so many scores. They also played for gold, and featherwork and sometimes played themselves away. The place where they played was a ground room, long, narrow and high, but wider above than below, and higher on the sides than at the ends, and they kept it very well plastered and smooth, both the walls and floor. On the side walls they fixed certain stones like those of a mill with a hole quite through the middle, just as big as the ball, and he that could strike it through there won the game, and in token of its being an extraordinary success, which rarely happened, he had thereby a right to the cloaks of all the lookers on, by ancient custom, and law amongst gamesters; and it was very pleasant to see, that as soon as ever the ball was in the hole, the standers-by took to their heels, running away with all their might to save their cloaks, laughing and rejoicing, others scouring after them to secure their cloaks for the winner; who was obliged to offer some sacrifice to the idol of the tennis court, and the stone through whose hole the ball had passed.

Of the substantial accuracy of Herrera's description there can be little question. Excavations in Mayan Yucatan, Honduras and Arizona have turned up ruins of massive athletic courts with such stone rings as the old historian describes. And in the case of the Central American one, the archaeologists depose that, close to the playing field, they found a temple of the rain god where sacred offerings of rubber balls had been sunk in a holy well.

One more of the Spanish chroniclers deserves mention — Juan de Torquemada, who in his 1615 *De la Monarquia Indiana* first described uses of rubber for other purposes than the fashioning of balls and the attaching of feathers to the person. His report out of Mexico contains a recognisable description of the *Castilloa elastica* tree, a description of tapping methods and a mention of numerous applications. The Spaniards in Mexico had found the tree milk of great utility in covering their capes, he notes, and his judgment on these first rubber raincoats of record is that they were, indeed, marvelously adapted to shedding water although they would melt in the rays of the sun.

The incurious homeland's ignoring of that bulletin is typical of an attitude which constituted the second of rubber history's two great mysteries. The first mystery, of course, was civilisation's failure even to notice rubber until it was shoved under its nose in the new world. Stranger, however, was its long failure to do anything with the substance once it was discovered. For this last neglect Spain and Portugal stand chiefly responsible. All of the great rubber lands of the Western hemisphere were in their possession. Portugal had Brazil and its Amazon lands overrun with the original and matchless *Hevea*; Spain the remainder of the *Hevea's* native land — the remainder of South America for that matter — and all of Central America and Mexico. Yet neither country made any attempt to import substance or manufactured articles; Juan de Torquemada's report, a report so well calculated to arouse a lively curiosity, fell on ears as deaf as those that had been turned to previous, less interesting bulletins.

So long was the versatile tree milk completely scorned that it was left for an adventurous Frenchman of the eighteenth century first to interest Europe in the subject. Generally, the name of Charles Marie de la Condamine is pretty well forgotten now but, no matter from what angle you approach him, he remains a scholar-explorer of considerable stature. For the first thirty-four years of his life he cut no such figure. He was, at the end of this period, a social climber who was making good in that difficult profession, a soldier with a reputation for being one of the foolhardy, a man of fashion, a frequenter of salons and trifler with verse. There are entries on the credit side of the ledger, however. Thorough Parisian the man may have seemed, but he was also a good friend of Voltaire's, a dabbler in chemistry, mathematics and *estrada*, a traveler who had made an adventurous journey into the Orient. And then he joined an expedition to South America. In invading that dark continent, his duty mission was to measure an arc of the meridian on the Equator for the Paris Academy of Sciences while another expedition under P. L. M. de Maupertius moved into Lapland to take similar measurements in the far North. Centuries before, Arab *estrada* of a caliph fired with the true spirit of Greek learning had arrived at the approximate size of the earth by measuring a degree of latitude on the plains of Mesopotamia while over all Europe hung a night of ignorance evoked by the true church. In a sense this was lost knowledge in the time of Condamine and Maupertius because of the uncertainty over the unit of measurement used. Something more precise was needed, and, impelling motive of the French expeditions, there was a great new controversy raging between the school holding that the earth is a sphere flattened at the poles and the rival professors maintaining that they infested a globe which lengthens in the direction of the polar diameter. Out of Lapland and Peru (that part which now is Ecuador) came finally, then, the measurements that established this world's size and figure and crushed the contentious Earth-elongators beneath the heels of the flatteners.

Condamine was seven years in performing his share of this task. His party landed at Guayaquil in 1736, journeyed bickeringly to Quito and, between the ranges of the Andes, settled down to workmanlike quarreling and workmanlike labors with crude pendulums and that measuring bar still identified in geodetic annals as the 'toise of Peru.' Early 1743 saw this work completed and the expedition splitting up. Those who had been unable to resist matrimony stayed on in Quito. Condamine and several others decided that they would head home by different routes in order to multiply their chances of making scientific observations by the way. For his own part, he settled on crossing the continent to Para by way of the Amazon. Francisco de Orellana, follower of Pizarro, had gone that way with fifty men so early as 1541 and emerged on the Atlantic coast with yarns of doing battle with lady warriors, from which tall tales the river's name derives. Pedro Texiera and nine hundred followers had performed the feat in reverse in 1638, and there had been others. Condamine knew, too, that there were Spanish and Portuguese posts along much of the route with priests to speed him on his way. It remained a hazardous stunt, nevertheless. He had a chance at a traveling mate, Don Pedro Maldonado, governor of the Peruvian province of Esmeraldas, but characteristically he arranged things so that this friend would start by a different and exploratory route of his own, the two keeping a rendezvous on the way. By land, by raft and by canoe, Condamine progressed and from May 11, 1743, to July 23 with only his changing relays of Indian guides to keep him company. With Maldonado, then, he journeyed on to Para, arriving on September 27. The governor took ship for Lisbon. Condamine decided to stick around a while and see what was to be learned. He returned to France, finally, in 1745; at a sitting of the Academy of Sciences he read a paper lengthy enough to be worthy of a filibustering United States Senator, repeated his tales in all of the salons, and, in the intervals of social life, managed to knock out six books in the next five years.

Had this true son of the eighteenth century not been an amateur geodesist, South America never would have seen him. Had he not been adventurous, his perilous descent of the Amazon would not have been made. If, however, he had been no more than a daring scientist with the usual one-subject obsession, the years in the Andes would have produced only the measurements of arc; the cross-country journey and the stays at Para and Cayenne on the East coast would have resulted in nothing. He was, however, one of those catholic minds of infrequent occurrence, a man as avaricious of all odd knowledge as Pietro Martyre was of all strange news, a scholar as keen to study plant life and figure out antidotes for poison arrows as he was to chart rivers or run a figurative tape-line around the earth. So was his stay near Quito of a multiplied richness, so was his crossing of the South American continent important because it was the first to be made by a man with a trained brain and a seeing eye, so were his months in *estrada* Brazil and French Guiana as productive as all the others he spent on American soil.

Most interesting of his natural science observations, perhaps, were his reports on rubber and on the *cinchona* tree. It was an appropriate pairing, considering that the rich commerce in both rubber and quinine would be lost to South America through transplanting to the East, considering that, without quinine, the white man could not have survived to supervise rubber-raising anywhere in the tropics. The notes on the *cinchona* tree have been called the first of scientific importance. And it is customary to say the same of the rubber notes. This, however, is a considerable exaggeration. Torquemada's reports are at least as precise and neither set is much more 'scientific' than the jottings of any of the chroniclers. The Condamine findings were of utilitarian importance, however, since they actually attracted Europe's attention. And they are additionally of great interest to the rubber historian because they probably constitute the first writings on South American rubber and certainly contain the first written mention of the *Hevea* tree. Also, it seems that Condamine was the first to send actual samples of rubber to Europe. And he tried his own hand at the manufacture of waterproof cloths and a rubber case for his quadrant.

As to the reports themselves; their sum total is a brief description of *Hevea* milk collection in Esmeraldas, of its use there as torches and for the coating of cloth sheets to be used 'for the same purpose we use waxcloth,' and of Amazonian manufacture of bottles, syringes, toys and boots 'which become quite waterproofed and when smoked acquire the appearance of leather.'

Not too seriously did France take these South American reports at first, but eventually they attracted the attention which Spain so noticeably had failed to give when Juan de Torquemada wrote as much about the Mexican industry more than a century before. Later, Condamine presented, too, the first really detailed observations of the method of collecting the *Hevea* latex. They were obtained for him by François Fresneau, a fellow Frenchman resident in Guiana, with whom he had come in contact in 1743. Fresneau, if we are to believe the French magazines of the last century, was a Lon Chaney character in real life, a man whose 'features, otherwise intelligent, had the savage character of the wild boar, a misfortune that caused him to shun society and finally led to his settling in French Guiana.' There, it is recorded, he still sought to avoid his fellow men and spent much of his time wandering in the forest, where his interest in botany led him to hunt for the rubber tree. Be that as it may, Fresneau would seem to have been the first European personally to observe native rubber-collecting in all its details. He wrote for Condamine a very good description of the *Hevea*, a report on its location, a minute account of tree-tapping and jar manufacture and a number of suggestions for practical application of rubber by Europeans. This, we should say, really was the first rubber paper of scientific importance. Nor was Fresneau content with advocating rubber tarpaulins, pump hose, diver's suits and air-tight bags for preserving food while at war or travel. He set about the making of various articles and sent them to Jean Frederick Phelypeaux Maurepas, a minister of state who had much to do with bringing on the French Revolution.

That Fresneau's gadgets and Condamine's samples of crude rubber, Condamine's own observations and his publishing of Fresneau's findings awoke a real interest in the tree product can be shown by listing developments. Before the century closed, it saw the first importation of Indian rubber products by Portugal, the first chemical investigation of the coagulated milk, the first patenting of an article made from it, the discovery of rubber in Madagascar, Mauritius, Penang, Sumatra and India. All these will be treated in their proper places in this volume. Meanwhile, it seems fitting to acknowledge that Charles Marie de la Condamine and François Fresneau stand at the dividing line between civilisation's long neglect of rubber and the first, slow, blundering exploitation of this now indispensable article of world commerce. They were the last of the explorers of the outer world to figure prominently in the history of rubber in the Americas, the forerunners of the scientists who were to write a great new story of discovery by taking rubber apart and reporting what they found concealed beneath the black, sticky, stinking surface of the strange substance out of the jungle.

The world measured by Condamine retains the same proportions today. It is a different world, however, because of the changes wrought by the 'resin' catching his fancy in remote Esmeraldas and the Para whose mucking with foul tree milk was of no interest to maternal Portugal for so many years, of little interest until it was too late for that nation of fortune-seeking navigators to attain to undreamed of wealth thereby.

2 Wild

2.1 Rubber River

Only as a novelty, interesting but of no commercial importance, did rubber reach Europe in the eighteenth century. To Para (Belem) at one of the mouths of the Amazon - the first Portuguese settlement in the far North of Brazil - went several pairs of royal boots to be waterproofed and returned to Joseph the King. In 1759 the government of Para presented him with a coat made of fabric covered with the new substance. For the rest, the casual traffic carried on by sea captains with a weakness for the curious was in rubber bottles of Indian manufacture. With these the scientists and experimenters toyed, from these the first simple articles of French and English manufacture were made. In 1800 some of the bottles reached the United States for the first time. Like those current in Europe, they had been obtained under difficulties, for Portugal's Brazilian policy prohibited all commerce and intercourse save with the motherland and sealed the great South American country as tight, almost, as Japan. Occasionally, vessels of nations temporarily in alliance with Portugal were permitted to anchor in the harbors, but passengers or crew could land only when escorted and watched by a guard of soldiers. One curious result of these oppressive regulations was that the Brazilians, loaded down with gold and diamonds yielded by the mines of the Southern states, were unable to buy ordinary household and agricultural essentials. Napoleon Bonaparte indirectly put an end to that. When he decided to move against Portugal, the prince regent, afterwards John VI, fled to South America with twenty thousand adventurers and court hangers-on, the only historical instance of a colony becoming the seat of government of its mother country. This was in 1808, and it was followed by the opening of Brazil's harbors to the commerce of all friendly nations. Twelve years later the first pair of rubber shoes arrived in the United States. They were gilded and had long, pointed toes. In 1823 a lot of five hundred pairs was brought into Boston. Minus gilt and stream-lining they were clumsy scow-like things, but they fetched high prices and a lively trade sprang up. By the middle of the century three hundred thousand to half a million pairs were being imported by the United States annually.

The Indian of the Para vicinity and especially of the islands in the Amazon's mouth supplied most of these, and he found the business to his liking. Working only during the dry season, he would collect his rubber and leisurely shape it into shoes unless there was wife or daughter to be assigned to this manufacturing. Completed and exposed to the sun for a day or two, the shoe would remain soft enough to receive impressions and during this time it would be figured with a pointed stick in such design as the Amazonian's artistry dictated. Tied together and suspended from long poles to keep them apart, they would then be toted to Para by foot or canoe. There the Portuguese merchant would stuff the still sticky brogans with dried grass to preserve their shape during the weeks in which they were set aside to harden sufficiently for export. For these articles he paid 10 to 15 cents a pair and this was satisfactory money to Indians with few ambitions beyond keeping themselves well supplied with the cheapest of firewater. Invariably this delivery of merchandise was under previous engagements and usually for advances received. Out of such petty juggling with dimes grew the vicious credit system which later made an actual slave of nearly every rubber gatherer, whether technically slave or free.

Simultaneously with this shoe traffic there developed, too, the exporting of raw rubber from the region around the Amazon's mouth, much of it going to the American ports of New York and Salem. This was in the form of sixty to eighty pound balls built up layer by layer over the smoke fires and, until the appearance and triumph of plantation sheets in the present century, these balls remained the principal rubber of commerce, the best of them coming to be classified as 'fine hard para.' Modestly, indeed, they first entered world commerce and so late as 1825 the total export amounted to less than thirty tons ('ton,' throughout this book, meaning the long ton of 2,240 pounds). By 1830 it had jumped to 156 tons and by 1840 to 388, although the price was down to five cents a pound in the United States at one time. The 1839 discovery of vulcanisation, essential the first process in modern rubber manufacture, now began to make itself felt. Exports of crude rubber selling at 18 to 30 cents a pound mounted to 1,467 tons by 1850. In the next decade prices touched 68 cents at one time, while exports of the native shoes faded and died in competition with the articles of factory manufacture. And in 1860 exports of the para were 2,670 tons.

As the demand for rubber grew, as prices rose, rubber gathering naturally came to be the only activity carried on in the Brazilian North East. Sugar cane was grown only in sufficient amount to yield a part of the supply of the booze called *cachaca* — sugar itself was obtained from the Southern provinces along with coffee, rice and cotton. So wholly was agriculture abandoned that beans, farina, maize - staples which could be raised in abundance and with little effort in a few months between rubber seasons — all came to be imported into a region which should have been wholly self-supporting. And this reliance on the world outside the fertile Amazon for all foodstuffs developed and persisted even as the iniquitous credit system. Rising rubber prices and freedom from growing his own food made no change in the Para Indian, however. He remained the same haphazard workman of shoe manufacturing days. Into the forests would go a family group or luckless bachelor and there they would live on provision of dried fish and farina (a coarse quality of tapioca made from the cassava or mandioca root and eaten dry) until they had collected such amount of rubber as they chose. Then it would be peddled in an eager market and the aborigine would retire to loaf and invite his soul until funds gave out and the returning dry season of four to eight months ushered in the period of collecting urucuri nuts and extracting rubber. Some hired out to Portuguese proprietors of district forests — collections of tree-to-tree footpaths circling about to begin and end at palm-tree huts - and made their rounds on a percentage basis. All of them put together - some 25,000 it has been estimated could not begin to keep up with the new demand even with the addition of rubber from the Middle East and Central America. The time was ripe for new methods. The coming of steam to the Amazon and the opening of the river to free navigation made it possible to put them into effect.

What the first real utilisation of the mighty water highway meant to the raw rubber industry can be appreciated only by considering a few facts about the stream and its countless tributaries. Nowhere in the world is there a river comparable to this Amazon with its drainage area of 2,722,000 square miles. The longer Mississippi - Missouri drains considerably less than half as large an area and the Congo little more than half; beside it the rivers of Europe are so many guttered trickles of rain water. River really doesn't tell the story. It is a congress of rivers; a titanic reservoir; more accurately yet, a great elongated inland sea. The Amazon proper varies from five to four hundred miles in width in the wet season and with its tributaries it drains more than a third of South America) an area equal to more than two-thirds of Europe. Sometimes and for long distances the main stream divides into two great rivers connected by a tangle of natural canals. And so huge are its principal tributaries that the river and its branches have an estimated total of ten thousand miles navigable by steamers below all falls. Tapping nearly all Northern and central Brazil) this river system reaches out) also) to drain half of Bolivia and Colombia, two-thirds of Peru, three-fourths of Ecuador, a part of Southern Venezuela. By a tributary of British Guiana's Essequibo, the Pirara tributary of the Amazon may be reached. Some fifteen miles from its source, Venezuela's Orinoco also bifurcates into a 2,150 foot wide natural canal running into the Rio Negro, thus affording a through water-way to the Amazon basin from another region.

Incomparably the greatest wild *caoutchouc* country in the world is that basin. To 2,250,000 square miles of it rubber is native and the accepted guess-work estimate holds that there still are some 300,000,000 untapped trees in the remoter forests of this region despite the hundreds of thousands of tons of coagulated latex that have gone down the river. Little rubber is found on the main stream itself, but virtually every tributary is loaded with it. The main belt of *Hevea brasiliensis* is a semicircle beginning near the city of Para and swinging South West through the Brazilian states of Para, *Amazonas* and Matto Grosso, then up through Matto Grosso, Northern Bolivia, *Amazonas* and the Brazilian federal territory of Acre until it reaches the Andean foothills in Peru. At either end the belt approaches to within one hundred miles of the main stream while near the center of the semi-circle its Northern limit is some eight hundred miles from the Amazon proper. Irregular in width, it is widest at the middle and throughout its entire span the *Hevea brasiliensis* areas are irregularly distributed: in the valleys, on the hills and along the banks of all rivers. Between this main belt and the Amazon it is found only along the rivers and in the flood lands of the river valleys; the theory that accounts for this and also for the Amazonian delta trees is that they sprang from seeds originally washed down from the uplands. Benthamia, the *Hevea* second in importance, is found in the upland forests North of the Amazon and along all tributaries, and tributaries of tributaries, flowing South through Brazil, Peru, Ecuador, Colombia and Venezuela into the main stream. So accurately does the big river divide the major *Hevea* varieties that south of it the *benthamia* is found only in a few isolated spots, north of it the *brasiliensis* is found only near the river-mouth and along the main course from west of Manaos to a point approximately opposite the mouth of the Teffe. As for the Castilloa, it rambles over most of the highlands between all the Southern tributaries at approximately one hundred miles from the main river, north up the Negro, and up any number of other streams further west. And without a great network of water highways this wealth of para and caucho rubber could never have been touched, for so closely matted is the jungle that land travel is impossible in most places.

The Amazon made the wild rubber industry, in other words, and the coming of steam and the rubber gatherers transformed the Amazon. Brief and disjointed was the history of the main stream prior to that time, and most of the tributaries had no history whatsoever. As Portugal, busy with colonial expansion in India and Africa, originally had ignored a whole American empire because of an apparent lack of precious metals, so independently and during its 1578-1640 union with Spain, it continued to ignore the Amazon while belatedly developing the rich agricultural possibilities of the coastal South and central portions of its territory. The very gateway of the river was not even occupied until the founding of Para in 1615, and that resulted not from the colonising urge but from the sending forth of an expedition to drive out the Dutch and English who had been encroaching on the Brazilian north. Separation of Portugal and its colonies from Spain found Para consisting of four monasteries and a collection of shacks housing a population of but four hundred. Already it was a headquarters of the Indian slave traffic, however, and a center of turbulent controversy between Jesuit and colonist. Two years before — in 1638 — Pedro Texeira had made that voyage up the river referred to in the preceding chapter and it was followed by a series of other such official explorations on the grand scale. After the explorers went the slavers and priests and then the traders and colonists. Twelve miles up the Rio Negro and a thousand miles up the Amazon, the city now called Manaos was founded in 1660 as a fort to protect the slave-hunting expeditions against the warlike Indians of that name. Single parties brought down red captives by the hundreds and the same decimation of aboriginal population went on at other points along the river. As far North as the Orinoco, these hunters traveled. In the meantime, the Jesuits were shepherding other thousands into the safety of the missions founded as the result of the labors of Father Antonio Vieria. Among the colonists Father Vieria aroused such enmity that they obtained his exile to Portugal and imprisonment by the Inquisition on heresy charges, but his work went on. Santarem, at the North of the Tapajos, five hundred miles up the Amazon, was founded in 1661; and by 1668 a member of the black-robed brotherhood had penetrated 1,400 miles up the main stream from Para to round up several of the docile tribes and found Ega near the mouth of the Tefte. Far up the Negro, the Society of Jesus established Barcellos, where, eventually, ten thousand Indians gathered. So arose the forts and the missions some of which became the principal cities and villages the Amazon has today. And Portugal by the early eighteenth century had gained a pretty fair idea of the main river and some of its more important tributaries; by 1756 it had explored a good part of the Madeira, greatest of all the tributaries, and founded Borba 150 miles from its mouth. All that it learned it kept to itself and the Condamine report comprised nearly the whole of Europe's knowledge of the Amazon at that time. Yet further knowledge would have availed it nothing, for the Portuguese isolation policy barred the whole river to foreign trade.

The Jesuits worked the Indians and sold their products abroad, but while the colonists killed them in the fields, the priests proved that under properly mild auspices the aborigines could both work and flourish. They had a real love for their charges; their kindness even included going light on religious instruction; and they bound the timid redmen to them by strong ties. In the end, however, their influence was a disastrous one, as we shall see. And it was a disaster that was not confined to Brazil for Spanish Jesuits had been as busy as the Portuguese brethren. Into country that would be parts of the rubber districts of Bolivia, Peru, Ecuador, Colombia and Venezuela they made their mission-founding way. There was, for instance, an important center on the upper Orinoco just above the canal leading to the Negro; others lay South of the Amazon and East of the Huallaga deep in Peru, scattered about the Beni river region of Bolivia. Many were founded by the Bohemian-born Father Samuel Fritz who began his great work on the Huallaga, Ucayali, Napo and other Amazon tributaries of the extreme West while Father Vieria was living out his last year on the Brazilian coast far South of the mighty river. Finally pushing into Portuguese territory, he was taken into custody, and armed bands were sent upriver to destroy his missions, which were regarded as outposts of Spain.

Usually, whether in Portuguese or Spanish territory, the Jesuits had the crown on their side, the colonial governments against them. In South Brazil as early as 1574 they obtained a royal decree giving them full control of all the Indians settled in their *aldeas* and restricting the colonists to the enslaving of hostile natives captured in battle. Launching their chain on the Amazon, they began with a 1654 order from Lisbon transferring the Indians from the jurisdiction of the governors to that of the Society of Jesus. Violations were plentiful, but the protection extended by the priests actually was real enough to decrease the supply of slaves to a point where the plantation owners had to start importing blacks from West Africa.

John V died in 1750, Joseph I ascended the throne and the iron Pombal emerged as prime minister. Pombal had no liking for the Jesuits and had at them at once. In 1755 he struck a blow at the Amazon missions by decreeing the emancipation of the Indians of Para state and of Maranhao to the south-east and by depriving the priests of all temporal power in these states. Those ordinances soon spread to the whole of Brazil. In Paraguay, meanwhile, the Jesuits were playing right into Pombal's capable hands. For a piece of Paraguay, Portugal in 1750 had traded Spain a piece of the land we know as Uruguay and the two nations then sought to have the respective populaces switch villages so that the subjects of each crown might be the same as before. Under priestly leadership the Indians, trained in European arms and discipline, rose in defense of their homes, and the troops of two nations had to be sent in to subdue them. That rebellion and a 1758 assassination attempt against the king which Pombal charged to the black-robed society gave the prime minister his needed excuses. From all Portugal the Jesuits were expelled in 1759, from all Portuguese possessions in 1760. From all Spanish lands they were expelled in 1767.

These were important dates in rubber history. We have said that the Jesuit influence on the Indian was a disastrous one in the long run. This was so because the priestly cherishing of these always weak ones reduced them to such a childlike state of dependence that they were completely lost when their guardians were snatched from them. It had happened that way after the razing of the missions in the South; whether adrift in the forest or sold into slavery, the aborigine was plunged into the worst state in his history. It was that way now in the North even though throughout Brazil the enslavement of the Indian supposedly had been abolished. Timorous, servile, helpless, the converts clung to the mission sites and became easy prey for trader and colonist or retreated back up the channels where descendants, raised in an atmosphere of abject hopelessness and incompetence, fell even easier victims to the rubber men.

From almost the beginning of the rubber trade, some latex had gone down from upriver. For years, however, the islands near Para and the swampy mainland fifty to seventy miles West remained the one district seriously dedicated to the business. And when the enterprise first spread to any extent it was farther up the Tocantins river, the mouth of which was near the border of the original district. Up the main Amazon the laying out of cacao plantations was the principal activity, although there was also some tobacco-raising. Here rubber was only one of several casually gathered forest products. The Indians peddled it to the traders as they did Brazil nuts, sarsaparilla or turtle oil, and it was probably in less demand than any of these.

Almost unbelievably were these aborigines thinned out along the main banks and the lower reaches of the great tributaries. Warlike tribes the Portuguese set one against another, sometimes taking a hand themselves as when they made common cause with Mundurucus against the nomad Muras. The 'civilised' Indians, as we have seen, decamped in some numbers following the Jesuit expulsion. Cruel trafficking in their hides and forced servitude swiftly cut down the numbers of the others. Diseases, including the mysterious fever which assailed them at the slightest contact with people from civilised settlements, swept off thousands more. Intermarriage with whites and half-castes cut a further hole in the number of pure-blooded survivors. In the comparatively civilised places, many of these last came, too, to know that there were laws protecting them against forced servitude and they quite naturally refused to work. So scarce did red labor become that Santarem, the Jesuit settlement which had become a slave center, and Obydos, in the important cacao district fifty miles away, turned to the importation of blacks. Farther upriver, however, there was still large scale slave-raiding as on the Japurá, where it really began in the late 1820s. On the higher reaches of that stream and beyond Brazil, incidentally, slave hunting by rubber men raged in the twentieth century.

Nominally, all of the Brazilian Indians had been placed under the protection of half-breed government appointees in the years following the expulsion of the Jesuits. These last had trickled back into South America with the 1814 revocation of the 1773 papal decree suppressing them throughout the world, but their day of power and influence was over. The new guardians of the Indians included directors of each of the important tributaries of the upper Amazon. Usually these half-Indian *mamelucos* used hundreds of their charges in their own private enterprises and were worse task-masters than the average white. Equally tyrannical on a smaller scale were the captains of *trabalhadores*, half breeds authorised to organise the scattered canoe-men of their areas so that they could supply travelers with paddle-wielders when needed. So thoroughly did they monopolise the men for personal ventures that it was a rarity when the loan of a canoe-hand could be wrung from them.

And while steam was conquering the world outside, Amazonia was actually allowing its means of communication to be slowed up by this scarcity of paddlers. Swift boats manned by a dozen of these redmen could make thrice the time of ordinary sailing craft, but it was the slow schooners of thirty and forty ton burthen that came to be practically the only link between the distant river towns and Para. Usually they were owned by traders living in these lost villages and trafficking with the natives up branch rivers still more remote. Natives from not too far up the river still would descend to

Para in their small canoes; Santarem, Manaos and other communities drew similar visitors, and occasionally some Portuguese merchant at Para would furnish a young countryman with merchandise and send him off to the interior to barter with the scattered population. In a strategic spot were these Portuguese when the rubber boom hit and they made the most of it. Agriculture they left to the *mamelucos*, water-work to the Indians, jungle-adventuring to the men of other nations. Shopkeeping and petty trading, these were the ambitions of most of the immigrants from Lisbon. Para fairly boiled with their mercantile enterprises, wholesale and retail. Even when they bobbed up in far-off, forest-surrounded Manaos, they would try nothing save setting up in trade. One store for every five residential buildings was the average there — stores carrying two or three hundred dollars worth of liquor and cheap wares and counting the day's business in coppers.

Dismal and forsaken was the whole panorama of river and tributaries until the steamboats and the rubber demand of the 1850s and thereafter wrought their changes. Para city was a weedy, dilapidated, over-grown village of swampy squares and unpaved or semi-paved streets. In 1820 it had attained a population of twenty-five thousand. A bloody revolt by half-breeds and Indians in 1835 had reduced that figure by ten thousand and now the survivors were battling a dreadful onslaught of yellow fever and had a scourge of cholera to come. Its Indians were all engaged in being absorbed into white and negro races or had been otherwise civilised, but, while free citizens, the aborigines had been deprived of their lands. South-west of Para and on the Tocantins river was Cameta, boasting five thousand residents of assorted breeds and the center of a rubber, cacao, and Brazil nut gathering district which claimed twenty thousand and admittedly was the most thickly populated part of the province of Para. Breves, on the island of Marajo, offered forty houses with Portuguese shopkeepers outnumbering Indian customers. From here on to Santarem, at the mouth of the Tapajos five hundred miles from Para, were but three tiny communities never found on any save the most inclusive of maps.

Santarem, important trading center, had three long streets ending where the jaguars roamed. On these and a few alleys clustered the tile and white-washed houses and the mud-heap hovels of 2,500 whites and Indians. In a radius of twenty-five miles it could claim another four thousand. And the town itself, remember, was the metropolis of all the Amazon, the largest and most active city on the mighty main stream from Para to the end of navigation thousands of miles away in Peru. Obydos, next beyond Santarem, offered 1,200 half-breeds scattered among cacao plantations and then came Parentins on the boundary of Para state and its offspring, the new province of *Amazonas*, a 730,000 square mile empire which the future 1858 census would credit with 55,000 souls. *Amazonas*, fresh from its creators, greeted the traveler with Villa Nova, seventy leaf-roofed mud hovels ruling a sparsely populated district containing not more than four thousand half-breeds and semi-civilised Indians and sending its rubber, balsam of copaiba and dried fish all the way to Para in exchange for European goods. Serpa, slightly larger, with two far smaller settlements and the small cacao and tobacco plantations dotting the coast opposite the Madeira, constituted all that was to be seen between there and the mouth of the Rio Negro.

Turning up the forest-clothed banks of this tributary, the voyager passed the homes of settlers hacked out on the wooded heights and came to Manaos, political capital of new *Amazonas* and real capital of the whole upper Amazon region. On the mercantile situation at Manaos we have touched. Politically, the influx of the army of salary-drawing, non-working officials had swelled the population to three thousand. Already it was suffering from a scarcity of the most necessary articles of food. Agriculture was completely ignored, the neighborhood not even producing sufficient mandioca for its own consumption. Foodstuffs came from Portugal, England and the United States while officialdom reclined on its pants-seats, Portuguese puttered about their alleged stores and the lower order devoted itself to the collection of the forest products. Its revenues, derived from export taxes, amounted to less than a fifth of its expenditures, and Para had to supply the remainder of the funds. Foreign food supplies, incidentally, reached Manaos some forty days after leaving Para in the six-month dry season of strong trade-wind and slack current, and only some three months after in the wet season of roaring current and no east wind.

On the Amazon and eighty miles from Manaos by the nearest road was Manacapuru, where citizens of the capital maintained summer homes. From there the voyager progressed for nineteen days without seeing a human habitation, often without seeing more than one vessel. There were a score of settlers' families in that 240 mile stretch to Quary, but all were located on the banks of lakes or inlets back from the main stream. After Quary came Ega, Brazil's one real settlement beyond Manaos, a village of 1,200 inhabitants — forty to fifty whites, five hundred to six hundred *mamelucos*, a number of blacks and mulattoes and a handful of Indians. Within a radius of thirty miles were another two thousand, perhaps. From Ega on to where Peru began some 1,900 miles from the Atlantic were just five so-called settlements, huddles of ruinous houses inhabited by a few hundred semi-civilised Indians and a handful of debased Portuguese traders, places so forsaken that not even a priest would consent to live in them. Over the Peruvian border, civilisation from the Pacific had left traces in the form of such tiny settlements as Iquitos and Nauta, but that was all.

And this was the world's mightiest river — not in part, but the whole panorama — when steam and the rubber rush woke it to life! And barren as our picture is, it gives, perhaps, an idea of a region more populous than really was the case. We have dealt with the main river, cited the comparatively dense population on the Tecantins and gone a short distance up the Negro to Manaos. Not the branch streams near Para, not the lower Negro, did the average tributary resemble, however. Up the more remote of them were only scattered Indian tribes that had never seen a white. Of the better known and more accessible ones, the Tapajos furnishes a fair example. At its mouth was the important city of Santarem, remember, and that mouth was but five hundred miles from Para. Yet all the Tapajos beyond a few hours walk from Santarem was unknown territory to the inhabitants of that community. And on the banks of this tributary were only a half dozen almost unnoticeable groups of half-breeds and baptised Indians before the country of the Mundurucu Indians, visited only by traders who came down occasionally to trade for mandioca, Tonka beans and rubber. On the tributaries of the tributary were here and there a *mameluco* settler or a cluster of Indians.

Then came steam and the rubber men to galvanise not only the Amazon but all these empty tributaries, all the tinier backwaters incredibly more remote and unknown. The decree opening the river to steam navigation was not signed until 1850 and it was 1853 when, by treaty with Peru, the subsidised Amazon Navigation Company placed three small wood-burners on the main stream. Peru ordered two wooden steamers from New York to be operated in connection with these vessels, but one never turned a wheel and the other made only a few trips to points above the end of Brazilian navigation at Nauta. The Brazilian move was made by the second and last of the sovereigns who ruled it as an independent empire and it was dictated by a number of coinciding events — the growing rubber demand; a belated recognition of the possibilities of developing the vast, unexplored reaches of the state of *Amazonas*; rumors of filibustering expeditions fitting out in New York to force the opening of the river. The result was the planting of colonies in the states of Para and *Amazonas* as per the navigation company's agreement with the government; the pouring into the city of Para of rubber, Brazil nuts, cacao from upriver, even the Peruvian mountain towns beginning to bring down their products by raft and canoe to meet the wood-burner. By 1857 seven steamers were on the river and new companies began organising to push the service up the Madeira, Negro, Purus and the smaller streams. Ten years later Dom Pedro II finally opened the water highway to foreign trade. Now Bolivia, Ecuador and Colombia had their needed outlet for forest products that had been going West over the Andes to be carried eight to ten thousand miles around the Horn before reaching the mouth of the Amazon on the way to the Atlantic ports of North America and Europe. Ocean liners from London and New York began nosing up the river to Manaos and to Peru's Iquitos, 2,300 miles from the sea. Steamers multiplied and shoved their way into rivers unvisited before save by canoe and *batelao*. Now that rubber was really wanted, it was possible to get it out.

In the world outside the Amazon, the American Civil War had given a sharp impetus to the manufacture of rubber articles other than boots and shoes and this was multiplied by the industrial expansion after the war. In 1865 sales of *caoutchouc* at $1.20 a pound were recorded. Out of the Amazon jungles came 6,591 tons in 1870 and it sold, nevertheless, at eighty cents to $1.07 a pound, so far did demand outrun supply. In all the hundreds of thousands of square miles of Para and *Amazonas* there were but 350,000 people - brown, white, black, mixed, slave and free - by that year and most of these were crowded on the seaboard. But now the leisurely mongrels of the districts around Para city weren't the only ones gathering rubber. Far in the interior the cinnamon-colored aborigines labored in dreadful bondage, were trapped and sold in a continuing domestic slave trade that was as terrible as the traffic in black pelts from Africa legally abolished by Brazil in the first half of the nineteenth century and actually and thoroughly suppressed in the early 1850s.

Into the Amazon had poured adventurers of all nations. Upriver they pushed, hunted out sites in rubber territory - sometimes far up the tributaries - and took possession. Next they descended upon villages of the helpless descendants of the Jesuit converts or of wholly primitive weaklings and with presents of liquor and factory-made trinkets lured them from their homes to work rubber. Whole communities would be abandoned, fields, grazing lands and all, as the natives were set to clearing the forest at the river-bank spot chosen by the exploiter for his estate. Then, after building home and other necessary structures for the master, they would be permitted to run up their own poor shelters off in the jungle. And footpaths would be cut from rubber tree to rubber tree in the forest, each serf being assigned a circuit. It was, of course, these paths rather than the land itself which constituted the chief value of a location. The paths, that is, plus the Indians using them.

At the beginning the exploiter would unload upon the eager aborigine firearms, musical instruments, fake jewelry, sewing machines of the cheapest manufacture. Typical of the fabulously priced but well nigh worthless goods foisted upon these jungle rubes was the 'tradegun' which came to be specially made in Europe for the Amazon rubber trade. Muzzle-loaders with wound-wire barrels, they would unwind and fall apart after half a hundred shots had been fired. Several factors were responsible for the excessive shoddiness of these articles. For one thing it was believed that any Indian coming into possession of a really good gun would feel himself so well off that never again could he be induced to work rubber. With the fragile 'trade-guns' it was necessary for him to fall to and earn a new one each season. Also, of course, the general theory of the whole commerce was to buy the cheapest possible goods and dispose of them at the most extravagant prices possible. Duly encouraged, the Indian sport would run up such a bill to be paid off in rubber that he would be reduced to slavery for life.

Always, of course, there were head-hunting tribes too warlike to be approached by the exploiters. Again there were others who, friendly enough to earlier traders, went wild on first experience of some wanton cruelty at the hands of the rubber adventurers. The Araras of the Madeira are a case in point. For almost a century traders from Manaos and Serpa had been regularly visiting them to distribute iron hatchets, cheap cotton cloths and other wares among the headmen and depart with sarsaparilla, turtle oil and copaiba balsam. With the first influx of the rubber men in 1853, however, these friendly souls changed overnight into demons who fell on canoes with club and arrow and indiscriminately massacred white traders and Indian crews. And for two or three years thereafter they held the river against all comers. In the case of these Araras, the legend is that they turned berserk after a rubber trader had fired into a family, killing the parents and carrying off the children as slaves. True or not in this particular instance, the tale is typical of latex-hunter treatment of the aborigines. And it turned others than the warlike against them, for as the fate of the workers enslaved by the *patraos* of the first upriver forest rubber districts became known, the timid among the Indians grew too shy to be coaxed with gaudy trade goods and gaudier promises. There was a labor shortage on the Amazon. That could mean resort to but one thing - the slave hunt. Arming their employees, the rubber barons would surround the chosen jungle village and close in on the defenseless Indians, preferably by surprise and with as little bloodshed as possible. If many had to be killed, however, it was regretted only because the number of work beasts captured had been reduced to that extent.

On their rubber beats these poor animals so illegally but thoroughly enslaved died like flies. Nor were the tin masks, the bull whips, the iron collars, the log-and-chain devices and the other frequently inflicted punishments of inhuman variety chiefly responsible for this. The treetappers were literally worked to death. Peculiar indeed was the constitution of the red aborigines of the warmer latitudes. By habit they were not even cultivators of the soil to any extent. Hunters, fishers, childlike irresponsible idlers they had been for generations. Strong enough they seemed — and were — when left to their own careless life of forest and river, but under regular, confined labor they could not hold up. Physically, the result was a serious undermining of the constitution; mentally, the loss of the will to live. And if this was true in the case of any sort of work systematically performed at appointed daily times, it was horrifyingly more so when the bondage involved toil and starvation in the swampy jungles and suffocating smoke-huts of the rubber estates.

In later years, the labor mainstay of the Brazilian 'industry' was a new class of workers brought in from outside the rubber lands, but in the other nations civilised and half-civilised Indians predominated. Augmenting their numbers were the *cholos* or *mestizos*, Indians with a trace of Spanish blood, and the wild aborigines of the forests. And in the twentieth century as in the four centuries preceding, these aborigines were subjected to organised slave-raiding of the bloodiest and most inhuman sort. By rubber men they were assailed and by slavers seeking not rubber but youths and girls to be sold to the rubber men. Slavers fell on friendly or hostile communities without discrimination. Rubber men usually had trade goods to offer and often it was only necessary to bribe the headmen to put their constituents to work. To convert the stubborn, however, or to re-convert the rebellious who repented of their rubber-picking, the good old technique was resorted to. Riflemen fired houses, caught as many inhabitants as possible, shot down the fleeing ones and often murdered the useless among their captives. Sometimes the prisoners were set to work in their own precincts, sometimes they were torn away and carried to strange regions where labor or women were in demand. The same policy of blood and iron applied when armed expeditions were sent out to overhaul such Indian groups as attempted to bolt after experiencing the life of rubber gatherers. And often, in the event of such flights, magistrates and other government officials took active part in capturing the fugitives and returning them to their bondage. In parts of twentieth-century Bolivia these things happened, in remote Brazil even, throughout jungle Peru from north boundary to south, but above all on the upper Ucayali and the Madre de Dios rivers of this last-mentioned nation, in the lands disputed by Peru, Ecuador and Colombia, in the lands along both sides of the line representing the Colombia-Brazil boundary and the Southern portion of the Colombia-Venezuela boundary.

In the closing years of the 1870s and opening years of the 1880s the Indians of Peru and Bolivia started to become acquainted with the rubber men as latex-gathering attained importance in those countries; breeds imported by wholesale from the north eastern coastal states, particularly from Ceara, were first injected into the Brazilian system; expanding industry touched off a new demand for rubber. The result was an Amazon exportation of eleven thousand tons in 1883, a doubling of production in thirteen years. In 1890 it was nineteen thousand tons out of a world total of twenty-nine thousand. By 1903 Amazonia alone was exporting more than thirty thousand tons of rubber annually. The motor era had arrived, and in 1910, beginning of the mass production of auto tires in the United States, world exports of wild rubber hit their all-time peak of eighty-three thousand tons. Over half of this came from the Amazon. For the five years 1909-1913 the South American river territory sent out an average of forty-two thousand tons annually. In its peak year of 1912 it exported over forty-four thousand tons.

Dominant, then, this original home of *caoutchouc* remained after more than a century of exploitation, dominant in the years when production in the other great wild rubber countries was at its height. And to far greater extent than Amazonian rubber dominated world production, Brazil, first source of commercial *caoutchouc*, continued to dominate the river basin's production. Of the 2,250,000 square miles of Amazon *caoutchouc* territory, 1,486,900 is in Brazil, and much of this is closer to the one outlet than are the rubber lands of any other South American nation. Some 424,600 of the square miles lie within the state of Para; 694,800 in *Amazonas* (which first passed Para in rubber exports in 1887); 308,800 in Matto Grosso (first exploited in 1900). The remaining 58,700 represent the whole of the tremendously rich federal territory of Acre, formally acquired by Brazil in 1903. A record year for these combined districts was 1912, and to the river's total they contributed 37,570 tons. And from non-Amazonian sources, Brazil shipped over four thousand tons in that same year. Dominant also Brazil remained in quality of product and price per pound received therefore. Over eighty per cent of the Amazonian rubber it sent out was coagulated milk of the wild *Hevea*; of this approximately sixty per cent was graded as 'fine hard para' — the best and highest - priced rubber known to the world. The rest of the Brazilian Amazon's exports consisted of caucho from the Castilloa trees, first heavily exploited there on both downriver and upriver tributaries in the 1890s; its non-Amazonian shipments represented manicoba rubber from the Manihot trees of Ceara and adjoining territory and mangabeira from the Hancornia, a fruit tree found in a number of Brazilian states. For all vast Brazil, the jungle product long was the second ranking export, accounting for twenty per cent or more of total exports. It was exceeded only by coffee, and in importance as a revenue-producer for states and federal government it outranked even that commodity. Trailing second to Brazil, both in total shipments and percentage of *Hevea* milk to the total, was Bolivia, where for a long time rubber constituted twenty-four per cent of the export trade. First trifling attempts at rubber exploitation were made along the Mamoré river tributary of the Madeira in 1864, but it was not until the 1880s that real development began. Bolivians pushed it in the Madre de Dios, Beni and Mamoré river regions comprising the present north of Bolivia; Brazilians in the vast tract north and west of the Acre river which then constituted Northern Bolivia, although mildly claimed by Brazil. And to such extent did they labor that output bounded to 290 tons in 1890, to some three thousand in 1898, to more than 3,400 for 1900 and again for 1901. And then came the sharp falling off which more than halved production by 1904. The mounting *caoutchouc* demand had brought the Acre territory dispute to a head and the smaller country had lost out. The trouble began with Bolivia's setting up a custom house on the Acre river for the collection of export taxes. The Brazilians, whose Cearense workers were gathering most of the rubber and shipping it to Manaos and Para, set up a defiant howl. Their government set up a renewed claim to the territory. Acre proclaimed itself an independent republic. Bolivia sent troops in and Brazil sent more. Assorted disorders followed until 1903, when the dispute was settled in the customary treaty manner, the larger country taking the rich *caoutchouc* territory, the smaller one trying to console itself with a measly indemnity of $10,000,000 and the promise of a Brazilian railroad that would aid it in getting its rubber to the world. Production in Bolivia's Territory of Colonias, South of the Acre, mounted by leaps and bounds in succeeding years, but not until 1911 did that nation equal the 1901 figure. Its all-time best showing, a total of 5,751 tons, was made in 1917, a world war year when all *caoutchouc* was in something of demand, although plantation rubber three years before had put the wild far out of the running. In that year the Acre territory had contributed almost ten thousand tons to the Brazilian total, in its best year (1912) some 11,500. Had it not lost that region of rich rubber forests, then, Bolivia would have been producing half as much rubber as Brazil during the great days of wild rubber. It still has, it is estimated, two hundred thousand square miles of the Amazon rubber area.

Interesting, too, it is to note that among the innumerable classifications of 'fine hard para' resulting from the general belief that quality of the product from certain localities differed, the highest ranking went to rubber from Bolivia, the Acre Territory and the Madeira river and that this always brought a fraction of a cent per pound more than the ordinary upriver 'fine', several cents more than the 'fine' from Peru, from the Tapajos river section of Brazil, from the islands near Para or from the other *Hevea* milk sources. As to the gathering and preparation of this latex in Bolivia, it went on as in Brazil. Brazilian Cearenses, too, were the laborers in the extreme North of the territory remaining to Bolivia after the Acre grab, but for the rest they were the native Indians or the half-breed *cholos*. As docile as dumb were both aborigines and breeds and, caught in a net of peonage laws or enslaved in regions beyond all law, they were exploited by masters even crueller than the Brazilians.

Caucho gathering there always had been in Bolivia, but, as in Brazil, it never came close to threatening *Hevea* supremacy. Peruvians were numerous among those who hunted it and eventually nearly all of the readily available Castilloas were destroyed in Bolivia as in Peru. Not only through Peru, northern Bolivia and far up many of the Brazilian rivers did the *caucho* chasing woodsmen of Peru penetrate, but all through the rubber regions of South America as far north as Central America. Added to its *Hevea* yield, *caucho* production made Peru third among the South American rubber countries. So late as 1912 and after thirty years of intensive Castilloa tree destruction, *caucho* amounted to a thousand tons, or more than a third of the Peruvian total, but rapidly indeed it has fallen away since then.

Ancient documents show that in 1853 some three tons of *Hevea* milk came down the Amazon in the first recorded shipment of rubber out of Peru, but it was not until 1882 that businesslike collection of *caoutchouc* began and it was the Castilloa tree that then was exploited. Speedily, annual yields of all rubber mounted beyond the two thousand ton mark to qualify as third national export although the whole forest region accounted for less than four per cent of the nation's population. Top year was 1915 with 3,347 tons of *Hevea* and Castilloa milk going out of the country, which possessed 225,000 square miles of Amazon rubber territory.

Of Peruvian rubber, 'fine' from around the headwaters of several of the streams flowing on into Brazil and Bolivia was among the Amazon valley's best, but a large proportion was 'weak fine' from *Hevea guyanensis* forests of the Amazon's Northern tributaries and particularly from the bloody Putumayo region.

Colombians, too, had a part in the Putumayo work and with Venezuelans they share the dishonor of having established some of the worst of the *caoutchouc* hells on tributaries of the upper Orinoco and Negro. Chiefly gathered by tormented Indians, shipments of latex from these areas cleared through Manaos, Brazil, or Iquitos, Peru, and were credited to these last countries since, in the absence of actual production records in the case of any wild rubber country, it is customary to regard export figures as equivalent to production. The four hundred and some tons annually averaged by Colombia throughout the great wild rubber years, the four hundred and some averaged by Ecuador and the two hundred and some by Venezuela, represent, then, only the Colombian and Ecuadorian *caucho* gathered west of the Andes and sent out of Pacific ports, only the Venezuelan rubber originating north of the perilous rapids guarding the upper Orinoco.

Altogether negligible were exports from the Guianas and the West Indies, but other parts of the Americas did have their periods of importance. Central America's lasted the longest. Caucho from its Castilloas and the poor rubber native to the *estrada* hemisphere entered the market in the 1850s as earliest competitors of the South American product, and the Isthmus held second position among world rubber producers until the rise of the African industry, then third position until surpassed by Mexico. It was a thrilling tale the *caucho* gatherers wrote throughout these middle Americas. Chiefly they were the breeds (Indian and Spanish with often a flick of the tar brush) of the coast towns, but West Indians, Spaniards and scattered adventurers of many a nation were included in their number. Cutting down the trees was one of three collection systems. The *cauchero*, appalled by the labor involved in chopping, used a system that killed the rubber-giver quite as effectively, however. Mounting into the branches on a ladder of vines, he would descend, cutting huge V-shaped gashes every two feet. Another idea was ringing the tree with a spiral cut, driving in an iron spout at the bottom of the spiral and receiving the milk in pails. In any case the heavy work of smoking was avoided by emptying the milk into barrels or pits in the ground and using an anciently known vine juice to accelerate coagulation. Early exhausting all of the Castilloas in the more accessible forests, these rubber hunters were forced to penetrate farther and farther into the unknown wilds. Nicaragua, greatest of the isthmian rubber countries, bred *caucheros* equal to Peru's, and in the later years it was no rare thing to find strays from their number as far south as the Amazon basin. Typical of their early explorations was the opening-up of the Rio Frio, Costa Rican tributary of that important highway into the interior, the San Juan river. About this unpenetrated stream clustered fabulous legends of golden cities and white aborigines, but more than counterbalancing the lure of these wonders were highly colored yarns of native ferocity. The tales sprang originally from the fact that the Spanish conquerors never had subjected these peoples and they thrived apace when Indian arrows wrote *finis* to later exploration attempts. So matters stood when the Nicaraguans began to eye this likely-looking stream. And not long was it before a party of these rubber men dared the adventure. They found no white Indians) and the red warriors they mercilessly mowed down with gun fire and drove deeper into the woods. And they found no settlements other than the familiar type of Indian village, but they did find and drain Castilloas in such quantity that their unlooked for return was a signal for a rush up the river by other parties. So ruthlessly did the invaders use their firearms that mere appearance of a boat soon was sufficient to send the Indians flying into the woods leaving their poor belongings and such luckless children as were lost in the shuffle to be gathered up by the rubber men and carried off downstream with the loads of *caucho*.

And the history that the *caucheros* wrote in Nicaragua was duplicated to some extent in Guatemala, Panama) Costa Rica, Honduras, Salvador and British Honduras, all of which produced coagulated Castilloa milk) in order of importance as named. In each, operations began with destruction of the readily available trees and settled down into a business of outfitting men in the communities that had developed into rubber centers and sending them into the deepest wilds on expeditions lasting several months. The merchants ran the risk of *cauchero* death in the forest and the commoner one of *cauchero* dishonesty in making for some other rubber center with his coagulated milk instead of bringing it in to the man who had fitted him out, but it was a highly profitable trade at that. It was killed by the blows that felled the trees of the interior. Production naturally slumped as the rubber forests were decimated, slumped right through the periods of greatest demand and highest prices for wild rubber. In some of these countries laws were passed specifying severe penalties for extracting latex by cutting down the trees, but these regulations, even if enforceable, came too late.

Mexico's original rubber export was also *caucho* from the Castilloa. A long rubber history is Mexico's, as our consideration of the Spanish chroniclers has shown, but only briefly — and in this present century — was it really an important rubber country. Wild, its Castilloas were chiefly in the states of Chiapas, Tabasco and Vera Cruz although they were to be found in the other Southern states as well. Despite its greater geographical area, however, this district always trailed various of the isthmian countries in production and export until it added plantation to wild *caucho* and leaped far ahead of all Central American countries combined. This happened coincidentally with the development of a new rubber source in the nation's far North near the United States boundary, and the result was that Mexico for a short time became the greatest rubber-producing nation the world ever had known, Brazil only excepted. Statistics tell the spectacular story. In 1900 the exports were 194 tons (to Central America's 774) — all *caucho* from wild trees. In 1905 they were 825 (to Central America's 639) of which 490 represented wild and plantation *caucho* and 335 were *guayule* rubber from Mexico's far North. In 1910 — Mexico's peak year - they were 17,481 tons! Of this record amount, 9,542 tons were *guayule*, while plantation *caucho* represented much of the remainder. The strange story of *guayule* and the frenzied story of the Mexican plantations belong elsewhere, however, and there is nothing in Mexico's wild Castilloa story to detain us.

It is worthy of note, however, that in Mexico as in all the *caoutchouc* territory outside of South America, wild rubber production either had faded away or was markedly on the downward trend even before the plantation product from the Middle East rose to dominance. Only in the valley of the Amazon were the wild rubber sources really inexhaustible and so Brazil was the only wild rubber country that showed a continually rising production from the beginning of exploitation until the plantation triumphed.

Despite continually mounting exports, there never was actual overproduction of any of its rubbers, however. As to 'fine hard para,' it was the one rubber always most in demand throughout the world and more especially in that greatest of all rubber-manufacturing countries, the United States. It was an Amazon monopoly in the days of wild rubber, and even after plantation *Hevea* milk appeared on the market, faith in the wild para was so strong that only necessity could induce tire makers to try the rubber from the Middle East. Never, therefore, did Amazonian rubber sell at anything properly to be called a low price until after the plantation product had changed the rubber world. In 1870, fine hard para sold at eighty cents to $1.07 a pound, as we have seen. Thereafter the lowest figure it ever hit was the forty to fifty-two cents of 1878 and this never happened again. Throughout the 1880S, the 1890S and 1900s it fluctuated around the sixty-, seventy-, and eighty-cent marks when left to itself, hit fantastic levels under manipulation - $1.25 in 1882-1883; $1.12 in 1898-1899; $1.50 in 1905-1906; $3.06 in 1909-1910. Usually it is 1910 that is spoken of as the time of the Brazilian rubber boom; less often the whole period from 1900 to the crash is cited. In a far truer sense, however, the history of South American rubber from at least the 1880s forward is the history of one long boom. The 1909-1913 period or, if you prefer, the 1900-1913 one was but the crest of a price and production rise that started its spectacular upward roll in the days of the first heavy industrial use of rubber. A few depressions wild rubber did experience, but they were of brief duration. Always when prices had fallen to a comparative low, there was a tumble in the value of the *milreis* which reduced local costs and, shortly, providence would provide some such fortunate coincidence as a war, a bicycle-craze or the invention and multiplication of the auto to quicken demand to a degree never dreamed before. Brazilian spending orgies of 1910 when that unheard of $3.06 price was reached differed no whit in kind from the wild squandering of previous decades, nor did the 1906-1910 stampede of tappers into the rubber lands differ from the one staged more than a quarter of a century before. When this book refers to the South American boom, then, it means the whole gigantic mushrooming and not just the more extraordinary price spurts which were merely a part of it. We turn now to the story of that mushrooming along the rubber river.

2.2 Jungle Boom

In part, the rubber boom story is one of 'black gold' multiplying the few Amazon steamers to fleets, crowding the river with ocean liners, sowing its banks with settlements, building Para into a dazzling metropolis and Manaos and Iquitos into roaring cities and carrying Northern Brazil to fantastic commercial heights. It is also the story of rubber sending small boats into all the malarial, death-breeding backwaters and penetrating the remotest reaches of the tributaries. Where the steamboats left off, the launches took up and where the launches came to an end, the canoe carried on. In the rush from Para that accompanied the beginning of the boom, the rubber-seekers had scattered up every main tributary. Up the branches of these tributaries they went in the years that followed, up the feeders of these branches, up the bifurcations of these feeders, shoving recklessly into the unknown so long as rubber was to be found and there was water enough to float a canoe. So far were they from a world they helped supply with its great elastic commodity that streams which would be unknown to government or mapmakers for decades were their familiar commercial highways. On the 'River of Doubt,' for instance, they had been located for twenty years before the eminent T. R.'s arrival. With all the penetration of remote places, however, with all the importation of labor, all the springing up of villages, all the swarming of men to the several boom cities, vast Amazonia so easily swallowed these new arrivals that the population average never even came close to reaching one inhabitant a square mile (an estimate of the early 1920s gave 1,500,000 as the sum total of those living within the 2,250,000 square miles of the rubber territory). Never was this jungle empire in any sense 'occupied,' then; it was but partially invaded.

Financing the invasion were the commission agencies and the great supply houses which had come into being at Para and Manaos with the first rubber rushes. These establishments, headed by the Portuguese shop-keeper of a few years before, extended unlimited credit both to *patraos* operating established rubber estates and to adventurers headed for new forests. The agencies bilked their clients on prices as they in turn mulcted half-breed serfs; although the prices which they paid made some of the jungle-invaders rich, they made all of the merchants richer. And at the top of the heap, eventually, were the exporting houses, which became the ultimate bankers of the 'industry' and piled up fantastic profits that dwarfed those of store owners and 'plantation' kings combined. Many were the stages through which rubber passed. From the tapper or *seringueiro* the *Hevea* milk went to the estate owner or lessee and theoretically was credited to the laborer's account. From the estate *patrao* it passed to the trader or *aviador* who had supplied the merchandise on credit and who himself had obtained the goods on credit from a wholesale house. This middleman made his profit by charging the *patrao* a 10 per cent commission on goods supplied and pocketing a bonus from the wholesaler for placing the order with him. From *aviador* to supply house went the rubber, then, and finally to the exporting firm through a commission-collecting broker. The *aviador* vanished as the boom collapsed; and long before that, of course, there were some wholesale houses exporting direct and a few such immense firms as Bolivia's Suarez Brothers which handled every stage in the marketing of their own production. But the complicated system above outlined is the one that generally prevailed throughout the Amazon in wild rubber's, wild days.

Exporter, broker, store owner, *aviador* — into these spots, then, swarmed the Portuguese first on the scene and, constantly, their numbers were augmented by new arrivals from home. Servile illiterates, clog-shod and with rear ends still aching from the bootings regularly administered as a part of the routine of Portuguese commercial life, debarked in Brazil with little save low cunning in their baggage. In a few years they would bounce from penniless apprenticeship or rudimentary retailing to supply store ownership, or perhaps to commission operation with fleets of flat-bottoms on the Amazon, estates in Portugal, parasite fortunes founded on the struggles of those who died in the jungle.

There is a great commercial saga in the story of the wiliest of the Portuguese traders, J. G. Araujo of Manaos, who rode the rubber boom to millions and became the king of kings in *Amazonas* with ships and trading posts and agents up all the rivers, who bought the rubber from the little kings of the tributary rivers and supplied them with their champagne and mirrors and music boxes, who still sends them their farina and dried fish today and remains the greatest power the Negro and the upper Amazon have known.

There is another such tale in the career of those amazing Bolivians, the Suarez brothers. Their sternwheel steamers, motor launches, *batalones* (cargo lighters) and canoes plied all the rubber rivers of their native land. At Cachuela Esperanza, between Riberalta and Villa Bella on the Beni, was their own town, built around headquarters and machine shop, and here all rubber coming out by that river paid toll to pass a half mile of rapids in push carts trundled over Suarez owned tracks. Next to the Beni, the Acre river was the principal exit and Cobija, accessible to flat-bottomed sternwheelers six months of the year, was its chief shipping point. To reach it from their establishment at Porvenir, the brothers hacked their own private cart road through 20 miles of jungle, transporting first with mules, then with motor cars and finally, in the era of rising gasoline and falling rubber prices, with oxcarts. All these were minor matters, however; the important Suarez feat was the acquisition of more than half the rubber *estrada* in all Bolivia! Their exact holdings were 21,265 of these circuits (over 16,000,000 acres) out of a national total of 40,000. Corporations Swiss, German, French, British and American, a number of them owning as many as 1,500 *estrada*, gave them what opposition they had, for of small individual producers there were virtually none. Like nearly all the real history of the Amazon, both the Suarez brothers' story and that of Araujo remain unwritten.

On Julio Cesar Arana of Peru there is more in print — chiefly British, Peruvian and American government documents. A peddler of Panama hats was fat Cesar when he started out; he ended up claiming 14,200,000 acres in the rubber forests of Colombia and Peru. Much concerning those acres of Hell will be found in the next chapter. Not so wholly as in Bolivia were the Peruvian *caoutchouc* forests tightly held by a few powerful corporations and individuals, but there were many important concessions, including Luis Morey's 400,000 acres and the Javary and Ucayali holdings of Israel & Co., the British operators of three-quarters of the steamers and launches on Peru's tributaries of the Amazon.

All the products of the luxury world these steamers carried up the Amazon: tons of pink and green glazed tiles for forest mansions, champagne, *paté de foie gras*, the most expensive of English biscuits, French lingerie to adorn the naked Indian wives of mongrel-blooded rubber 'kings' on rivers far beyond steam. Pianos, even, were carried by the steamers to the end of navigation, transferred to rafts, hauled a thousand miles farther up stream — and, once delivered, were never played.

At all the principal rubber estates the steamers made stops, and elsewhere two shots of a rifle were sufficient notification that a passenger wished to come aboard. Nor were these all of the halts. Even when well on the way up a tributary, it was occasionally necessary to wait a day or so for a sudden rise permitting further journeying. Other navigation difficulties make quite a story. Some of the Amazon's branches are among the most crooked in the world and there are stretches over which no ship would proceed on a moonless night. Sandbanks and submerged logs were a common menace and, even when no accidents were met with, it was far from a rarity for steamers to fail to escape low water. Surprised upriver after too long a stay, they would have to come tearing back down, ignoring all cargo waiting on the wharves and lucky if they did not ram a sandbank and hang rotting and rusting for months until the next rising of the water.

As to the independent plantation *patraos* served by these steamers, they were of all varieties. And in Brazil their number was legion. Here there had been no such wholesale seizure of astounding areas of public lands by a few individuals as went on in Bolivia and Peru. The great majority of these *seringals* ultimately fell into the hands of the commercial houses of Para and Manaos, true enough, but this was a case of none-too-eager acquisition dictated by the debts piled up before the crash. In rubber's boom days the comparatively small holding was common in Brazil. Men of all nations were the proprietors, but the greater number were Brazilians of more or less Portuguese blood; their serfs were the wiry breeds from the coastal states south-east of Para. Many had come from the same sort of social strata as their workers, a few had risen from the actual tapper ranks. Even more ignorant than the merchants, they were as strong-willed as the city men were cunning. Starting often with no capital save their inherent ability as brutal drivers and enslavers of their countrymen, numbers of them ended as paper millionaires. Sometimes they carved out or acquired their own *seringals*, sometimes they leased from the actual proprietors. Variety was added by British, American, French, German and Swiss corporations erected as fronts for wily South Americans hot for expanded credit or actually organised to take a flier in Brazilian, Bolivian or Peruvian rubber extracting. The *bona fide* foreign concerns sometimes employed managers of their own nationality, sometimes left everything to a native go-getter. Often the conference table boys in London or New York knew nothing at all concerning the men directly in charge of their interests; there are records of such corporations being represented by South American dolts who could speak neither Portuguese nor Spanish and had to call in assistance in making up their reports since they were unable to keep accounts or even to write. Under almost any conditions, however, there was big money in the business — even though few indeed were the Brazilian *seringals* not mortgaged to the commercial houses. During all of the boom years cash as well as merchandise was easily obtained on advances to be covered by future deliveries of rubber, and the profits on sales of goods to enslaved laborers constituted a gold mine in themselves.

And it was worth a fortune to live in the rubber lands even as an idling proprietor. Slack dry season currents of numerous tributaries never visited by the Amazonian trade winds meant stagnant water in regions of stagnant air, and one of these regions differed from another only in the degree to which it was fever-dominated. At worst, entire towns — and comparatively important ones — were afflicted with malaria. And everywhere, main stream included, hookworm reigned. So universally did it ravage and enfeeble the dwellers in the forest and country districts of Para that the entire state had an 80% prevalence. In *Amazonas* 96.4% of the populace was infected. And so prevalent was leprosy that Para authorities once counted over two thousand victims and probably missed at least as many more. But if climatic unhealthiness and the neglect of elementary hygiene offered little choice, climatic annoyances did vary greatly. To foul water, death dealing insects and complete disregard of sanitation, all universal in the Amazon country, some of the upriver districts added a vapor bath atmosphere in which biscuits sprouted whiskers, salt refused to remain solid, sugar dissolved into syrup, a gun loaded overnight could not be fired in the morning and even gun powder in canisters turned liquid after several days.

Varying, of course, were the degrees of isolation of the *seringals*. Some, as on the lower Madeira, had steamboat service the year around. Others, not too far up smaller tributaries, had it during the six months of the dry season but were cut off from the world half the year around. Numbers, remote from any settlement, received no mail save by courtesy of some wandering rubber hunter or explorer. And with rubber prices and Indian attacks almost the only topics of conversation among the p*atraos*, yet they were always months behind in their knowledge of figures quoted on 'black gold.' Still other spots there were that no white man or breed other than the rubber exploiters themselves ever had seen. So ignorant of the world were all of the Indian tappers and many of the breeds that they had no slightest comprehension of what the white man wanted with the jungle stuff they so laboriously gathered, had no more idea of auto or auto tire than of the New York Giants, the philosophy of Schopenhauer or the obligatory chastity of Harry Lehr.

Little more can be said for both their masters and their masters' masters, however. Greedy as were exporters, dealers, 'plantation' operators, sharp as were export houses in gouging supply stores, supply stores in gouging rubber estate *patraos*, *patraos* in gouging workers, the whole trade was based on ignorance so profound that the Amazon valley never housed a dealer or producer able to calculate the production cost of rubber there. All that any of them knew was that it had to be less than the price paid the estate owner which in turn had to be less than the price received by the dealer supplying the *patrao* which in turn had to be less than the price obtained by the exporter. On this basis, then, the rubber bubble swelled and ballooned and rolled forward. We have mentioned one special oddity of the gamble — the fact that nose-dives in the value of the *milreis*, although fatal to many a Brazilian business, were always a gift to the rubber dealers. At one end was the rubber exporter being paid in gold in the markets of New York, London, Paris and Hamburg; at the other were laborers paid, when at all, in paper. The workers who were paid received no more paper when the rate of exchange was low than when it was comparatively high, the unpaid serfs consumed no costlier beans and mandioca, for on these necessities money market antics had little if any effect. In the exporter's case, then, low prices for rubber in the occasional depression periods were compensated for by the larger number of *milreis* his gold dollar would buy, and because of this he could afford to carry the plantation operators with lavish advances based on expectation of the next rise in rubber. And in good times or in bad, the *patrao* was able to maintain his Manaos or Para mistresses in the style to which they were accustomed, to make his trips to Europe and the usual round of vice resorts or cures, to levy on the supply houses for the expensive watches, furniture, diamond rings and costly attire beloved of his kind, and to figure on settling for all this at some vague time in the future. To such rubber tappers as were not absolute slaves, high and low rubber prices made a difference — but a slight one. When things were at their height, no *patrao* would allow the men to leave his 'plantation' store without having canned goods, liquor, perfume, patent elixirs, fake jewelry and gimcracks of all varieties foisted upon them at ridiculous prices. When prices were down, the tappers found difficulty in obtaining the bare necessities of life. Either way, however, the majority remained caught in the debt trap.

Naturally, it was up the far reaches of the smaller tributaries that the really picturesque rubber 'kings' flourished. Here any ruffian who could enslave an Indian tribe, lure hapless half-breeds into the rubber lands, or dominate the fugitives from justice all too common in these shadowy boundary areas might become a feudal ruler beyond all law save that of lash and gun. Not of the trip-to-Paris type were most of these *patraos* but brutes well content to be lost to all the world save the dreadful one of their own making. To these despots, too, the launches brought luxury cargoes, but there was no temptation for them to ride the boats back down. Some were men marooned in the jungles by the price on their heads; some were Spanish, British, German or American wanderers driven by inner necessity to remain beyond a civilisation of which they had had more than a bellyful; some were native to wild places or were blacks sifting down from the West Indies, Panama and the Western coast towns of South America in whom the atavistic jungle urge was overmastering.

At the end of the Isana tributary of the Negro in Colombia ruled a Corsican killer who had escaped from Devil's Island off the coast of French Guiana, who had found wild sanctuary in the forests, subdued and enslaved a settlement of fugitives from 'plantation' hells on half a dozen rivers, brought the Indians under his heel and become a rubber king in as lost and lawless a region as a trackless continent owns.

Up the equally remote Uaupés, the explorer Gordon MacCreagh found a black ruler of rubber days still bossing that water highway years after rubber from South America had ceased to be in demand. His Indian subjects had fled up remoter creeks; decay had overtaken his lath, plaster and tile palace with its nude ceiling decorations painted by imported artists; his patent mechanical band instrument had fallen apart; and the king lived in a part of his aforetime rubber shed surrounded only by a half a hundred immediate relatives of his numerous wives. He was still lord of that stretch of the river, though, still waiting for wild rubber days to return or the demand for rubber's cousin, balata, to have the same effect.

In all the upper Amazon it was considered that the adventurer who first went up or down one of the smaller rivers and set the Indians to rubber gathering acquired ownership of stream and aborigines thereby. For another to enter that water highway was 'piracy' punishable by being shot at sight. To take away any of the first corner's Indians or even to enter into any sort of business relations with them was 'theft,' and such crimes bred reprisal expeditions and private battles reminiscent of the baronial warring of medieval times. Nor did the white, black or mixed despots fall beneath poisoned arrow or rival's bullet only in the remotest forests. Many a proprietor of the more civilised *seringals* was done in by his brutalised slaves, and there were others who died in the boundary battles of *patrao* against *patrao*. Among the exploiters invading new territory on the larger streams, the custom was to stake out their claims by cutting a mark on river-bank trees at either end of the desired stretch. This served well enough to establish river frontages, but lines back from the water often had to be marked in blood. Usually this happened when rubber barons worked back from parallel rivers and collided. And it kept on occurring even in the days of title deeds from the government, for these documents usually stated no exact depth but permitted the properties to run back until they reached an unknown part of the forest or any part already being worked.

Rubber thefts constituted another potent source of combat. Somewhere in an estate claiming hundreds of thousands of acres, trees would be raided and the boss would hand out the Winchesters to his headmen, fill a couple of boats with serfs armed with trade guns, and move against a near neighbor thirty or forty miles away. Especially common were seizures of the coagulated latex. One chunk of rubber is so much like another chunk: and letters stamped on with hot irons were so easily changed or removed that establishing title to any lot of the stuff was well nigh impossible — save when that establishing was done by a man looking down the sights of a gun. It was one *patrao* raiding another's shed or collaring his boat load that touched off these wars, but that was far from an end of the stealing. Overseers credited workers with half of what they brought in and disposed of the remainder on their own account. Serfs would hide out a part of their collections to trade with the wandering Syrian peddlers who braved jungle, river and lash-law of the *patraos* to carry their trinkets to God-forsaken spots where they were good for 'black gold.' From the very sheds of the importers on the Para wharves the stuff vanished, and en route to manufacturers and even in the United States as well.

With the *caucho* hunters it was customary to establish title to trees by cutting a distinctive insignia into the bark and to return to fell them only after the wanted number had been so blazed. Times there were when these symbols of ownership were disregarded, but it was at the risk of life if caught in the act. Easier was the lifting of the coagulated rubber that had been left in holes in the ground to harden. *Caucho* exploiting companies solved this with warehouses and guards, which reduced their theft problem to the amounts which workers and overseers succeeded in getting away. Among the wandering independent Castilloa hunters theft remained a matter for blood-letting, however, since there were always numbered among them rogues who would steal another's accumulation with no more compunction than they felt in planting a bullet in an Indian or disease in a raped Indian wife.

The whole melodrama of rubber-gathering in Amazonia is a story that can never be told, of course. It is a history written in substantial part by lone men and small parties adrift in the rubber forests; it is a narrative composed of thousands of individual tragedies and terror stories, one which could be properly recorded only by an omniscient author with access to the separate details of each of those obscure, unrecorded lives, with no space limitations upon his output. We can suggest, can cite highlights, that is all. A complete Amazon River anthology would deal, remember, not only with the tough adventurers hunting likely centers for *Hevea* tapping or trailing the Castilloa up the Purus, Jurua and the cannibal-infested Javary rivers in Brazil and through Peru and Colombia. It would tell also of the greenhorns on the same missions. It would recount the desperate sagas of criminals fleeing from justice on the other side of the Andes, of those who bolted the *seringals* with or without first committing rape or murder and found themselves not only up against impassable jungle but up against it without proper weapons, tools, provisions.

Included under these subdivisions would be the stories of blow-gun aborigines ambushing white intruders; of man-eaters castrating foolhardy rubber traders and fattening them in pens until the proper stage of plumpness was reached; of whites torturing Indians, burning their hovels, ravaging their planted fields because food or paddlers or rubber gatherers had been refused — or for no reason at all. There would be the chronicles of those queer pioneers, rubber men with wives and children who became permanent settlers squatting along the jungle rivers so remote their very existence was unknown to church or state. There would be all the individual stories of fortune hunters dashed to death in careering rapids; of native women ganged by half-breed parties; of white killing white over a load of 'black gold' or brown woman; of monstrous diseases fastening upon men in return for the 'social' sicknesses they brought into the jungle. And most tragic and frightening of all, the tales of the bewildered and fear-shriveled lost ones tortured by hunger, thirst, fevers, relentless insect clouds, deluging rains, staggering through clutching swamps and flesh-ripping underbrush until reason and humanity vanished and it was a mad animal that death pulled down.

That is a hint of what went on in the theatre of hairy spiders, stinking orchids, blood-sucking vampire bats. It is a story of adventures in the far corners where no two individual dramas were alike. With everyday laboring life in the jungle it is possible to deal in mass terms as we have done in recounting the fate of the earlier Indians in the 'industry,' as we will do in sketching the labor story of later Indians and imported half-whites. It is possible, too, to paint in broad strokes the rubber boom's effect on the cities where crawled the little wise ones, the merchants, officials and bloated parasites. We must record, too, the immense civic improvements in the cities of Para and Manaos as the states of Para and *Amazonas* gathered in millions in rubber taxes. And there were projects far from the rubber country financed by the Brazilian federal government's tax on exports out of its territory of Acre. One thing all these improvements had in common — their power and works were confined to the cities, a revealing comment on the governmental attitude towards those whose dreadful travails produced every last *milreis* of the wealth. Out of the millions upon millions in rubber revenues, not one lonesome coin went back into the immense region that supplied it all. It was, perhaps, the largest scale slaughter of the layer of the golden egg ever witnessed on this foolish globe. Never a *milreis* for sanitation work. Never a thought of trying to drain the rubber regions, where death's mosquitoes swarmed. Never an attempt to clear a portion of the jungle, encourage crop and cattle raising in these reclaimed areas, assure the Amazon valley of an indigenous food supply and cut the preposterous living expenses of the exploited toilers among the trees. Had engineers, medical experts and plant authorities been turned loose, had cheaper transportation been supplied, untouched and impenetrable rubber regions could have been opened up, the number of permanent dwellers in the far districts could have been multiplied, immigrants able to thrive in reasonably healthy torrid climates could have been introduced, modernised methods of cleaning and coagulating the gum could have been taught, government-assisted plantations could have been developed, if necessary. Then, if the *Estrada* competition threat ever had developed, Brazil could have met it with a rubber free of foreign matter, an adequate and efficient labor supply, a transportation service imposing no excessive cost penalty. Eventually Brazil did talk of doing those things, but the goose was already dead.

That it could ever die was conceived by none while Brazil's Coolidge era lasted. It was Para city, of course, that first really felt the boom. Rubber carried it to a quarter million population. Entrance and exit for the whole Amazon valley, headquarters of the Amazon Navigation Company, it battened upon the latex trade and soared to commercial domination of all Northern Brazil. Not long did Para state have a monopoly of *caoutchouc* production, not at the height of the boom was it the leading producer, and yet its capital remained one of the two great markets and shipping points as the smoky hams from a thousand tributaries of the Amazon came floating down the river to the vast zinc warehouse sheds on Para city's wharves. There they were cut in half (exposing the rudimentary deceit of the simple-minded *seringuero* who had introduced rocks, axe heads, nails, wood blocks and inferior rubbers to add weight to his ball), graded, packed in cases and swung aboard the ships which eventually were visiting Para to the number of over 1,600 a year.

Exorbitant living costs hit this metropolis early. Already we have seen how agriculture was abandoned in the original rubber lands of Para state, how later the luring of the upriver Indians from their cattle and their cultivated fields stripped that part of Brazil of its food-producing lands. To this impoverishment of Brazilian productivity rubber contributed further by robbing the coastal states neighboring Para of their agriculturists and cattle raisers — men who should have been given government aid in their fight against peculiar rigors of nature instead of being permitted to flock away to leave their bones in the Amazon, their home states depopulated. Also contributing to the food shortage was the fact that with the abolition of slavery in 1888 (previous legislation had suppressed the slave trade but not slavery), great numbers of farm workers forsook their fields. All of these developments combined to make living costs exorbitant in rubber's boom towns, worse than exorbitant in the remote rubber districts. Para city was until as late as the middle of the nineteenth century one of the cheapest places of residence in the Western hemisphere. With the abandoning of agriculture and with rubber money pouring into the city, however, living costs rocketed to four or five times the New York City figure. And the farther man went up the Amazon, the higher mounted these prices. Yet so helplessly dependent on store food did the Manaos area's labor become that even in recent years balata gatherers have been known to travel for a month or so to reach regions rich in the trees (and potentially rich in agricultural products) and to give up and return after a few weeks' work because of failure of supplies to arrive from the capital of *Amazonas*. In the boom days, rubber workers so remote from the Atlantic as Bolivia ate tinned sardines, salmon and tuna on the banks of fish-filled rivers. Fruit, too, they knew only in cans. So even with sweet potatoes, a crop that could have been grown in the Amazon with little more effort than is required to turn this page. Nor was the economic penalty the only one paid by the river rubber lands for failure to produce food locally. Through that failure they brought upon themselves the dreaded scourge of beriberi.

Labor scarcity in cities as well as farm regions was another consequence of the reign of wild latex. Soldiers of fortune from all points of the compass, men toughened by farm and ranch life, were far from comprising all of the rubber adventurers. Portuguese youths a cut or several above the ordinary strata were drawn to Para by the El Dorado publicity, grew dissatisfied with store or office pay and routine and jumped counters to head up the Amazon on their own. So powerful was the lure of the upriver regions at the crest of the boom in 1910 that the tram service given Para by rubber temporarily was revoked by that same agency. So many of the motormen had lit out for the *caoutchouc* forests that operation of the lines went very thoroughly to pieces. Citizens demonstrated in the Latin manner by stoning the cars and, as annoyance mounted, burning them. Mobs numbering thousands joined in the raids and bonfires, traffic was entirely suspended and the cavalry had to be called out. Order finally was restored by the military, but bringing back the motormen from the rubber forests was beyond their power and tram movements remained exceedingly irregular.

It was an era of frenzied finance, public as well as private. In Brazil, as in the other South American countries, export and import taxes were and are the principal source of government revenue. Import taxes were uniform throughout the nation with the greater part of the total going to the federal government. Export taxes were left to the states and they collected with both hands. On rubber the great producing states of Para and *Amazonas* slammed levies (revised weekly in accordance with prices) which, combined with municipal and other charges, ran to a third of the value of the 'black gold' in boom days. Para city officialdom reveled in such a spending orgy as not even South America had seen before. Millions in graft were tossed about, millions went into boulevards, electric lights, telephones, street cars, schools, hospitals, sanitation, water systems, harbor works - an eighteenth century city was transformed into a glittering modern metropolis. Rubber gave it its opera house, its theatres, its swarming cafes, its hundreds of gambling houses. A rubber metropolis it was; bounded on one side by the bay and its great stone quay jammed with iron liners from all the oceans and with upriver craft bringing down the latex balls from hundreds of streams unknown to the geographers; bounded on the other sides by the forest pressing in where the pavement ended. A city of millionaires and lepers, of carrion-eating birds performing the duties of 'whitewings,' of streets hung with advertisements of syphilis cures and of hotels renowned among the vice traffickers of three continents. To Para, in those days, flocked the adventurers and the harlots of all the world. From there the men in pursuit of rubber fortunes scattered all through the valley of the Amazon. And the women in pursuit of the men with rubber fortunes divided to occupy Para and Manaos. Manaos, as a city in more than name only, goes back merely to the rubber boom; as a modern city just to the early years of the twentieth century. A federal port of entry complete with custom house, a rubber market exporting direct to the United States and Europe without reshipment at Para city, it ballooned in importance as Acre and Matto Grosso production developed, was even with Para city as an export center at the height of the boom (1910: Para, 16,687 tons; Manaos, 16,680), has led it two to one in the days of languishing wild rubber and dominant Middle East plantations. So late as 1902, however, it was still a mushroom metropolis with but two noteworthy public edifices, a church and an asylum; its port was only a river anchorage. Steam had helped make the mushrooming possible by cutting the time from Para to five to seven days instead of six to thirteen weeks. Rubber did all the rest. It skyrocketed the population past the 75,000 mark. It attracted $40,000,000 in foreign funds to build a stone quay and floating wharves connected by floating bridges, an arrangement made necessary by the rise and fall of a river the levels of which vary by as much as 30 feet and more. It poured into this capital of *Amazonas* an export tax revenue that in one year hit $8,000,000 at the rate of exchange then current. Millions for graft there were, then, and millions for stone and marble government buildings, monuments, schools, opera house, museum, library, zoo, public gardens, meteorological observatory, electric lights, water supply. Boom Manaos, it has been said, displayed more luxury for its size than Paris. It was one of the world's chief markets for the costly things, a city avid for the most expensive wines, silks and furniture obtainable, the greatest diamond buying community on any continent. In every street were jewelry stores, on the grimy fingers of down-at-the-heels clerks to whom the rubber boom was but a phrase, glittered diamond rings worth hundreds of dollars, diamond rings the absence of which marked any man as a confessed and pauperised failure. A thousand miles from anywhere, Manaos gave birth to a modern brewery, spawned suburbs and linked them to itself with the best tram service in South America. By way of contrast, it offered over two hundred lepers, most of them roaming at large. The Manaos opera house (estimates of its cost range from one to five million dollars) was built by the city fathers, who adorned the lobby with paintings by an Italian artist and stocked the house with a company of Italian singers. Half of the vocalists promptly contracted yellow fever and turned up their toes. Their comrades beat a hasty departure. That was the end of opera, but plenty of diversion remained. There were the sidewalk cafes, the gaming halls, the jeweled prostitutes, bar after bar. And to this pleasure city stampeded every rubber picker of the region who had finished the season with sufficient means, for only in the wilder regions was debt slavery universal. In boom Manaos, then, the top level of society presented the spectacle of Brazilian new rich lighting their cigars with three-figure *milreis* notes while the bottom level offered champagne-drinking Indians. These two fantastic figures were the seal and symbol of the times.

Of the Brazilian cities other than Para and Manaos, Santarem probably was the one most affected by the rubber boom. Outlet of the Tapajos river latex, it was a port of call for all steamers, but dealt always in diversified products and rubber had less to do with its expansion than was the case with Manaos. One of its mass increases in citizens it owed to the Union victory in the American Civil War. Numbers of Confederate soldiers, devout believers in slave-holding, signaled the end of hostilities by heading out of the United States rather than live under Yankee rule. Many of them landed in South America. Those who picked Santarem did better than most, but their great gesture netted them little enough in any event. Within two decades after they had established their cacao, sugar-cane and tobacco plantations, slavery was legally abolished in Brazil. It was a bitter pill to swallow, but most of them stayed on in Santarem. Today one or two of the ancient rebels still survive among their swords and souvenirs. The Southern flags are gone, though — conquered by the Amazon climate - and not all the numerous descendants of the original Confederate colony are of unmixed blood or able to speak English.

In Peru, Nauta originally had seemed cast as the town of destiny. It was in rubber territory, it was the final point of call of the Brazilian Amazon Navigation Company's steamers and it was the port out of which the Peruvian government's steamers supposedly made their runs. And then one day the river decided to change its course and Nauta was a mile up the sort of creek for which the sporting world of the United States long has had a peculiarly graphic name. That left Iquitos without a rival. And if Manaos owed nearly all to rubber, the Peruvian city owed everything. From an 1863 population of 431, it boomed to 10,000 in 1905, on to a city boasting 15-20,000 permanent residents, a city which every day greeted, fèted and robbed an army of whoopee-seeking visitors outnumbering the whole populace of before the boom. Over 2,300 miles from its ocean, it became Peru's second seaport, topped only by Callao on the Pacific. On its wharves ocean liners from England and America dumped goods to be distributed by smaller craft among all the upstream tributaries. And steamed back down with fortunes in rubber, fortunes so huge the government cut alone inspired one revolution after another, the patriots seizing the customs house wealth and withdrawing at their leisure weeks before the news could cross the mountains to Lima. Such antics eventually gave it a garrison of a full regiment, made it the base for a flock of river gunboats. A colorful city it was, this capital of the department of

Loreto — Peru's largest and remotest state — a city of Peruvians, Brazilians, Indians, Moroccans, Turks, Arabs, Syrians, Japanese, Chinese, Greeks, a city where, well into the present century, lath-thin, flea-bitten scavenger pigs swarmed in all the so-called streets, rats challenged all corners on the garbage-littered waterfront, and leaning out from the bedroom window to aim chamber pot contents in the general direction of the middle-of-the-street drainage ditch, with or without regard to innocent passers-by, constituted the one gesture in the direction of sanitation. It remained a city of thatch and tile roofed adobes, a city without a water system or any sewerage system other than the ditches in the center of the streets; but Southern Hemisphere go-getters did achieve two partially-paved thoroughfares, a three-story hotel, a narrow-gauge steam road for hauling freight the length of the metropolis (property of Julio Cesar Arana), electric lights, three newspapers, numerous warehouses and cafés and a system of floating docks constructed by the British steamship line which gave this farthest inland of all ports its connections with Europe and America and made it one of the Amazon's three direct export cities. Here was headquarters of Israel & Co. and of Julio Cesar Arana. Here traffic in the Indian youths rounded up by the forest hunters of men was open and notorious; here flourished traders who extended credit to Peruvian estate operators on no other security than the bodies of rubber serfs, for most of these customers were squatters and failure meant that the Iquitos lender had nothing save workers to seize in satisfaction of the account. And here 2,260 tons of *caoutchouc* went aboard the jungle-invading ocean liners in 1910.

In Venezuela, San Fernando de Atabopo, at the confluence of Orinoco, Guaviare and Atabopo, rose to importance and to bloody notoriety as the rubber capital. Never housing as many as 1,000 inhabitants in the village proper, it was yet the center and the distributing point for a rich latex-producing district shut off from much traffic with the North by dangerous rapids which could be passed only by Indians in small craft, in communication with the regions of the remoter Amazon by a portage between Atabopo and Negro, the Casiquiare canal linking Orinoco and Negro. Here every article brought in from Ciudad Bolivar on the lower Orinoco represented transportation by 19 distinct carriers; every machete, every square of cloth whether from the North or the Amazon sold at a price increase of a 1000%. Enslaved Indians gave their lives as raft-men, canoe-men, rubber tappers; debt-shackled half-breeds of the latex forests passed from hand to hand by sale or by armed seizure of authority over the region. All officialdom was in the rubber business; the very governor of the territory used one hand in operating as a dealer paying no duties, the other in collecting export duties in hard cash from his business rivals. The sequel is easy to imagine. On the night that Colonel Tomas Fúnes took office by the easy process of murdering the incumbent, 70 whites died in the one massacre. It was only the first of a series of murders and atrocities. Fúnes butchered his accomplices, exterminated competition and harried the Indians with his rubber-collecting tyrannies until these Caspar Milquetoasts of the forests were lashed to such a pitch of desperation that they fell on San Fernando one night and, to their own profound and frightened astonishment, accounted for 50 victims before stampeding back up the Ventuari.

Bolivia, though second to Brazil in rubber production, was as one with Venezuela, Colombia and Ecuador in having no great Amazonian market and export city. In the disposition of its rubber revenue, it differed from Brazil, too. Here there was no thought of public improvements in any communities; here *all* of the governmental cut went into the pockets of the politicians. Threadbare scoundrels took office without a silver *boliviano* to their names and piled up takings of thousands of dollars in a year's time. Rubber money pouring into the Beni region North of Santa Cruz de la Sierra, birthplace of the Suarez brothers, boomed that squalid flyspeck on the map and drove its population far past the 20,000 mark despite its lack of water communication or any other adequate transportation service. It was the commercial capital of all *estrada* Bolivia, the playground for all those on holiday from the rubber, fever and sudden wealth belt of the Beni. Yet it emerged from the boom still a straggling frontier community ordering its goods one year and receiving them the next, still without railway or highway connecting it with the world, still ten to twelve days by mule track from Cochabamba on the West, 20-24 days from Puerto Suárez on the Paraguay to the East. It had, however, the memory of one long debauch of crime, drunkenness and sex play bringing it to a level where case-hardened US missionaries, inured to the roaring wickedness of Para and Manaos, labeled it the one place in the New World most heartily abandoned to profligacy and prostitution. Women far outnumbered men and all of them were kind; *chicha* (a dynamic drink prepared by boiling and fermenting maize after it had been thoroughly masticated in the human mouth) was plentiful and cheap. The Santa Cruz citizen drank deep, loved often and envied not the bureaucrats and merchants who pocketed the gold. Only when he went to jail did he ever have cause to regret that attitude. Selling prisoners to the rubber barons of the Beni at one thousand *bolivianos* a head was one of the perquisites that went with the office of *Intendente de la Policia* and rarely, indeed, did one of these bartered jail birds live to see Santa Cruz again. Cochabamba, in Western Bolivia, was Santa Cruz all over again. Cutting in on the Amazon trade through the little river ports on the Mamoré river tributaries, it was helped by rubber to a population of over thirty thousand. It maintained, it has been estimated, over 1,400 bars for the maize and saliva addicts.

The real rubber towns of Bolivia as distinguished from such rubber men's playgrounds as Santa Cruz were Cobija on the Acre; Riberalta on the Beni, near the mouth of the Madre de Dios (a mud church plus rubber-stacked warehouses in high-fenced compounds); and the border community of Villa Bella where Beni and Mamoré unite to form the Madeira. It was a railroad that brought decay to Villa Bella; brought into being Brazil's Porto Velho, modern city of roundhouses and shops, docks, sewerage system, hospitals, hotel and copper screening next door to the ancient and notorious cesspool of a town called Santo Antonio. From Matto Grosso's Guajara-Mirim on the Mamoré down to Villa Murtinho, in Brazilian territory opposite Villa Bella, and from Villa Bella and Villa Murtinho down the Madeira to Santo Antonio is a series of twenty roaring rapids often not navigable even by canoe. Yet through all or part of these rapids went not only the rubber from Bolivia's Beni and all its important tributaries and from the Bolivian Brazilian Mamoré but also that from Bolivia's Abuna and Brazil's Guaporé and Jacy-Paraná; went, in fact, virtually all Bolivian rubber save that finding its exit by the Acre — and a good portion of the Matto Grosso production as well. And getting these thousands of tons of *caoutchouc* to the world was an epic task indeed before the coming of the steam road. For the Indians and chiefly for the ones in bondage to the Bolivian transport agents of Villa Bella were reserved the job of putting the 'black gold' through the rapids and onto the steamers (Santo Antonio is accessible to ocean liners). The assignment decimated the population of that portion of Bolivia. There is a record of one, three-boat expedition from which but three men returned. Drownings of half a crew of 20 to 30 were not infrequent and rarely indeed did a crew come back with a full roster. This irked the rubber kings somewhat. More bothersome were the losses of 'black gold' in the river (a third of it, it has been estimated) and the fact that it took six weeks to run down through the rapids, two to three months to return from Santa Antonio with merchandise. A short railroad around the rapids seemed to be the answer. In 1871 and years before the rubber boom in these parts, Colonel George Earl Church, New England engineer, soldier, war correspondent, explorer and geographer, had formed a couple of companies and tackled the job after placing an $8,500,000 bond issue (net proceeds $4,266,430) in London as special agent of Bolivia. A leading authority on the Amazon was Colonel Church, but his construction contractors quit after a two year massacre, complaining that the country was a charnel house and that given command of all the capital in the world and half the population, the construction of the railroad would remain an impossible task. That reaction is understandable. Of Santo Antonio a Brazilian authority on tropical diseases once said, 'There are no natives of this place; all children born there die.' Bolivia defaulted on the bonds in 1875 and by a judicial settlement in 1879 the bondholders got $225 per $340 bond. That same year a company was formed in Philadelphia to build the railroad and 941 men sailed from the United States. Eighty died in a wreck off the coast of Carolina, 140 fell victims of disease in the charnel house, one was killed by wild Indians. Of how many died fleeing by canoe or raft there is no record, nor were any statistics kept on the mortality among the native workers. The final conquerors of the impossible, American engineers in the employ of an $ 11,000,000 British concern registered in Maine, took up the task in 1908 in pursuance of the conditions of the 1903 treaty under which Bolivia had yielded up the best part of the Acre lands. Sanitation measures copied from those found successful on the Panama Canal gave them their victory.

Old rails these engineers found serving as posts and even as a sidewalk in Santo

Antonio. Buried in the jungle they discovered a locomotive formally christened 'Colonel Church' and dating back to the ill-fated 1879 expedition. Fixtures were gone, firebox and flue were shot and there was a tree growing from the smokestack. Exhumed and repaired, 'Colonel Church' was placed on the rails and has been in regular use since.

An especially interesting angle of the railroad construction work was the decision to import eucalyptus ties by ocean liner direct from *Estrada* to Porto Velho. It was easier and cheaper than getting the wood out of the forests through which the right-of-way passed. The fact that much of the local timber had a higher specific gravity than water and could not be floated was one of the factors contributing to this oddity.

One of the costliest lines in the world is this bijou Madeira-Mamoré and the one imposing the highest tariff, a sliding scale based on the prices of rubber. As to its cost in suffering and death, there is a popular saying that every tie represents a life and this saying vividly expresses a truth that remains a truth whatever its mathematical accuracy or inaccuracy. Ironically enough, sanitation experts and engineers scored their final triumph just as the wild rubber trade was coming to the beginning of the end. Villa Murtinho was reached by the railroad builders in 1912, Guajara-Mirim (226 miles from Porto Velho) in 1913. According to the treaty with Bolivia, the Mamoré was to have been bridged and the tracks extended another 61 miles to Riberalta, thus eliminating the portage of rubber at Cachuela Esperanza, but the work was suspended after a portion of the roadbed had been completed. Today only one train every two weeks runs over those rails figuratively laid on human bones. It, too, is the symbol of an epoch whose larger labor story is the subject of our next chapter.

2.3 The Cearenses, the *Caucheros*, the Infidels

It is by a series of accidents that an incredibly expanding industrial world always has been kept supplied with the raw rubber that is one of its prime necessities. One of the most notable of these occurred on the threshold of that factory era which first put upon the rubber producing countries a strain to supply the demand. More than enough *caoutchouc* the Amazon had, but labor to extract it was another thing. The Ceara accident of the late 70s supplied that.

In the north-east of Brazil lie the 57,000 drought-wracked acres of this state of Ceara. It is just south of the Amazon valley, which is one of the wettest places in the world, yet Ceara is almost as often a desert as a land habitable for human beings. Its rivers are small and vanish in the dry season; the scant soil of its interior does not hold moisture; and, when the supposed rainy season of four to six months fails to produce rain, drought, famine and pestilence rack the land. Any of its droughts would be historic in almost any other inhabited land. In its annals, however, are half a dozen droughts so devastating as to be historic even in Ceara. One of these befell in 1825, another in 1844-1845. And the greatest of them all was timed exactly to coincide with the Amazon's first great need of men.

There were something under a million residents — small farmers, cattle raisers, hardy toilers — in the state when the drought of 1877-1878 smote the land; there are only a few more than a million today. Half a million people died in this one outburst of starvation and pestilence. And the land was swept as clean of live-stock as though the Lord had wiped it as a man wipes a dish, wiping it and turning it upside down. By man, too, the interior was wholly abandoned for that time being. On blistering feet, the people of the exodus surged East to the coast; driven by furnace-winds, by rampant smallpox, by the strength of delirium; stumbling over dead, dust-coated bodies; straggling past blasted bush, withering trees, dead rivers and occasional stagnante, fœtid pools until they reached the safety of the sea. There over a quarter of a million shook the dust of Ceara forever from their feet. Some went far South to toil on the red-earth coffee plantations of São Paulo. The most - and they were the luckless - took ship for the Amazon.

Here were exactly the workers that the rubber exploiters needed. An Indian-Portuguese-Negro breed as hardy as the pure Indian is weak, they had grown hardier than ever through their years of warring against relentless natural forces in their own sun-burned, grudging land. Tough as they were they perished by the thousands in the Amazon — from fevers, malnutrition, insect venom. And the thousands who survived found themselves trapped in a dreadful bondage from which they looked back to desolate Ceara as a happy land compared to this dank, poisonous jungle exile where they drudged out their days with never a hope of balancing credit against debit on the estate owner's books.

Every year after the one of the great exodus saw more men of Ceara and neighboring Maranhão being brought into the Amazon, the soaring rubber prices and mounting demand of the 1906-1910 period resulting in an especially large influx. Few were used in the region around Para city and not many more in the other rubber districts of the lower Amazon. Natives of the states of Para and *Amazonas*, a mixed race in which the black had come to predominate, were sufficient to work the *caoutchouc* in these areas. In the islands, especially, it was a miserable starvation existence eked out between dripping jungle and muddy, mosquito ridden side channels. Here, and on the nearby mainland the majority of the rubber gatherers were independent squatters or small individual landowners; on the lower rivers where large estates were the rule the comparative closeness to civilisation insured against workers being subjected to all the feudal tyrannies flourishing in the lands beyond the law. It was into the lawless lands, into the newer, richer and remoter districts far up the river, that the Cearenses were taken. In the Madeira river zone of *Amazonas* and Matto

Grosso they predominated; up the Purus, Jurua, Javary and other Amazon tributaries between Manaos and Peru; in the Acre (which they largely settled and developed) and in Northern Bolivia. And, until the collapse of wild rubber, almost none of them ever came out. The few who did became, unquestionably, a curse to their own race. They were wealthy by the standards of their countrymen and it was their example that enrolled other hundreds when the agents of the rubber plantations arrived on their annual trips to drum up recruits. In a drought year, these enlisters would find all the men they needed among the sufferers arrived in the coast towns. If there had been no drought that year, the tempters would cross the plains and mountains to the interior and line up as many as possible by contrasting the riches waiting in the Amazon with threats of new droughts to come in Ceara.

It was on the river steamers out of the rubber capitals that first disillusion came to the Cearense long lapped in brightly colored promises reinforced by cash advances. On these boats he traveled third class, 'cattle passage,' hundreds packed in the lower deck. These stinking quarters he shared not only with comrades in misery but with all the other cattle and with four-legged porkers. The other pigs - the two legged ones - lolled about on the upper deck and slopped themselves with liquors that represented the purchase price of more than one of their serfs stacked between the livestock and the firewood below.

At journey's end, the exile found a group of buildings set in a clearing hacked from the river-bank jungle and out of reach of the waters in time of flood. With one of these structures he would have no concern. That would be the home of the proprietor. The other of the two biggest buildings would dominate his new life. This was the warehouse-store. His introduction to it came when he lined up to draw necessary farina, dried fish, sugar, trade-gun, ammunition and the tools of his trade hand axe for tapping the trees, tins and basin for collecting the latex, earthenware jar for smoking. Cost of these made wholly certain a bondage already virtually sure as a result of the advances received from the recruiting agents. To those first advances a 20% commission had already been added. To the plantation store he had to deliver all rubber and there he had to do all his purchasing.

On everything, of course, he was cheated shamelessly. For his rubber he received such book credit as the master chose to give him — he had no check on the weighing, no knowledge of current prices and no way to enforce such prices had he the knowledge. For every purchase he paid prices fantastic even beyond the usual fantastic prices charged in the earth's wild places when booms are on. Clothing, firearms were as shoddy as those foisted on the Indians. Food, due to careless and overmuch handling on long hauls in a tropic climate, was uniformly bad. And for everything in his store the *patrao* had himself paid excessively. Often the cost delivered at the plantations was thrice the landed price at Para or Manaos. Then, too, interest was charged by the commercial houses at the rate of 10-20% and it began at date of invoice and ran until sale of the rubber delivered in return. There were estates on remote Amazonian tributaries where the up trip with the goods required as much as eight months, the down trip with the rubber another three. The further away from Para or Manaos, the more the *patrao* paid and the more he charged his *seringueros*. And a crowning irony of the whole system was the fact that most of the foodstuff transported up the Amazon at so great expense could have been grown by the *seringuero* at his very doorstep. In the rubber-gathering season, however, he had no time to hunt or fish, much less engage in any sort of agriculture. And at no time did the employer want him to be scratching the ground. That would have eliminated the profit on the food sold at his store. So long as the *patrao* could peddle imported farina at a profit, then, he had no desire to see his serfs raising their own, and so long as they could supply their needs at his store they had no desire to tackle the light labor of planting. Such was the vicious circle.

As to the work life of the *seringuero*, it was a horrible thing. Only because he found his paths already laid out for him did it differ from that of the first Indians inducted into the 'industry' in its early upriver days. To spot the rubber trees, open up the jungle and layout the circular beats of one hundred to two hundred trees there had come into being a special class of workers corresponding somewhat to the timber-cruisers of the American North-west. Paid at approximately the rate of a dollar a tree, these matteiros were the aristocrats of Amazon labor. With the tappers they had nothing in common and sometimes a forester even turned overseer and became the tyrant directly over the rubber collectors. Under any conditions, however, actual overseeing of labor was a nominal thing. The *patrao* had no need to check his men on the job; he had a sufficient check when they brought in the results of their labors. If the man produced the required amount of rubber, well and good. If not, the plantation operator shut down on the quinine, alcohol, canned goods, ammunition, patent medicine, cotton clothing and other luxuries and reduced the food ration to a starvation minimum. Almost invariably this turned the trick. If not — if a tapper continued to report with rubber insufficient to cover his food — then he might starve or die in such other manner as he saw fit.

Season's production for one man was expected to amount to 450 to 500 pounds in the oldest districts, from 800 to 2,200 in the upriver areas where the trees had not been tapped so long or where virgin *Heveas* were available. On the average, 61% of this was fine hard para; 11% was the medium or *entrefina* produced when the latex coagulated in a clotty condition or the smoking was incomplete; 28% was coarse or scrap (*sernamby*) resulting from natural coagulation and including the scrapings from cups and containers and the drippings adhering to the trees. And woeful indeed was the lot of the serf continuing to show up with too large a proportion of *entrefina* or *sernamby*.

There were ways of beating the quantity checkup, of course. A man could spend pleasant days in his hammock and then make up for them by bleeding the higher parts of the trees, by kindling fires at their bases to hasten the flow. This meant the killing of the tree, however, and ultimate discovery often resulted in death for the offender. And so it was that most of the tappers stuck to the prescribed round of work so long as they were physically able. It began, in the case of a new beat or *estrada*, with building his own leaf-thatched palm hut close to his forest route and as close as possible to a source of water. If it was an established *estrada* the hut left by the previous victim of fever or beriberi was there and ready for occupancy. As the season opened, many days had to be spent in clearing the paths, for so fast does the jungle reclaim its own that the footway of a few months gone would now be true bush. With this cut out, with the temporary bridges over swamp and stream replaced, he would be ready for his true work.

Each day began at 4 AM, for the hotter the hour the slower drips the latex. Rolling from his hammock before dawn, the half-alive *seringuero* would make and down his cup of coffee, light his hurricane lantern, gather up axe or machete, gun and tin cups and start out in the night on his long round.

Five to six miles it was and twice each morning it had to be covered, once to cut the tree bark and affix the tin cups and once to empty the cups into a two-gallon container. Nor was this ever the sort of stroll in the woods the Northerner knows. The paths were almost imperceptible trails daily invaded by lianas, creepers, razor-edged grasses, thorns that could and would nail shoes to feet; always defended by mosquitoes as poisonous as ants, by carnivorous ants as fearsome as tarantulas, by deadly snakes ambushed in the leaves underfoot, by blood-drawing flies, hornets, spiders, scorpions of a deadliness unknown to temperate latitudes. Insects laid their eggs in the slightest scratch and every rubber tapper became a walking worm farm. He could and did avoid the streams and ponds and fermenting flood-deserted patches of water where shoaled the flesh-eating, toothed caribe fish — the poisonous stingrays — the paralyzing electric eels. He could not avoid swampy ground where *Heveas* grew. In these swamps, then, men and leeches worked together, men tapping trees and leeches tapping men, ulcers on over-worked human trunks matching swollen, tumorous-looking masses of scars on overworked trees.

Malaria, too, was always on the job and, it is probable, caused more deaths in the Amazon than almost all other causes combined. It is in the dry season of rubber collecting that the jungle is unhealthiest and no rubber worker escaped the fevers generated in the damp heat and rotting vegetation. To the tapper quinine was a luxury - often he endured the fever without it and always, while a victim, he worked as usual save during the actual attack. So it was that fever-ridden tappers fell in the swamps and died under a blanket of flies. Or the dread beriberi fastened upon them and they swelled, these rubber gatherers, as though they themselves were but rubber bladders under inflation. Legs, chest, face ballooned, while burning pain crept through the body. Sometimes the maddened man would swing his machete, cut off the affected leg and die a gangrenous death that was the easiest way out.

Always he was an easy mark for disease. Poor and insufficient food was one contributing factor; drinking water foul with decaying animal and vegetable matter another; the unsanitary hovel in which he lived a third. Chills took him in the swamps and the consumptions and other chest diseases acquired on his morning rounds of the forest he aggravated by his afternoons in the smoke houses. A dreadful device, the smoke hut. Necessity of obtaining a perpendicular smoke giving a maximum heat dictated building the smallest possible shack and seeing that there was the least possible ventilation. Necessity of keeping the heat even required a continual blowing through the hole at the base of the smoke-concentrating cone during all the time that the pole on which latex was poured to coagulate was revolved over its apex. Often the job lasted until sundown. And, huddled in this kennel, sucking up the dense, oily smoke, groggy from asphyxiating gases, the rubber picker sacrificed his lungs, his general health and sometimes his eyesight.

The off-season was as bad. It was an annual flood the man of Ceara had to contend with now instead of an annual drought, and it meant imprisonment for several months. Marooned in his stilted hut with an army of insects, he passed week after week in hammock hibernation waiting for torrential rains to cease, for waters to recede sufficiently to permit him to set foot on earth and resume work. This rotting inaction and the outbursts of malaria that always followed the recession of the flood waters were among the principal factors undermining his health, but the isolation was less burdensome than might be imagined. The tapper was accustomed to it, for, even in the 'dry' months, his was one of the loneliest lives in the world.

On an estate where trees were tapped daily there would be a man in each center; where the every-other-day system prevailed each man would work two of the *estrada*. Day after day passed in solitude, misery and monotony, broken only by the weekly or monthly visit to headquarters to draw supplies and down drinks with the other enslaved wretches caught in the jungle trap. Sometimes two or three tappers shared a shack together; again a man would have wife and family; but the general rule was otherwise. Bestiality and other perversions were common in the spots remote from handy Indian women of the semi-civilised tribes. And to these depraved hermits, visitors were the one thing in the world most to be dreaded, for they could have but one mission — the bringing of death.

In general, as we have noted, the 'wild' aborigine of the Amazon was as gentle as he was timorous, but there were warlike tribes. And when not warring among themselves, these looked to the rubber estates for their victims. Usually it was a case of waylaying the solitary tapper and filling him with poisoned arrows. Huts they set on fire; what there was to destroy they destroyed and then departed with the slain man's head as trophy. Among the true head-hunters this was preliminary to peeling skin from the head, stuffing this mortal envelope with hot sand and boiling it down until it became a suitable ornament for any hut — a fist-sized reduction preserving in miniature the face as it saw death coming. Estate headquarters too the combative Indian tribes were known to raid, and far upriver trading villages. Chiefly, however, it was as assassins of the lone tappers that these peculiarly bellicose natives of Amazonia figured in rubber history. Plantation *patraos*, incensed at having hired hands shot out from under them, staged punitive expeditions; but the warrior villages hidden in the heart of the jungles were hard to come at and little was accomplished. And so in time it came to be accepted that it was easier to replace the slain tapper with a new serf than to bother with profitless campaigning.

And the wild Indian took his place with wild beast, with poisonous snake, with the swamp fevers and the strangling atmosphere of the smoke house as a menace to the sorry life of the rubber tappers. Their compensation for facing these dangers, for the lonely monotony of the life-sapping labor? Food, poor and insufficient, liquor in the form of cachaca; that was all. Not even the dream of wealth and a triumphant return home was left them. It faded; the recollections of the comparative happiness of their lives in harsh Ceara slipped from them and they became brutalised, unthinking, living for their cachaca bouts, with no aspirations, no hopes, nothing to distinguish them from the other jungle animals.

Escape? It was something they thought of constantly during their early years in these hells. Inevitably it was something they learned to forget. In all directions save the river the jungle blocked them completely, implacably. Boats? They owned none and the headquarters craft were chained up to guard against any contingency of seizure. The miracle sometimes happened, however, and a man got away. Then, if jungle or jungle waters did not take his life, there was small chance of his evading the master of some other plantation. And whether put to work at the new spot or returned to the old made small difference save that a flogging could be expected in the last event. In Brazil, it was true, the *seringal* proprietors had no actual legal hold on their men and the law had no punishment for desertion or failure to pay debts. But in the upriver rubber districts this was a pure technicality, for there were no authorities to protect the tapper and the only law was the law of gun and lash. And not Brazilians alone, but a sprinkling of Portuguese and other foolish whites from Europe found themselves in bondage far from civilisation and unable to communicate with a representative of any government. They toiled as slaves; they watched their own physical and mental disintegration; and unable finally to return a profit to their masters, they were left to die lonely deaths never reported to families and friends overseas.

In Bolivia, Peru and the other rubber countries, the tappers, white, red or mongrel, were treated even more brutally than were Brazilians and in these nations the very laws of the land forbade any peon's leaving service unless he had paid his indebtedness in full or could find another proprietor willing to take him on and assume responsibility for his old debt. When this last was resorted to as a way of escape from some particularly intolerable rubber estate, the new master added a 20% commission to the debt and the serf found himself chained just that much more securely to slavery for life. Incidentally, it was only when he sought such a transfer that he ever was informed of the supposed amount of his indebtedness. In Brazil accounts, real or imaginary, were rendered to the tapper at the beginning or end of the season, but in the other countries there was little pretense of keeping books save as the selling of debts made some sort of farcical calculating necessary. And usually the sale of slaves (for that is what the sale of debts amounted to) was a business transaction between the masters rather than the result of a chattel's request.

Some of the Castilloa exploiting was done by concession corporations or squatter 'kings' of the remote places who changed their bases from one spot to another in their own empires as areas were milked dry. Some of it was done by migrating proprietors who would abandon exhausted estates and take to the rivers with family, *peons*, goods, seeds and domestic animals in search of unclaimed forests where they would settle down to fell rubber trees, build homes, cultivate plantations and) perhaps) found a new village. And in these cases, of course) the *cauchero* often had one master for life. Always, however, much of the *caucho* gathering was the work of small gangs brought, stolen, enlisted or enslaved by wandering adventurers. The customary procedure in setting up a main camp was simply to furnish each man with a week's supply of farina, sardines and ammunition and send him into the jungle to spot and mark his trees. When the serf returned he would take on supplies sufficient for months, head for his first tree and throw up a temporary shelter. First bleeding the roots with his machete, he would next have at the trunk with his axe and, after felling and bleeding this tree, would move on along his carefully blazed trail to the next victim. The rubber would be accumulating and hardening in tree incisions and holes in the ground during all this tour; later it would be washed and rolled into balls or tied into blocks. Each Castilloa yielded forty pounds or more of dry rubber and a season's gathering by a *cauchero* ranged from the one thousand pounds of the procrastinator up to the three to four tons an exceptional man could collect in a few exceptional spots. From 2-3,000 pounds was considered a satisfactory amount for a good *cauchero* in a good district to tote into camp at the end of the season. It was a worse life than that of the *Hevea* tapper, for the *cauchero* worked in even wilder places and in greater isolation. No weekly or monthly trip to store and fellowship was there for this jungle laborer. If he ever went to the expedition's base, it was apologetically and of dire necessity and when master or overseer occasionally looked in on him in the jungle it was an unpleasant business of kicks, blows and other expressions of vigorous dissatisfaction with the amount of *caucho* gathered. At the end of the season he might be carried off to a new location, might be sold to another adventurer or into one of the permanent hells of the remoter forests. Few, indeed, were the *caucheros* who ever saw home again regardless of the short or long time it took them permanently to deposit their bones on the jungle floor. And whenever one figured he had cleared the amount of his debt and waxed presumptuous enough to request a settlement of accounts, the settlement commonly took the form of a murderous flogging laid on with whips of *balata* or twisted tapir hide.

Peruvians there were who apparently hunted *caucho* for themselves; but actually they worked for the Arab running the village trading store instead of for concession company or wandering exploiter. Somnolent villages these were for most of the year, waking to roaring life and trebled population when the rainy season brought back the men and their rubber. The usual procedure was to leave wife and children dependent on credit extended by the Arab or Moroccan merchant, to line up a concubine to go along into the bush, and to return after months of toil and privation to tear into the trader's liquor and, under its influence, exchange rubber for impossibly priced goods of the sort that only a drunken half-breed would be seen carrying home.

Among the gang *peons*, escape attempts were common, of course, and the initial get-away was comparatively easy. But even when a *peon* was not fleeing at the instigation of, and into the hands of, a new exploiter, even when he had store of jungle lore, the fugitive never attained to freedom. In the wild spots anyone on the loose belonged to the first rubber forest owner or master of *caucheros* who picked him up and even those rare individuals who had been released with documentary evidence of debt satisfaction were likely to be captured, relieved of their papers and put to work pending a supposed 'investigation' that never was made. And there were guards on all those rivers down one of which a fugitive from the remotest tributaries eventually had to come.

From the headquarters estates of the large exploiters, the *caucho* hunting squads were sent out under foremen and sometimes to great distances. Occasionally one of these functionaries worked up a runaway of the whole party, promising to share the sales proceeds of the stolen rubber and let the men go free. What he did, of course, was pocket the cash and sell the dupes to other masters.

Even worse off than such serfs, however, were the *infidels* of the rubber 'industry' — the wild forest Indians enslaved either by Castilloa hunters and their *peon* followers or by those fortunate conquistadores who stumbled upon remote areas where *Heveas* and savages flourished together. New exploiters were as cruel as the old, their greed and ambition had grown apace with the world's ever-increasing clamor for rubber, and barbarities multiplied in direct ratio to the waxing of their demands.

The atrocities in 'the Putumayo' perfectly exemplify this. Constituting the largest single exploitation of the wild redman in the twentieth century and minutely and impregnably documented, they well may serve as the horrible example of that modern slavery. In a number of the boom years, the slaves of this one squatter's domain accounted for as much as a fourth of the total Peruvian production, and the crimes of this largest of the jungle hells constitute an orgy of bestiality without an exact parallel. The specific atrocities, of course, were duplicated in any number of lost forests of the upper Amazon and it is well to remember this if you want the complete picture of South American latex gathering at its worst. And as with Indian exploitation and torture, so with the other two main narratives that go to make up the complete Putumayo chronicle: the tale of jungle war between exploiter and exploiter, the recital of the manner in which a forest squatter's slave enterprise was transformed overnight into a banker-endorsed corporation trading under the flag of a civilised nation and offering its shares to the people of that nation over the impressive names of silk-hat directors who couldn't have known less about the rubber lands if they had been located on the moon. It is because none of the three narratives is unique that each is so instructive.

Briefly the story of 'the Putumayo' is the story of something under 4,000 tons of rubber extracted in the years 1900-1912, rubber that brought some 5-7,000,000 dollars on the London market and cost little more than the lives of a great many thousand Indians. Actually, the history of *caoutchouc* collecting in this area goes back into the last century (19th) and is still being written today, but the important chapter covers only the dozen years cited. Like many a generally accepted label 'the Putumayo' is as inaccurate as it is handy. The inaccuracy we shall deal with immediately and thereupon feel free to drop quotation marks and employ the tag without twinges of conscience. The crimes of 'the Putumayo,' then, are actually the crimes of the Igaraparani, Caraparaná and Cahuinari. The first two of these rivers rise in the equatorial jungle near the Japura (Caqueta) watershed and flow south-east to join the thousand-mile-long Putumayo at points 400 and 600 miles, respectively, above its junction with the Amazon in Brazil. The Cahuinari is a southern tributary of the Japurá. All together the squatter's domain they drain comprises some 12,000 square miles (just the area of Belgium) roughly extending from the Japura on the north to the Putumayo on the south, although none of the rubber and almost none of the hell was raised on either. This area has formally been a part of Colombia for not quite a decade. In the years with which we are principally concerned, however, it was but one piece of a no-man's land actively claimed by both Colombia and Peru. Peru's idea was that it owned everything as far north as the Japurá; Colombia's that it owned everything as far south as the Napo (which is a considerable distance south of the Putumayo). The political fuss had little real bearing on the atrocities of the Igaraparaná and adjacent precincts, but it did serve as a very welcome alibi for both nations in the day of investigations heard around the world.

Actual ruler of this Belgium sized empire during most of its ghastliest years was not Colombia, not Peru, but one fat, yellow, sciatica afflicted Peruvian bearing the appropriate name of Julio Cesar Arana. Its pioneers were Colombian *caucheros* who came down the Putumayo as Castilloa destroyers and settled on the Caraparaná and Igaraparaná in the late 1880s or 1890s (the records disagree) after discovering the promising *Hevea guyanensis* forests and, even more promising, the innocent, unspoiled and defenseless people occupying them. These were the Huitotos, a frail, dark, Asiatic-featured race split into a multitude of clans not one of which could muster more than two hundred men for the family fights and raids after women which furnished diversion without too much bloodshed. Easily conquered by advances of tinned beef, beads and trade guns, they were set at tasks not difficult in themselves; gashing the trees and collecting and washing the naturally coagulated latex was all that was involved. Of poor quality was this rubber, but its abundance, and above all the abundance and docility of the free labor, made the exploitation profitable and shortly the original conquistadores were followed by numbers of other Colombians.

The state of comparative amiability did not last long, however. The exploiters were there not to become permanent settlers but to amass wealth as rapidly as possible; the increase in their numbers and the expansion of their wants and their ambitions necessitated the collection of more and more rubber; the very abundance of potential labor encouraged the wanton using up of red lives since there were so many more replacements to be had. Enticing of the Indians gave way to the business of rounding them up in bloody forays; cheating them of such rubber as they brought in gave way to the fixing of mandatory quantities to be delivered on pain of death or torture; violence and oppression continually increased in intensity, depravity and volume as random small-scale activities gave way to larger and larger operations.

It was the Colombians who began the gory work, but it was only with Cesar's entrance upon the scene that oppression entered upon its worst stage; it was only with his rise to control through a variety of interlocking partnerships that it took on wholesale proportions; it was only with mere control giving way to absolute ownership that horror mounted to the ultimate heights. A geographical accident had brought the Peruvian into the picture. From back in Colombia supplies could have been obtained only after difficult and expensive journeying over the mountains. Downstream on the Amazon were the rubber markets and it was easiest to have the luxuries for the masters and the barter goods for the Indians sent by steamer from Iquitos, 1,200 miles away. There Cesar, who started his commercial career as a barefoot peddler working out of the town of Yurimaguas on the Huallaga, had been an established trader for a decade when he made his first purchase of Putumayo rubber in 1899. By 1901 he had the trading and navigation monopoly of the Igaraparaná-Caraparaná, and was advancing credit to all of the Colombians. Soon nearly all of them were heavy debtors compelled to take him in as ruling partner. By 1904 he was buying them out (chiefly by releasing them from debt) in the name of J. C. Arana & Brothers, capitalised at $400,000, while as Arana, Vega & Company he sent Peruvian (and some Colombian) gunmen into the forests of the Cahuinari region to subdue the more warlike Andokes, Boras and Ocainas Indians with faggot and Winchester and machete. Always there had been a certain amount of fighting between the various rubber exploiters but from now on it assumed really serious proportions and was marked by atrocities almost as dreadful as those visited upon the Indians. Usually the battles were over claims that the Colombians had been stealing Arana Indians or poaching on Arana preserves, for by now the Peruvian's representatives were claiming title not only to the aborigines actually in their service but to all the as yet 'uncivilised' ones as far north as the Japurá. And so successful were they that by 1907 all but three debtor-rivals had been muscled out or forced to sell. Juan B. Vega of Iquitos, a Colombian but a partner of Cesar's in pre-Putumayo days, was paid off in that year, and the Peruvian Amazon Company, Ltd., of London, England, capitalised at $5,000,000, took over the whole 8,000,000 acres — three sections on the Caraparaná excepted — as successor to J. C. Arana & Brothers. In the early months of 1908 the job was completed. Under a fake scare of invasion by Colombian troops, a Peruvian gunboat and an Arana steamer loaded with Peruvian soldiers were sent up from Iquitos, capital of the department of Loreto, and joined with the British corporation's gunmen in attacking and seizing the largest of the remaining Colombian rubber stations. The Arana mobsters shortly thereafter descended on the last two rivals and a wholesale massacre of captured men completed the conquest of the Putumayo. Prominent in this massacre, incidentally, was Bartolomé Zumaeta, a brother of Cesar's wife. Later on he committed the serious social error of publicly raping the wife of an imprisoned Boras *cacique*, the youthful Katenere, while this Indian looked on from his vantage point in the stocks. Katenere escaped, captured several Winchesters from Arana gunmen, hunted up Zumaeta and killed him. Thereafter he warred both on whites and the Indians who submitted to them and numerous expeditions were sent after him. One did succeed in capturing another of his wives who was immediately imprisoned at a station prepared to receive Katenere when he came for his woman. The avenger came as expected, came at dusk with his red riflemen, and was shot down. Unfortunately, the Putumayo produced but two or three like him during all of its years under rubber's birds of prey.

How many of the other kind infested the land when Cesar began to gobble it up at the turn of the century, it is impossible to say. Optimistic promotion estimates of Cesar, his partners, associates and subsidised Peruvian government officials for the pre-exposure years of 1906-1908 ran from 40-70,000 thousand, with the number of those actually subdued and working rubber being set at 30-40,000. Supervising these laborers, it was claimed, were not less than 1,500 'armed vigilantes' (also referred to as 'civilised employees'), and the number of rubber collecting centers or sections was set at 45. Commissions of inquiry in 1910-1911 found that there were 10,000 Indians and that 4,000 of them were at work in 20 collection sections under 600 gunmen of whom

400 were aboriginal youths trained since boyhood as oppressors of their own kind. Ignoring births, were 30-60,000 of these reds actually plowed under in the course of that five-year plan? Hardly. Official investigators have set 50,000 as a conservative population figure for 1900, however, and, leaning over backward in the desire to be cautious, we'll say that there certainly were 30-50,000 Indians around at that time. Something happened, then, to more than 20,000 constituents of Cesar's in the first decade of this present Christian century. What that was can be briefly summarised. Some of them ran away, most of them died from starvation, exhaustion, smallpox, putrefaction of the flogged carcass, severed heads, bullet wounds, bonfires and other diseases of the Putumayo.

Cesar, who visited the Putumayo only in the key years 1901, 1905, 1908 and 1912, took no personal part in such bloody goings on, of course. Nor did his brother and partner Lizardo Arana (also appropriately named, for Cesar has revealed him as a sort of lounge-Lizardo who was 'too fond of amusing himself and did not stick to business'). Nor did his partner and brother-in-law, Pablo Zumaeta of Iquitos, though this gentleman could point to two brothers killed by the Indians while serving as slave-drivers. Nor did his partner and brother-in-law, Abel Alarco, reputed former tree tapper who swanked about as managing director in a London setting until his unfortunate connection with the attempted bribing of a journalist named (believe it or not) Horace Thorogood. Nor did the baron, the baronet and the bankers who drew their fees as English and French directors of Cesar's British corporation (*one* of them did own a few shares of stock). All that these gentlemen did was engage the services of some of the worst guerillas in the upper Amazon, load them down with armaments and turn them loose on defenseless Indians with the explicit understanding that their pay would be in direct proportion to the amount of rubber they succeeded in belting out of red hides. There is a reservation here in favor of the European directors. Incredible as it may seem, the evidence points to their actually having been such super-stuffed shirts that they never tumbled to the fact that they headed a corporation engaged in slaving and wholesale murder.

A number of motives had impelled Cesar to register as a British company. For one thing there was a report that Colombia had granted an American syndicate a concession to work the area and the business of trading under a European name would be definitely useful in case of complications. For another he was pinched for capital and the partnership had reached the limit of its borrowing. And, last but not least, was the desire to have a whirl at coaxing dollars from the investment and speculation public. To England he went because his rubber always had been consigned there and because his banking connections were with London. European protective coloration he assumed on the day of registering his company. Credit to the tune of $300,000 he wangled from a London bank on the strength of the same transformation. And though the first and only public offering of shares came in the panic year of 1908 and was a colossal flop, it gave Cesar an additional $500,000 to work with, some $25,000 to $40,000 of this being supplied by the lambs and the remainder by the silk-hat financiers of the City who had underwritten a half million of the issue. Not bad at all for the one-time barefoot bargainer of Yurimaguas. Especially when one considers that the whole expense of underwriting and promotion — $150,000 — was borne not by the vendor partnership but by the new corporation. And this whole corporate structure, nominally capitalised at $5,000,000 and with $3,400,000 in shares issued ($2,750,000 going to Cesar and partners as a part of purchase price) was founded on little more substantial than the bones of Indians. Arana properties other than the Putumayo were involved, but the Putumayo was overwhelmingly the principal asset, so much so that its book value was set at $2,964,140. And in the Putumayo, J. c. Arana & Brothers and successor corporation had a dubious squatter claim on the land and a slaver's claim on the bodies and souls of its inhabitants — just that and nothing more!

Incidentally, such squatter title to the Putumayo and its Indians as existed (the documents transferring the interests of the original Colombians) remained vested in J. C. Arana & Brothers as trustees instead of being transferred to the company. And Pablo Zumaeta was fixed up with a power of attorney from the company which assured his being able to do things with rubber and estates without waiting the formality of directors' action - should such fast work become necessary. And Señora Leonora Arana, wife of Cesar and sister of Pablo, was down on the books for a $300,000 claim which the silk-hat directors looked on as her interest in the business; a capital liability, that is, not a debt liability which the company was undertaking to meet. Cesar permitted them to think that for quite some time. Meanwhile, Alarco ran the London office, Zumaeta ran the Putumayo from Iquitos, Lizardo ran Acre river properties from a branch office at Manaos and the commuting Cesar ran Zumaeta, Lizardo, Alarco and Alarco's innocent English successors in person or by remote control from such favorite spots as Paris and Biarritz. On one of his trips - the 1908 voyage home after formation of the English company - an odd juxtaposition occurred when Cesar, at dinner, was seated beside a fellow passenger named Roger Casement, who also was returning to South America. They weren't interested in one another then.

In floating the British corporation, the aborigines of the Igaraparaná and Caraparaná did far more than their share. Their duty it was to pile up the record of astounding output increases in 1904-1906 which were featured so prominently in the Peruvian Amazon Company prospectus.

From the time of Cesar's first turning his attention to the district, rubber production had climbed in this sufficiently startling fashion: (1900) 35,000 pounds; (1901) 120,000; (1902) 272,000; (1903) 445,000. Then the pressure really went on and the Arana slaves bled and died to compile this truly convincing advertisement : (1904)

757,000 pounds; (1905) 1,038,000; (1906) 1,420,000. It was by extending rubber collecting into new forests and stepping up the output of established centers that this was achieved. A Babson graph of the Putumayo would show crime mounting in a steady curve in the wake of production and belatedly slumping off in quantity although not in ferocity even as the rubber yield diminished to 1,384,000 pounds in 1908, 879,000 in 1909, 699,000 in 1910. The last three years of J. C. Arana & Brothers (1904-1906), then, and the first two years of Peruvian Amazon Company (1907-1908) marked the height of the Putumayo atrocities.

Actual perpetrators of the crimes were the copper-colored 'white' chiefs of the sections - nearly all Peruvians - and their staffs of enforcers. The company's insatiable desire to obtain the greatest production in the least time was inculcated into these chiefs by a system of paying them no salary whatsoever, their remuneration being a flat sum per arroba (33 pounds) of rubber or a percentage on the gross yields of their territories. These ran to 20%, 25% and even 50% in some cases, notably among the Arana-founded sections in the Boras and Andokes country where the worst atrocities occurred. With these more warlike Indians subdued and with the necessary record for the London prospectus safely compiled, the high percentage contracts were steadily eliminated, although percentages remained the system of payment. Too exclusive a concentration on quantity had resulted in an excessively poor quality of rubber and this became matter of concern once the English corporation had been floated and the attempt to foist shares on the public had been made. And so it was that, with the fattest contracts bought up or stepped down, output and atrocities fell off together. Almost unbelievable was the criminality even in its period of decrease, however, and some of the worst of the Putumayo horrors occurred within a few months of the arrival of 1910 investigating commissions.

Only to the managers of the Putumayo's two divisions, the Igaraparaná and the Caraparaná, were the section chiefs responsible. In turn responsible only to Iquitos headquarters, these two gentlemen relied on a percentage of the profits of their districts as principal source of income and their interests were one with those of their superiors and their murderous inferiors. Colombia's original joke 'authorities' had all been rubber men, and the later Peruvian 'police' and 'judicial' appointees were either Arana district managers or section chiefs. And the few score 'troops' kept in the territory after the joint army-navy-civilian maneuvers of early 1908 were under strict orders to confine their activities to garrisoning the Colombian station they had seized and the two Arana administrative seats of La Chorrera, at the rapids 220 miles up the Igaraparaná, and El Encanto at the mouth of the Caraparaná. Under no circumstances were they to poke their noses into the rubber-gathering centers or otherwise interfere in matters which did not concern them.

On his own bone-pile, then, each section chief was king. And in addition to drawing dividends on the intensity with which he drove and slaughtered in order to pile up production, he was especially recompensed for long trips after runaway Indians, these poor wretches being chased as far as six to twenty days' journey into undisputed Colombian territory North of the Japura. This system of rewards for huntsman work was extended also to include the half a dozen to a score of salaried breeds and blacks assigned to each section chief as assistant terrorists. Additionally, some few of the mongrel trigger men drew small bonuses of so much per arroba while the negroes were spurred by the demonstrated fact that they themselves were not immune to punishment of lash and stocks should the chiefs find cause for dissatisfaction with their Legree work.

The system of using armed bullies as Indian slave-drivers was as old as the Putumayo rubber 'industry.' The half-breed *peons* or *racionales* (so called to distinguish them from the completely ignorant forest Indians) brought in by the early Colombians were the first such musclemen and this class of Peruvians, Colombians and a very few Brazilians continued to be imported for the filthy service. The negroes, British subjects imported from the West Indies by Alarco in 1904, were an Arana idea and not a particularly brilliant one in the final analysis. Augmenting the breeds and the black Barbadians were the red *muchachos* who drew no pay and lived in special Indian houses at the section stations. The educating of these boys and youths to butcher and torture the other redmen was also an old story in the Putumayo. Often they became almost as vicious as their masters, for their promotion gave them the use of Winchester rifles and made them free to rob the laboring Indians of any possessions that took their fancy. Ordinarily they were employed against tribes other than their own and for them it can be said that their worst crimes were committed under direct orders from their superiors. And that sometimes they turned on the masters and killed.

In considering these masters, the 1936 comparison that comes readily to mind is with the New York and Chicago gangsters, their mass murders and their jesting ways of putting fire to the feet of their bound captives or cutting notches in them with Boy Scout axes to the accompaniment of anecdote and belly-laugh. Imagine a Schultz or Diamond or Capone mob turned loose to run a land where the adult population was physically as well as mentally childish and you will have a very good idea of what happened in the Putumayo under Arana's enforcers.

As practiced by the monsters who subdued the Andokes and Boras Indians and opened up their country for Cesar and partners, the conquering process followed a basic pattern. Locate a communal house, fire it, capture as many as possible and shoot all the others. Chain the captives, march them into the new station (beheading the laggards along the way), keep them in chains or stocks to soften up for a month or so, and send out the survivors to pick rubber. Repeat until the desired quota of workers is obtained. Nor did this technique die with the end of station-founding, for as enslaved Indians were slaughtered, replacements had to be rounded up from among those still running wild. Standard as the recipe was, however, it was subject to infinite variations, the individual section chiefs adding such torture ingredients as wanton impulse dictated or as they had found useful in reducing these animals to abject fright. Nor was it only in rounding up the untamed that mass murders were staged. Recaptured runaways, whether Boras or Huitoto, were dragged from the prisons under the station houses and shot down by tens and twenties by way of example to others. Expeditions were sent out to exterminate clans consistently failing to produce as much rubber as ordered, sometimes with instructions to bring back fingers, ears and heads of victims as proof of performance. Squads of men, women and children were deliberately starved to death in stocks often so jammed that captives could be fastened by but one ankle although the Peruvian specialists built 20 and 24 holers only. In a wholesale massacre of Indians who foolishly had allowed themselves to be enslaved by Colombian 'poachers' near the Japurá, children were jammed head first down the post holes dug by their parents in connection with building a house for the whites of the wrong nation. During one attack on a settlement of fugitive Indians, 18 small children had their brains dashed out against the trees when enraged hunters barely missed their elders. At La Chorrera itself, 30 captive Ocaina Indians were clubbed, shot and burned to death in one day.

In the punishment of individuals for running away, for encountering search parties and being unable or unwilling to furnish information, for failing to work rubber, or for falling short of the quota by too wide a margin, ingenuity ran wild. The lucky were those, juvenile or adult, who had their heads instantly severed with the machete. Children were dismembered limb by limb, adults left to die with arms and legs cut off. In the field men and women were hung from trees and cooked over bonfires of dry leaves. At the stations they were drenched with kerosene and set ablaze, were bound hand and foot and thrown into the flames. Men were clubbed between the legs until they died, were chained up before marksmen who competed at shooting off their privates. *Muchachos* and their masters snatched babies from the arms of offending mothers and threw them in the river or wrapped them around the trunks of trees.

Men, women and children were set at latex gathering. As there was little time to cultivate crops around their own communal houses in the bush, starvation was a regular guest and times there were when they were not even able to furnish food to the *racionales* and *muchachos*. Then the minor terrorists themselves had a taste of subsisting on wild herbs. These gunmen actually had duties additional to subduing wild Indians, pursuing ones who had fled the territory and assisting in the flogging and torturing at the stations. A chief occupation, in fact, was supervising the bringing in of the rubber every 10-20 days. Between these times the Indians gathered latex in the forest entirely without supervision for the chief was not concerned with how the stuff was collected, but only with its being delivered in the specified quantities on the specified day. On the day when the *puesta* was due, he would send out parties of *racionales* equipped with *muchachos* and with lists of the red workers to be collected. First the *cacique* or *capitane* would be located and if all his people did not appear, the *muchachos* would be sent out to summon them. If some still did not show up, the *capitane* would be hung up by the hands or worked on with a lash to make him confess where they were hiding. When the gunmen then had collected all who could be found, the entire party, those who had their rubber and those who had failed to accumulate it, would be marched off to the section house. Men, women and children made up these caravans and in many cases it meant a journey of 10-12 hours, always burdened and always through difficult jungle all the way.

Then began the separating of the sheep and the goats. Good Indians were allowed a mouthful or two of beans and rice before being started back to their rubber forests. The others knew what to expect. Above five these Indians could not count (20 seems to have been the limit for the Einsteins among them) and they had no idea of weights, but they did learn to know when the needle on the scale fell short of the mark indicating their *puesta* assignment. And so humble were they that numbers, on seeing this, would lie down on the ground, stretch out their arms and await their floggings. This was standard punishment, although Indians often were shot or had their heads lopped off with machetes for no more than failure to produce the specified amount of rubber. The wallopings, all administered with a twisted lash of tapir hide, were dished out for being as little as a pound of latex shy, and the amount of the shortage dictated the number of stripes. Always they were laid on with such a will that one authoritative estimate had 90% of the Putumayo's working population bearing scars of such sessions. Whatever the accuracy of that, these badges - and they were presented at administrative centers as well as at the section stations — were common enough to win a Putumayo nickname—'marca de Arana' (the mark of Arana).

The really terrible hidings went to those rubberless wretches who had hid from the *muchachos*, although others less 'guilty' shared in them, too, on occasions when the chief happened to feel that way. Such punishment also was mingled with burnings, mutilations, deliberate starvation and other reprimands administered for such major crimes as attempting to flee the territory. Pegged out between four stakes, fastened in the stocks or cepo, hung in the air from a cross pole by arms tied behind the back or jerked off the ground at a whipping-post by a chain around the neck, the victim would soak up 50-200 strokes of the kind that cut flesh and drew blood every time. The floggers worked in shifts, relieving each other at 25 strokes or so, and Indians sometimes were deliberately flayed to death in this manner. If not left in the stocks, the survivors of the worst lashings were turned loose to die on the way home. Backs and buttocks literally torn to pieces, they could be left for the maggots. They putrefied and died, then, so many of them that whole sections sometimes stank from the rotting flesh of dead and dying redmen. As one of the least squeamish of the Barbadians phrased it, 'you could not eat your food on account of the dead Indians lying around the house.' Nor was appetite helped by the sight of dogs worrying the carcasses to pieces and dragging the limbs about.

When the corpses became too much of a nuisance for the section chief, Indians were ordered to pile them in heaps and set fire to them. At the river stations they usually were tossed into the water. Children of eight years or younger were immune from neither lash nor *cepa* and women were flogged not only for bringing insufficient rubber themselves but for the sins of husbands who failed to come in. And if a chief did decide some juvenile worker was too tiny to be given the tapir-hide, the alternative would be to have the mother beaten before the child's eyes.

Tortures to one side, the average Indian's hardest time came, twice or thrice a year, with the *fabrico* — delivery of the rubber of preceding *puesta*s to administrative centers or to the nearest shipping point. These journeys of entire rubber-working populations 'injure them so much that many of them fall ill and die' reported 1911's reform manager of La Chorrera district, and this at a time when some changes for the better had been made. His estimate was that hundreds had perished in previous years' treks from the distant sections and the testimony of those who had witnessed such safaris would indicate that he did not err on the side of overstatement.

Fever, starvation, exhaustion, these were the enemies that pulled down the overburdened staggering along the jungle paths. From many of the inland stations the laden journey was of days' duration. One required five full days until in 1908 a launch was placed above the Igaraparaná rapids in order to eliminate the necessity of toting overland all the way to La Chorrera. Even with this change, three days of 12 hour marching remained; thus, considering previous journeys from homes into section station and the return trip, these Indians were away from home for well over a week. And they subsisted on the country as best they might in the intervals of their killing labor. A minute portion of rice and beans at the section station and at journey's end was all that the company contributed. Those who had a bit of cassava bread in the larder at home brought it along. Those who had none foraged for roots and leaves and berries along the way and on some of the forest routes their progress was blazed by the branches, creepers and even trees torn down in the desperate search for something to stave off the hunger pains. Numbers journeyed as much as 20 miles into the section station with their current *puesta* contribution, there to have the burden increased to 140 pounds or more for the final lap that sometimes was all of 45 miles. Children of 12 to 13 were loaded down with more than their own weight in rubber. Nor was it unusual for women to come to journey's end carrying the added load of a baby born on the road and coated over with rubber latex to keep it warm and to protect it from the insect swarms. And all of the marchers — men, women and children, starved and half-starved — had to battle their burdened way through matted tropical vegetation, over fallen trees and across the swamp lands, for there were no roads and no bridges. Those who collapsed were kicked to their feet and set going; those who died were left at the roadside. Armed *muchachos* and *racionales* were scattered through their ranks to keep them moving, other gangsters brought up the rear and a day behind followed section chief and additional gunmen to make sure that no pack animal succeeded in slipping his burden and getting away home.

Perhaps the most horrible feature of the whole ghastly business was the fact that it was matter of common knowledge throughout the upper Amazon, matter of official knowledge in the capitals beyond the Andes almost from the beginning. Specific atrocities were the town talk of Iquitos, general charges of murder and slaving appeared in the reports of missionary priests of Loreto department, found their way into documents submitted to the Peruvian congress and issued as government publications. And once the Peruvians began to work on Colombians as well as Indians, details of massacres appeared in the press of Bogota, in the press of Manaos (where Colombia had an unusually efficient consul), in official correspondence between Bogota and Lima. To an Iquitos editor, however, was left the task of first marshaling the case with every possible circumstances of name, date, location and extent — all backed up by eye-witness affidavits of former Putumayo workers, some of whom had escaped, some of whom involved themselves as coerced perpetrators of crimes described. Benjamin Saldaña Rocca was the name of this always obscure and now forgotten crusader who 'had at Cesar in Cesar's capital.' Whatever his motives, the courage he displayed should have made his name forever memorable in the journalistic history of both Americas. Friendless except for the futile folk below the salt, without financial backing of any sort, impeded by the authorities at every turn, he waged the unequal combat through the pages of *La Sancion* and *La Felpa*. Not newspapers of regular issuance were these, but mushroom affairs that sprang up in 1907 and died in early 1908 — 'scandal sheets' we would call them. All that Rocca had to say in them he backed up with a legal action which he finally forced on the attention of the Iquitos courts, along with a counter attempt to have him indicted. Now the airing of these charges was the last thing that the rulers of Iquitos wanted, and from this hot spot judicial authority escaped by the shameful dodge of declaring itself incompetent to act because of the Putumayo boundary dispute. Outmaneuvered in the courts, busted, and with the good fight waged so far as it was in his power to do so, Benjamin Saldaña Rocca vanished over the mountains and is said to have wound up pounding the Peruvian equivalent of a beat for one of the newspapers in Lima.

Not in the long run was it a futile fight, however. As the editor pulled out of Iquitos, there pulled in one W. E. Hardenburg, an engineer from the United States who had come blundering down the Putumayo from beyond the Andes just in time to get caught up in that early 1908 assault of army-navy-Arana forces upon the Colombians.

Smarting under personal ill treatment and appalled at what he had seen while prisoner of the Peruvians, Hardenburg gathered up the Rocca material, collected further affidavits on his own hook and made for London. There the daily press, fearful of libel, shied away from his attempts to donate to them the story of the Putumayo; at last, in a muckraking periodical, *Truth*, the American found his mouthpiece. In September, 1909, it shot the works. With questions popping in the House of Commons and with the Anti-Slavery and Aborigines Protection Society working on Foreign Minister Edward Grey, such pressure was exerted against Peruvian Amazon's directors that after months of denying, squirming and fencing they consented to organise and send out their own commission of inquiry in July, 1910. In marked contrast to the Peruvian Amazon board's determined fight against any kind of investigation, and its flat refusal even to receive Anti-Slavery society officials, was the attitude of another rubber concern which had no British connections whatsoever. A letter printed in a London newspaper (and apologised for the next day) had charged that there was another Putumayo on the Suarez properties in Bolivia. Immediately, one of the Suarez brothers wrote to the Anti-Slavery group offering to pay the expenses of any investigator it cared to send out. That stand, combined with the society's inability to find the letter writer, sufficed to acquit the Bolivians.

In London on leave at this time was Roger Casement, British Consul-General at Rio and a world figure since his earlier report on rubber atrocities in Africa. Delegated to conduct a separate investigation for the Foreign Office, Casement traveled back to South America. Arana's importation of those black Barbadian subjects of Britain's king was the loophole by means of which this official entered the Putumayo. Signed statements of the Barbados men, some of whom admitted their own participation in the crimes, confirmed the worst of the atrocity stories and were from day to day accepted as substantially correct by the current manager of La Chorrera. With the Peruvian Amazon Company's commission this gentleman sat in on most of the questioning and he turned down invitations to have the accused called in to confront the negroes. The upshot was that Casement, after weeks of interviews and investigations on the scene of the horrors, confirmed everything alleged by Rocca and Hardenburg. So did the company commission which investigated independently of Casement a part of the time and which drew up its report without having seen his. With the cooperation of the State department of the United States, Sir Edward Grey's office began bearing down on the Peruvian government in January, 1911. Peru, with full access to Casement reports otherwise held secret, elected to stage a probe of its own instead of making immediate arrests. Into the Putumayo went Judge Romolo Paredes, a conscientious official who emerged in July, 1911, with 237 secret apprehension warrants and a 1,300-page report confirming everything turned up by Rocca, Hardenburg and Casement and adding to their findings. Iquitos — if not Lima — was playing horse with the diplomats of London and Washington, however, and virtually no effort was made to apprehend even those criminals who had not fled by this late date. Save for one retired section chief picked up at Iquitos, all of those taken into custody were minor offenders. None of these were punished and numbers of the same stripe remained at work in the Putumayo itself. As to the section chiefs and their principal lieutenants, they chiefly scattered into Brazil with 'their' Indians. Some of these Huitotos were sold as slaves in the Acre territory, others were set to picking rubber for the Putumayo 'fugitives' in forests close to the border. Those section chiefs abandoning the business wound up in such diverse spots as Lima, Callao, the Argentine, Barbados and New York City. Similarly farcical were proceedings against the Iquitos masters of the criminal syndicate. Against Pablo Zumaeta a judge of first instance issued a warrant of preliminary detention which police deliberately refrained from serving. The superior court annulled the warrant and dismissed the impertinent judge from office. Pablo sued the judge for 'revealing public documents,' and was successively elected president of the chamber of commerce, head of a benevolent society and member of the Municipality. Similarly, the order for Cesar's arrest was barreled by the superior court and the traders of Iquitos united to throw a banquet for the hero returned from London and his hired hand, Rey de Castro, Peruvian consul-general at Manaos. Disgusted at the course of events, England in July, 1912, made public the Casement reports and its two-year correspondence with Peru and now, for the first time, the Putumayo scandal was heard 'round the world'.

Meanwhile and throughout all the uproar, the slaves of the Igaraparaná and Caraparaná continued getting out the rubber for Cesar, liquidator. From the day of Peruvian Amazon's inception, Cesar, Alarco and Zumaeta had ridden it straight for bankruptcy. With the end of the corporation comedy in sight in early 1911, Zumaeta used his power of attorney to slam on a $300,000 mortgage in favor of his sister and Julio's wife. This made the señora the preferred creditor. With credit automatically shut off, rubber shipments tied up, and between $10 and $15 remaining in the treasury of the $5,000,000 corporation, the puppet board thereupon legislated itself out of existence by naming Cesar as autocratic liquidator. In such capacity his principal duty was dealing with that claim of his wife's.

Such little matters as the mortgage and the investigating commission's report were kept from the handful of small stockholders in England, but everything was brought to light at 1912-1913 hearings of the committee appointed by the House of Commons to inquire into the responsibility of British directors in the matter of the atrocities. It was in the course of these hearings — and this seems to have been completely forgotten — that Cesar himself admitted the occurrence of most of the crimes alleged. This was in connection with his attempt to prove that they occurred without his knowledge and his only reservation in the matter of the most hideous ones was an 'exaggerated in description' generality. Cornered on the witness stand, he could not cite one specific instance of exaggeration and after 'binding' himself to send such information to the committee, failed to do so. The committee found that the English directors had not been parties to any overt act, but that they had been 'culpably negligent.' Further, it came to the conclusion that 'Senor Arana, together with other partners in the vendor firm had knowledge of and was responsible for the atrocities perpetrated by his agents and employees.' English law had no power to punish the Peruvians, of course, but on the petition of the minor shareholders the courts ordered the compulsory winding up of the company and Cesar was bounced in favor of an official liquidator. Eventually Arana picked up the pieces, however, and continued to operate under the style of Peruvian Amazon Company.

Exports in later years have followed the trend of world rubber prices and demand, in one year (1919) mounting to the new record of 1,709,000 pounds in response to the impetus of the war period, and two years later falling to 69,000 pounds (little more than one-half of the 1901 figure). One tentative effort there was to interest American capital, but it got nowhere. Approved by Colombia in 1925 and by Peru in 1927 was the 1922 treaty establishing the Putumayo as boundary save for a narrow strip up against the Brazilian border which gives Colombia a corridor to the main Amazon. Peruvians interested in Arana's company were about the only ones affected by the major portion of this cession, but the Amazon port of Leticia in the corridor area was seized by Loreto volunteers so recently as 1933 and only restored to Colombia when the League of Nations intervened in 1934. Undisputed is Colombian sovereignty over Igaraparaná and Caraparaná, but the extent and effectiveness of its control is highly dubious.

There have been no official reports on the condition of the Igaraparaná-Caraparaná slaves since late 1912. In 1923 a rubber survey commission representing United States departments of commerce and agriculture went through the principal Amazon regions, however. Cesar, by then, was a Peruvian senator representing the department of Loreto and, according to the commission, owner of 'definitive title to 14,200,000 acres of land in the Putumayo region, but extending across to the Japura in one direction and the Napo in another,' land inhabited by 'over 10,000 Indians of whom 3,000 may be considered as actual or potential laborers.' Concerning labor's treatment in the Peruvian rubber lands, the commission notes, 'There would appear to be very little ill treatment of labor in these regions, and the stories of former 'atrocities' are now greatly discounted by the calmer and more unprejudiced judgment of those who were actually familiar with conditions.' The committee, incidentally, went no closer to the Putumayo than Iquitos. And in the introduction to its report, it gracefully declares: 'Among the many private individuals whose kindness we take pleasure in acknowledging are ... Sr. Pablo Zumaeta of Iquitos'.

2.4 Congo King

GREATEST OF ALL the wild rubber kings was a man named Coburg. First and last *caoutchouc* monarch thrown up by the African 'industry,' he looms over the members of the South American royalty as these towered above, say, the leading latex controllers of Central America. The Suarez brothers, we have seen, came to operate half the rubber estates in Bolivia, second ranking producer of the Western hemisphere's Southern continent. Their African rival accounted for a half of his continent's rubber production in some years of this century, from a third to a fourth in others; got out more rubber than the whole of Bolivia, more wild rubber than any nation in history with the exception of Brazil and of Mexico in its top years. Julio Cesar Arana started with a load of Panama hats and pyramided it into a personal estate the size of Belgium. The man Coburg started with an idea and pyramided it into a personal estate a fourth as large as all Europe; into architectural extravagances that made the marble boasts of Para and Manaos look like the efforts of a Lem Putt in comparison; into the greatest rubber fortune ever known — and this includes industrial barons, inventors and share gamblers as well as the exploiters of the trees. Multiply Arana's commercial cunning until it becomes the financial prowess of a Morgan, multiply his ability to control associates into a genius for manipulating whole populations, multiply his paltry millions of profits by many tens and you have Cesar Africanus. Multiply the Putumayo's area by seventy-five, multiply its minimum of 20,000 deaths and flights by hundreds and you have the Dark Continent's equivalent of the Igaraparaná-Caraparaná sector. It was with the development of this private property that African rubber first bulked importantly in commerce; it was with the manipulations of Coburg that raw rubber, now a potent source of international jockeying, first became matter for world diplomatic manoeuvers.

Historically African rubber dates back to the 1760s, when *caoutchouc*-yielding plants were discovered by Pierre Poivre, French intendant of Mauritius, and by M. Coffigny, nosing about in Madagascar. Mainland rubber does not enter the record until 1805. In that year the French botanist Palisot de Beauvois classified a West African vine as a member of the dog-bane family and christened the new genus '*Landolphia*' in honor of Captain J. F. Landolphe, privateer who had participated in the one French expedition into the gruesome land of human sacrifices ruled by the black priestly theocracy of Benin city. That territory in the Niger basin became a part of Britain's rubber-producing colony of Nigeria and varieties of *Landolphia* scattered from Senegal on the North to British South Africa qualified as the continent's leading rubber yielders.

By the natives rubber never had been known or used. Once the white traders on the coast had shown them what was wanted, however, the knowledge spread into parts of the interior as, long before, had been the case with the European diseases and the plants out of the Americas (cassava, peanuts, maize, sweet potatoes). To an American missionary the traders themselves were indebted for the first serious directing of attention to the West African *caoutchouc* vines. Reverend T. L. Wilson, established on the Fernand Vaz river in present day Gabun, did this pointing out in the 1850s. During the same period French-American Paul du Chaillu was making the first exploration of the country back of the Gabun *estrada*. Wilson, rediscovering the rubber vine, was considered something of a benefactor. Du Chaillu, rediscovering the gorilla after more than 1,300 years, was set down by the world as an unmitigated liar. The judgments were equally erroneous.

From about 1860 on, then, a small amount of rubber went out of the West coast settlements and trading 'factories.' It wasn't even a noticeable export until 1870-1875, and it amounted only to 220 tons by so late as 1878. From these worst of the old slave shores the Portuguese had been loading black cargo since the middle of the fifteenth century; Spanish, British, Dutch, French and Danish joined in as the necessity for repopulating the New World arose. Not until 1808 was the British slave trade prohibited and not until 1836 (two years before the abolition of slavery in British possessions) did the last European nation declare the business illegal. Smuggling of slaves had gone merrily forward, of course, but it hit something of a snag in the early 1830s when Britain and France entered into an arrangement for a mutual right of search within certain seas. And it sustained a really serious blow in 1845 when these two nations combined to maintain squadrons off Africa's West coast. Even then a considerable bootlegging went on until after the American Civil War, but for the whole last half of the nineteenth century the important African slaving problem was an internal rather than an export one.

And so it was that commerce in forest products came into being chiefly because traders from Senegal to Angola were looking for other things than black meat to deal in. British consular reports of as late as 1877 note that the 'factories' of these traders all were more or less worked by slave labor, that tortures were inflicted by the Portuguese and that the murder of one or more slaves by white men was almost a daily occurrence in 'the Congo.' It was in Portugal's Angola that the illegal slave trade chiefly hung on and that there was direct penetration of the interior. Here, a considerable distance South of the main Congo but within the Congo basin, half-caste slavers long had ranged from the coast to what is now Northern Rhodesia. North of it, the dealers in ivory, palm oil, gums, peanuts and, less importantly, rubber (which then and always was inferior to the South American product) hugged their few coast locations, trading directly for products of the immediate vicinity and making advances of goods to neighboring natives who journeyed inland to barter with other blacks or in turn give them merchandise on credit to be traded to tribes still deeper in the forests. In Gabun, for instance, the British merchants who chiefly fostered trade were under a French political authority which was confined to the Gabun river *estrada* itself. Lured by their liquor, kettles, umbrellas, muskets and looking glasses, coast natives shoved in to the West and South. Born traders were all of the Central African natives, traders they had been among themselves long before the white man came. Ivory and gums meant little in their lives; rubber nothing at all. But for booze, cloth, cooking pots and snuff boxes they did have an abiding lust. Whatever the difficulties of ivory-hunting, gum and rubber collecting practiced when and as one feels like it is no onerous task, and thus it is easy to see why in African eyes it was the white man who was the sucker swapping valuables for trash.

On the Congo as on the Gabun, direct trade was confined to the *estrada* and only through black intermediaries was 90 miles in 1483, but there had been virtually no attempt to exercise sovereignty over it. So late as the 1870s, Portugal's nearest settlement on the coast was Ambriz, a good 125 miles to the South as the crow flies. Inland was San Salvador, some 60 miles South of the nearest stretch of the Congo. To the North France's foothold on the Gabun *estrada* was 500 miles from the Congo in a direct line. International were the settlements on the Congo *estrada*, collections of Portuguese, Dutch, French, British and German 'factories,' each flying the flag of its own nation and claiming that country's protection but possessing no actual political character. Most of the traders were located on the North bank, the principal settlements being Banana, immediately within the river's mouth, and Punta da Lenha and Boma, respectively, 30 and 60 miles up the 60 mile *estrada*.

The Amazon of Africa is the Congo. World's second largest river in volume and in area drained, it has a length of 4,600 miles and main stream and affluents have over 6,000 miles of navigable water. Nothing at all of this was suspected by its discoverers, however, nor by the traders who came after. About 93 miles from Banana is the foot of a series of rapids through which the river plunges for 200 miles in an 800 foot drop. It was these cataracts separating lower and middle Congo which shut off navigation and, it would seem, curiosity as well. Not until 1816 was there a real exploration attempt, the British Government sending out an expedition under Captain James Kingston Tuckey, R.N., who managed to pass some of the rapids and reach a point 170 miles up the river. There a fatal sickness claimed the leader and 16 of his followers and the attempt was abandoned. Later on several other tries were made with even less success and so it was that for 60 years more 'Tuckey's Furthest' remained the highest point on the Congo known to the white race. Up the line of cataracts from the lower river, the Ba-Congo blacks carried goods of the European merchants and returned with slaves or forest produce obtained from the Batekes who were settled about the lake-like expansion (now known as Stanley Pool) at the head of the rapids and who acted as middlemen for the tribes voyaging down by canoe from farther up. Ramifications of this commerce extended to considerable distance, but without adding to civilisation's knowledge of a 4,600 mile river which, in nearly 400 years, had been explored for just 170 miles. On the upper, or Lualaba-Congo, Arab slavers were located, but they had pushed across from the East coast, had no idea of what river they were on and no intentions of following it through. David Livingstone, working through this South Eastern area in 1867-1873, thought that he was tracing the head-streams of the Nile. 'I have no fancy to be made into 'black man's pot' for the sake of the Congo' was his attitude. From Livingstone's own reports, some of the English armchair geographers were able to arrive at the conclusion that the missionary-explorer was at large in a river system belonging to Congo rather than to Nile, but actual proof was lacking. It was to settle this question, to complete the exploration of Lake Victoria, to do something about filling in the immense blanks on the map of Africa, and to 'show' those who had sneeringly doubted his story of finding Livingstone that Henry M. Stanley, the journalist baptised John Rowlands, launched his second African journey.

From the Congo's mouth draw a straight line across the continent of Africa and it will hit the East coast island of Zanzibar. It was from there that Stanley started in November, 1874, with three white assistants and 224 Mohammedan blacks. He followed even less of a straight line than does the horseshoe-shaped river, for in all he traveled some 7,000 miles to reach Boma from Zanzibar, going first North-west to Lakes Victoria and Edward, then dropping South to Lake Tanganyika and discovering its outlet (one which connects it with the Congo system, incidentally) and finally striking North-west across country until he hit the Lualaba-Congo at a point where only Livingstone and the Arabs had been before him. Here, in the heart of the continent, he decided to follow this river to whatever ocean was at its end. Only one white companion remained alive at this time, but with this assistant, his remaining blacks and a party of Arabs under the slaving chief Tippu-Tib, Stanley struck out down the Congo on November 5th, 1876, the leader by boat, the great body of his followers marching through the forest. On December 28th the Arabs reached their limit and turned back, the others continuing in a fleet of canoes first captured from battling cannibals and then purchased from them. A week later they sighted the first of the cataracts (Stanley Falls) which separate upper and middle Congo and are 1,000 miles above Stanley Pool. On June 3rd, 1877, the rapids below Stanley Pool took the life of the leader's sole white comrade. On July 30th, 'Tuckey's Furthest' was reached after more disasters by water and almost incredible portages up cliffs and through jungle. Knowing now beyond all doubt that he was on the Congo, Stanley abandoned the boats for an overland march. On August 4th he sent a message ahead to Boma asking help: '... You may not know me by name; I therefore add that I am the person who discovered Livingstone in 1871.' The aid came, and on August 12th, 1877, the explorer reached Boma, nearly three years after he had set out from the East coast. Those bare dates naturally give little hint of the tremendous odyssey which the seven months on the Congo constituted. The wrestling's with disease, the combats with the upper and lower river cataracts, the weeks of daily battles with fierce blacks catching their first glimpse of a white man — it is not with these we are concerned, but with the mere fact of Henry M. Stanley's completing that journey. Now for the first time the immense commercial possibilities of the Congo basin dawned on certain of the European nations. Before the end of the decade the scramble for territory was on and a new era in the history of West and central Africa had opened.

First to catch fire at Stanley's exploit was Leopold II, King of the Belgians. Right at the boat were two of his emissaries when the returning explorer reached Europe at Marseilles. Stanley shoved on to his native Britain, however — he had his books and his newspaper stuff to write and he wanted to turn over this newly discovered territory to his old countrymen. Back home he was a hero all right but only until he started trying to give this land away. Britain was having all it could handle in Egypt; the Congo sounded like quite a burden to take on and suspicion grew as its sponsor waxed ever more rhetorical about its golden prospects. The end of all Stanley's campaigning, lobbying and speech-making was failure; and then he turned to Leopold.

Cramped by his tiny German duchy of Saxe-Coburg-Saalfeld, Leopold's father had escaped from it to wind up as elected king (second choice) of newly independent Belgium after just falling short of becoming ruler of Britain (as prince consort or prince regent), Mexico and Greece. A rare opportunist was Leopold I, a marrying gent as smart as a Mdivani, and a modern-minded ruler who played a leading part in making Belgium one of the most highly industrialised nations on the continent of Europe — and one of those paying the lowest wages. More importantly, he sired his superior to carry on. A razor-witted and unscrupulous trickster was the oldest of the Coburg boys, a master of gargantuan and profitable buffooneries, an implacable hater, a mover of heaven and earth in pursuit of his ambitious ends — in short the machine-age Ulenspiegel of machine-age Flanders, the great Beggar in modern dress complete with trade mirror, and dividend-conscious owl. As Duke of Brabant and then as King he found himself almost as confined by Belgium as his father had been by his German acres. In Europe there was no room to expand. Mozambique—Portuguese East Africa — seemed to offer some possibilities, the Philippines more. Leopold would found a chartered company like the old East India companies, lease the islands from threadbare Spain and show the world some first-class exploiting. The offer was made and rejected. Back to Africa turned the Coburger, nothing definite in mind yet, just a general idea that an alert king ought to be able to snare something pretty juicy when the time came for partitioning that huge continent. On September 12th, 1876, then, Leopold opened in Brussels an international African conference of geographers and explorers. It was a good hunch, it gave the ruler of the Belgians a splendid chance to impress himself upon world consciousness as a Great Humanitarian interested in having lost, forsaken, heathen Africa prettied up with civilisation, progress and Christianity. It was a better hunch than he knew for, all unknown to Leopold and the world, Henry M. Stanley even then was closing in on the upper reaches of a Congo he was to introduce to the world as a great commercial highway. What Leopold talked at the conference was a sort of international league to establish coast bases opposite Zanzibar and at the Congo's mouth and lay roads and sprinkle posts and settlements through the interior, the funds to be raised for the world body by national subcommittees. Out of it came the Association Internationale Africaine (AIA) dedicated to converting the Negroes of Central Africa and spreading trade and industry among them. There was, of course, an international commission with Leopold as president but the national committees were slow in coming into being and even slower in coming across and it remained a nebulous venture, one looked upon as the quixotic idealism of a pious dreamer. In June, 1877, the international commission gathered at Brussels for its first and last meeting, reelected Leopold as president and gave birth to an AIA flag. And while it deliberated, the long lost Stanley was battling the rapids of the lower Congo and drawing ever closer to Boma.

By October the AIA was virtually dead. In October, Stanley's cablegram announcing the tracing of the Congo broke upon a startled world. Now that Leopold had his goal he could be trusted to work out his route to it. That first attempt to flag the explorer at Marseilles was but an opening move. King bombarded reporter with invitations, watched anxiously the Stanley attempts to interest British statesmen and capitalists, kept an eye on the disappointed man who packed off to Switzerland to write his book, netted him at Paris through the medium of emissaries, signed him to a contract at Brussels. There was no idea of establishing a Belgian colony, the Englishman was told (and this was perfectly true). Stanley was only to investigate economic possibilities, civilise blacks, and found stations on a generous salary and bonus plan — all this on behalf of some sort of philanthropic international society. Leopold molded him into the desired shape, swore him to secrecy, headed him for the Congo's mouth in early 1879 and went back to his international wire-pulling. Already the AIA was being brushed aside. Stanley was to fly the ALA flag from his stations; he agreed, however, to induce the native chiefs to sign away their territory to the Comité d'Etudes du Haut Congo, and it was as leader of a Comité d'Etudes expedition that the puppet hero went out. A commercial company nominally capitalised at $200,000, this new outfit had been organised in the palace at Brussels by the innocent explorer, an Englishman, a couple of Dutchmen and a posse of Belgians headed by the trader-king and a banker who fronted for Leopold and posed as chief angel of the venture. The moribund ALA did have several farcical national committees, but the new concern was no more than Leopold. And especially was this true after it was wound up under the Comité d'Etudes name and brought back to life as an all-Belgian venture styling itself the International Association of the Congo 'founded by His Majesty the King of the Belgians for the purpose of promoting the civilisation and commerce of Africa, and for other humane and benevolent purposes.' Purposely confusing were these assorted associations with vaporous outlines, and it is small wonder that governments, even, were ignorant of just which one they were dealing with; what if any were the differences between them; and what either one, let alone both, really amounted to. Actually the AIA now consisted of a flag.

For the International Association of the Congo, however, Stanley and other agents labored mightily in sewing up some 500 native chiefs in 'treaties' stripping them of their sovereignty, their property, their all. Stanley also sweated at building roads and founding trading stations during his five years as a Coburg servant. Never, it seems, did the gallant and honest Bula Matadi, 'the rock breaker,' quite figure out what everything was all about. At times this was vastly irritating to Leopold, especially when the man went ahead with his road-building knowing that France's Count Savorgnan de Brazza was racing towards Stanley Pool from the North but not appreciating at all what this meant. Usually, however, inability to tumble to what was up was a very valuable quality in one whose irreproachable character so well qualified him to front for the schemer of Brussels.

Meanwhile, as Stanley and successors were carving stations out of the African forest, the Ulenspiegel of Brussels was spinning a State out of thinnest air. In the beginning his trouble was that the nations now competing for African acquisitions brought up his old fine phrases to use against him and solemnly treated his associations as the philanthropic, scientific and commerce-encouraging bodies they had professed to be — bodies which under no circumstances could be viewed in the light of a sovereign State or States. What Leopold had to have was recognition of his enterprise as a State by just one important nation. That would give him some sort of ground to stand on.

He achieved his end by taking the reformers and politicians of the United States of America into camp. The tool in this campaign was 'General' Henry S. Sanford, former United States minister at Brussels and one of the two Leopoldian couriers who had sought to enlist Stanley at Marseilles. With the two-plank platform of free trade and wholesale conversion, then, Leopold won the Protestant Missionary Society, President Chester A. Arthur, the State Department and Congress. The ballyhoo referred only to the impressive-sounding AIA. The International Association of the Congo was confused with this one and referred to as the AIA in a number of documents including State department communications and Senate committee reports, but this was straightened out in connection with the formal exchange of declarations which constituted the American recognition of April 22nd, 1884.

That was all that Leopold needed; but he needed it desperately by then, for Great Britain two months before had signed a treaty recognising Portugal's historic claims to lands running North to include both banks of the Congo *estrada*. In England Leopold was already at work and shortly had public opinion in his capacious pocket. On the continent, the Coburger Ulenspiegel wooed France by sending it, the day after American recognition, a promise of preferential right to purchase Association possessions should circumstances ever compel their surrender. Bismarck, which is to say Germany, currently was interested not in colonial possessions but in free trade for Germans — and in having France absorbed in African rather than European activities. Here was sufficient support. England, under fire at home and abroad and even in Portugal, abandoned the treaty. Bismarck granted recognition. An International West African Conference was called at Berlin late in 1884, the very time at which Roger Casement, a twenty-one-year-old 'black Protestant' from Ulster, embarked on his first voyage to Africa as a purser in an Elder, Dempster & Company steamer. Casement returned from the trip in 1885 fired with a determination to travel in the Dark Continent.

Leopold in February, 1885, emerged from the Conference (attended by 'General' Sanford as a United States delegate) with 900,000 to 1,000,000 square miles of that continent, emerged just a bit dissatisfied because the stupidity of Stanley had cost him all that North bank territory allotted to France, because it had been necessary to allow Portugal the Southern share of the Congo *estrada* and a couple of seaports North of the new Congo Free State's Banana region. With no army or navy to back him up, without so much as a square inch of Belgian toehold in Africa, with the Portuguese and the British having done all the discovering and exploring, the King of the Belgians, by spending a few million francs, had been able to cut himself in for one of the richest million miles of African territory — a triumph of adroit wire-pulling and wholesale propaganda work without equal in world history.

And what he had gained only Leopold knew — if, indeed, he fully realised it himself at this date. Not a Belgian colony, of course, for when the parliament of that nation gave its necessary assent to his becoming 'the chief of the States founded in Africa by the International Association of the Congo' it was with the emphatic reservation that 'the union between Belgium and the new state of the Congo shall be exclusively personal.' No desire to take up the white man's burden had the then stoutly parochial Belgian populace. By the reformers, missionaries, aborigine-protectors and anti-slavery groups of the world, it was clearly understood that that noble philanthropist, Leopold, was no more and no less than a guardian placed over the Congo blacks, a trustee named to protect their interests, an appointee responsible to the Powers who had selected him as their representative in carrying out an international goodwill project that marked a new high in world progress towards the brotherhood of man.

Freedom of trade had been the one dominant theme of the West African Conference. To this Leopold had, perforce, to subscribe and to the prohibition of import duties, although he did succeed in defeating the move to ban export duties as well. Cautiously, then, he proceeded at first in his new dominion. It was a matter of placing steamboats above the rapids separating lower and middle Congo and above Stanley Falls, projecting a railroad around the lower river cataracts, arranging for regular steamship service with Europe, multiplying his stations, coming to terms with Tippu-Tib (who had raided the Stanley Falls station out of existence), recruiting among the fierce black tribes beyond Stanley Pool to form the nucleus of an army — in short, creating a real State in place of the imaginary one that had attained recognition. All of the butchering that occurred in the first half-decade of Leopold's rule was connected with subduing warlike tribes and with casual raiding of villages to seize men for the army; this rough treatment the plenteous dark populace could survive. Far more sinister was a first-year decree declaring all 'vacant lands' the property of the State, although it was not at once interpreted and neither native rights in the forest nor native trade with assorted Europeans was interfered with during the operation of the original five-year plan. It was in these years that the white man penetrated the interior and pointed out the rubber-bearing vines to tribes not familiar with the latex or the European demand for it. In 1887, as a result of this work, 30 tons came down the river from above Stanley Pool; in 1890 some 130 tons.

One of the first concerns to seize the opportunity of trading directly with the blacks beyond Stanley Pool was the Sanford Exploring Expedition organised by 'General' Sanford and M. Georges Brugmann in 1887. Roger Casement was among those who went up the Congo for Sanford in 1887-1888. Thereafter he headed for the United States to lecture about the wonders of Africa, and the Sanford properties at Matadi, terminal point of navigation on the lower Congo, were taken over by the Société Anonyme Belge pour le Commerce du Haut-Congo, newly organised commercial adjunct of the railroad construction company and related Belgian concerns. To pilot one of its steamboats on the middle river, the Société enrolled a Polish sea captain, Teodor Korzeniowski. As fellow employees he and Casement met in 1890 at Matadi. Simultaneously, a young Parisian, Edmund Dene Morel (born and baptised Georges Edmond Morel-de-Ville) was giving ear to the entreaties of a mother whose brother-in-law had died in the Congo. He had, perforce, to turn down a chance at a pursership on an Elder, Dempster & Company ship operating between Antwerp and the African river. And thereafter he crossed over to his mother's native island and became an Englishman and a clerk in the Liverpool home offices of the line just as the onetime purser, Casement, saw his Polish friend off on the overland journey to Stanley Pool. Six months of the Congo and the Belgians was enough for Korzeniowski. With him he carried away undigested material for two stories, Heart of Darkness and An Outpost of Progress; a jungle illness that was to plague him for the rest of his life; and a memory of Roger Casement that became a passage in a letter from Joseph Conrad to R. B. Cunninghame-Graham and an immortal portrait. The subject of this word-picture was lost by the Société to the British Consular Service, which placed him first in Nigeria and then moved him all over the map of Africa before returning him to the Congo. Of the pre-1891 style of free trading in which Casement participated, a picture has been set down by a missionary, who thus describes the founding of the rubber traffic in the territory that was to become the domain of Coburg's 'Abir' trust:

'It is interesting to hear the Bongandanga people tell of the beginning of the rubber trade. How wonderful they thought it was that the white man should want rubber, and be willing to pay for it. How they almost fought for the baskets in order to bring them in and obtain the offered riches. But they say, 'We did not know, we never understood what it would become in the future.'

The thing it became in the future was crystallised in a Congo proverb: *Botofi bo le iwa* (rubber is death).

As yet the Coburger was but laying the foundations for this aphorism. He was a good waiter when necessary, and it required a good waiter to keep on pouring money into Congo wars, administration and improvements while taking little out. Into that black hole went Leopold's profits from shrewd speculations in Suez Canal shares and other enterprises, as well as millions of francs belonging to his crazed sister and ward, Charlotte, sometime Empress of Mexico. With Belgian citizens too wary to buy his Congo bonds, with banks as coy, with stock exchanges of the Continent refusing to have anything to do with this African madness, the king signed notes of hand, pawned his civil list, piled up debts it seemed he never could pay. Very well, Belgium should pay them; there were statesmanlike ways of tapping its pockets. For this move Leopold had paved the way in 1887 by making it clear that the preferential purchase right granted France could not, of course, be opposed to the rights of Belgium. In July, 1889, he was able to manipulate the reluctant Belgian parliament into kicking in $400,000 towards the cataract railway. The next month he made public a will bequeathing to Belgium all his 'sovereign rights' over the Congo after his death. On the strength of this Ulenspiegel-style 'gift' within a year he obtained a $5,000,000 no-interest loan from his unwilling Belgian government. Leopold pledged himself to incur no further financial liabilities without the assent of the national Chambers, and Belgium was to decide in 1900 whether it would annex or demand repayment. How it could conceivably hope to collect from its king was not specified. Nor how Leopold could be held to his pledge of no further borrowing.

He broke this last almost at once, secretly borrowing $1,250,000 from A. de Browne de Tiege, banker and former Member of Parliament for Antwerp, on conditions which assured the creditor's coming into possession of a chunk of Belgium's Congo 'patrimony' five times as large as Belgium in the event of non-payment.

Meanwhile, Leopold, beginning to consider himself pretty well dug in, was putting pressure upon the European traders on the Congo and drawing battle lines for a showdown with the Arabs who had been raiding middle as well as upper river. In connection with this last he staged a conference of the Berlin Act signatories to the accompaniment of a press and pulpit hullabaloo about wiping out the slave trade. Net result was what he desired: the right to levy import duties in order to obtain money with which forcibly to put down his Mohammedan competitors in the ivory business. The war on the higher Congo broke in 1892, lasted for two years which saw the use of cannibal warriors by both sides, and ended in the complete smashing of Arab power. This was a commendable deed — if we consider it separately from its aftermath — and it is only fair to emphasise here that it was no earthly paradise Leopold ultimately turned into a super-efficient shambles. A dark and bloody ground always, it had boiled with tribal wars long before the Arabs came and in the districts to which the Arabs never penetrated. Black slaughtered black in the competition for wives and for involuntary honor guests at the cannibal feasts; black sold black in the indigenous meat and servitor markets. As for the Moslems from the East, they killed, plundered and burned in the wholesale fashion when they fell upon villages; murdered the wounded; butchered dragging captives without compunction as they marched bound victims to the coast. Numerous as were the all-Negro combats, however, they were waged with primitive weapons by participants meeting on pretty much of an even basis and were a routine peril in Congo life — something that menaced the individual or an entire small community, but cut no notable gaps in the population as a whole. Leopold, repressing both cannibalism and the blacks' own slave traffic to considerable extent, went on to subject the 'rescued' Negro tribes to a new continuous exploitation worse by far than the old intermittent slaughters, worse, even, than the exploits of the Arabs. On a smaller scale the whole picture is easier appreciated, perhaps. For Leopold substitute that other sufferer from sciatica and the instant wealth itch, Julio Cesar Arana; for the dissension-torn blacks, the Putumayo reds who also warred among themselves; for the Arab invaders, the original Colombian exploiters of the Caraparaná and Igaraparaná. That will give the approximate setup. Where Arabs and Colombians slaughtered, Leopold and Cesar depopulated; where pioneer exploiters harried parts of jungle empires in a comparatively small way, their successors extended their activities into all the forests, put wholesale murder on a business-like basis, and trapped survivors among once-carefree aborigines in a living hell of systematic starvation, overwork and torture that made mere sudden death a positive pleasure to endure.

As in the case of the Putumayo, the amount of depopulating accomplished cannot be accurately arrived at. During the late 1880s assorted travelers and administrators offered population estimates ranging from 20-30 million. In 1911 the official return was 8,500,000. Even today there is no wholly satisfactory and comprehensive figure, but we do have an individual enumeration of 7,727,316 announced in 1923 and it is presently agreed by all hands that actual population at most is eight to ten and a half million. Making generous provisions for possible 1888-1889 over-optimism and 1911 under-estimating, then, it is still certain that something happened to some millions of Leopold's black children. The diagnosis is the same as for the Putumayo: starvation, exhaustion, diseases, murder, tortures and flight. Direct agents in the acts of violence were the 14,000 or more blacks eventually enrolled under the white officers serving Leopold, Sovereign, plus the thousands of Negro riflemen (10,000 perhaps) who came to be employed by the rubber concessionaire companies dominated by Leopold, businessman. Decrees promulgated by Leopold were the cause of their carrying out the slaughter and, again as in Arana-land, the zeal of the white chiefs of enforcers was assured by making their remuneration dependent upon the quantity of marketable stuff they squeezed, flogged and shot out of the natives. The result was the same. Murder and atrocities followed the rubber production curve so that we see the crimes of the Congo steadily mounting through the 1890s to climax in a burst of horror and then subside in volume and finally disappear with Leopold and the wild rubber business.

It was in the year 1891 that the great white father of the Congo blacks really got down to business by placing in effect the theory that 'vacant lands' belonging to the State included every inch of soil save those actually lived upon or 'effectively' cultivated by the Sovereign's niggers. Application of this interpretation meant, of course, that the wealth-producing forests comprising nine-tenths of the Congo passed into the possession of Leopold, who was the State. The subjects were left with their villages and the patches of tilled lands where they raised their food. Ivory, rubber, gum copal, any and all products in demand by the world, became a Leopoldian monopoly. This did more than withdraw the entire Congo Free State from foreign trade; it denied the native even the right to make use of the forest abundance for internal commerce or internal industry.

To bring this home to his administrative staff the Coburger on September 21st, 1891, caused issuance of a secret decree (it was kept from the world for several years) which laid down as the one paramount duty of officials the taking of 'urgent and necessary measures to secure for the State the dominal fruits, notably ivory and rubber.' It was swiftly followed by a series of regulations specifically forbidding the natives to sell ivory and rubber, threatening traders with prosecution as receivers of stolen goods should they buy from black poachers, enjoining each and every native from even leaving his village without a special permit. So loud an uproar went up from established business, including Belgian, that the King had to pull in his horns a bit and the result was that private traders were allowed to remain on the short stretch of the lower Congo, in some narrow strips of territory along the middle Congo, and in the basin of the Kasai tributary.

Save for these comparatively trifling exceptions, Leopold now had all the wealth of the Congo in his hand. He obtained the necessary manpower to work it by the imposition of ' taxes' upon his black subjects. These 'taxes' were exacted by the concessionaire companies just as by the government and, whether in government or concession areas, actually were limited. Officials or commercial agents decided what amount of rubber or other produce they wanted brought in every week or fortnight or month, district commissioners went through the formality of working out the amount per native, and it thereupon became official for that area. To this spectacle was added the further oddity of Government and companies dishing out arbitrarily chosen trinkets or insignificant amounts of brass wire currency in exchange for 'taxes' whether in labor or in kind. One might almost have suspected that this was slavery and robbery under the guise of forced commercial dealings rather than a legitimate transaction between taxpayer and government. As a matter of fact, even the Congo courts discovered that it was illegal (in 1903 and after eleven years of operation) and thereafter the requisitions were 'legalised' by a decree limiting the labor tax to 40 hours a month to be remunerated at the 'market wage.' In practice the new rule was administered by setting down any desired amount of rubber as the quantity which the taxpayer should be able to collect in 40 hours and charitably giving him the whole month in which to produce it in the traditional weekly or fortnightly or monthly installments. After as well as before the 'reform,' then, the dispossessed wild man toiled dreadfully and continuously seven days a week and four weeks a month for no more than the figurative receipting of his tax bill plus such trinkets as the State or its concessionaires chose to throw in.

From the beginning Leopold insured necessary 'energetic' and economical cooperation of his on-the-spot officers by instituting a sliding system of bonuses under which the less paid the native the more paid the official. To steam up the impressing of native enforcers for the tax-collecting army, a bonus was allotted not only for every adult recruit 'whose stature exceeds 1 metre 55 centimetres,' and every youth 'whose stature is at least 1 metre 35 centimetres' (the Aruwimi River pygmies were being discriminated against, it would seem) but also for every male child 'at least 1 metre 20 centimetres in height, and sufficiently strong to be able to support the fatigues of the road' (page Mussolini!). And a special fund was set up for distribution among officials who equaled or exceeded the quotas assigned for their districts. The children were sent to 'camps of military instruction' where they were manufactured into soldiers under the direction of officers from the Belgian Army. And they, as well as their elders, were chiefly obtained by newly systematised raids upon blacks and purchase from blacks.

The best of Leopold's Congo Governor-Generals resigned in protest against the oppressive decrees of 1891 and 1892. The caliber of his successors can be judged from the fact that they were willing to hold the office. In the ranks of their subordinates were scoundrels as filthy as any of the section chiefs of the Putumayo. That this force would administer a rubber empire, the Coburger had not originally foreseen. Logic originally had turned his eyes to Africa because it was the one still unpartitioned continent. It was as a gambler, however, that he founded his first association, rushed Stanley at the drop of the cablegram, sunk his francs in the Congo before he had title to it and before its wealth had been really demonstrated, and threw borrowed sums into a costly and futile 1891-1894 attempt to extend Free State boundaries by grabbing the upper Nile. Prior to revelation of the Congo's navigability he had been wagering that there was something extremely worthwhile in the unknown interior of Africa. After the revelation he was wagering both that he could trick his way into possession of this particular river and that the area was rich enough to payout. And after he had possession, the commercial and speculative world saw him as a sucker with a white elephant on his hands, a pachyderm tusked with valuable ivory but still far from likely to earn its keep.

Even in Leopold's own thoughts, rubber was seldom present in the early years of ownership. He imagined in ivory and conceived in tusks in building his dream-fortune; rubber was just one of a number of forest products promising supplementary revenue. In the 1880s it was hardly even considered; in the edict years of the early 1890s the references were to 'ivory and rubber' in that order. And it was partly over the pachyderm product that the costly wars with the Arabs were fought. In this attitude Leopold was wrong, of course. To pull him through the critical years, to realise for him the tremendous profits for which he lusted, the need was for some product not only strongly in demand but readily available and, over a reasonable period of years, inexhaustible in quantity. Rubber fulfilled all these conditions, but the Coburger, on the basis of insufficient information, failed to realise this. So late as 1894-1895 we see a definite wavering in his faith in his Congo luck. Revenues were rising but debts had accumulated even faster, and with M. Brown de Tiège's $1,250,000 claim falling due, the busted King made known its existence and apathetically settled back to let the hurricane roar. It did. In the Congo unpaid cannibal troopers were in riotous revolt, in Brussels once-subservient legislators uproariously bawled that only on condition of immediate annexation would Belgium take over the Congo debts and end this folly of lavish African spending. Leopold himself had assured that this was the form his defeat should take if it became inevitable. Only briefly, however, did the kingly gambler hover on the verge of giving up his Congo. In his determined and successful comeback thereafter his greatness as a speculator is shown. Mustering all his resources of intrigue and propaganda, he sold his constituents the idea that giving him the money to pay Brown de Tiège would be cheaper and more satisfactory than assuming the Congo load, that if they kicked in now it would be for the last time.

Payment of Brown de Tiège did no more than carry the King over an immediately impassable hump, but his luck was in and the rise of Congo rubber speedily gave him the needed cash and credit. In 1886 the Free State had exported but $32,000 in rubber. For 1888 it was but $52,000 out of an export total of half a million dollars, principally ivory. For 1895 it represented over half a million dollars or better than a fourth of the two million total. The rubber demand was rising, rubber prices were booming and onto the Antwerp quay poured a flood of rubber out of the Congo. For 1896 it represented $1,300,000 or more than half of the $2,500,000 export total and the Sovereign was able to float a bond issue in France and Belgium. This year was the turning point and after it came the deluge: $3,170,000 (out of $4,430,000) in 1898 as Leopold's huge rubber concessionaire companies really got under way and the titanic nine-year job of building a 260-mile railroad around the lower Congo's cataracts was completed; $8,000,000 (of $9,500,000) in 1900; nearly $9,000,000 (of $10,000,000) in 1901. For Leopold's eleven juicy years, 1898-1908, the value of rubber exports was $80,000,000 as against a mere $25,000,000 for all other products. In nine years, 1900-1908, *caoutchouc* production in Congo Free State totaled 46,090 tons, the record year being 1901 and its 5,954, the lowest 1908 and its 4,487. For Leopold's four best years the total was 22,782 tons; for his last four years of Congo ownership 18,584. For the first four years of the Congo under Belgium (headed for the last three of these by nephew Albert) the total was but 13,855 tons although these years were the four greatest wild rubber years in the world's history. Exhaustion of man and rubber vine in some areas tells the story of the decline in production from 1903 to 1909. That same reason plus the gradual abandoning of the Leopoldian system tells the story of the continuing decline which ended in complete smash as the World War and plantation rubber from the Middle East combined to wreck the African industry.

It is only when we compare imports with exports that the extent of the thefts from the Congo natives begins to shape up. In the developing of such wild regions as West central Africa, imports naturally should exceed exports. They did and to vast extent —but not in the Coburger's lands. With a steadily falling population, we yet find that the Congo Free State's 1902 exports (of a diminished market value per pound) have a value more than three times that of 1897's while its imports are some $600,000 less than in the earlier year. For the eleven years 1898-1908 imports are just $50,000,000 as against that $105,000,000 exports. Over 75% of the imports into the Congo was for administrative, military and public works purposes - all not only useless to the blacks but worse than useless since all imposed heavy burdens upon them.

From well before the turn of the century, therefore, the aging Leopold (he was 60 in 1895) had no further financial grief. He did have to worry about the Belgian attempt to come into 1900 possession of the Congo as provided for in that contract he had personally framed.

With every resource at his command the King fought off the annexationists, beating them down in 1901, shaking them loose in 1906 when the whole world rang with denunciations of Congo atrocities and demands for his removal, staving off the inevitable until almost the end of 1908. And when he finally yielded, it was to collect his price and give up a land gutted of men and rubber, tied up in exploitation grants and loaded with unnecessary debts on some of which the bankers would rake in an extra profit of 28% representing the discount on these loans to the King.

It was African rubber, then, that made Leopold even as Leopold made the African rubber industry. It rescued him from bankruptcy. It started him on the road from jitney kingship of a pint-sized constitutional monarchy to a great fortune. And then it carried him all the way along that road and to a place among the financial titans of the time. The fact that he had played ivory to win shows only that he was not completely infallible. The important point is that Leopold created a system for exploiting everything of value in his Congo park. When he placed the grab decree in effect and turned loose his soldiers, the machine was able to extort ivory or increasingly-important rubber indiscriminately. And it proved capable of handling all the rubber of the raging boom days.

For purposes of practical exploitation, the Sovereign handled parts of his million-mile estate in three different ways. The territory along the main river and nearly all the North Eastern portion of the whole Free State (this is what remained after breaking off pieces from year to year) was administered as a *Domaine Prive*, later called *Domaine National*. The rubber and ivory of these districts the State itself (Leopold) gathered, exported and sold. Sums so derived were to be used for Free State improvements and administrative expenses (including the expensive military expeditions into the upper Nile) but it is almost certain that they did far more than that. One investigator has claimed that total realisation of *Domaine Privé* rubber and ivory sales ascertained for two of the fifteen years and partial realisation ascertained for six, sum up to nearly $10,000,000 more than Leopold's 'estimated' total receipts for those eight years.

Other rubber areas the Sovereign controlled in partnership with court cronies and financier friends who floated companies on the Belgian stock exchange. In most of the concessionaire trusts Leopold awarded himself a half interest, in all of them a heavy interest either as shareholder or recipient of part profits (this in addition to the special tax on all products they collected). And with these shares the one-man State had control of the market in the time of Belgium's rubber boom. Greatest of the latex and revenue producing monopolies were the Anversoise (organised 1892) with all rubber rights in 12,000 square miles North of the Congo and in the North-west part of the Free State; the Abir (organised 1892) South of the Congo and the Anversoise lease; and the Kasai (organised December 31, 1901), South-West Free State.

It was in 1898 that the *caoutchouc* concessionaire companies really got down to business. That was the year in which Abir (Anglo-Belgian India Rubber Company) was reconstituted as a purely Belgian venture with nominal capital of $200,000 and paid up capital of $46,400. Of its 2,000 shares of $100 each, Leopold pocketed just half. In six years this company made a net profit of $3,600,000 and each share of a paid-up value of $23 received $1,675 in dividends. And what happened upon the Antwerp Stock Exchange was a caution. On the basis of exchange quotations, Leopold's 1,000 shares with a paid-up value of $23,200 were worth $3,590,000 in 1899; $5,005,000 in 1900; $3,160,000 in 1903; $940,000 in 1906, the year of Abir rubber vine exhaustion and tumbling output. The Anversoise had a capital of $340,000 divided into 3,400 shares of $100. Of these shares the Sovereign held 1,700 and, additionally, he collected 15% on the company's profits beyond a fixed figure. Profits for 1900-1902 were shot to pieces by the necessity of subduing a tribe of fierce rebels, the Budjas, but for the three quiet years, 1898, 1899 and 1903, the net profits totaled $1,800,000. In early 1900 each share stood at $2,746 so that Leopold's holdings amounted to over $4,600,000. Revolt and falling production knocked those figures into collapse, of course, but with the Sovereign sending in his troops to take over in place of the company's less efficient gunmen, profit taking was restored and by 1906 value of Leopold's 1,700 shares had been shoved up to $2,380,000 - more than half the record figure on the exchange. Originally, as we have noted, the Kasai basin had been left open to trade. Late in 1901, however, the grudging monarch stepped in and caused an amalgamation of the fourteen Belgian and Dutch firms there established. The new trust had a capital - fully paid up - of $201,000 divided into 4,020 shares of $50 each, exactly half being held by the Sovereign. It was directed by an administrative council in which the King's appointees had the controlling vote. In the four years, 1902-1905, the outfit piled up a net profit of $3,658,400, some 1,200 tons of rubber having been ground out of the niggers in the last of these years. On the exchange Leopold's 2,010 shares were worth $6,432,000 in 1906. So late as 1908 the hundredth part of a share was being quoted at over $20. At the time of annexation this trust area was yielding net profits of $1,600,000 a year and producing immensely more rubber than any other section of the Congo.

Why did the Sovereign cut others in on the graft? There were a variety of reasons. For one, some of the concessionaire companies were formed when the Congo was not yet paying its way, when it was not yet certain that it would pay its way. Leopold needed financial assistance then, and the granting of concessions was one way to get a little. For another, had the King hogged all the profits the King alone would have had to face the storm of denunciation certain some day to break. By taking in selected bankers, statesmen and officials of his Court he acquired devoted and influential allies. And by passing out stock exchange tips he multiplied them many times over. Financiers, journalists and high public officials of Belgium and other countries were let in on the good thing and became whole-souled Leopoldians as dividends of 300 to 800% sent quotations soaring and touched off a frenzy of Belgian speculation in tenths and in hundredths of shares. Shares of a nominal value of $50 changed hands for as high as $3,200; others with a face value of $100 brought over $5,000. And even before those heights were reached, the master plotter succeeded in putting over the concessionaire system in French Congo, as later will be related. That, it must be admitted, was one of his master entrenchments, for the result was that a Government of France stood by him. For a brief time, too, he succeeded in persuading Germany to follow suit in the Cameroons, but that scheme didn't quite work out.

To Leopold State and Leopold Concessionaire must be added Leopold Coburg. Some 112,000 square miles of chosen territory - an area 10 times as large as Belgium -

Leopold Coburg reserved as a personal patch of ground, inalienable from himself and his heirs forever regardless of what went on in the way of Congo annexation. Carved out of the *Domaine Privé*, this Crown Domain in the rich-in-rubber central part of the Western Free State was created by decree of March, 1896, and embellished with mining rights to additional lands in 1901. Its very existence was denied by the royal liar and all his creatures until 1902 and none of its revenues passed through the Free State budget either before or after Leopold set up a Foundation of the Crown in 1901. In deepest secrecy both rubber shipments and profits were shrouded, not even estimates being furnished Belgium, the Congo or the world. Through all the clouds of concealment burrowed one investigator, however, and this sleuth, Professor F. Cattier of Brussels University, was able to show the world that for the decade 1896-1905 Leopold's Crown Domain profits from 11,354 tons of rubber had been $14,270,000, an average of nearly $1,500,000 a year!

Nor does the lion's share of the profits on $105,000,000 worth of exports end the tally of the amounts Leopold Coburg sucked out of juicy Africa. To meet a claimed deficit of $5,500,000 in the 1891-1905 Free State operations, he saddled Belgium's patrimony with debts of $26,000,000 and pocketed the difference. In 1906 he borrowed $6,000,000 as Leopold Free State, immediately made it over to Leopold Crown Domain and thereafter made away with the whole amount. That was by way of cushioning the blow of threatening annexation. In that year, also, he plastered unencumbered areas with mining and rubber concessions in order that these rights might not pass to Belgium. By stipulating impossible annexation conditions, he held off action until November 28th, 1907, and then compromised by signing a treaty of cession with the Cabinet which still permitted him to retain the Crown Domain. Parliament turned this down and more profitable Leopoldian months passed before he acknowledged the inevitable.

This acknowledgement was the triumph of Edmund Dene Morel, viper nursed in Leopold's own bosom. The Coburger's carriers, Elder, Dempster & Company, had created a special Congo department and the French clerk (he became a naturalised Englishman in 1896) was placed in charge. About 1895 Morel began making frequent journeys to Antwerp and Brussels to interview Free State authorities and to discuss affairs with company agents and it was in connection with his job that he started to study West African affairs. To this man, businessman by force of circumstances, economist by predilection, Free Trader by profound conviction, trade statistics and ship manifests told a story and an appalling story indeed. While still indirectly a Coburg employee, then, Morel launched his journalistic assault on the concessionaire regimes in both French and Belgian Congoes. The fellow was at no pains to conceal authorship and naturally it soon became apparent that this just wouldn't do. In 1900 the company and Morel parted and the latter set up as freelance journalist and founder-editor of the *African Mail*.

To take Leopold apart there was needed an analyst of world affairs as keen as Leopold himself, a fighter as bitter and relentless, an even abler publicist. In the man Morel were united these qualities. For what they were worth, he used the howls of the Congo to the last faint squeal, but the thing that made him really formidable was that he took up where those concerned only with the immediate woes of the aborigines left off. It was Morel who showed the world that depriving the native of all rights in the soil and the products of the soil inevitably had to result in just such atrocities as were embodied in the unbelievable crime stories sifting out of the Congo. It was Morel who pinned that system squarely on Leopold and not on the poor brutes mowing down niggers in the African forests. It was Morel, free trader and economist, who stirred up commercial England as Morel, bloody hand waver, stirred up the reform crowds. It was he who built the bonfires under the tail coats of the hesitant British Foreign Office; who organised and made effective use of the missionary testimony that started flooding in once the soul savers had Someone To Write To; who brought into being the Congo Reform Associations which became so potent and especially in Congress-memorialising America; who put that able Parliamentarian, Sir Charles Dilke, on the right track in England; who made plain to Emile Vandervelde the economic monstrosity of the system he had been courageously denouncing as a humanitarian; who converted Pierre Mille, first to crusade against Leopoldianism in France; who touched off the thundering broadside from Mark Twain which, badly as it reads today, was tremendous in its effect when that effect was needed.

Very good use Morel made of Consul Casement's Congo report, *Africa No. I (1904)*, one of the most famous White Books in England's history, but the generally accepted idea that this was the first real expose of the Congo horrors is without any basis in fact. In African as in later Putumayo affair, Casement's findings were the first to be issued in an official government document and it was the official character of his report that was responsible for a whopping uproar. Casement had been back on the Congo only since 1901 and made his first trip upriver in late 1903 under direct instruction of a Morel-bedeviled British Foreign Office briefly out to get the goods on a Leopold who had been making great mock of it. Returning to London in December of that year he busied himself for some months with the Irish affairs that ultimately would result in his wearing a hemp collar. Africa he never saw after 1903.

Even to skeletonise the history of Morel's campaign would require a chapter. It climaxed with a world storm blowing as Belgian sentiment swung against the King, the United States made strong representations, and Edward loosed a diplomatically worded but unmistakably businesslike speech from the Throne. And so Leopold had to step down, although from behind the scenes he would continue to dominate Congo policy. And for finally suppressing the Foundation of the Crown and surrendering an impoverished Crown Domain along with the rest of the Congo, he demanded and received $10,000,000 payable in 15 annuities from the Congo Government 'in testimony of gratitude for his great sacrifice in favor of the Congo created by him.' Supposedly the King turned over all of the Free State and Crown Domain property. Into the portfolio of the new Belgian Congo did go those shares in the concessionaire companies. For some reason, however, the newly named Minister of Colonies (drafted from a minor concessionaire company) did not check over the properties to determine whether everything had been transferred. This was to be matter of interest at a future date. Meanwhile the Belgian Government was to expend $9,000,000 in completing some of the King's 'works of public utility' and there were other concessions including one by which Leopold retained usufructs of such 'public works' as the European palaces, chalets and royal parks acquired with Congo revenues and now accepted by Belgium without payment of a single franc to its new colony.

Into the purchasing of these magnificent properties in Belgium and France, the King had dumped some $6,000,000 of the rubber pounded out of his tortured niggers. Into additions to the Laeken palace had gone another six million bloody dollars from Africa. Appropriately these embellishments included both a new chapel and hothouses where plants from the Congo bloomed. The black slaves paid also for the establishing of a $6,000,000 museum at Tervueren in Belgium, for a $1,000,000 triumphal arch at Brussels, for a hippodrome at fashionable Ostend, for Belgian roads, buildings and forest prettifications of all kinds.

Not yet, however, have we traced all the rubber dollars that came out of Leopoldland. There were the millions that went to the prostitutes, the prostitutes of bedroom, press, book world, lecture platform, political world and international law who tickled the old goat's vanity and did his business for him. In palaces on the Riviera and in Laeken he installed one of these, the one who bore him the three bastards of his seventies — or so he believed. Through the United States, England, France, Italy, Germany, the world, flooded the propaganda given birth to by the less comely of the others.

And still there were millions. Into imperialistic, financial and industrial speculations of all kinds they went — whenever the King couldn't induce the other participants to put up all of the money. Continually he was rebuffed by propositioned kings of the blood royal, but among the kings of finance he had little trouble in finding friends, associates and partners. Of their number were: Baron (by grace of Leopold) Empain, Belgian banker and self-made multi-millionaire; Sir Alfred Jones (knighted in 1901), the steamship company apprentice boy who became a member of the Elder, Dempster shipping firm carrying Coburg's red rubber from Congo to Antwerp; Thomas Fortune Ryan, dry goods errand boy who rose to be a power in Tammany Hall, milked millions from investors in scandalous New York traction enterprises and single-handed beat out the Harriman and Hill crowds in the scramble to gain control of Equitable Life and its immense funds available for flotations and promotions of all sorts. Entertaining the flattered Horatio Alger boys from America at one or another palace was routine business with Leopold, and while most of them shied away from Congo propositions, he considered it definitely worth while to have the moneyed interests of the States favorably inclined to him.

In China, Leopold took Li Hung Chang into camp, piled up profits with a Franco-Belgian syndicate which built the Peking-Hankow railroad, and would have made more save for impotence of his wiles and blandishments when exercised upon the elder Morgan. In another venture involving American associates Leopold lost out. With the American rubber shoe manufacturing trust, the United States Rubber Co., Sir Martin Conway, the British explorer, and F. W. Whitridge, husband of poet Matthew Arnold's daughter and sometime receiver for Ryan's busted Third Avenue line, the King of the Belgians and the Congolese got in on a concession under which Bolivia granted the best of the Acre rubber lands. That one had to be abandoned when Brazil took the Acre from under the noses of the financiers. An interest in the Congo was offered by the King to Charles Ranlett Flint, founder of U. S. Rubber, but this appropriately named gentleman thought the price just $ 15,000,000 too high.

Tom Walsh, the Irish immigrant who made one of Colorado's big gold strikes, was a Leopoldian social conquest with a deaf ear for the King's prospecting propositions. Tom Ryan, however, was sent by a couple of easy pushes right into the Congo. He came up with diamonds!

The Ryan deal involved two of those concessions slammed through in 1906 when Leopold was making certain that as few Congo rights as possible should pass to Belgium if annexation came. There were other angles, too. Congo rule was under extremely heavy fire in the United States at that time, Leopold was fighting back with the claim that it was all a lot of Protestant jealousy over Rome's achievements in Africa, and Ryan and his first wife were among the principal benefactors of the Catholic church in all the Americas ($20,000,000 they gave, approximately). Add the fact that Ryan and others controlled a process for Mexican guayule rubber extraction applicable to some of the Congo rubber vines. Add the fact that one of Ryan's rubber associates was Senator Nelson W . Aldrich, wealthiest and most influential political figure in the United States. There you have the reasons why the King of the Belgians summoned the Tammany Hall committee member from holiday haunts in Switzerland. Together with the usual Belgian banking associates they hatched the American Congo Company as a subsidiary of the Ryan-Aldrich Mexican exploitation corporation. It was granted 60 year rights to collect rubber and other products in some 4,600 square miles on the main river just above Stanley Pool. Also evolved was the Société Internationale forestière et minière du Congo (Forminiere), with a 99 year monopoly on all mines discovered within six years in a district covering half the Congo. All Foundation of the Crown mining rights were included in this deal. American Congo abandoned African *caoutchouc* gathering as unprofitable after several years — one reason being the inadequacy of the vine supply. In 1921 it surrendered its forest monopoly rights in return for full title to 386 square miles plus 99 year exclusive rights to any mines discovered in additional territory. Forminiere's American engineers found diamonds on the Kasai, something of which Leopold lived to hear. In Forminiere, as in American Congo, the King-State had received his usual half of the shares, this time in the names of both National Domain and Foundation of the Crown. Subscribers to one of the additional fourths included Coburg, private businessman, and the Société Générale de Belgique, the financial concern in on all his ventures. To share the one-fourth American interest, Ryan took in Aldrich, Daniel Guggenheim, John Hays Hammond and Harry Payne Whitney.

So much, then, for samples of the later ventures of Leopold as money king. What the specimens do not fully indicate, perhaps, is that the majority of these enterprises were profitable. The question, then, is not only what became of all the bloody rubber dollars but of the new dollars they brought home. 'The Duke! the Duke in Brussels! Where are the strong boxes that have wings?' Wings they had and whence some of them had flown was disclosed after Leopold's death on December 17th, 1909.

Inside of a month, millions in Congo loot were found hidden away under the names of companies and secret concerns and, above all, in the Foundation of Niederfulbach, furtively established under German law. In his will Leopold devised \$3,000,000 which he claimed to be all his property. Actually, and with all his colossal spending, he left a discoverable (by sleuthing) estate of nearly \$16,000,000, nearly all held out when supposedly surrendering all Congo assets and revenues. Embezzlement is the word for it among commoners. To the Niederfulbach foundation, created just before he had to dissolve the Crown foundation, he had transferred properties and investments to the value of some \$9,000,000. Stocks and bonds were of 35 industrial companies located all through the world. Into them, with other funds, had gone that \$6,000,000 borrowed by Leopold Free State and transferred to Leopold Crown Domain as previously recounted. Of what the man took out of the Congo over and above properties handed over on annexation and hidden away in the just mentioned magpie nests will never be known, for Leopold was extremely careful about destroying all his books. As we have seen, the German foundation deception was discovered almost at once. The funds remained deposited in the Banque Nationale of Belgium, the daughters Leopold had sought to disinherit went to law, the courts in 1913 decided that the stocks belonged to the State, and into the Colonial Treasury they went as the last hank of red tape was unwound a decade later.

With this was completed the dispersal of the greatest fortune ever built by rubber. To round out the picture of its founder, however, it is necessary to note that he it was who stubbornly accomplished fortification of Antwerp, Liege and Namur over strong objections from within his nation. And that the Belgian universal military service measure he long and persistently advocated was enacted as a farewell tribute while he was on his deathbed.

He usually had a pretty fair idea of what was coming, did Leopold. And there he had the edge on the prophets of the press. The specific reference is to one extra-emotional and moralising Belgian newspaper writer who, we may guess, never fully recovered from the shock of witnessing the manner in which his own forecast worked out.

In 1909, year after the Free State annexation, year of Leopold's death, this journalist considered the still existing Congo rubber horrors in the columns of Le Patriote. 'God will punish Belgium' was his finding as to the future.

In 1914 came '*Me und Gott.*'

2.5 Outposts of Progress

The story of Africa's rubber monarch has been told. This is the story of the subjects, of the Congo blacks on whose stinking backs the tinpot King at Brussels mounted from virtual bankruptcy to the imperial magnificence of rubber sovereignty and a place among the real rulers of the world. Leopold himself never saw these blacks. Oddly enough, neither did Edmund Dene Morel, their emancipator. Leopold, Duke of Brabant, had traveled in three continents during the 11 years before ascending the throne in 1865, but his African visits were limited to Egypt, Tunis, Algiers. Morel's one African visit took him closer to the horror scene — to Nigeria — but the visit was not staged until 1910, two years after Leopold lost the Congo, a year after Leopold's death.

Black subjects of the controversies between these two experts never heard of either, of course. They did know 'Bula Matadi,' not Stanley now but the State and one of the most execrated names in that land of witch doctors. It was Bula Matadi who lodged those extortionate rubber demands; Bula Matadi who dressed up their old tribal foes, armed them with rifles and cap guns and turned them loose. For Bula Matadi they sweated and died; against Bula Matadi's arms of precision they opposed their spears and arrows. No such sissies as the Putumayo reds were the Congo blacks, but stout fighting men who rose again and again and, even with their poor excuses for weapons, sometimes gave the Belgians more than they could handle. The battling began with the river natives who refused to gather and deliver rubber, ivory and foodstuffs, unless recompensed by desirable trade goods as in the past. It spread inland as the horrible collecting system inaugurated on the Congo was pushed into those regions and it raged throughout the whole of Leopold's reign.

From Stanley Pool to the *estrada* frontier the general picture is of a main river well nigh licked clean of blacks, a rubber hinterland with here a spectacle of frenzied and bedeviled latex gathering, there a battleground for guerilla warfare, the other way a patch of howling wilderness (wilderness furnished by the Sovereign, howls by the handfuls of survivors). In all quarters the atrocities occurred, in the two 'Domains,' the concession areas, in some of the few upriver unreserved spots as one or two small trading companies imitated the concessionaires and drafted black gunmen.

On soldier- and laborer-recruiting raids led by white officers troops 'regular' and irregular shot down defenseless blacks, struck down babies with gun butts, finished up with cannibal banquets. On punitive expeditions from major stations and from lost one-white outposts in the bush they went out unsupervised by white officer or agent to wreak vengeance on villages negligent in feeding rubber to Leopold and food to his gunmen, wood choppers, laborers and station slaves. Smoked hands they brought in by the basket in proof of orders obeyed and to demonstrate there had been no undue waste of ammunition. Sometimes male sexual organs were required, but if it was a showing of hands no questions were asked even if they had been sliced from children. In each of the rubber-gathering villages one or more of these cannibal riflemen was stationed as a sentry living on the people and driving all adult males into the forests to glean *caoutchouc*. Kings in little were these tyrants from strange or hostile tribes, and they flogged and raped and murdered not merely with impunity but in secure knowledge that Bula Matadi demanded this tribute. The statute books the cannibal tax-gatherers could not read, but they did know that sentries who failed to obtain sufficient rubber were removed and degraded.

Nor was the white personally guilty of no more than murdering with firearms under the guise of war, of watching his cannibal troopers kill and mutilate and quarter. In his very stations slaves were forced to eat their rubber when it was badly prepared (a Congo court once held that subsequent illness and death of such victims could not be attributed to this disciplining). Chiefs were degraded in the eyes of their subjects by being compelled to drink: from the latrine, were chained or tied up in ropes for weeks at a time to await ransoming or death from ill treatment. Women were flayed with hippopotamus-hide whips until they ran blood and urine. Sentries who had dared steal rubber and dream of bootlegging it to traders were tied to stakes and left under the vertical rays of the sun, naked, without food and water, tongues hanging out. Niggers who brought in less than the required amount of rubber were shot down by irate whites.

Classed with the chain-gangs as a regular institution, hostage houses were to be found at every State or concessionaire station and outpost. The idea was to bring in chiefs of tribes and the fathers, mothers and wives of able-bodied bucks who thus would be spurred to rescue them with latex no matter what the difficulties encountered in achieving the arbitrarily designated output. Ransomed from the pangs of hunger and from the stench of weeks of accumulated filth, prisoners were unable to stand. In just one of the Abir's houses, deaths of three, five and ten a day were the rule. Often a rubber post was some distance from the nearest food town and many times it was as much as an official could do to feed self and black soldiers. When that happened, the always half-starved prisoners were left to perish.

Ordinarily, however, magistrates considered the detention of women the 'most humane form of coercion' employed in Leopoldland. It was. And when it didn't work, when men (in virtually exhausted rubber regions, sometimes) were considered a bit dilatory in showing up to redeem their women, the armed mobs of cannibal enforcers were sent out as coercers by massacre and atrocities. When they left, the luckless villagers were further penalised by fantastic fines which could only be met by selling some of their number as slaves.

So much for raids, atrocities and hostage houses. In taking leave of them, it may be well to underline, at risk of repetition, that the root of nearly all these horrors was the latex from root, vine, bush and tree. King of the African lianas was *Landolphia owariensis* followed by other species of that and the closely related *Clitandra* genera and then by the genus *Caprodinus*. The *Landolphias* furnished the red rubbers, the *Clitandras* the black. Starting in life as shrub-like growths, most of these plants put out shoots provided with numerous branch tendrils which, coming in contact with tree branch or other support, keep right on going until they reach the top. It was the collector's job to haul down these climbers, stretch them out on the ground and start carving incisions every few inches or cutting the vegetable cable into pieces. A peculiarity of these vines is that conduct of their latex varies with atmospheric conditions determined by localities or seasons. Sometimes the milk would simply fill up the cuts and coagulate, a process hastened by sprinkling with lime juice or salt water. After four or five hours the rubber then could be collected by pulling it from the notches and rolling it into balls. In cases where the milk flowed freely at first, it was caught on a leaf and spread out in a thin layer to coagulate, or smeared on the body and afterwards pulled off in long strips. Latex that continued to flow freely was caught in leaf cups placed beneath incisions or collected in troughs or holes in the ground in which lengths of vine were placed upright. Mostly, juices of wild fruits then were added to hasten coagulation although with some of the milks a long boiling was necessary.

In grass and short scrub regions, moreover, there had developed *Landolphia* varieties which put out subterranean stems rising every 50 feet or so to a small tuft of branches. These immense worm-like stems, the early Congo blackbird pulled up for yard after yard. The rubber was in the bark and the process of getting at it began with exposing the vine in the sun to dry and cutting it into workable lengths. Then it was necessary to beat these sections with wooden mallets on a log to remove the bark. Thereafter the bark was hammered until it dropped away in a red dust, leaving the rubber in thin cakes. In the Kasai region especially, the thud of mallets pounding out rubber became the new Congo drum and it was to this monotonous accompaniment that hopeless villagers dragged themselves about between ruinous houses set amid the piles of red dust and the overgrown and neglected cassava fields.

Native industries, the forging of ironware, the making of baskets, the fashioning of weapons; native husbandry and cattle-raising; native social life - all now had been swept away by the necessity for concentrating on the rubber work. No time was allowed for cultivating the ground, for hunting, for fishing. And even with all able-bodied men prowling the rubber lands, many a native tribe could only meet its exorbitant latex tax by impressing women and children into the service and leaving huts to fall apart and plantations to return to jungle.

Starvation and overwork, it is probable, killed more hundreds of thousands in the Congo than any other cause. Sleeping-sickness and smallpox were second and here again rubber was at the bottom of it all. Before the white man's exploitation, diseases ran their course in a single village or congeries of settlements. Under the reign of rubber, epidemics spread across the face of the land with the wholesale deportations of women and the moving about of black armies. And it was because of the wretched condition of the slaves, because of exposure and lack of food and lack of rest that they so easily fell victim to sleeping sickness as well as to the lung and intestinal diseases directly arising from the rubber district life.

Flight was common, of course, and before France's Congo lands were switched to the concessionaire system by the scoundrelly politicians at Paris, some 30,000 of Leopold's children are specifically known to have crossed over. Into the British lands, too, they escaped and there were many who fled or sought to flee from Coburg rubber districts to other Coburg regions where no vines grew. Altogether, however, the thousands of escapes were but a single white drop in the red lake of millions in population decrease brought about by starvation, overwork, disease and murderous raiding.

Not in itself should the rubber work necessarily have been a dismal and horrible thing. It was the Leopoldian demand for excessive quantities produced on rigid schedule which made it that. And this was as true under colonial setup as under Free State. Belgium had taken over the whole Leopoldian setup, lock, stock and gun barrel. The Cabinet was a Leopold crowd, eight of the 14 members of the Colonial Council were his nominees, the vice president of that Council was the president of Abir, the Colonial Minister was drafted from another concessionaire company, the Congo laws remained, the administrative force in the Congo remained unchanged, the first budget set up an estimated revenue which included, among other things, \$3,000,000 in public domain rubber 'taxes.' Morel, however, had not relaxed in his crusading with Leopold's defeat. Now he kept the protest meetings seething while Vandervelde roared powerfully in the Belgian Chambers; England withheld formal recognition of the annexation; the United States sent a sharp dispatch. How long reforms would have been delayed had Leopold lived it is impossible to say; the probability is that nothing less than an ultimatum from one or more of the signatory Powers would have turned the trick. It came to the very edge of that, as it was. Parliament, reconvening in February, 1910, found England boiling with an indignation greater than it had shown over any foreign affair in the recollection of the legislative veterans. In a March 10th speech in Parliament scoring the Foreign Office for its weak-kneed attitude (construed as resulting from fear that stern insistence on action would drive Belgium into the arms of Germany), Sir Charles Dilke voiced the general sentiment by calling attention to the current bill for the greatest army and navy expenditures in England's peacetime history and demanding that the weight of these armaments be placed behind its diplomacy.

Twelve days later Belgium announced graduated abolition of the State's forest product monopoly in the national domain. It was effective July 1st, 1910, in the *estrada* part of the Congo; July 1st, 1911, in the former Crown Domain and one other area; July 1st, 1912, in the remainder. On May 2nd, 1910, the rubber tax and all such labor taxes were abolished. In January, 1911, the Government effected a compromise with the Kasai company, which accepted the regime of free trade and converted itself into an ordinary trading concern in return for receiving stock formerly held by Leopold and having a value at the time of $2,500,000. In May it obtained abrogation of Abir and Anversoise rubber purchase contracts by granting them full property rights, subject to rights of third persons native or non-native, in lands where these now merged concerns today operate palm plantations, palm oil mills and trading posts dealing for copal, palm nuts and other native products. Other monopolistic concessionaire rights were scaled down when possible, although all of the colony's most valuable lands, an area ten times the size of Belgium, still remain in the hands of corporations holding them as full property and without payment of rent to a Government which also is a shareholder. This is one souvenir of Leopold. The colony's public debt burden (none of it ever assumed by Belgium) is another. And in a land so short of workers as the Congo has been since Leopold's devastation, Government partnership in industry has meant the survival of forced labor to some extent; recruiting pressure, however, stops short of violence and the workers, once dumped down across country, are as well treated as any in Africa.

Even Morel was satisfied. May of 1913 offered the unusual spectacle of Leopoldianism's severest critic lecturing in Brussels on tropical African government by invitation of the university. In July the Congo Reform Association held its last meeting in London and closed up shop. Secret diplomacy, into which he had obtained so educational an insight while hounding the Foreign Office on the Congo matter, was now the focus of the ex-clerk's absorbed interest. His outspokenness during the World War landed him in a Pentonville cell on a trumped-up charge. He came out a full-fledged Labourite, went on to Socialism and in 1922 was elected Member of Parliament for Dundee, defeating Winston Churchill. He died in 1924 and today is virtually forgotten although so early as 1907 the voice of the American people, the Floyd Gibbons of the day, Mr. Richard Harding Davis in person, was confidently predicting Morel's burial in Westminster Abbey when death should scoop him in.

By Leopold, this clerkly conqueror of a most powerful king was variously pictured as anti-Catholic, blackmailer, agent of British imperialism or agent of jealous British trade interests. The last of these claims has certain merits and French concessionaire apologists wisely have stuck with it instead of switching over to the pro-German charge as their Belgian brethren have done. Morel campaigned as Free Trader, not as an aborigine protector; his assault on the French concession system led to France's paying damages to English trading companies robbed and evicted from Gabun and to the reestablishment of the traders in 1911. Also, the French point out, these companies were the principal contributors to the $15,000 purse presented to him by his admirers in 1911. Leopold paid three times as much for one book by an American lawyer (Morel wrote a dozen assailing the concessionaire system), paid Colonel Kowalsky 100,000 francs for no more than getting a Congo defense letter from Leopold to President Roosevelt and to the press at the time Morel was in the United States presenting a memorial to T. R., addressing a meeting at Boston and becoming involved in an acrimonious controversy with Cardinal Gibbons. All in all, the $15,000 hardly seems enough.

It was Leopold, as we have already mentioned, who brought about the concessionaire regime in the French Congo. Approximately the size of the Belgian Congo, this African possession (officially known since 1910 as French Equatorial Africa) is composed of the colonies of Middle Congo, Gabun, Ubangi-Shari and Chad, the first three of which are in rubber territory. Middle Congo stretches along the North bank of the main river and its great Northern tributary, the Ubangi, and is the political subdivision immediately between Belgian Congo and Gabun. Ubangi-Shari, interior colony, adjoins both Middle Congo and Belgian Congo.

Here the rubber concessionaire system arrived later, waxed as horrible in intensity but not in extent, hung on far longer, and claimed some tens of thousands of lives. Discussed in 1891, it was abortively applied to parts of Congo and Ivory Coast by secret 1893 decrees at a Colonial Under-Secretary, howled down and revoked when the concessions became known in 1895. By 1899, however, the Coburger had got in his work. Impelling him were a variety of reasons. The colossal increase in Free State rubber output, the enormous premium at which his shares stood on the Antwerp stock exchange, the now frenzied speculation in anything having to do with Congo rubber, all these had turned French eyes to the most neglected of its colonies. Should there be an influx of French capital, Leopold foresaw a railroad from French coast to French bank of Stanley Pool threatening his transportation monopoly with the interior. That he most emphatically did not want. He did want to get his own hooks into the rubber North of the Congo and he did want to see his own Free State position strengthened by the spread of the concessionaire system throughout all West Africa. The job, then, was to drive through the system but to see that the concessions fell to his Belgian band. And so it was done — by financial and political intrigues involving corruption and cowardice, and over the strong protests of the French Congo's local administrators. And so well was it done that later-day investigators found that five-sixths of the total $11,845,000 capital of the 40 odd concessionaire companies came out of Belgium. The concerns had nominal French heads, but the Belgians on the boards of administration represented the controlling factor and it was the Belgian system of exploitation that was applied. How these outfits maintained their hold on successive French governments and more especially on changing Colonial Ministers is no mystery in view of the venalities of French politics. In this connection it is especially interesting to note that the *chef de cabinet* of one such minister subsequently became the director of six rubber concessionaire companies.

All together these companies originally were granted a third of the French Congo, an area larger than France, to exploit for 30 years. At the end of that time they acquired title to such of the land as had been developed by cultivating a fifth of its surface with such crops as rubber or a tenth of its surface with foodstuffs. What they did, of course, was fall to work raking in the wild rubber and ivory without attention to development. To the government they paid land rent (amounting to less than 1% of the value of their capital) and in their enterprises the Government shared to the extent of 15% of the profits. To the concessionaire companies belonged all of the products of the soil, it naturally was decreed in Leopold-emulating style. Merely depriving the native of his rights to forest valuables wasn't the same as putting him to work, however, and in 1902 the Government took action to supply the requisite slave labor. The whole French Congo now was blanketed with a head tax payable in rubber or other forest produce which was to be received by administrative officers and handed at a nominal price to the company upon whose territory they had been gathered. To the natives went not one gewgaw, and the administrators, playing in close to the companies, had these black simpletons gather *caoutchouc* to ten or more times the amount of the tax. This difference was kept by the concessionaires, of course. The remainder of their rubber they obtained by forced labor recompensed by the salt indispensable to native diet or by such valueless and usually unwanted trinkets as were distributed to the Free State slaves.

How this worked is exemplified by one concession in the lower French Congo; there natives were forbidden to make salt in order to compel them to buy it from the company, which would sell only for large quantities of rubber. Deprived of it, the blacks sickened and died.

Everywhere the system's results inevitably were the same as in Leopoldland. The force of government arms — fierce Senegalese soldiers commanded by French officers — was behind the tax impositions; the companies themselves maintained the ferocious slave-driving sentry system invented across the river, and government and company forces united for punitive raiding. First to rise were the blacks of the maritime zone who had been accustomed to barter their rubber with now dispossessed British traders and to receive merchandise that at least satisfied them. To the upper French Congo European trade had not yet penetrated, but when the rubber demand came in the form of year-round slavery it met first with bewilderment and then with the usual pitiful and powerless insurrections. In the whole history of the French Congo under the concessionaires there is just one deeply satisfactory episode. That is the 1904 rising on the upper Chari where 27 tyrants were killed and eaten. Their skulls, stuffed with rubber, were found by the punitive expedition. All that mars the perfect felicity of this incident is the fact that the 27 were black guards and not white superiors. A number of such superiors were rubbed out, however, in successive revolts in the Ogowe, the Sangha, the Loyaye and elsewhere.

These whites drew commissions on the rubber profits and were the foul type customary to such exploitations. Some came, in fact, from the Abir inferno to the South. As to their governmental running-mates these were a new breed insofar as French colonial activities anywhere in Africa were concerned. They came in with the concessions and to their number were added some members of the old regime corrupted by the consequent putrefying of the administration.

'In the Ubanghi-Shari,' wrote Savorgnan de Brazza, 'I have found an impossible situation, the continuous destruction of the population — purely and simply.' Fully to understand the force of that, you must conceive of Stanley (who never saw the Congo after 1889) returning to the Free State and discovering what Leopoldization really meant. Of the great line of African explorers was de Brazza, and to him France owed its very possession of its Congo lands. It requited this founder of its Equatorial empire by retiring him from the Governorship before installing the concessionaire system, and then recalling him from retirement to head an investigating mission intended to be used as a false face before the world. So completely had the politicians underestimated the courage, the integrity and the ability of this man that his mission must have spelled the end of the concessionaire system had not monopolist luck been riding high: de Brazza died on the way home.

It was a series of particularly atrocious crimes occurring in the upper Shari that made the sending out of a mission obligatory. In the early years France heard only of the crimes in the maritime zone, and since this was the region in which the dispossessed

British traders were concerned, it was easy to attribute all the charges to the plottings of these interested parties. Until the concessionaires began their penetrations, the interior was unknown to all save the de Brazzas and when the concessionaires went in a veil of secrecy descended behind them. From behind that veil officers of the old regime sent out the gruesome true story, but of these reports direct to the Colonial Office in Paris every last one was thoroughly suppressed. Then came the trial of a couple of 'indiscreet' administrators on ghastly charges which resulted in five-year sentences. News of this got home in 1905 and in the denouement some of the suppressed reports were dragged into the light. Now for the first time the populace of France learned that it had a Leopoldland of its own, learned that the concessionaire-collecting of rubber had meant lash and torch, rape and mutilation, natives stood up as targets for revolver practice, used as black guinea pigs to test the efficacy of dynamite cartridges. Panic-stricken, the usual venal government of the day had no choice but to attempt using de Brazza as a front.

For four African months de Brazza lived in a nightmare, a new Dante touring a country of the damned which he had himself created — although in all innocence. It was he who had sewed up this land in 'treaties' with the native chiefs, he who had persuaded these populaces to accept French 'protection,' he who had ruled them justly and well — and then unwittingly had turned them over to the scoundrels of the *caoutchouc* Inquisition. And it was this man who on departure from France was furnished with secret instructions from the Government in which it was urged that his report should make it clear that the system in effect in French Congo was not identical with the Free State one! In his first report from the scene he told the diplomats that he could not do what was demanded of him. Uncompromisingly, he informed them that the conditions he found could only be explained by the system and not by individual acts of brutality. From that moment the whole weight of French secret diplomacy was thrown against the continuing attempts to call a conference of Berlin Act signatories to act on the case of the Free State. It had known it possessed the same abbatoir as Leopold; now it knew that it faced the prospect of its presence there being established by the man whose word on French Congo subjects carried more weight than that of all others.

'Ruin and terror,' 'terrifying depopulation,' 'universal exodus,' these are the phrases that recur in de Brazza's messages home. To say that witnessing of these horrors had a part in bringing about his death is to venture into the realms of the melodramatic. That, however, was the view of Felicien Challaye who went with him through the devastated Congo and who was with him at the end. We give Challaye's words for what they are worth:

'The fate of the Congo troubles him more than his own. When he has the strength to talk it is of the Congo that he speaks. From these sinister discoveries M. de Brazza suffered in the deepest recesses of his soul. They hastened his end.'

It was at Dakar in Senegal that de Brazza died in September, 1905, on his way home. Had he lived it is certain that the French Government would have gone to almost any extreme in attempting to suppress his report, but it is probable that the political tools of the financial ring would not have succeeded. As it was, Challaye gave them more than enough trouble and he had no such reputation as de Brazza's to throw into the fight. He and other members of the Congo mission did have documents for a report that would have made the Casement paper look like an Elsie book in comparison. They were forbidden to prepare it, were mercilessly assailed in the kept Colonial press, and finally were compelled to hand over their evidence to a committee appointed by the Colonial Minister and inclined in the system's favor. So overwhelming were the documents that even the report of this group was suppressed by the Ministry on the grounds that its effect internationally would be prejudicial to France and that its publication would lead to actions at law by the concessionaire companies which had threatened to take proceedings against the Government!

'Blackmail!' a deputy called this last as the Chamber characteristically engaged in tumultuous three-day debate over the suppression. In that 1906 debate, deputies with sources of information dragged out more than enough of atrocities equal to Leopoldland's but the demand for the production of the whole of the de Brazza evidence ultimately was voted down. Challaye, however, used much of it in a series of articles later published in book form and his writings and speeches, together with those of Pierre Mille, Anatole France and others, did much to formulate and keep alive a substantial and extremely articulate opposition to the whole concessionaire system. In the year of Chamber debate, too, he organised a league for the protection of the Congo natives which performed invaluable services. The one immediate result of the tragic de Brazza mission had been the appointment of traveling inspectors by the Government and while their reports were officially suppressed as fast as they came in, Challaye's society was able to get hold of a number and give them to the world. In these 1906-1909 reports the Company du Congo was accused of burning 20 villages and capturing hostages (fine of $40 in the local courts); the M'poko Company was charged with 1,500 murders in its concession; the Fernan Vaz Company was cited for causing the 'exodus and revolt' of the natives by its 'coercion and slavery.' Against the Lobaye Company a dossier deposited with the court at Brazzaville on Stanley Pool listed 153 counts of 'crimes and delinquencies' on the part of its agents. Other companies, Sette Cama, Brettonne, Haute N'Gunie, Bavila, were similarly accused. One and all the inspectors urged cancellation of charters, prosecution not of individual agents but of the corporations counseling 'practices against which humanity protests.' Not one concessionaire company was proceeded against, not one charter was annulled — but the inspectors themselves were suppressed when the concessionaire boards objected to the leaking of the reports.

So aroused was public opinion that reforms had to be made, however, and so the rubber tax (never so integral a part of the picture as in Leopoldland) was abolished in this year of 1909. A year later the Government began its attempts to cut down the colossal area of the original grants by negotiating with the various companies to surrender their exclusive collection rights in these areas for full property rights in much smaller ones. This met with success in a number of cases, since corporations had bitten off more than they could handle and freehold land remaining theirs forever looked better than monopoly rights to territories which they would lose in 1929 through failure to develop. Of particular interest to us, for reasons later to be disclosed, is the first such agreement entered into by the government. This was a 1911 settlement arrived at with 11 concessionaire companies which a year before had associated themselves as the Compagnie Forestière Sangha Oubangui (CFSO) holding rubber rights in a total of 37,400,000 acres. Each of these companies agreed to give up the concession expiring in 1929 in return for full ownership of 22,000 acres plus the exclusive right for a 10 year period to gather the rubber on the originally conceded territory. At the end of this period each company could acquire full property rights over lands in the area it had put under cultivation and for another 10 years would enjoy the exclusive right to gather rubber in an area 10 times the planted acreage. The land coming under cultivation at the end of this second period (1931) would likewise become the freehold property of the companies. Rents and 15% of the profits would continue to go to the Government. In 1920 this combination still was influential enough to wring from the Government an amazing extension. For surrendering its rights to rubber throughout its former concessions it received exclusive rights to exploit rubber for a period of 15 years (ending with the close of 1936) in a vast area covering part of the territory ceded to Germany in a 1911 deal and recovered as a result of the World War. Through the whole of this territory the gathering of rubber by blacks or Europeans not authorised by the company was forbidden under penalty of arrest by company agents, seizure of rubber and suit for damages. It was also provided that the company should receive at the end of 1936 four times the amount of land it had cultivated for the six years preceding, that it could obtain a renewal of the entire concession if it annually exported 500 tons of plantation rubber in addition to the forest product and that if this plantation figure was not attained the concession still could be renewed upon a proportionately smaller scale.

Nothing at all approaching the Free State's spectacular rubber record was ever achieved by the French Congo, it should be noted now. The Colony had far less rubber to begin with, and far fewer native inhabitants; transportation, save in the case of Gabun, was far more expensive; the Government was far less helpful than Leopold, being far less vitally concerned; and among the more than two score companies there was no such complete and efficient exploitation as where one great army served one despot — the Coburger. To explain early lack of profits there is also the usual period of expensive conquest which had to be gone through with before the dividends could start rolling in. Until 1904, then, only the few of the concerns operating in the old free trade area of the maritime zones made money. Eight of the others folded, something not exactly surprising to promoters who floated their companies in the Antwerp market on the crest of speculation in wild rubber shares and got out far ahead of the game. So quickly did the French Congo boom fizzle, in fact, that in 1902 there was not a single transaction in its rubber shares, with the exception of two companies. Trade itself increased, however, from $2,600,000 in free trade, 1899, to $4,200,000 under the concessionaire regime of 1904. And thereafter profits began to come in as numerous amalgamations produced larger and stronger concessionaire companies while liquidation took care of still others that had served their purpose by enjoying a brief but sufficient run-up on the stock exchange.

Rubber production figures show that atrocities, while individually as intensive and horrible as in Leopoldland, yet were not nearly so widespread or numerous. The 1900 production of 645 tons was doubled by 1904 (1,229 tons) and tripled by 1906 (1,924 tons), but this last was the French Congo's high mark in the era of booming wild rubber throughout the world. The 1911 and later reducing of paper areas marked no lessening of severity in crushing and subjugating the slaves, however, and along with the still-continuing amalgamations which gradually reduced the number of companies to 13 there was evolved a more efficient setup. Sunk by the flood of plantation rubber and by the advent of the World War, production was revived in the later war years of great demand and in 1917 hit its all-time top of nearly 3,000 tons (Belgian Congo: 2,800 tons for the same year). Thereafter it gradually subsided, while timber, palm-oil and cocoa came forward as the colony's chief exports and rubber regained importance only as Britain's plantation restriction scheme of the 1920s caused the wild rubber demand briefly and partially to revive.

Governmental cruelties are far easier to liquidate than commercial ones, however, and, even with rubber production a comparatively minor matter, the old atrocities have hung on here as they have not in the Belgian Congo. The distinguished French author, André Gide, traveling in these regions on a semi-official mission in 1926, came out of the CFSO concession area with a report on 1924-1926 reprisal expeditions, floggings, tortures, starvation, hostage imprisonments, and hopeless slavery which reads like a de Brazza letter of two decades earlier. For their rubber these slaves were being paid one franc a kilo, while in adjoining free trade regions the British traders were paying the natives 10 to 12 francs a kilo. To gather 10 kilos of rubber in the CFSO area meant spending a month in the forest, often a week's march from any village. Far more profitable and immensely easier would have been the gathering of palm nuts near the town, but since the concession was for rubber only, the natives were harried and terrorised into bringing in the specified monthly amounts of latex. And still helping to carry out the reprisal work of concessionaire agents paid in commissions on rubber was the French military! Gide brought about the prosecution of one particularly inhuman agent — but he was under no delusion that this constituted breaking up the system. What has happened in the CFSO's lost forests since 1926 only the company itself can tell, for there have been no subsequent reports and a world adequately supplied with plantation rubber has no interest in the atrocities that still may be going on in odd corners of Africa and South America. Always there is some sort of market for this fractional amount of wild rubber which continues showing up in raw *caoutchouc* statistics, and, where slave labor is involved, it doesn't take much of a rubber price to mean a profit of a sort to the slave drivers.

Concessionaire slave-driving, it is worthy of note, is something that even such heavy-handed colonisers as the Germans couldn't quite stomach. In the Cameroons (the territory between Nigeria and the French Congo), Leopold stampeded them into trying it, as noted in the preceding chapter. First of these rubber concessions was granted to a Belgo-German group in 1898 and involved nearly 18,000,000 acres; the second was granted to a similar outfit in 1899 and involved 11,000,000 acres. A year later the Government drastically restricted their privileges and definitely went out of the business of making such grants. By 1905 the largest original concession had been cut to some 3,700,000 acres by the scaling-down system later adopted in both French and Belgian Congoes. The other group was obliged to recognise native rights, to respect the principle of free trade and to carry out a number of other obligations looking to the development of the country. It did not live up to these obligations and its rights expired in 1913. Reduced properties of the concern earlier in the field were sequestrated by the French during the World War and its rubber monopoly was abolished when France was given a mandate to administer the greater part of the colony (the remainder is under British mandate).

Necessary to the piling up of Leopold's fantastic profits was the concessionaire system, but not to the mere getting out of rubber, as the history of the Cameroons clearly shows. In non-monopoly areas, native middlemen multiplied until in 1910-1911 there were over a thousand of them buying rubber from fellow blacks and peddling it to some 50 dealers operating 230 stations. A fifth of the size of French or Belgian Congo was this Colony, yet it boosted its production from 538 tons in 1900 to 1,130 in 1906, passed the French Congo in 1910 (Cameroons: 1,927 tons; French Congo: 1,632 tons), and hit the high mark of 2,832 tons in 1913, in which year rubber represented half the total value of exports. The war wrecked the business and it hasn't amounted to much since. Like all African states, it has some rubber plantations; but here we witness the peculiarity of blacks owning a number of them — a policy inaugurated by the Germans and since encouraged by the French, who confer upon leading native planters membership in the Order of Agricultural Merit. Where the proprietor is a chief he often has been known to send out his police to conscript laborers, who thereafter serve for little or no pay. Sometimes they escape, again they prefer to remain, since their ebony exploiters usually allow them to work as they please and at least they live according to native customs and are freed of most Government exactions while under such protection. The sex lure, too, plays a part in the arrangement, for in some districts it has been the custom among the chiefs to farm out a surplus wife to each new employee. With dowries coming high and with the chiefs marrying most of the women, the black proletariat can work or go without. So they work.

An attempt to introduce the concessionaire system into French West Africa was supported from Paris but defeated by the combined opposition of local administrative officers and powerful trading firms. Governor-General Ballay concisely summed up the administrative repugnance to the proposal when he told the home superiors, 'It means that you must place a soldier behind every producer.'

Five of the eight colonies in this Federation — French Guinea, Ivory Coast, Niger, Senegal and Dahomey - were rubber producers and in about that order of importance. And here it was shown that administering and developing colonies in the best French tradition (imperfect, admittedly, but as good as Africa can show) was not inimical to the business of rubber gathering. French Guinea alone got out 2,000 tons in one year (1912) and in its best year (1910) the whole Federation passed the 4,500 mark. Taken as a whole, the French rubber possessions - French West Africa, French Congo and the island of Madagascar — went around the Congo Free State in 1904 and thereafter led the African parade. Their record was the 7,304 tons of 1910.

Of individual political units, runner-up to Leopoldland was Portugal's Angola (including the Kabinda enclave North of the Congo) which passed the 3,000 ton mark in 1904, 1909 and 1910. Here, again, the story is slavery without palliation, but in this instance rubber can't be blamed for the situation. It was from Angola that the slaves for the Brazilian sugar plantations and gold mines were exported, and until as late as 1850-1860 the black man was the only product of the land held in esteem from the commercial viewpoint. And when, with Brazilian merchants leading the way, attention was turned to the forest wealth in rubber, palm oil and wild coffee, it was only in combination with slave labor that these new industries appealed. An inversion of the old slave trade — sending in the overseers to sweat the blacks on their own ground instead of shipping the blacks to lands where the overseers awaited them — fairly describes the forest and plantation exploitation of the concessionaire lands and Angola. By raid, by kidnapping and by purchase the rubber labor of the Portuguese colony was acquired, but in spite of all the bloodshed and cruelty involved in the captures and the subsequent slavery the situation was not so bad as in Leopoldland. And for one reason: the machinery was less efficient.

Collectively, the Portuguese rubber colonies — Angola, Mozambique (Portuguese East Africa), Portuguese Guinea (where the sheep give no wool and the goats give no milk)- had their best year in 1909 (4,134 tons), surpassed Belgian Congo in that year, in 1910 and in 1915. The usual *Landolphia* and *Carpodinus* vines supplied the most of the output, but Angola additionally boasted the oddity of a rubber-bearing tuber the shape and size of a small turnip and resembling a yam in color.

In all of Angola's history there is probably no story more interesting than that of the exposure of the twentieth-century slave trade in full swing between this mainland and the Portuguese-owned West coast cocoa islands of San Thomé and Principe near the equator. Rumors of this horrible trade and of the death-trap nature of the islands reached the London firm of Cadbury Brothers, cocoa manufacturers importing a part of the product. Representations to the Portuguese government followed, but when a year passed and the reports still were coming in, these business men approached their colleagues in the industry with a proposal to send out an investigator. American firms refused, but three other British concerns joined in and Joseph Burtt was dispatched on the mission in 1905. His exposures and those of Henry W. Nevinson, noted journalist representing *Harper's magazine* (United States), resulted in immediate boycott by the British companies. That action, probably unique of its kind, suggests how speedily the rubber terror reigns could have been ended. Had the British, French, German, Belgian and Dutch manufacturers who consumed the rubber from Putumayo and Congo felt any moral responsibility in connection with the origin of this bloody stuff, they could have worked reform overnight by the boycott weapon. Never in any country has a rubber manufacturer taken such a stand, however, and the fact that virtually none of the Leopold or Cesar *caoutchouc* reached the United States is no tribute to American industrialists, who happened to be supplied with a Brazilian, Bolivian, Mexican and Central American product concerning which they made no inquiries. And in a great many of the African colonies it is probable that conditions under which rubber was produced were far better than in South America. Among these must be listed Britain's chief wild *caoutchouc* lands, Gold Coast and Nigeria.

Gold Coast is of peculiar interest to us since the 1883 native discovery of a rubber tree there signaled the first real entrance of tree latex on the African scene. Previously only poor rubber from one or two species of Ficus had been included in Africa's exports. The new tree — *Funtumia elastica* — yielded a product of good quality, however, and its presence from Gold Coast to Lake Victoria (that is in nearly all the West and central rubber lands of Africa) made it of substantial commercial importance. Like the *Castilloa* and unlike the *Hevea*, the *Funtumia* yields all its available latex at a single tapping and no response is obtained by reopening the original cuts or by making fresh incisions above or below them immediately afterwards. Two or three times a year is the limit of tapping, therefore, and the result was destruction by impatient gatherers as in the case of the *Castilloa*. Either cuts were made too deep and too far up the trunk (incisions often were carried to a height of 50 feet and more) or the tree was felled at once.

Not at first did the *Funtumia* discovery send Gold Coast exports rocketing, but as

Leopold's success in the Free State turned trader attention to the sticky stuff, the British blacks responded with a fine frenzy of tree destruction. Production soared 2,000 ton mark in 1897, hit a record of 2,672 in 1898 and then fell steadily away as the trees gave out. Only 679 tons were exported in 1901, only 714 in 1902. Then came the opening up of new forests in the interior and production bounced back to 1,792 tons by 1904. In the very midst of this revival there was a peculiar one year slump as statistics show: (1907) 1,585 tons; (1908) 792 tons; (1909) 1,234 tons; (1910) 1,439 tons. One fetish priest of the Ashanti region caused this halving of production. Bounding out of the bush to herald the coming of a new god who would exalt the blacks over the haughty whites, he specifically prophesied that anyone collecting rubber would be transformed into an antelope on the messiah's arrival. Says an official report: 'This imposter succeeded in paralyzing the local rubber trade. He was eventually arrested and detained in Kumasi and his fetish fell into disrepute.'

Gold Coast today produces half of the world's cocoa, not an indigenous product anywhere in Africa, but one which figured in the shift sending *Hevea*, *cacao* and *cinchona* to the *estrada* hemisphere and coffee to the West. Due to the enterprise of illiterate black farmers unaided by European capital, this amazing cocoa industry was built in the short space of 25 years, a development which graphically illustrates the wrong done in treating the Congo blacks as wild beasts instead of men capable of learning ways of peace and industry.

Another illustration of the capacity of the African native to progress along indigenous lines is furnished by Southern Nigeria, where the Bini took readily to wild rubber collecting and the laying out of communal plantations once the bloody priesthood of Benin city had been overthrown. Star rubber-producing district of the lands now known as Nigeria was, however, the old separate Colony of Lagos. With discovery of *Funtumia* there, exports jumped from three tons in 1894 to 2,263 in 1895. Seventy-five per cent or more of the trees fell in the slaughter of this and later years. For all Nigeria, including Lagos, the twentieth century high was the 1,533 tons of 1906; the low (considering pre-war years only) the 454 of 1902. Twentieth century high for all British colonies and protectorates having indigenous rubber (Gold Coast, Nigeria, Sierra Leone, Gambia, Kenya, Uganda, Nyasaland, Northern and Southern Rhodesia, Anglo-Egyptian Sudan and Zanzibar and Pemba islands) was the 3,639 tons of 1905.

Completing the African *caoutchouc* roll call are: Tanganyika (formerly German East Africa and now administered by Great Britain under mandate); Togoland (former German West coast Colony now under British and French mandate); Rio Muni (tiny continental portion of the Spanish Colony of Guinea); and the Republic of Liberia. None of these produced wild latex importantly, but of Liberia we shall have something to say under rubber plantations later on.

The continent as a whole produced over 15,000 tons in 1900 and thereafter never fell 12,000 until 1914 brought the World War to Africa. Record years were 1906 (20,539 tons) and 1910 (20,136 tons) and only twice in the 1904-1913 period was the yield less 18,000. From well over a score of different political units came this rubber, as we have seen, but the full horror of what *caoutchouc* collecting meant to Africa can only be appreciated by remembering that the Congo Free State hell alone accounted for nearly a third of the total 1900-1908 production. And that of the 1900-1912 total, Congo Free State accounted for more than a fourth, Congo Free State plus concessionaire-ridden French Congo for more than a third, the two Congoes plus slaving Angola for nearly a half.

Far more gruesome than the tragic history of wild rubber gathering in South America, then, is its history in Africa. So gruesome, in fact, that in recent centuries only the overseas slave trade itself can be cited as a comparable wholesale atrocity. Terrible as have been some of the military barbarities committed by Britain, Germany, Italy and the other European nations in nineteenth and twentieth century imperialistic campaigns in Africa, these were episodic where the rubber atrocities were continuous. Whether the object was white settlement or political control, the business of grabbing territory did not run to mass murder save when original conquests were being effected or rebellious districts were being subdued and 'punished.' These were ventures with the character of permanence and self-interest alone dictated stopping short of unnecessarily butchering off the populace. In the case of rubber exploitation, however, there was no single demand or series of demands which the native could buy peace by accepting. Climatically the Congo's chief rubber areas are such that the white man could not have settled and raised families, even had he wanted to. Nor did European financiers or their changing field agents have any least concern for the ultimate fate of either lands or inhabitants. They were exploiters with the one aim, as we have noted, of getting out the greatest possible amount of rubber in the least possible time, and so their demands were incessant and incapable of satisfaction. As against imperialistic campaigns lasting a few months or a few years, then, we have commercial spoliation involving 20 years of continuous warfare in Leopold's land, 10 years at least in the French part of the Congo. A record far worse than that of the militarists, far worse than that of the continent-partitioning statesmen, this is what capitalism as exemplified by the rubber industry has to answer for in the case of Africa.

3 Plantation

3.1 Seed Snatch

On one side is the Bay of Bengal, on the other the Gulf of Siam and the South China Sea. This is Southern most Asia thrusting down towards the equator in a goose-neck peninsula. The neck itself is a part of Siam, the head and the beak are that jumble of colonies and federated and unfederated protectorates loosely known as British Malaya. West of it across Malacca Strait lies the long length of the Island of Sumatra, which runs on to the South and East and breaks off in a tail of subsidiary Dutch islands the closest and largest of which is Java. East of Malaya, East of Sumatra and North of Java is the squarish bulk of the Island of Borneo, mostly Dutch, but including also another jumble of British protectorates. This is the new world of rubber, and its capital, its Para, Manaos and Iquitos rolled into one, is stinking Singapore, a yellowman's city on a dot of an island at the Southern extremity of the Malay Peninsula, halfway between India and China.

In 1910, we have recorded, the wild rubber lands hit their all-time production peak of 83,000 tons out of a world total of 94,000. To the 83,000 figure the exploited portion of the 2,250,000 square miles of Amazon *caoutchouc* territory contributed more than half, the continent of Africa a fourth, Mexico most of the remainder. In each of these areas the figure represented virtually the maximum; in Amazonia, a maximum dictated by available labor.

Two years ago — 1934 — the world hit its all-time rubber-producing peak — 1,019,000 tons. To that record Brazil, Africa and all the other wild rubber lands contributed just 1.4%. From British Malaya, that 52,500 square mile tip of the Malay Peninsula, came nearly half — 467,000 tons to be exact. From Sumatra, Java and Dutch Borneo of the Dutch East Indies came 380,000. From the British sphere of influence in Borneo came 29,000, from Siam's share of the Malay Peninsula 18,000.

For 894,000 tons of the world supply we have accounted, then, without leaving the one peninsula and the three of the Sunda islands mentioned at the beginning of this chapter. Nor do we move very far from them to account for most of the remainder. Adjoining Siam on the East is the medley of protectorates and colonies known as French Indo-China. From here, and chiefly from Cochin-China Colony, the tip of the Indo-Chinese peninsula across the Gulf of Siam from the Malay Peninsula, came 20,000 tons.

From a casual glance at the map, you would say that the Tenasserim division of Lower Burma (British) shares the upper part of the Malay Peninsula with Siam. Geographers say that this isn't so and that the peninsula begins just where Tenasserim leaves off. At any rate this gives you an idea of how close Burma's rubber lands lie to the new heart of the *caoutchouc* world. In official returns they are teamed up with the rubber lands of British India's extreme South-West and the 1934 total was 12,000 tons. Just South and East of the tip of India and across the Bay of Bengal from the Malay Peninsula is Britain's great island of Ceylon. Its share of the record total was 79,000 tons.

That leaves just 14,000 tons to be accounted for. Brazil supplied 9,000 of these, Africa 3,000 and the remaining 2,000 came from all over the world, from the Philippines to Mexico and back to the cannibal island of New Guinea.

Immensely more than Brazil ever dominated wild rubber, therefore, Malaya and its neighboring islands today dominate a production stretched to so vast a figure that little more than two decades ago it would have been regarded as a fantasy of the trip-to-the-moon variety. How this came about is now our story. To the story, rubber indigenous to these parts of the world contributes very little. And yet save for its presence originally in India and Burma and Malaya and Sumatra, it is unlikely that planting attention would have been turned to this area and it is for this reason that we deal with it briefly.

The formal history of rubber in Asia and Asia's islands begins, then, with the swing of a cutlass in the Penang brush. Immediately off the coast of the Malay Peninsula is the island of Penang, immediately opposite what then was headquarters of one of the most notorious bands of pirates infesting the East Indian seas. Target for the swung blade, however, was nothing more formidable than one of the innumerable arm-thick vines blocking an exploring party's way into the forest. The vine spurted milk, some of the milk dried upon the blade and James Howison, Esq., inspecting and experimenting with the coating, was surprised and delighted to find something possessing 'all the properties of the American *caoutchouc*.' So was rubber discovered in Asia in the 1790s, a mere two and three-quarters centuries after the men of Cortez gazed upon it in the halls of Montezuma. Surgeon of Penang was this Howison, and a servant of the greatest of the East India companies, the English one chartered by Elizabeth in 1600. The island had been leased by John Company's renowned Captain Light from the Sultan of Kedah in 1785. Virtually uninhabited, it was made a penal settlement. Today it houses the number two rubber export market of British Malaya and is, approximately, the geographical center of the whole island, peninsula and mainland world of British, Dutch, French and Siamese plantation rubber.

It is the *Hevea brasiliensis* tree about which that world now revolves. What the vine was Howison was unable to determine, nor was that at all surprising since he had hold of a new genus as well as a new species. Furnished with a specimen from Sumatra in 1797, the eminent botanist William Roxburgh, MD, christened it '*Urceola elastica*.' Experimenting with its juice, he too held that it really was comparable to '*caoutchouc*' from America and was not another 'triflingly elastic' false alarm such as the milk of Asia's sacred fig tree under whose shade Vishnu was born and Gautama Buddha became endued with his divine powers. By 1810, however, Roxburgh, first really scientific classifier of rubber plants, had his attention called to an infidel cousin of *Ficus religiosa*'s eventually found to be most important of all Asiatic *caoutchouc* sources. First spotted in the Indian province of Assam just South of the Himalayas, *Ficus elastica* actually was distributed from there through Burma, Penang, Sumatra and the other Sunda islands. A huge tree in the tropics, this rubber tree is familiar to all United States citizens of 30 years or older as the stunted exile once to be found decorating every pador. Whether the natives of Assam and Burma ever made any use of its latex is open to question but certain it is that if they employed it at all it was to exceedingly small extent.

As an article of commerce Asia's *caoutchouc* dates back to the late 1830s and the idea of trading in it seems to have struck England and Assam just about simultaneously. From 1838-1839 periodicals we learn that by the time a London company's offer of a premium of $250 for the first hundredweight of such rubber reached India 'the great demand existing at home for the article had been heard of and large quantities were already on shipboard, compared to which the hundredweight stipulated for was but a molehill to a mountain!' A molehill to a South American mountain was all the wild rubber ever to come out of Asia, but by the middle of the century latex from *Ficus elastica*, assorted *urceola* vines and numerous other lianas and trees of Sumatra, Java, Borneo, Singapore, Penang, Burma and the wild elephant valleys of Assam was a familiar article to the European trade. Despite optimistic beliefs of discoverers and christeners of the first varieties noted, none of these plants exuded a latex rivaling the rare old juice from Para, however. Resinous was all of this rubber, and its comparative quality can be deduced from the fact that it brought but five to eight cents as against the 18 to 30 commanded by para in the first half of the nineteenth century.

Only at odd intervals were the combined Asiatic exports of any consequence during the later years of the nineteenth century. Two such occasions were 1872-1873 when 1,095 tons came out of India and Burma and 1897 when 1,288 tons cleared through Singapore. It was three years after this last date, incidentally, that the first shipment of *Hevea* milk came out of the Middle East.

It was an old dream, this idea of transplanting and cultivating the American rubber tree, a dream as old as Europeans' knowledge that *caoutchouc* was yielded by plants native to the *estrada* hemisphere. In writing we encounter it for the first time in the March 23rd, 1791, issue of *The Bee* or *Literary Weekly Intelligencer* of Edinburgh. Therein, James Anderson, LLD, FRS, FASS, etc., argues for importing the seeds into 'our West Indian islands, or the Cape Verde islands, or along the coasts of Africa, where there are such extensive tracts of uninhabited country laid waste by the depopulation that our destructive trade in slaves occasions.' He continues, 'What a difference there would be in the state of the inhabitants of that unhappy country, were they to be taught to cultivate the arts of peace, and to enrich themselves by industrious labour, instead of those cruel wars fomented by our miserable trade in slaves.' Edinburgh Jamie could not foresee a Leopold, it would seem. In another sense, however, he showed himself a shrewd and far-sighted prophet, indeed. 'Could this juice be had in abundance so near to Europe,' he notes, 'it might be sent hither in its fluid state in closed casks or bottles, so as to be here manufactured for the purposes it were fitted to answer.' Early as he was, a French chemist was a year ahead of him in calling for importation of noncoagulated milk; but industry has taken up the idea only within the last 15 years.

Howison, writing in 1798, spoke of the possibility of cultivating plantations of his vine discovery or of transferring the American trees to India and he recommended the province of Bengal as site for a vine experiment. Thomas Hancock, one of the great men of rubber invention, was a plantation advocate of the 1850s. His idea was to try cultivating the Amazon trees, the *Hancornia* of Brazil and the *Urceola elastica* vines in West and East Indies. And so early as 1855 his correspondent, Sir William Jackson Hooker, director of the Royal Botanical Gardens at Kew, was on record as pledging any assistance in his power to parties disposed to make the attempt. Sir William's son and successor, Sir Joseph Dalton Hooker, friend of Darwin and one of those who induced him to publish his views on the origin of species, made good on that pledge in the 1870s.

Meanwhile practical planters were doing something — but not much. In 1830, for instance, *Castilloa* seeds from Guatemala were distributed to Cuban planters and there were later small plantings not to exceed 100 acres in all. Of record is a shipment of 1,600 pounds of *caucho* in 1924. In 1864 the Dutch planted some eighty acres in Java with *Ficus elastica* and these trees were tapped for some years after 1881. In 1865 Joaquim Antonio da Silva, living on the River Guama, 12 miles above Para, had 20,000 young *Heveas* planted on a low island in the stream. To save labor they were set out near the circumference of the island and the tidal ebb and flow destroyed most of them. In the extreme South of Mexico, one Romolo Palacios pioneered in a big way by planting 100,000 *Castilloas* in connection with *cacao* near San Benito and close to the Guatemalan line in 1871. A fire swept this property and it never was replanted. In the same general region, General Sebastian Escobar stuck out a million trees in 1872 and Don Matias Romero, former Mexican diplomatic representative at

Washington, tried his hand with 100,000. After Escobar's trees had matured rubber was taken from them for a number of years, but the chief business of the estate was grazing, and interest in the *caucho* lapsed. Romero's plantation was not a success but he retained his faith in the idea and published a book upon it later on. In England at the time of these first Mexican experiments in cultivation, interest in transplanting the American trees had spread to the high places and it was this that resulted in rubber plantations as we know them today. And wholly on the right track was one man at least, for in a paper published in 1869, James Collins, curator of the museum of the Pharmaceutical Society, called not merely for acclimatisation of rubber trees in Britain's *estrada* possessions, but claimed that the *Hevea* was superior to all others. Collins drove home his argument by referring to the successful introduction of the *cinchona* into India and Ceylon a decade before. Clements Markham, assistant secretary of the India Office, was basking in a knighthood conferred upon him by Victoria for having personally supervised the kidnapping of the quinine tree. Cordially disliked in Colombia, Peru, Ecuador and Bolivia was Sir Clements for the part he had played in that business, but Brazil was to have even more reason for detesting the name. Interesting himself in rubber, he entrusted to Collins the bringing together of all existing information and the publication of the latter's report in 1872 may well be considered the beginning of the project. Embodied in the report was much valuable information gathered by Richard Spruce who, in Condamine fashion, had been studying the *Hevea* and communicating his information before as well as during the time he chased the *cinchona* for Markham. Equipped with this document, Sir Clements enlisted the aid of the Duke of Argyll, Secretary of State for India in the Gladstone cabinet. Kew was notified of developments and Lord Granville, Secretary of State for Foreign Affairs, was asked to instruct the consul at Para to obtain seeds. Brazil supposedly was strongly averse to the exportation of *Hevea* seeds; that proved no insurmountable obstacle, for by May, 1873, Collins was in touch with one Farris, who had landed from Cameta with 2,000 seeds, 'some quite fresh.' England bought them for $27 spot cash. Only a dozen germinated and half of these were taken to Calcutta in September by Sir William King, director of the Royal Botanic Gardens there. The climate of India did not prove suitable for these first *Hevea* trees brought alive to the East and they shortly died. Propagation by cuttings taken from the remainder was started at Kew. And from there Hooker wrote the India Office that he had a correspondent, Henry A. Wickham, a planter at Santarem on the Amazon, whom he would ask to send seeds. Wickham proposed raising *Hevea* seedlings as the seeds lost their germinating power so quickly. Nevertheless he would furnish seeds for $50 a thousand and the India Office agreed to this price.

In such businesslike manner, then, enters an important figure in rubber history, the chief physical agent in the transferring of the *caoutchouc* world from one hemisphere to the other. Without this vain fellow the change nevertheless would have been effected; but the fact remains that it was he who actually performed the feat, and there is no denying the presence of mind he displayed and the workmanlike way in which he carried out his commission. It was as a youth of twenty that Wickham first came out to the tropics in 1866, traveling through a part of Central America and witnessing *caucho* collecting at first hand. Three years later he invaded Venezuela and decided that this was a country no self-respecting young British prig could care for. Native labor was unreliable, the ruling class was 'despicably corrupt' in sexual matters and the self-exiled Confederates just in from Dixie were 'anything but well-selected, respectable men, and for the most part, the offspring and refuse of the long civil war.' So off to an island in the upper Orinoco packed Henry to work rubber with an English companion and half a dozen Indians. This was his first experience with *Hevea* milk. Starting down to Para with his rubber in 1870, he had a good chance to look over the Negro and Amazon territory on the way and picked the triangle between the Amazon and the Tapajos just behind Santarem as the base for his future operations. His return there and his correspondence with Hooker brings us to the close of 1874.

Less successful than Farris was Wickham in his first attempts to reverse the hemispheres. Kew received a few seeds from him in July, 1875, and a few more in August. All failed to germinate. Between these shipments the India Office acquired some 378 through its own channels and passed them on to Calcutta. Their fate was the same. Tired of fooling around, officialdom then sent Robert Cross, another of Markham's aforetime *cinchona* collectors, to Panama and he returned with 176 *Castilloa* seedlings which were dispatched to India the next year. In March, 1876, it was decided to send him to Brazil for *Hevea* seeds and he actually took off in June, a week or so after the forgotten Wickham's triumphant arrival at Kew with more seeds than Markham and Hooker ever had dreamed of. Cross was not called off, probably because of the chance that the Tapajos importation again would disappoint. This time, however, Henry had fooled them all. It was the wholesale character of his 1876 seed snatch which brought success and its story is one of unexpected opportunity smartly seized.

Up the river had come the first ocean liner of a new Liverpool-to-Upper-Amazon service, Captain Murray in command. Invited to a dinner on board the *Amazonas*, Wickham and the other planters of Santarem were entertained by the two gentlemen in charge of inauguration of the new line. Henry admired the blue lights in which the ship was dressed up, participated in a hearty attack on a first-rate supper in the saloon, and departed ruminating not on *Hevea* seeds but on the excellence of that chow. Some days later, however, surprising news came down the river and now for the first time two pictures clicked into one in Wickham's mind. Up the Amazon the two gentlemanly masters-of-ceremonies had vanished with everything aboard the vessel. A liner, an English captain and not a stick of cargo for the return voyage to Liverpool. And in the high forests up the Tapajos the *Hevea* seeds were just then beginning to ripen. Wickham hardly could have asked for more nor did he think of doing so. He had no cash on hand and there was not the slightest chance to realise any on his incipient plantation, but he did have that commission from the India Office. Boldly, then, the 30 year old opportunist wrote Captain Murray chartering the *Amazonas* on behalf of the Government of India and setting a date to meet him at the junction of Tapajos and Amazon.

Piling into a canoe then, he went 50 miles up the tributary river and, gathering together as many Indians as he could find, struck back into the forest. With this crew he daily ranged a rich *Hevea* area, packing the heavy loads of seeds down to a village of the aborigines. There, as Wickham tells it:

> 'I got the Tapuyo village maids to make up open-work baskets or crates of split *Calamus* canes for receiving the seed, first, however, being careful to have them slowly but well dried on mats in the shade, before they were put away with layers of dried wild banana leaf betwixt each layer of seed; knowing how easily a seed so rich in a drying-oil becomes rancid or too dry, and so losing all power of germination. Also I had the crates slung up to the beams of the Indian lodges to insure ventilation.
>
> 'I was working against time. It was true that the seed would still continue to ripen and to fall from the trees for another month or so, but it would be inexpedient to risk the vitality of some thousands I had succeeded in securing. The rendezvous with Captain Murray of the *Amazonas* would soon fall due at the river mouth, and if I missed that, when and how another such opportunity? I had got to look upon the heavy oily seeds in their dappled skins as become very precious, after having packed them down so many long days tramping across the forest plateau, and so lost no time in getting them carefully stowed under the tolda of the canoe, and starting away down stream, duly meeting the steamer, as appointed, at the mouth of the Tapajos.
>
> 'I found Murray crabbed and sore from the experiences with his rascally supercargoes. It appeared they had given instructions to land the whole of the trade-goods with which his ship had been freighted ostensibly for purchase of incoming rubber season crop at the town of Manaos. He was then to anchor his ship at the boca of the Rio Negro and 'await orders,' they meanwhile to dispose of the goods and to advise when they had got together sufficient rubber in order to load ship for the return trip. The time becoming unaccountably long, he landed, and on making inquiry he could only learn that the goods had indeed been disposed of but no one could give any information as to his two supercargoes, and so found himself left with an empty ship on his hands.
>
> 'For my part, as the fine ship sped on her way with my prospective *Hevea* so far safe aboard, slung up fore and aft in their crates in the roomy, empty forehold,

I became more and more exercised and concerned with a new anxiety, so as not much to heed Murray's grumpiness. We were bound to call in at the city of Para, as the port of entry, in order to obtain clearance for the ship before we could go to sea. It was perfectly certain in my mind that if the authorities guessed the purpose of what I had on board we should be detained under plea for instructions from the Central Government at Rio, if not interdicted altogether. I had heard of the difficulties encountered in the Clements Markham introduction of the *Chinchonas* in getting them out from the Montana of Peru. Any such delay would have rendered my precious freight quite valueless and useless. But again fortune favoured. I had 'a friend at court' in the person of Consul Green. He, quite entering into the spirit of the thing, went himself with me on a special call on the Barao do S—, chief of the *Alfandiga*, and backed me up as I represented to his Excellency my difficulty and anxiety, being in charge of, and having on board a ship anchored out in the stream, exceedingly delicate botanical specimens specially designated for delivery to Her Britannic Majesty's own Royal Gardens of Kew. Even while doing myself the honour of thus calling on his Excellency, I had given orders to keep up steam, having ventured to trust that his Excellency would see his way to furnish me with immediate dispatch. An interview most polite, full of mutual compliment in best Portuguese manner, enabled us to get under way as soon as Murray had got the dingy hauled aboard.

'Now fairly away, I could breathe freely, and soon had the hatches off with the open-work crates slung up on lines fore and aft in the air and free of danger from ship's rats. Again blessed with fine weather, I was able to keep the hatches off all the way over.

'I got Murray to put me ashore at Havre, and there posted over to Kew, saw Sir Joseph Hooker, so as to enable him to dispatch a night goods-train to meet the ship *Amazonas* on arrival at the Liverpool docks.

'June 1876 was a time of commotion at Kew, as they were compelled to turn out orchid and propagating houses for service, and to make room for the sudden and all-unexpected inroad of the *Hevea*; but Sir Joseph was not a little pleased. The *Hevea* did not fail to respond to the care I had bestowed on them. A fortnight afterwards the glass-houses at Kew afforded (to me) a pretty sight-tier upon tier-rows of young *Hevea* plants — 7,000 and odd of them.'

Wickham in later years served as inspector of forests in British Honduras and inspector of forests and commissioner of Crown Lands under the Government of India and explored parts of Malaya, *Estrada*, New Guinea and the Pacific islands, studying native rubber plants and making planting experiments. Inventor of two rubber smoking devices which failed in the market, he never ceased roaring the virtues of the smoking process and denouncing chemists and others responsible for plantation rubber made with coagulants 'like junket' and 'baked like unto toffee in vacuum driers.' Other great and troubling griefs were the 'fallacy' of 'stinting' on space for what is 'by very nature a large forest tree' by planting more than 40 in an acre and planter willfulness in practicing excision (a paring away or gouging of outer bark so as to open the latex vessels) as against 'clean sharp incision' with ax or other sharp instrument in the traditional Amazon manner (which covered South American trees with excrescences and killed them in what should have been their prime). His seed snatch had made him the supreme authority on all things pertaining to rubber and rubber trees, he figured, and every planter, botanist and chemist departure from South American methods was all wrong. Paying no attention to him, planters and scientists went right along improving on Brazil and on nature and increasing and then multiplying tree yields while Henry continuously whooped dire predictions of root failure, loss of vitality, bark disappearance, exhaustion and other calamities which never occurred. When the Amazon belatedly began aping *Estrada* tapping methods he piped down on his crusade for incision (without ever admitting he had been wrong) but to the day of his death he never relinquished the smoke fixation. He was a 'practical' man, was Henry. That accounted for his success in the seed grab. It accounts also for his being wrong in nearly every argument with a plantation world progressing and improving through newfangled scientific methods instead of abiding by the rules laid down by the Indians of the American jungles.

Late in coming were the honors and emoluments for the good job of seed transporting, but through no fault of Henry's. He was an inveterate writer of letters to editors, he delighted in identifying himself by such sonorous appellations as 'sometime commissioner for the introduction of the Para (*Hevea*) Indian Rubber Tree for the Government of India' and he could and did refer to his own disjointed, repetitious and often inaccurate account of his feat as 'a thrilling narrative of resource, initiative and organising capacity, crowned with complete success.' In 1911, three years after the publication of this book, he was presented with $5,000 and an annuity by the Rubber Growers' Association of London and the Planters' Association of Ceylon. In 1920 he was knighted by George V. On his eightieth birthday, 1926, he received a $25,000 present from Edgar B. Davis, American rubber and oil magnate renowned for carrying generosity to the limits of eccentricity, and $40,000 from the Straits Government. He died two years later.

As to the seeds Henry brought out of the Tapajos, the present writers are agnostic enough to doubt that anyone ever 70,000, and this is the figure that Wickham himself used in 1927. As to the 7,000 plants which, according to the 1908 book, he saw in the glass-houses at Kew, this was no more than an optical illusion for the records show that there were just 2,397 of them. All the other seeds failed to germinate.

Because of the demonstrated lack of suitable botanical gardens in India it had been arranged between the India and Colonial Offices that the seedlings should be sent to Ceylon to be cultivated and propagated for distribution to Burma and other hot, moist districts of the Indian empire. In August, 1876, then, 38 cases containing 1,919 seedlings were started from Kew to Ceylon with W. Chapman, a gardener, officiating as chaperon. At the Heneratgoda gardens there, conditions proved so favorable for their thriving that years later Wickham was able to have himself photographed with one of the original kidnapped seeds now grown to a record tree which had given 95 pounds of dry rubber in 1909, 55 pounds in 1910, 67 in 1911, 86 in 1912 and 68 in 1913 before being retired to stud and used only as a seed bearer.

In the month of the Ceylon shipment 18 seedlings also were sent to the Dutch island of Java and a few seeds to Singapore island and Perak (one of the Federated Malay States under British protection). On these the India Office neglected to pay freight and delivery was delayed until too late. In September 1876, a hundred plants, propagated at Kew from seeds whose source is not of record, were added to the Ceylon collection.

In November of the same year Cross returned in triumph from Brazil with 1,080 *Hevea* plants plus *Cearas* and *Castilloas*. Kew gave away 680 of the *Heveas* and retained 400, of which 3% survived. Whether any of the plants propagated from these *Hevea* seedlings went to Asia is subject of dispute. According to an account in the *Kew Bulletin, Additional Series VII*, however, 100 of these plants were sent to Ceylon in 1877, the year in which Kew also sent 50 *Heveas* to India and Burma. Ceylon made its first distribution in 1877 by sending 22 trees to Singapore, and Henry N. Ridley, director of the Singapore Botanic Gardens after 1888, once went on record with the opinion that these probably were Cross plants. He added, 'It was from these 22 plants in the Singapore Botanic Gardens and Perak that three-quarters of the cultivated plants of *Hevea brasiliensis* have sprung.'

Accepted at face value, this would mean that Cross rather than Wickham deserves the title of 'father of the plantation rubber industry'; but the total testimony sums up to the probability that it actually is from Henry's seeds that the majority of plantation trees are descended. The probability also is, however, that Cross as well as Wickham trees spread across the face of the Middle East and the whole record is proof positive that, had Wickham never lived, Brazil's rubber industry yet would have been moved overseas. We have gone into detail, then, in order to dispose of the generally held and often printed myth that the creation of the plantations was conceived, advocated and executed solely by Henry A. Wickham. As to the Singapore trees, they bore seeds in 1881 which were distributed throughout Malaya and British Borneo. Two years later Ceylon sent 28 plants to India and some few to Burma. From these lands, thereafter, the industry spread to the Dutch East Indies, French Indo-China and elsewhere. It was a long time doing so, however.

Perhaps as widespread as the legend of Wickham's omnipotence is the conception of the *Hevea* and the *Hevea* alone being rushed to the East expressly to replace devastated coffee plants and of an eager army of planters whirling into action to set out the seedlings and reap 'white money' just as fast as they could tear up other crops and knock over jungle. Actually it was a case of India's Government importing all varieties of rubber plants to be tested in competition with the native ones, and it remained an experimental business for decades. Nor was it merely an elimination contest in which *Hevea* competed with all other rubber trees; it was also an elimination contest in which the British and Dutch East competed with almost the whole of the tropical and sub-tropical world, South America alone excepted.

In the history of wild rubber we have given the *Hevea* the place of prominence because it accounted for nearly all the wild to begin with, for a third to a half in all the boom years when the forest trees and vines of the Americas and Africa were pouring their latex into the world market. Under Middle East plantation auspices it has been developed to a point where it accounts for about 97% of the multiplied and re-multiplied total, and so it is that we have been so attentive about tracing the careers of each batch of *brasiliensis* seedlings that went out to the Orient. To deal in detail with all of the prior, simultaneous and subsequent importations of rival rubber plants by the East; to deal in detail with all of the prior, simultaneous and subsequent plantation experiments in other lands would be a weariness and a waste of reader time. To outline briefly and broadly the whole panorama of plant exchange and experiment is necessary, however, in establishing that today's preeminence of *Hevea* and Malay-Sumatra in this rubber world is no accident. Favorite at the start the *Hevea* had been because of its being the source of the world's best rubber, but many were the disappointments it caused. And, in the reaction that set in, it traveled under a handicap for years.

Ficus elastica was so assiduously experimented with in India that the Government had over 2,000 acres of it set out in plantations in Assam before 1892. From there it was imported by Lagos for a trial on the West African coast in the 1890s. In 1897 it was introduced into Egypt by Kew. Belgian Congo experimented with it in this century. In Angola it became the tree second in popularity. The Dutch saw it as the most promising plantation tree for Java in 1900, preferred it above others in Borneo until after 1910.

The Ceara or Manihot was tried literally all over the world. First American rubber tree to show flower and to produce its latex in the Middle East, its seeds were distributed by Ceylon to Burma, India, and Singapore in 1878. From Kew in 1877-1878 living plants went not only to Ceylon but to India, Java, the Fiji Islands, *Estrada*, to Jamaica and Dominica in the West Indies, Zanzibar off the East coast of Africa, Trinidad off the North coast of South America. It was sent into the Philippines in 1905-1906 and the United States Government made extensive experiments with it in Hawaii prior to and during that time. It was planted in Nicaragua, on the Island of Mauritius, on San Thome. It was introduced into every rubber country on the African continent. In Tanganyika the Germans had planted 112,257 acres of it to 1,010 of all other rubber trees at the end of 1913. In British Nyasaland, three-fourths of the 10,562 acres of planted rubber area in the same year was *Ceara*. In Angola it was the big favorite, seven to ten thousand acres being planted.

Castilloa had its chance abroad in Ceylon, Java, India (Cross reported to the Government that it would give a larger return than any other plant in India), San Thome, Principe, Portuguese East Africa, Nyasaland, Belgian Congo, Kenya, Zanzibar, Uganda. At home it was planted late in West coast Colombia and Ecuador, early in Mexico and Central America.

Africa's indigenous *Landolphia* vines were given a plantation tryout in Leopoldland and Sierra Leone. The native *Funtumia* tree was planted in nearly every Dark Continent rubber land and especially in Nigeria where the government-encouraged communal plantations of the natives numbered 602 in 1911. Kew obtained specimens in 1895 and this African was given the usual opportunity in Ceylon and elsewhere in the Middle East.

As to the *Hevea* it went, prior to 1900, not only to Ceylon, India, Burma, Malaya and the Dutch Indies but also to Nigeria and the Belgian Congo in Africa and to British Guiana in South America. In each case, incidentally — and this was true later as well — the seedlings came out of their adopted home in the Orient, and not from Brazil. After 1900 came the *Hevea* trials in all of Africa, Florida (United States), British Honduras, Mexico, Nicaragua, Costa Rica, Dutch Guiana, Trinidad, Tobago, Haiti, Hawaii, New Guinea, Samoa, Fiji, Solomon Islands, the Philippines, Hainan off the coast of China and where have you. In many of these places soil and climate were unsuitable, others never got into large scale production because Middle East supply had caught up and passed demand or because labor was lacking or because of land laws which forbade the acquisition of vast holdings by any individual or corporation. Out of a number of them small amounts of para still come but as in the case of the New Guinea shipments to *Estrada* are too minute to contribute more than a fraction of a fraction to world trade. In all of them, however, there was sufficient planting to show that the cultivated *Hevea* nowhere grows and yields so well as in Malaya and Sumatra.

Wherein *brasiliensis* triumphed over all other plantation trees has yet to be recounted. To begin with, the others, with the exception of the *Ceara*, had the disadvantage of being tappable but two to half a dozen times a year. *Ficus* had the additional disadvantage of being extra long in coming into bearing and of being so free-branching that no more than 40 could be planted to an acre. Also it was a capricious and unpredictable creature giving a pile of rubber one year and scarcely any the next. *Funtumia* required even more time before being ready for tapping — in some of its home precincts up to twenty years. *Castilloa* yielded no rubber at all in many places where it apparently grew well, and but a few ounces a year on any plantation. *Ceara*, like *Hevea*, can be tapped regularly. Unlike the *Hevea*, which demands absolutely correct conditions, this hardy native of the drought lands grows any place where there is enough dirt to fill a fingernail, yields rubber in most localities. Again it scored on the *Hevea* by being tappable at two to three years of age as against the five to six of the *brasiliensis*. Six ounces a year was about the best it would yield, however. With *Hevea*, before the days of seed selection and bud-grafting, averaging four or five pounds a year at an age of 10 years or less, and with individuals running up to 30 pounds or more, the fate of *Ceara* on a competitive basis can be easily seen. Growing and yielding in lands where *Hevea* would not, there was, of course, some profit in fussing with it so long as world rubber demand was ahead of supply and prices rode high. Tanganyika was the star instance of this and even there the Germans were contemplating uprooting their trees and planting the areas with sisal hemp when the War broke.

Why then was this *Hevea*, this king of planted rubber trees, so long in establishing its supremacy? *Hevea* was so long in establishing its supremacy over the plantation world for the same reason that the plantation world was so long in really digging in to start overhauling the wild: official and farmer ignorance of everything connected with rubber trees. Was this not a sorry state of affairs? It was not, it was the greatest break this *caoutchouc*-consuming world ever had. Had the British Government's botanists at the *Estrada* gardens and experimental stations and their planter constituents been well versed in Amazon lore, had Wickham's report to the Government of India been broadcast as a Bible instead of bureaucratically being filed away to gather dust, things would have started off on a Brazilian basis. This would have meant a moderate success, and the heavy odds are that widely spaced trees and reckless and wasteful slashing in *seringuero* style would have prevailed; as a consequence, the dawning motor era would have been handicapped by a rubber shortage and prices beyond the wildest dreams of the greediest *patrao* in South America. All the wild rubber lands of the world plus all the plantation acreage that could be laid out and worked on Wickham lines would fall far short of supplying world needs. Needed was the very plantation system that came into being in the East, and it came into being because men, knowing nothing whatever about the *Hevea*, blundered and experimented and tested through failure after failure until they hit on the methods of planting, of tapping, of coagulating, that gave the sort of results we know today. That they succeeded just in time to provide supply for the automotive demand must be attributed to the same *caoutchouc* providence which decreed that Henry Wickham should go steaming down the Amazon with his *Hevea* seeds in the very year that Dr. N. A. Otto of Cologne launched the internal combustion engine upon its world-conquering commercial career.

It was not the trees that died, but the trees that came through to fine, lusty growth that gave the most trouble to the planters of the Middle East. Locked in the bark was a wealth of milk, but so insoluble seemed the puzzle of how to get it out that it was pretty generally concluded for a time that the milk just wasn't there. First of the *Estrada* trees to feel the steel were five in the botanical gardens in Ceylon, tapped by Dr. Trimen in 1882 when they were six years of age. The yield was 2½ ounces of rubber. Disappointed in this insignificant amount, the scientists gave the trees another five years, then had at them with chisels, cutting 40 or more V-shaped incisions right through the bark and its latex tubes, through the cambium (paper-thin seat of tree growth) and into the wood. The yield, allowed to dry in tears on the tree, was about 1½ pounds in the case of each plant.

Not until the bark had healed were the trees assailed again and thereafter tappings every other year became the order of procedure. At every point within reach, tapping was tried, even at 30 feet from the ground by means of ladders and scaffolds. Typical of the sort of advice that went out to planters was the 1888 circular of the Ceylon Forest Department gravely advising them not to tap *Hevea* until it was 10 years old. And the later one mildly encouraging further commercial planting on the strength of this finding: 'A yield of over 10½ lb. of first-class rubber from a single tree in six years fully warrants a belief that the cultivation of large plantations would be profitable. Nor is there any reason to suppose that the trees would not easily bear tapping annually.'

Such bulletins naturally killed just about all the little enthusiasm for the *brasiliensis* that had been generated among planters who, freely furnished with seeds and plants but with no reliable information, had themselves been assaulting the Brazilian immigrant with axes, cutlasses and every other engine of destruction at their command. Accustomed to the free flow of milk from *Ficus elastica*, the few drops of a para seemed to them a sorry showing, indeed. Ceylon's planters originally had turned to *cinchona* to replace the coffee ravaged by the leaf disease of 1875-1880; the collapse of the *cinchona* boom seemed to suggest a switch to rubber, but the disillusioned planters wanted little part of it. Electing to try tea, they set out *Cearas* and *Heveas* chiefly to shade this crop.

So much for the nineteenth century history of *Hevea* in Ceylon, its original home in the East. It was in Malaya that the beginnings of the success story were written and it was due as much to botanist Ridley, before mentioned, as to completely suitable soil and climate, that this happened. In 1889, year after his arrival at the Singapore botanic gardens, Ridley first tapped the trees there. Interspersing his important experiments with attempts to interest planters, he met with no success in this latter respect until 1895 when R. C. M. and D. C. F. Kindersley started the first practical estate on a commercial scale in Malaya, where coffee had failed because of the high price of labor as compared to other lands producing the berry.

It was important that Malaya, though it fussed around with interplanting of *Hevea* and other crops, wasted no time fooling with other varieties of rubber trees. One thing that Ridley was careful to explain to the planters was the necessity of 'calling the rubber.' The free flow of *Hevea* milk was to be obtained, he had discovered, by making a first or preliminary cut and returning to it to shave a thin slice off the edge. It is this daily (every other day in more recent practice) reopening of the original wound which constitutes the foundation of plantation success. 'Wound response,' the modern term, also was noted by experimenters at Ceylon and Perak botanic gardens in the 1890s. The other important discovery of the period was achieved by John Parkin. As scientific adviser to the Ceylon gardens, he developed the acid process of coagulation.

The tapping knife and the careful removing of a shaving of bark without penetrating into the cambium, as against the *seringuero*'s careless hatchet hacks. Clean and efficient acid coagulation as against the dirty and uneconomic smoking methods of the Amazon. In those comparisons is patent a considerable part of the plantation triumph. In the true sense that triumph was achieved before the twentieth century rolled around. On the face of things, very little had yet been accomplished, however.

In 1894 the first plantation rubber had been exported from the Middle East — 3.1 tons of hardened *Ficus elastica* juice out of Java. In 1900 there were four tons of *Estrada* plantation rubber reported in world statistics. That fractional advance, so far as the world could see, represented the sum total of what scientific effort had accomplished in the 24 years that had elapsed since the *Amazonas* had pulled into Liverpool with its load of seeds. Nor was there much of a story to tell for the first few years of the new century. With Malaya and Ceylon now in the business in a small way, Sumatra, too, turned to interplanting a little *Ficus* and *Hevea* with tobacco as coffee prices went to pieces in 1902 in the face of the huge increase in Brazilian output. By 1905 the four tons of *Estrada* plantation rubber had grown to but 174 (104 Malaya, 70 Ceylon), however, and the number of acres planted to rubber on this peninsula and this island was but 50,000. Average yields of the period promised a possible ultimate output from this acreage of no more than 6,800 tons.

3.2 Gold on Trees

It was in 1905 that Britain, fast getting nowhere after nearly 30 years with *Hevea* seeds in its pockets, finally took the jump into large-scale rubber growing. Nor is jump quite accurate. Literally it was a case of reckless and complacent Brazil booting the tight (in two senses) little island into the competition that so soon and so completely would end with Northern South America out on its feet.

In spendthrift North America, meanwhile, the first rubber plantation speculation frenzy in the history of the world already had run more than half its course. The thing stemmed back to 1890, when destruction of wild rubber trees began to give rise to reports that the untamed *caoutchouc* lands were heading hell-bent into exhaustion. Of numbers of latex countries this was true, but so little of its wealth had Brazil tapped that, alone, it could easily have supplied the world until that world went automotive. Nevertheless, the United States Government asked its consuls to gather available information on the reputedly nearing shortage. In 1897, William McKinley, former representative of rubberised Akron, O, in Congress, moved into the White House. Inspired by the supposedly approaching rubber famine and by news of planting experiments going on in various spots around the globe, McKinley in a 1899 message to the Fifty-Sixth Congress advised *caoutchouc* cultivation in the nation's new imperialistic possessions — the Philippines, Porto Rico and Hawaii.

By this vigilant action the President innocently performed a very kind service for promoters steaming up the floating of stock companies to get into the business of *Castilloa* planting that had started to attain sizable dimensions in Southern Mexico and Central America. By early 1900 the plantation 'investment' madness was sweeping over McKinleyland and the United States Department of Agriculture's attempt to cry it down was so much expectorating against the whirlwind.

Hundreds of stock companies were formed and all of them issued glowing prospectuses describing the great redistribution of wealth just at hand. Many of these corporations were started in good faith; many were outright frauds. Most entertaining prospectus of the lot, perhaps, is the 47 page pamphlet of the Florida Rubber Culture Company dedicated to cleaning up by selling *Hevea* trees to the new plantations springing up in Mexico. Its cover screams the announcement: 'The Gold which grows on the trees in the California of South America to be transplanted to South Florida. Secretary Blaine's foresight of the immense value of the rubber industry explained by Ex-Consul Kerbey.' Inside we learn that this outfit's capital stock had been placed at $100,000 which 'seems trifling compared with that of the $50,000,000 of the Rubber Combine' (United States Rubber Co., then popularly reputed to be sewing up all the output of the Amazon). Reading on, we learn how Major J. Orton Kerbey had just 'explored' Southern Florida along both coasts and into the Everglades and had tracked to their lairs three species of rubber trees 'indigenous to South Florida': *Ficus elastica*; *Castilloa*, 'the same as the Central American variety'; and 'also a rubber plant similar to those found on the islands of Cuba and Jamaica.' Conditions of soil, heat and moisture of Florida all were similar to those on the upper Amazon, the Major was quoted as saying, and it further was recorded that this daring adventurer was in darkest Amazonia scooping up *brasiliensis* seeds and that shortly young *Heveas* would be on sale at the company's nursery near Orlando.

Of the plantation companies formed, the first 145 showed an authorised capital of $74,539,400, owned or had options on 1,665,646 acres of Mexican and Central American land, and actually planted some 32,000,000 *Castilloas* and a sprinkling of *Heveas*. In a single Department (Palenque) of Mexico's Southern state of Chiapas, it is estimated that nine to ten million trees were set out by 18 to 20 stock companies, mostly American, between 1898 and 1910. The craze to invest in rubber farms struck all classes. The Obispo Rubber Plantation Co., incorporated in New Jersey in 1901, had for president no less a personage than Thomas A. Edison. The New York Teachers' Plantation Co. was organised in July, 1902, to sell stock exclusively to school instructors of Gotham. The Police and Firemen's Mexican Plantation Co. was organised in May, 1904, for the exclusive benefit of Chicago flatties and smoke-eaters, although the bars were later taken down and Uncle Sam's letter carriers and postal clerks were admitted.

Typical of the many purely fraudulent enterprises was the San Miguel Plantation Co., which paid substantial dividends from 1900 to 1906. When the Post Office Department caught up with vice president Talton Embry and secretary Hiram Rose, the promoters, in 1908, it proved the dividends had been paid out of the sale of bonds.

One of the more notorious scandals was that of the Ubero Plantation Co. and the Consolidated Ubero Plantations Co., which had their start in Indianapolis at the hands of William D. Owen, a former secretary of state of Indiana and a three term congressman. The center of financial operations was transferred to Boston and the company soon began paying dividends, supposedly through the sale of short crops. The Post Office Department in issuing fraud orders in 1905 said: 'The plan of Owen and his colleague Barges in both these companies, was to secure the names of prominent men to create an appearance of stability, and then by alluring literature and fraudulent dividends, to give the scheme the appearance of a profitable enterprise. The development work done was solely for the purpose of misleading visiting directors and investors. All the funds of the company were diverted to the pockets of the promoters through the medium of selling companies.' The prominent men referred to were Arthur W. Stedman and Frederick C. Hood, rubber manufacturing barons who were named president and vice president, respectively. These two sent a well known scientist to investigate the project. He returned with such ecstatic reports that Stedman became suspicious and his further investigation resulted in the fraud order. Owen escaped to Europe but Barges was found guilty of conspiracy and of 73 charges of larceny.

In the depression years 1907-1908, the United States citizen really began to snap out of his dream of North American rubber wealth and the awakening was accompanied by failures, bankruptcies, the loss of millions of dollars. Not until after 1910, however; was the large-scale planting of trees abandoned — the best of proofs that a good many of these ventures were honest ones.

In Central America as in Mexico it was the *Castilloa* that almost universally had been favored although there were some plantings of *Hevea* seedlings brought back across the oceans from the Middle East. And Costa Rica experimented with *Cearas*, although unwittingly. These trees were what came up after 'genuine Brazilian rubber seeds' had been sold in this unsuspecting land during the height of the boom. In both Mexico and the nations to the South the Castilloas on some plantations were tapped to so late a date as 1920 before they were cut down to make way for bananas or the jungle was permitted to come in and claim its own.

In Mexican exports of *caucho*, which does not include *guayule*, of course, is written the rise and decline of its plantation industry (remembering that four or more years passed between planting and tapping): (1901) 186 tons; (1904) 303; (1907) 4,616; (1910) 7,939; (1913) 3,816; (1914) 286! The fault, then, was not that the planted *Castilloa* did not produce rubber. It was - and this has been the case with every tree save the Hevea - that it did not give enough latex per tree or per acre to justify the cost of planting and maintaining, save when rubber prices were at their very highest. Three or four times a year is the most that the *Castilloa* can be tapped and each tree gives but 2½ ounces to considerably less than a pound of rubber per annum.

Also complicating the Mexican picture was the fact that many United States citizens sent South to start plantations knew absolutely nothing about the thing they were trying to do. Plus the fact that labor, besides being scarcer and costlier than in the East, was far more intractable. On a comparative basis the rubber tappers of Chiapas actually were veritable Ryans and Morgans compared to the coolies of Asia. Indians recruited from the interior mountains of Mexico were housed, given the use of land to cultivate and were not required to work full time for the plantation. They were paid 37½ to 50 cents a day (D. S. money). Laborers imported from Vera Cruz by the operators who planted the trees on a contract basis or by the plantation companies were required to work exclusively for the plantation and received no perquisites beyond a house. They were paid from 50 to 62½ cents a day, however.

Nor was costly labor the only Mexican drawback from a capitalistic exploitation standpoint. Grasping local and federal governments were relentless in demands for taxes, special charges and duties. Towards the last the death of the industry was hastened by the banditry that drove the workmen away or forced them into the army, that plundered and destroyed the plantations, and that sometimes murdered the Yankee owners and supervisors.

None too distinguishable from slavery is Mexican peonage at best, but atrocity charges involving Chiapas rubber exploiters came to light only once. That was in February, 1910, when the *American Magazine* broke out with an article, 'Barbarous Mexico: Rubber Slavery in the Mexican Tropics,' by one Herman Whitaker, a writer who had spent six months on a *Castilloa* plantation. It bristled with graphic pictures of 'prison huts' for the 'herding and guarding of men, women and children at night,' of peons so frenziedly toiling that 'their pantings could be heard at a hundred yards,' of an armed force of planters who fell upon an unarmed band of escaping workers and 'simply strewed the ground with lopped fingers, hands and arms that were raised to protect heads from the keen machetes.' In estimating these accusations it must be remembered that they came at a time when Congo atrocities were an increasingly lively topic and within five months after Hardenburg's publication of his Putumayo expose. And the kind of evidence proving Free State, French Congo and Aranaland horror stories to the very hilt is wholly lacking here.

A good guess is that the worst of the Mexican atrocities were performed upon the pocketbooks of Yankee speculators. For many of the English the same thing was in store, but the North American automobile saved and rewarded great numbers of them. Up to and into 1905, raising capital in London for such dubious business as rubber growing was an extremely difficult matter, and this it was that had held back the expansion of plantations in Malaya and Ceylon where the coaxers of *caoutchouc* from imported trees really had been learning their new business. In 1905, as we have seen, 174 tons of tame *Hevea* milk came to market. First noticeable shipment from the Middle East this was, but in itself it bore no seeds of any boom. Here was where Brazil extended the helping hand. Confident in the power of its own overwhelming domination, it put on the squeezer and ran rubber prices up to $1.25–$1.50 a pound. With that it signed its own death warrant. The spectacular price rise, the fabulous profits realised on those 174 tons out of Malaya and Ceylon were all that were needed to bring capital into the field. All along West coast Malaya the tapioca, pineapple and gambier plantings went down before the advancing line of *Hevea* seedlings. In Ceylon profitable and successful tea began to edge back towards the high lands in a retreat that would soon be a stampede as rubber took over the low country. In Java and in East coast Sumatra the phlegmatic Dutch roused from their tobacco dreams to squint across the straits and see what had bitten those British planters. Soon they were following suit in a leisurely, smaller scale way.

Thanks to Brazil, then, 100,000 Middle East acres were planted in rubber in 1905, just twice as many as for all the years preceding. In 1906 the new planting was 200,000 acres. In 1907 it was again 200,000, in 1908 again 200,000 - and this in the face of a United States panic rocking the whole world of commerce and cutting rubber prices by two-thirds. A total acreage of 750,000 was in rubber in 1908, then, but not yet had the real plantation boom hit England. Nor was Brazil yet paying much attention for, with none of the 1905 or later trees in bearing, the 1908 Middle East production was a mere 1,796 tons (1,402 from Malaya, 390 from Ceylon, 4 from British Borneo) - just 2.7% of the world total. Blithely and blindly, Brazilian government and speculators now proceeded to execute the wild *caoutchouc* 'industry' by running up 1909-1910 prices to an all time high of $3.06. Out of the East had come but 3,386 tons in 1909 and 7,269 in 1910; into the soil of the East, on the other hand, went nearly the equivalent of all the rubber trees then planted - 300,000 acres of them in 1909 and 400,000 in 1910 — principally in Malaya and Sumatra. It was 400,000 again in 1911 400,000 in 1912; in the three years 1913-1915 it was 250,000 annually.

Prior to 1909 it had been a case of the moneyed man figuring out a good one. A certain amount of popular speculation in rubber companies wild and plantation there had been, but the complete extent to which it had vanished during 1908's hard times was well shown in our account some chapters back of the fate that overtook Julio Arana's Peruvian Amazon issue. But the boom that hit about the middle of 1909 was a veritable speculative insanity that swept an entire nation, scooping them all in from the lowliest clerk to the eminent Charles M. Doughty of *Arabia Deserta* immortality. Anything connected with rubber sufficed, any *caoutchouc* flotation whether it involved Middle East, South America or Africa. The published figures of companies already operating were all that anyone saw; the warning voices of those who knew anything at all about rubber were lost whispers in a mob roar of greed that howled down the proposal that the London stock exchange be closed for a week or so while the frenzy subsided. This suggestion was one that came at the height of the madness in 1910. It well might have been made as early as July 1909. From July 1st to November 1st, 77 companies were organised with a total capital of $31,453,868. Between February 19th and March 20th, the shares of 10 plantation companies chosen at random rose in market value 37½ %.

A specific example should be of interest. During January the West Jequie Rubber Estates, Ltd., was floated to take over the Jequie Rubber Syndicate, Ltd., a *Ceara* rubber plantation in Southern Brazil. The day after the company was advertised 298 subscribers received an average of 2½ % of the shares they applied for and 561 subscriptions representing 734,511 shares were returned.

Here are some of the 1910 London dispatches (displaying an astonishing ignorance of rubber). From the New York *Sun*:

> 'The boom in rubber shares which has excited the Stock Exchange for some time has reached unexpected dimensions ... The demand is due to the public realising that rubber can be profitably cultivated in parts of the world where it is not yet grown, such as in Ceylon and the Malay states.
>
> 'Numerous companies have been formed and some have already declared big dividends. Each day sees riotous excitement in the section of the exchange devoted to trading in rubber shares. Brokers are making huge profits. The public is buying furiously. Many persons who bought shares some time ago and sold out at a handsome profit are now rebuying shares at double the former price.

> 'New companies continue to be floated. The subscription list of three such companies will open tomorrow and they doubtless will be closed before the advertised time, as has been the case with many others.
>
> 'The brokers and clerks are becoming physically worn out. They rarely leave the City, snatching short spells of sleep at hotels.'

From the New York *Tribune*:

> 'It is a maddening revel of speculation by the multitudes who are investing their small savings in rubber and oil shares. Rapid profits are made when two shilling shares rise to 50 or 70 shillings, and fresh investments are made in new issues.'

The suddenness with which the rubber industry ballooned up from obscurity into the City light reacted, of course, in far Malaya. There, as in London, fortunes were made and lost overnight as worthless shares circulated among all classes. Plantations that never did and never would exist were bought and sold. And the gay girls of Singapore's Malabar street gambled in rubber stocks as did their sisters under the skin and across the seas.

In London, Sir Frank Athelstane Swettenham, former governor of the Straits Settlements, had a word for it all. 'A great many companies have been floated in which I would have thought no one outside a lunatic asylum would have invested a shilling,' was his succinct comment.

Sir Frank knew what he was talking about. Very few of the other experts, whether pessimist or optimist, did. Few of them realised that the *Hevea* had conquered the plantation world, none of them that the plantation world had conquered the wild. Here, for instance, is the eminent rubber authority, William Wicherly, FRHS, writing in 1911 (*The Whole Art of Rubber Growing*): 'Regarding it from a commercial point of view, the *Hevea* does not occupy the pre-eminent position it held only a short 10 years ago. At that period its status from every standpoint was apparently unassailable. But in the interval a great deal has been learned about its great rival and relative the *Manihot Glaziovii*, and also concerning the *Castilloa* ...' Even then the Dutch were wholly abandoning *Ficus* and *Manihot* for *Hevea* while in Mexico and Central America the *Castilloa* plantings were being handed back to the jungle.

Wicherly also did an impressive amount of figure juggling to show that Middle East plantation production for 1912 could not possibly be more than 12,000 tons and that it was extremely doubtful whether it would pass 10,000. '... and, the optimists notwithstanding, it will certainly be many years before the Mid-East can hope to compete seriously in tonnage with the yield of wild *caoutchouc*.' In 1912 the Middle East produced 30,133 tons of rubber, in 1914 it tore around wild to represent 60.4%

of world production (plantation 73,153, wild 48,052). By 1916 and with those 1909-1910 plantings in bearing it more than doubled itself (plantation 158,993, wild 51,086). Three years later it had doubled itself again (plantation 348,574, wild 50,424).

Leopold Coburg's 11 juicy years of rubber profits had been 1898-1908. Taking up where he left off, fortunate plantation investors had the same number, 1909-1919. The break in rubber prices which carried them from $3.06 at the 1910 peak to an average $2.06 for the year, an average $1.41 for 1911, an average $1.21 for 1912, an average 82 cents for 1913, had, of course, wiped out the paper millions made by the waiters, bobbies and apprentice boys who had paid fantastic premiums for their shares. And thousands lost every shilling they had invested when the purely fraudulent or lunatic companies turned up their toes with or without ever having produced rubber. There remained, however, a number of soundly conceived companies and a far greater number of wildcat corporations which went on to pay such dividends as the average man collects in dreams (460 British plantation companies were operating as 1920 arrived). The thing that saved the wildcat outfits and converted them into smugly respectable and bloated affairs was the motorisation of America. The business of putting the United States on wheels carried annual automobile production from 63,500 in 1908 to 181,000 in 1910, to 356,000 in 1912, to 543,679 in 1914, to 1,525,578 in 1916. Rapid growth of the automobile industry had been foreseen, of course, but no such extravagant multiplication as this. And without such extravagant multiplication the world would have been so surfeited and glutted with rubber that plantation companies would have been exploding sky-high all over the Middle East. As it was, any wildcatter had only to produce rubber in any quantity to make more money than could be had by trimming the public on the market. America produced 85% of the world's motor cars and consumed 70% of its rubber; Britain supplied the rubber. This was the picture for many years.

Nevertheless, rubber production so closed in on demand that the average yearly price per pound slid off to 65 cents in 1914 and 1915, to 72 cents in 1916 and 1917, to slightly less than 49 cents in 1919. The American crackup of 1920 brought it to 35 cents and thereafter it was never to see 30 cents again save in the three artificial price establishing years 1925-1926-1927. Even in careless 1909-1919, production costs in the East were but 25 cents a pound, however, so it can be seen that huge profits were made during 1914-1919. Naturally, the almost unbelievably rich gravy went to the plantations first in the field and especially those that had been set out in 1905 or earlier. With the trees not producing until five years of age and not reaching free-flowing maturity until several years after that, the investors of 1910 had but short years to get in on the big payoff and those of 1913-1914 hardly got in at all. Of the amount sunk in British companies that immediately went under, there is no estimate reliable enough for citation. Those that survived had in 1920 an actual capital invested of

$505,000,000 (as against $130,000,000 for the Dutch companies, $241,000,000 for French, Belgian, Japanese, American and all others investing in Middle East rubber). And as widely as steel or railroad or industrials are distributed in this country are rubber shares held in Britain.

When the boom ended, it is interesting to note, those *caoutchouc* corporations that had been coining vast fortunes for a decade were in little better shape than those just ending their entirely profitless five to seven year period of waiting. Enormous gross profits had been blown in enormous dividends and shareholder bonuses. Retained in the business were little more than those amounts used in opening up new acreage, for obviously this Midas stuff would last forever and very few companies deigned to set up reserves.

Of the kind of dividends that fed the flames of the 1910 frenzy, it is sufficient to note that one company in that year paid a dividend of 375% on capital invested. In the nine years, 1911-1919, Pataling Estates earnings amounted to 2,025% of the capital invested; Batsu Caves Co. 1,515%; Selangor Estates 1,427%. Eight leading companies exclusive of these earned an average of 811% of the capital invested. Vallambrosa Rubber Co. paid dividends in 15 of the years 1906-1923 and they amounted to 1,386½%. In addition, its area under rubber was increased from 930 to 3,292 acres — all out of profits. In 1910, earings of 27 companies chosen at random from among thise then operating amounted to 53% of capital issued, dividends to 46%. For 1915 earings of 49 assorted companies were 32%, dividends 27%. For 1919 earnings of 51 miscellaneous outfits were 30%, dividends 26%. And this despite the fact that much of the area owned by the groups representing each of these years was immature or not yet in the full-bearing stage. Instructive it is, too, to take a group of 51 and find 21 of them paid dividends out of profits in 1920 and 22 in 1921 and 46 in 1922 — these being the panic years that had Malaya yowling for the government aid which was supplied by John Bull near the end of 1922 at the expense of Uncle Sam. Also, the dividends of those so paying were large enough to equal 3% of the issued capital of all 51 in 1920-1921 and 6% in 1922.

This was achieved by increasing yields per acre while simultaneously cutting upkeep, administrative and labor costs all along the line. 'All-in-all' cost of producing rubber was knocked down to nine to twelve cents a pound by these means and today ranges from 3½ to 8 cents, with five cents about the average. Today, then, the plantations after profiteering in 1925-1926-1927 and selling below cost in the early years of the present decade are pulling in a sweet profit from sixteen cent rubber they once would have considered bankruptcy bait.

In this ability to cut costs as much as necessary, the East always has had it on the Amazon as it has had it also in the matter of cleaner product, transportation modern and standardised (if for longer distances) and freedom from grafting governmental assessments running into ridiculous figures. Temperamental inability of the South Americans to get down to earth was perfectly demonstrated by the ruin that overtook them as plantation production surged past wild in 1914 and drove prices down to 65 cents. Never, as we have said, was there a dealer or producer able to calculate production costs in Amazonia. For years, however, it had been an accepted truism that if rubber sold below 70 cents a pound in New York, there would be no inducement for tapping the wild *Heveas*. As things worked out, that guess was shown to be very nearly correct.

It was in late 1913, with the East fast overhauling Brazil, that the day of reckoning for those decades of inflated credit extension arrived and the result was a tumbling and crashing of the existing system that spread black ruin through the whole North of that prodigal republic. And this within three years of the day of $3.06 rubber, within two years of rubber averaging over $1.40, the very year after rubber averaged over $1.20.

As Para city was the first to feel the boom, so was it the first to be overwhelmed by the panic. Supply and exporting houses began demanding real money instead of rubber in payment from *aviadors* and refused to ship more goods. *Aviadors* in turn called for cash from the *patraos* up river. In the final washout the wholesale house assets proved to be enormous debit balances with the *aviadors*, the *aviador* assets proved to be huge debit balances with the *patraos* — all paper profits containing not a *milreis* of actual value. Old established houses commenced to fall and as one after another slipped into liquidation the mortality list grew to scores in Para alone. Over $20,000,000 was involved in this city's failures. Pereira, Bessa & Co. for $4,700,000, B. Antunes & Co. for $3,500,000, Mello & Co. for $3,000,000: so it went.

Commercial Manaos toppled in the wake of Para and starvation clamped down on the jungles up the tributaries as the supplies ceased coming in. By raft and by canoe the liberated Cearenses came floating down the Purus and Jurua to Manaos in an exodus from the Acre. Swarming out by way of the Javary, the deserting *seringueros* robbed, plundered and destroyed in an appropriate farewell to the hell they had so long populated. Elsewhere the tappers remained in bondage but, with downward dropping rubber prices wiping out their last faint hope of ever working off their indebtedness, a mingled lassitude and defiance fell upon them. Fishing and hunting for their food, they made no rounds of the *estradas* save when the overseers were upon them. It became necessary to have one of these drivers daily in each center, for in the bosoms of the *patraos* still welled the wishful belief that this was a passing depression and that prosperity was just around the corner. By now the most of the estates were in the hands of the remaining or reconstructed commercial houses, but the merchants had no desire to administer rubber properties in person and so the former owners leased or rented them, paying a percentage of the *caoutchouc* or a specified weight. And while rubber prices in the foreign market came to represent too narrow a margin of profit to warrant production, they still hung hopefully on and still turned out this that no longer was 'black gold.' With credit relations reestablished to shrunken but still excessive extent, Para and Manaos were making a good paper profit on merchandise sold to the estate lessees and these gentlemen were pocketing a similar imaginary take on the re-sale to their *seringuero* serfs. Over whole vast areas the rubber industry was sustained by this kind of play-money transactions for many years and, locked in the jungle, slaves toiled away in a bondage that returned a profit to none and was a drag upon all Northern Brazil rather than the conjectural benefit it once had been. Conjectural, we say, for by striking a balance on the history of booming 'black gold' in the Amazon, one finds Brazil's loss through the degeneration of the transplanted Cearense stock far outweighing its gain by way of marble monuments, empty opera houses and vanished boodle.

For long, then, Amazonia's production at or near a loss held close to the old boom figures, falling only from 1912's record of over 44,000 tons to less than 42,000 in 1913, to less than 38,000 in 1914. There was a production rally in 1915, a slump in 1916, a War-dictated rally to over 40,000 in 1917, a fall to 28,000 in 1918, a post-war rally to 40,000 in 1919 and then the long and virtually continuous drop. At the 72 to 48 cent range of before 1920, the 'industry' had just been able to hang on. At the 68 to 13 cents of 1920-1924 it died; it returned to brief life in 1925 as Great Britain's price-raising scheme brought dollar rubber, and relapsed permanently into the grave as effects of that stimulant speedily wore off. An average of well over 20,000 tons a year it got out in the 1920s and in 1925 passed the 30,000 mark, but with the 18 to 2½ cent price of the 1930s it has slipped 60 years back and is where it stood in 1870 (6,000 tons in 1932 and 9,000 in 1934). In the readily accessible (as the Amazon goes) wild trees still remains latex enough for annual yields of 50,000 tons and much higher, but it isn't worth getting out.

The steps taken by official Brazil to prevent this debacle were as funny as *Charley's Aunt*. When it first began to realise that the East really was more or less in earnest about this *Hevea* planting business, it whanged down with a rigid embargo on exportation of rubber seeds (belated) and *urucuri* nuts or specimens (timely) and settled back to chuckle over the idea of anyone's trying to produce first quality para without benefit of the magic smoke of the fetish. Among the *patraos*, as time went on, the cheering and universally held belief was that all *Estrada* trees would die, or living would give no latex, or, yielding latex, would not surrender it in marketable quantity or quality. American manufacturers supported them in this idea by shying away from the suspect plantation stuff and rating it at a figure markedly below that paid for 'fine hard para.' Whereupon the Brazilians requited these boosters by running up prices until the factory men literally were chased into the open arms of the Orient's planters.

About 1911, then, officialdom began to think it was time for something to be done. Up the rubber land tributaries, however, many a *patrao* still was serenely waiting for word of the certain *Estrada* crop failure, glad tidings for which some of them are waiting to this day. The governmental action took the form of a grandiose *Defesa da Borracha* (Defense of Rubber) scheme announced by decree of January 5, 1912, and brought into being with complete regulations on April 17th. This was, indeed, a honey. It not only hung up premiums for planting rubber trees and opening rubber factories, but provided for running railroads all over the Amazon Valley, opening all the tributaries to year-round navigation, establishing a chain of experiment stations for the study of *Hevea* and *Ceara* trees, reducing Acre territory export fees and getting together with the state governments with a view to lowering their larceny rates also. Nor was this all. Immigrant hotels were to be erected; cattle ranches, packing houses, dairies, farms and fish salting enterprises established; and, crowning touch, a rubber industry exposition would be held every three years at Rio.

Needless to say, very few of these elegant projects ever were put into execution. Immediately upon appropriation of $2,400,000 a commission was established to spend it. This Brain Trust thereupon set up a central bureau at Rio — as far as possible from the rubber lands — and equipped it with a yearly payroll of $92,400. Monthly salaries of the bureau's rubber defenders ranged from $1,600 for a superintendent to $300 for a paymaster. Seven branches and experimental stations, each with an annual payroll of $18,600, were then opened. Next came a series of contracts with private wizards, including $180,000 a year for a secret coagulation process and $210,000 for a plan to introduce sanitation into the Amazon. The next year the whole thing was abandoned with a mere $1,768,400 having been tossed away in salaries and contracts.

And so it is that in Brazil, home of the *Hevea*, the planted *Hevea* never has had a fair trial. This is being remedied now, not by Brazil, but by a Mr. Ford of Detroit. Aside from Ford's, no Amazonian 'cultivation' venture of before, during or after the Middle East rise to dominance deserves the name 'plantation.' Intermittently, Brazilians and Peruvians would decide to have a try at this game and the result was plantings ranging from a few trees up to as many as 20,000. Without exception, however, the trees were stuck into the forest to be overwhelmed in the Amazon's vegetable warfare or rammed into unsuitable soil or lined up in picket-fence style and never thinned out. In growth and in rubber yield the best were far behind their expatriate brothers of half a world away and for years few have even been tapped.

On the Amazon has arrived the ruin and decay which it asked for and which it deserved. The rubber-tapping which goes on is not in the richest spots but the more accessible ones. Up many of the rivers there is no activity other than Brazil nut gathering. On the verandas of tumble-down estate buildings, listless one-time *patraos* stir from their hammocks only to inquire hopefully for the rubber quotations when the ever rarer steamers come in. High on the remote streams, kings black and kings more or less white still hold to the law that none may travel on 'their' rivers without their permission. But since none now has any desire so to voyage, the enforcement problem isn't what it used to be. It is the gatherers of *balata* alone who preserve the true Amazon tradition in the lost places. This gum, an inferior *gutta-percha* chiefly used in machine belting, never has been sufficiently in demand to warrant scientific cultivation. And so it is that it is hunted today in the forests of Colombia, Peru, Bolivia and up-river Brazil by half-breed balateros corresponding to the *Castilloa*-trailing *caucheros* of yesterday. There the story of lonely toil and privation, of raids on Indian women and exploitation of Indian bucks still is being written as in the old days.

Ghost towns shrouded in a graveyard atmosphere are the rubber-made villages on the main stream. Upriver spots that still draw large type on the maps have moldered away to collections of a few houses and these few collapsing into rickety abandonment due any day to inspire some Rio wit to suggest that the time has come for giving the Amazon back to the Indians or the Jesuits.

Gone, too, is the roaring life of the boom cities. No ocean steamers call at Iquitos, now shrunken to a population of 9,000 or less. Its port works are in collapse, its warehouses are empty, its $4,000,000 imports of 1910 had dwindled to $94,000 by the 1921 of port closing; what it has for export it sends by river steamer for transshipment to sea-going craft at Manaos or Para. Its celebrated three-story hotel and the other properties of conjectural value are included in the holdings of a non-Aryan who descended on the metropolis as a $20 a week clerk and became a power in the rubber days. Cows graze on grass-grown boulevards in fulfillment of a familiar North American political prophecy and Cesar Arana's freight railway (valued so highly in that London prospectus) now serves principally as a vehicle for stately joy rides by Peruvian aristocracy on Sunday outing bent. Such is Iquitos.

In Manaos linger more remnants of vanished grandeur — empty opera house, groups of unfinished buildings begun at great cost and left in a state of undress as deflation hit. Out of Manaos go such comedy cargoes as hides, horn, dried fish and pissava fiber and there isn't a diamond ring in a carload of its reds, blacks and mongrels. The commuters are gone from its suburbs, the animals from its zoo and the life from its cobblestone thoroughfares. Such is Manaos.

Para, still claiming some 230,000 population, has an empty opera house, a pile of pink and blue houses stocked with mosquitoes and women, a shoal of backing and filling open street cars. It has also 300 automobiles (85% Ford), six motion picture theaters and a dream of rubber — principally to be supplied by Henry Ford. All the money it has seen in recent years is what Ford has spent; its glittering night life has degenerated into 148 licensed gambling houses which hardly make the light bills. The six months annual water scarcity fills in the gaps in conversations about rubber and for excitement there is an occasional visitation of yellow fever. Such is Para.

Larry, turn the crank.

The hand-colored slide now depicts that two-by-four island (actually 27 by 14 miles) hugging the South coast of the Malay Peninsula which a decade or so ago flung out a protecting arm in the form of a causeway road and railroad across the Strait of Johore. The railroad runs on to Bangkok, Siam. The causeway blocks the strait, protected farther East by the newest of Britain's major air and naval bases, hacked out of the island jungle in the last half dozen years and complete now with giant guns, munitions depots and an 885 foot floating dry dock equipped to receive battleships and battleships alone. This is Singapore Island, outpost of commercial and imperialistic Britain, 8,300 miles from London, outpost of commercial China, 1,440 miles from Hong Kong. The Western island is coconut groves, the central island is rubber plantations and the Southernmost tip is Singapore city, rubber—and tin-exporting capital of the globe.

'Crossroads of the world' it is called, most important halt on the trade route to the Far East, meeting place for ships bound for all corners of the earth. Little more than a century ago, city site and island were a crocodile wallow, a mean jungle, a huddle of Malay fisherfolk infesting its fringes. Para by then was already such rubber capital as the times needed.

Founder of the city was Sir Thomas Stamford Raffles who in 1819 obtained a trading factory concession from the Sultan of mainland Johore in behalf of Britain's East India Company, then established on Sumatra, where Raffles ruled in its name as lieutenant-governor. The Raffles aim was to foil ambitious Dutch attempts to obtain complete trade control over the archipelago and it wholly succeeded on that day in 1824 when Britain bought the island from the Sultan for $77,500, a transaction almost comparable to the white acquisition of Manhattan. No city could have had a finer father. Where Para was founded by slavers, Singapore was founded by a man censured by the English courts in 1824 for his 'precipitate and unauthorised emancipation of the Company's slaves.' It was the career of Raffles that Sir James Brooke, who would be the first of the line of white rajahs of rubberised Sarawak, was emulating when he sailed in 1838 to do a little battling with the pirates of Borneo on his personal account.

A half million population Singapore has today, and through the city passes three-fourths of the rubber grown in British Malaya; rubber from Burma, Siam, Indo-China, Dutch East Indies, British North Borneo and Sarawak for sale or auction or transshipment to New York, London or Amsterdam; more rubber than all South America ever has seen. This is, perhaps, the world's most cosmopolitan city, governed by the British, overrun by sleek and sassy Chinese, but including also a medley of

Dyaks, Javanese, Tamils, Sikhs, Siamese, Persians, Armenians, Jews, Annamese, Japs, all the Orient, all the breeds and races and divisions of Malaya and India known to ethnologists, geographers and historians.

Lost in the shuffle, the sullen Malays stick to their huts among the coconut groves or loungingly infest the same sort of shacks in a city featuring the pagoda-like residences of the wealthy merchants from Cathay, the handsome teak-wood (a dirty trick on the predatory white ants) homes of the European and American representatives of oil and rubber interests, the business buildings owned by Chinese, Oriental Jews and the whites out of another world. Some 70,000 the Malays of all varieties number, some 56,000 the Indians and oddments of all kinds, some 7,600 the Europeans. And over 380,000 the Chinese.

Shipping point for the world's rubber and tin, it is also the distributing center for all the goods coming into Malaya, a city with exports of over £60,000,000 and imports of over £76,000,000 in 1929, with total exports-imports of £74,000,000 in the 1931 of microscopic tin and rubber prices. Singapore River is as jammed with sampans as a Detroit parking lot with Fords; Singapore Harbor with liners, coasters, junks, fishing craft of every cockeyed Oriental size and shape and state of seaworthiness. Six to seven thousand commercial vessels is the number clearing annually as against the 1,600 of Para at its most riotous.

As a haunt of glamorous vice, pre-war Singapore had a reputation overshadowing boom Para's, but white officialdom and the more than Rotarian spirit of the successful yellow man have transformed it into a city with no more than licensed opium dens and mean street wenches to match against impoverished Belem's licensed gaming dens and alluring women. These things are Singapore today: modernised mercantile establishments; latex tanks, rubber godowns (warehouses) and rubber milling plants; an ocean swimming pool fenced off with concrete shark protection; Chinese amusement parks, Chinese golf-fiends and Chinese Boy Scouts; pneumatic-tired taxis, pneumatic-tired rickshaws and pneumatic-tired bullock carts; Malays and Chinese airing their caged, imported singing birds on foot or in rickshaw. Occasionally, of course, it offers the spectacle of a kris-carrying Malay running amok and clearing the streets as effectively as a kill-crazy gangster shooting up the Loop in Chicago. Late in 1935, for instance, there was a minor epidemic in which nine persons lost their lives. And, interestingly, recent years have offered the spectacle of Chinese going wild in this hitherto exclusively Malay manner. In each case the yellow stampeders came from families more than a generation in the tropics, a suggestive fact which hints that whites, too, may be joining in the game one of these decades.

Humid in the Amazon manner, this is no city for white residence and so it is that the mandarins, the shoguns, the barons, the big shots of commercial, industrial and mercantile life are the striving Babbitts out of Hongkong, Shanghai and beyond.

Foreign money is represented only by hired hands; Chinese capital is right in Singapore, and Chinese manufacturers, contractors, bankers and rubber plantation owners. Here are second and some first generation members of yellow laboring families that climbed into the millions even as the Portuguese shopkeepers and the barefoot Spanish-American peddlers made the same ascent in Brazil, Peru and Bolivia.

Most interesting of the Chinese industrialists from a *caoutchouc* standpoint is Tan Kah Kee, son of a Singapore merchant prince, who years ago got into the manufacturing business and expanded his enterprise into a rubber shoe factory sizable enough to consume over 2,000 tons of crude a year. A price-cutting king worthy of mention in the same breath with the barons of Akron, he sewed up the archipelago business and reached out to compete in English and Canadian markets. He was into the banks for over £1,000,000 when they banged down on him two years ago. The factory is still closed today, the several thousand relatives of the headman have scattered to other fields and the aged Tan Kah Kee lives in past glories.

A going Chinese concern also of interest is the Lee Rubber Co., through which small rubber buyers operate. According to Singapore gossip this is financed by Mitsui Bussan Kaisha, Ltd., Japanese trading company backing numerous Chinese who buy from small estates throughout Malaya, Borneo, Sumatra and Java. Mitsui, with six vessels of its own calling at Singapore, also has control of the Kokusai Kisen Kaisha, which has seven. In these ships, all of which travel through the Panama Canal, half of the rubber leaving Singapore for the United States has been carried in 1936.

Common are the Chinese rubber plantation company chairmen; the Chinese 1% commission on the turnover; the Chinese owners of rubber mills which exist to clean, dry and otherwise prepare for export the wet and dirty native rubber which comes piling in to Singapore by bullock cart and stinking native boat. Importance of this branch of the business can be seen by the fact that 160,000 tons were turned out in 1935 by the rubber mills (including one on the Kallang River owned by an American tire corporation which houses its representatives in a Firestone Park section named after its worker allotment on the edge of Akron, O). Of all the rubber that went out of British Malaya in 1934 a total of 32½% was exported by mainland shipping agents and export dealers, 23% by Singapore and Penang export dealers, 22¼% by local representatives of American and European rubber manufacturers and 22¼% by the Singapore rubber mills.

This rotten smell of processing rubber it is that dominates the variegated odors of the new *caoutchouc* metropolis as the smoked—ham atmosphere of the Para wharves hung, not unpleasantly, over that onetime rubber capital. The source of the stuff that stinks in the Singapore mills and adds its quota to the musty smell of the Singapore docks we shall now consider - and specifically the contented coolies who supply this milk.

3.3 Brown Man, Yellow Man

Hevea brasiliensis at home we have described. *Hevea brasiliensis* as it principally appears overseas presents a far different picture. It is again a sea of trees stretching away in all directions as far as the eye can follow. These are all of one species, however, and there is more than the uniform appearance of the trees to betray the guiding hand of man — man the destroyer and rebuilder hacking holes in such Malayan and Sumatran jungle as would not be out of place on the Amazon and in these clearings bringing into being his own artificial forests. Evenly spaced with the distance between tree and tree in a row the same as between row and row, a geometric design repeats its pattern over and over with the effect of an immense formal garden running away to the horizon. Planted by rule 200 to an acre, nursed along, thinned out to save only the 80 heaviest rubber producers per acre, individually numbered and recorded in a Who's Who religiously kept up to date, these pampered members of the vegetable world are sturdier, more productive, of far more utility than any of the rugged individualists of their own age scattered about Amazonia.

This is the development that the world owes to the British scientists and planters who worked out the problem of making the imported *Hevea* flourish and yield milk while the Dutch, supposed masters of all things having to do with tropical planting, failed with *brasiliensis* and placidly and pointlessly experimented with *Ficus* and *Ceara* in the intervals of bringing in the tobacco. Some one to two hundred pounds an acre where soil is exhausted or conditions poor, 600-700 pounds on conservative tapping methods in the best of the inland rubber areas, an average of 400 pounds for mature trees in areas considered good: these are the sort of annual yields the English achieved and are achieving.

As to what the economic life of those trees will be no man knows, for the industry still is far too young to have found out. Planted under proper conditions and not tapped to excess, the tree will go on renewing its latex-secreting bark indefinitely. And it gives larger and larger yields until about its thirteenth year. That is an approximate maximum, but when the decline sets in has yet to be established. Where the yields of some of Malaya's 36 year old trees have declined, indications are that poor treatment rather than age is responsible. We have mentioned that a star performer among the original Wickham trees in the Ceylon botanical gardens gave 68 pounds of rubber when tapped during its thirty-sixth year. That was in 1913. Of more practical interest is a 1935 report from one of Malaya's oldest estates — one in existence since 1899. In its 36 year old area the 1934 yield was 719 pounds an acre from half the acreage, this estate being modern enough to have adopted the post-boom usage of tapping and resting the trees in alternate years.

So much for the achievements of the first generations overseas. As to the younger generation it has surpassed and will continue surpassing anything ever dreamed of in connection with its forebears. This improvement the world owes to the Dutch as it owes original development to the English. The years 1877 to 1910 the Dutch required for pulling out of their lethargy but once in the *Hevea* business in earnest their scientists were but six years in developing and but two more in beginning to put into effect the greatest discovery ever made in connection with rubber planting. This is bud-grafting, first successfully accomplished by Bodde in 1916 and systematically tested by Cramer and van Helten, who published their first records in 1918. It is based on the theory that the high-yielding properties of a known efficient tree can be impregnated in another by grafting, and large scale commercial demonstrations of the proposition have been convincing, to say the least. By this grafting of pedigreed dormant buds on unproved young trees it has been shown that there an average yield of 650 pounds an acre is obtainable in the second year of tapping. This steadily increases to 1,200 pounds in the sixth year and runs to 1,300 pounds or more thereafter. Since the operation is not practical in the case of trees much more than a year old, it is only when new plantings or replantings of unsatisfactory areas are being made that it is called into use. Even so, however, it already has become an extremely important factor in rubber production. Objectors to it on the ground of cost (as much as $60 an acre for the dormant buds alone) argue for selecting seeds of high-yielding trees and using them instead of ordinary trees. Not yet extensively tested, this practice remains a theory, but it would seem that combining both systems — using selected seeds only and then bur grafting the seedlings — would doubly assure rubber yields of three or more times those the planting world had grown accustomed to before the Dutch worked their magic.

Turning now to the physical collectors of these yields, it is well to note, perhaps, that it was not altogether in irony that we spoke of milk from contented coolies. Among the hired hands on the estates, today's underpaid but more or less decently treated brown men are fairly well satisfied with things in their passive, undemonstrative, unexpectant way. In the breasts of the money-minded yellow men still wells the divine discontent, of course, but many of them are, in any event, better off than their fellows at home in China. Also to be remembered is the fact that the native rubber which gives the large plantations such stiff competition in the world market is produced by thousands upon thousands of Orientals when and as they feel like it, and in periods of high prices Straits dollars, rupees and guilders come to them in quantities they never had thought possible before. Naturally, their holdings of a few acres each have little in common with the large plantations in appearance. It might almost be said that they represent a compromise between the glossy estates and the wild *estradas* of the Amazon. Here the *Heveas* are irregularly thrown in 200 or so to an acre, and allowed to stand. Here, in contrast to the estates which once cleared the ground of all weeds and still cut out all harmful ones and periodically cut back the others, hardly any care is given. To such extent are the weeds permitted to grow that often the tappers have difficulty in getting from tree to tree. Yet due to the thick stand of trees and the more intensive tapping, native owners, interested not in the future but in today, get out an average of 500 pounds an acre as against the 375 of all estates. All that they invest is the labor of selves and relatives, and there is nothing to prevent them from raising their own foodstuffs, leaving profit from rubber as so much gravy.

The time element, of course, played an important part in keeping *caoutchouc*'s record in the island and peninsula world as clean as it has been. The Middle East has its familiar history of European conquest; but the rubber industry, not really arriving until the twentieth century, came too late to share in any responsibility for this bloody work. Chiefly the history of this world is of a Malay influx which overwhelmed or drove into the remotest parts the aboriginal pygmy Negritos; of sea-roving Malays, gunpowder-and cannon-equipped, engaged in ceaseless warfare as rajah battled rajah and all of the despots heavily oppressed their own people; of slavery, cannibalism and head-hunting among those Malays in the bamboo-spear and blowgun stage of barbarism. Came then the Portuguese with their oppressive cruelties; the heavy-handed Dutch with their inhuman labor taxes and their inhuman pertinacity in collecting them; the British with their slave-holding and exploiting and punitive expeditions; and the French, who were no better than the others. Dutch drove out Portuguese; subdued or failed to subdue natives who rose again and again against them; executed and even tortured English as well as killed them in battle. Here the English exploited the natives; there they ganged up with them to wipe out the Dutch; all over peninsula and archipelago they intrigued and warred and filled their pockets. Meanwhile, from every island save Java, the Malay pirates harried the commerce of these Christian intruders.

Generally speaking these things were long over, however, by the time of *Hevea*'s arrival. It came first, as we have seen, to the British colony of Ceylon, which until 1916 was the second in importance of the new rubber lands and since then has been but a poor third but incomparably far ahead of those next in order. Not Malays but a Dravidian type out of India had been the dominant race on this island, and rubber's arrival found their last revolt against the British long in the past, their earlier experiences with Portuguese atrocities and Dutch extortions happily forgotten. Until just recently these Sinhalese would not themselves toil on European estates, and so it was that black Tamil coolies from the arid lands of South East India had to be brought in to work coffee. And naturally they provided the *cinchona*, tea and rubber labor as the successive switches to these crops were made. Chiefly in the South-West of the island are the 606,000 acres of *Hevea* trees, and they are far from being a European monopoly. Sinhalese planters took readily to rubber culture after having scorned *cinchona* and tea; such wealth did they acquire thereby that social upheaval, political reforms, and assorted caste and religious riots developed from the situation. In 1915 new and jittery British colonial officials used blood and iron methods to suppress blazing brawls between Buddhists and Mohammedans and a legacy of bitterness has survived. As a rubber country Ceylon must now stand still, for almost no land suitable for *Hevea* growth remains. Nor are its average yields up to those of Sumatra and Malaya, this chiefly being due to the fact that so great a part of its plantation soil has suffered from previous cultivation. In Malaya the islands and islets and the scattered bits of mainland comprising Britain's colony of Straits Settlements amount to but 1,246 square miles. Its protectorate over Federated and Unfederated States has meant a very real improvement in the condition of the natives, however. They remain under their own rajahs and chiefs, but British officials ride herd on these headmen to maintain a peace, security and freedom from revenue-devouring despots previously unknown by the peasantry. Splendidly proud and splendidly indolent are the Malays of the peninsula and the finest thing that can be said for the English administration is that it has permitted them to lead their own lives. On the plantations owned by Europeans, Chinese and Japanese, the toilers have been Tamils, Chinese and imported Malays from Java (1935: Indians 185,479; Chinese 72,690; Javanese 24,640; all other 9,605). Nor has it been a case of the choice lands being stolen from the peninsula Malays as they went about their business of loafing, fishing and cultivating their own small plots of foodstuffs. Rubber first took over developed estates previously devoted to other crops, but the later and by far the greater part of the *Hevea* milk farms have been carved out of rank jungle of no more use to natives than to whites. And for the Malay too proud to work for the whites, but not too torpid to ignore the chance to pick 'white money' on his own, there is opportunity to acquire the kind of small holding he wants either inside or outside of the Malay reservations. Of the 1,261,074 planted acres in small rubber holdings, 693,591 (or 55%) are held by Malays in average patches of 4.2 acres, 378,322 by Chinese in average chunks of 18 acres and 189,161 by other Asiatics in average areas of 17 acres. Not too unfavorably does this acreage compare with the 1,890,888 planted acres in estates of which 73.7% is held by European corporations (and one American), 19% by unincorporated Chinese 3.3% by Indians, 3% by Japanese companies and individuals, 66.7% by Malays and 33.3% by others. Of the total planted area, then, 55% is held by assorted Asiatics, 45% by whites.

Native rulers as well as progressive peasants and pushing immigrants are in the rubber business, and so as early as 1908 a collection of success stories about Malays who had made good took great pains to point out that Ibrahim, genial and enterprising Sultan of Johore, is of their number. 'He takes a close personal interest in the administration of his country, but even the active supervision of the various State departments does not absorb the whole of his energy, for he finds time to superintend the management of several rubber estates of which he is the owner.'

All along the West coast are the heavily important rubber plantation lands and in none of them have there been murders of British agents, reprisal expeditions or outstanding feats of piracy since the 1870s. Additional *Hevea* territory includes parts of the interior along the railroad lines and the North-East coast. Here the last important trouble was in 1895, when there were outbreaks in interior Pahang which required an irregular expedition into the East coast refuges of the rebels led by Sir Hugh Clifford, famed author and friend of Joseph Conrad. Not by any means is all of the suitable rubber land in Malaya now in bearing or planted or even specifically alienated to *caoutchouc* (as must be done before it can be bought). Future planting, if any, will be done in the unplanted land already reserved by estates or small holdings, it seems certain. Without any planting, however, mere coming into bearing of present young trees, of which 170,000 acres are bud-grafted as 30,000 acres of bud-grafted trees now in bearing and return to bearing of areas now in resting, promise that normal production (which is far under potential) will be some 570,000 tons by 1940 (in 1935 it was 484,650).

As against Malaya 6% of planted area in bud-grafted trees, the Dutch East Indies and French Indo-China have 10% and 28% respectively, another factor which shows how hard it is going to be to hold swelling rubber production in. Should there ever again be a state of rubber demand outrunning production (which seems improbable), the Dutch have, too, more qualified *caoutchouc* lands not yet devoted to that purpose than any other nation. It is Sumatra, its leading latex island, that supplies it with this insurance, for many parts of Borneo are not well adapted to the *Hevea*, and Java, like Ceylon, has planted nearly all of the suitable territory. Total *caoutchouc*-planted acreage of all the Dutch Indies is 3,500,000.

It was in Java that the worst story of commercial exploitation in the whole Middle East was written, and this not so long before rubber arrived. Of all its holdings in the East Indies, Holland really 'developed' only this one island before the 1880s, by forced labor, extortion and terrorism comparable to the later Leopoldian exploits in Africa. The cleanup began in 1864, however, the worst features of the system were eliminated by 1872 and the final abolition of forced labor was achieved in 1885. This sordid tale is no part of *caoutchouc* history but rates passing mention here because in reaction to these crimes the Dutch became some of the better colonial administrators at large in an imperfect world. Today's *Hevea* tappers have benefited by all the suffering on the plantations of pre-rubber days, then, and it is to this that they owe especially their status as independent planters aided by a Government going far beyond anything ever done in Malaya to encourage such native independence. To achieve this status, however, they are forced to leave overcrowded Java and to voyage to underpopulated, under-developed Sumatra or Borneo.

Even more than the peninsula Malays the Sumatrans are averse to working for the whites. Rare fighting men they are, however, and in this year of 1936 much of the interior of the land still is unsubdued. In both seventeenth and nineteenth centuries the Dutch actually were driven off the island, and not until 1851 did they obtain a really secure foothold. In Achin, Northern end of the island, they established supremacy on the coast only after a 1873-1903 war estimated to have cost a quarter million lives and a quarter million dollars. In the West central part of Sumatra are the Menangkabau Malays who claim descent from Alexander the Great and who used to boast of a powerful kingdom which represented Malay civilisation at its peak. Not among such gentlemen as these are rubber coolies to be found. Java, with a population of 35,000,000 to Sumatra's 5,800,000, has more than enough to spare, however. As to the military work in Sumatra, it has not been pursued on behalf of the rubber interests. It is on the coasts subdued in precaoutchouc days that the best of the *Hevea* lands are to be found and along the North-East (but South of Achin) coast that most of the plantations exist.

In Java the plantations are distributed through all sections of the island. And of all the new rubber lands it is only here that indigenous labor is used by the estates. Tamils and Chinese in Malaya, Javanese in Sumatra, Tamils in Ceylon, Chinese and Javanese in British and Dutch Borneo, so runs the setup. Cochin-China's rubber workers are the same sort of Annamese as inhabit that colony, but principally are drawn from the protectorates of Tongking and Annam. The semi-wild Mois of the interior mountains (who seem to have been the indigenous people of all Indo-China) and some Javanese also are used. As to Siam, it relies not on the Siamese but on the assorted laborers of Malaya. And South-West India draws on South-East India as does Burma.

With one exception, therefore, the coming of *caoutchouc* has not meant toil on the plantations for inhabitants of the new latex lands. Nor were they otherwise affected by the innovation in the earliest years. By 1912, however, the Asiatic holdings (Chinese, Japanese and the Malay potentates' plantations lumped with the small acreages) had reached a fourth of the Middle East's planted whole. By 1925 this was a third. By 1930 brown and yellow man production had bounced to 54% of the total and since then (barring the later five-nation rubber restriction act to be discussed later) it has varied with rubber prices. Unlike the estates which have to keep shoving out their product regularly, the majority of natives just don't tap when there is no price incentive. Pretty much peculiar to Sumatra, Malaya and Borneo are the small holdings, incidentally. Here again it was a case of the British originating and the Dutch later surpassing them. What attracted a great many of the Malays was the chance to obtain rubber lands in the Malay reservations for $6 to $30 an acre as against the $120 to $180 an acre for small holdings outside the reservations. It was not until about 1916-1918 (native holdings in all the Middle East had reached a million acres in this last year) that the Dutch began to take a hand, but when they did, it was with free land, free seedlings and expert advice. In Java by then the *Hevea* areas were all in the hands of the Dutch, British and American corporations. If the Javanese couldn't go into business at home, Sumatra was at least the next best thing. And such Sumatrans as cared to get into the game were welcome. Seeing a rare chance to take rubber markets away from the British in the 1920s, the Dutch fostered this development intensively and so it was that the Asiatics drove around the whites. With world production outrunning demand, the inevitable result has been a new discrimination against the brown and yellow small holders by British and Dutch governments trying to take care of the alarmed owners of European estates. So strongly entrenched is small-fry production, however, that it is going to remain an important part of the Eastern setup no matter how much the Governments repent of the good they have done.

Lumping rajah, sultan, Chinese and Japanese estates with the European and American ones, plantation production so far has always been in front of the independent efforts, however, and it is on these holdings that the greater part of the Middle East rubber labor story has been written. It was a sordid story of the greed and callousness of men white, yellow, brown and black in the earlier years, but improvement has been consistent and today it can fairly be said that for the last 15 years or more the underdog of the Middle East has been substantially better off for rubber having come to these lands.

This was not always true. With all labor, save Javanese working in Java, once recruited under the one- to three-year indenture system, with all rubber land statute books then studded with penal provisions for such offenses as contract breaking or absconding while in debt to the estate, with the filthy professional recruiters working the famine lands of China and India and the backwoods of Java to prey upon misery and credulity, with whites knocking their serfs and debt slaves about at will, with the opening up of the forest lands loosing wholesale malaria among imported laborers — with all this on the record, plantation history in its first decades is not too distinguishable from what went on in the less obscure and less dreadful parts of the Amazon. Frequently it is, of course, a case of balancing starvation at home against suffering abroad and it also must be considered that the coolie indenture system was in flourishing existence long before rubber and continued to supply tin, tea, tobacco, coffee and other slaves as well as the *caoutchouc* victims. That the rubber boom brought into this slavery thousands who otherwise would have escaped it is not to be questioned, however, for there would have been no such unparalleled (and deadly) leveling of jungle, no such gargantuan planting, no such extraordinary demand for labor in connection with any other crops that might have been tried.

It is from the Madras Presidency of South-Eastern India that the black Tamils were and are drawn. To this poverty stricken breed, crop failure meant death or migration. Whenever there was a poor monsoon, then, famine and the recruiting agent triumphantly traveled the land together. The recruiting was done in small lots by men of their own nationality — *kanganies* sent into India by the managers of the Ceylon and Malaya estates where they were employed. In theory it was a patriarchal system, the *kangany* supposedly being the head of the family group and being held responsible for the debts of all the laborers he brought in. Employers liked it because the joint debt responsibility made all gang members vigilant in seeing that all others worked, and in coffee days it is said to have functioned fairly well. The booming demand for rubber and tea workers proved too much for *kangany* morals, however, and they went into large-scale recruiting outside their families, into chiseling advances which never reached the laborers and into piling up huge debts against the dark laborers on estate books. Individuals could not break a contract without being pursued, collared and tossed into the hoosegow, but the *kangany* could pull his gang out of one plantation and take it to another (which also took over the debt) by giving a month's notice. Using this as a lever, the Indian exploiters of Indians pried cash from the estate managers and higher and higher were the debts chalked up against poor devils in bondage both to alien overlords and under-officers of their own race. When Ceylon in December 1921, finally passed a law making all advances non-recoverable and abolishing the penal clause in work contracts, a total debt from *kanganies* and laborers of 14,000,000 rupees had to be written off by some 800 estates concerning which statistics have been collected.

Malaya's record in the matter of Indian labor is better than Ceylon's. It abolished the debt system as early as 1907, but, while claiming to have abolished the indenture system in 1910, it did not ditch the penal clause in labor contracts until 1921. India itself took belated interest in the poorest of its sons by the 1922 passage of an emigration act designed to prevent the revival of indentured labor and generally to protect the exploited ones. Malaya and Ceylon the next year passed labor ordinances in conformity with this act and since then both have maintained Government departments to supervise and control all recruiting of unskilled workers in the dry land to the West. Rising wages followed on worker freedom and the Tamils today are far better off in Ceylon or Malaya than at home. In India these agricultural laborers are irrevocably shackled to the pariah class or the lowest castes. Away from home, the offspring of the untouchables attain to respectability — sometimes to native prominence — and the widely established estate schools for laborers' children constitute an important route up. Free housing greatly superior to the mud shacks of Madras, free medical attention, obligatory sanitation - this is the estate picture today and the record would seem to indicate that even in 1917 and earlier, living standards on the plantations were far in advance of India's. Although there is a Governmental setup for distributing the cost of Tamils among all those who import them, the proprietarial attitude still exists and penalties for the enticing away of labor are provided for by Malayan statute.

Swifter tappers are the Chinese, especially when on piece work (400 to 600 trees a day against 300 for the Tamils) but being less docile and having a better idea of the business and what they should be paid, they are less popular than the Madras clods. It was because these last are too frail for the heavy work of felling and clearing jungle that such work went to the Chinese and Javanese. And it was chiefly in weeding, drainage upkeep and similar work that the Chinese continued to be employed in Malaya, save for the unhealthy districts where the Tamils refused to go. Once any region acquired a bad reputation for malaria, the word spread through South-East India so swiftly that no more of the Madras boys were to be had. To the yellow men malarial risks were all in the day's work, however, and in they went as intrepidly as any of those white heroes of tropical medicine hymned for sacrificing all for dear old science.

In connection with the Chinamen, the Malayan abolition of the out-and-out indenture system did not come until 1914. Recruiters worked for the lodging-house keepers of Hong Kong and Swatow and chiefly sifted through the starving provinces of Kwangtung, Kwangsi and Fokien. 'Kidnappers of young pigs' was the Chinese designation for these good-will ambassadors. To the planter of Malaya the cost per young pig was $20 to $34 (United States). Of this about $10 represented transportation and other expenses and the remainder was the flesh broker's profit. Contracts generally were for 300 working days at 2½ cents a day and food. The $10 passage money had to be repaid to the employer and there was no obligation upon him to repatriate the laborer. Mathematics says that under such a setup not even a Chinaman could be clever enough to escape after 300 days and mathematics is right. What happened was that the recruit or sinkeh stayed on as an old hand (*laukeh*) at $4.50 a month.

Today's Chinaman sometimes figures in a triple play from homeland broker to Singapore broker to the employer, who pays all expenses and such additional profit as the current labor market dictates. Again yellow contractors employed by the plantations will recruit direct, with the estate paying passage and expenses only. Work agreements are limited to a month and the Protector of Chinese is authorised to fix the top amount which any coolie may be liable to refund to his employer on account of importing expenses. On the job the workers toil directly under a contractor and look to him for pay. Often enough this dignitary lights a shuck away from the rubber lands with the advances intended for his men, but the plantation managers continue to find the Kwangtung and Fokien hands too inscrutable to be dealt with directly. Or too sharp at bargaining. Even on piece work it's a case of paying the total for the rubber to the contractor, who makes the individual distribution himself. And pockets a cut, for his boys are always in debt to him. Tamils, Javanese, these become individuals to the whites driving and babying them, but the Chinaman always remains a composite, anonymous figure. The Tamil especially needs much looking after. He has to be forced to sleep under a mosquito net; if left alone will exercise no care in the preparation of his food, will drink from the handiest death-breeding stream or stagnant pool. To lesser extent the Javanese also must be policed in these matters. On the Chinaman, however, there are not even reliable death statistics. When he decides that he is seriously ill he leaves the plantation and makes for friends. And if he leaps the dragon gate thereafter, the estate manager often never hears of it.

In Java itself all rubber plantation and other labor is free, but the Javanese tapper going to *Hevea* estates elsewhere in the *Estrada* world still does so under three-year indenture with penal provisions. In the days of booming rubber the recruiting agents for the Sumatran estates were brown-skinned city slickers working out of the Javanese ports of Batavia, Surabaya and Samarang and what those spell-binders did to the country boys was a pity and a shame. From the poor but easy life in his native village the innocent youth was lured away by rosy pictures of the gold and maidens then in process of redistribution in some Utopia not specifically identified. Most engaging and unsophisticated of all those who came to labor for king *Hevea* in the Middle East was this unspoiled rural character. The dark Tamil knew where he was going, and so did the lemon-colored rookie from South China. But the lad from interior Java had never been out of his village, had never ridden on a train, had never seen a white woman, a city, a steamship. These things and stranger he now gazed upon but it was only a swift, weird interlude between rural riverbank home of the past and jungle home of the future. Taken on the railroad to a terrifying metropolis, he there bound himself to three years of slavery by signing papers concerning which he was told nothing. Before he could look around, he was being goaded into fierce labor in the unbroken or newly broken jungle.

In the days of clearing jungle there always were the native headmen — the *mandurs* — hurrying them along; they in turn were badgered and driven by *tuan besars* frantically anxious to get their share of the money flowing into the Middle East. Dosed with quinine, in fear always of the white man's hospital, the preparation of their food depending on what few women lived in the compound, the brown men found their only pleasure in the feverish gambling on the bi-monthly pay day, *Hari Gagi*, and the holiday which followed, *Hari Besar*. No matter how determined the individual was to save for that hoped-for return to his homeland, the urge to gamble usually overcame his slight resistance. And on the day after *Hari Besar* the majority went back to work as penniless as when they first arrived. Contract renewals in those days were for three years. Later they came down to 18 months, and, as planter associations opened their own headquarters and supplanted the professional recruiters, provisions were made for sending back to Java any workman who within a month after arrival in Sumatra declared that he did not know what he was doing when he signed his contract.

Efficiently supervised by the Government are both recruiting and working conditions now, and corporal punishment is strictly forbidden. That the indenture still is something with teeth in it and not just a name is shown, however, by the fact that in Sumatra and Indo-China the working hours are longer and the labor more 'efficient' than in any other Mid-East land.

How, then, does the state of the Javanese in Sumatra or other exile compare with his condition at home? If he has his own small rubber holding, he is better off, of course. As to the plantation serfs it is, perhaps, true that in Sumatra a majority of the recruits have either been drawn from starvation-poor agriculturists or the dock hands and warehouse workers of the Batavia (population 290,000), Surabaya (250,000) and Samarang (150,000) slums. Mostly bad boys and 'undesirables' were these last, and numbers of them were wanted by the police at home. Sumatra was preferable to prison in their eyes, then, and the authorities were perhaps a bit more willing to have them in the first place than the second.

A chief argument of the Dutch defenders of the indenture system has been the great improvement wrought in the physical condition of the migrants, and there seems to be no question that food, housing and hygiene standards of the exile island are superior to those of the Javanese homes. As to our deluded country boy of old who had rice and contentment in his native village before the recruiter's arrival, he at least has become not unhappily resigned to the new life. The first renewal was obtained by the persuasive mandur who dangled rich promises before him and shamed him out of the idea of slinking home with nothing to show. The Javanese would dig in then with renewed determination to save his wages. Again he would gamble them away. And the re-signing for period after period usually continued until there was little desire to return to a home which by then would have seemed far stranger than the plantation. Women had come in, families had been started and this once queer life seemed natural. Then, too, the work grew lighter. Clearing days were over and tapping and caring for rows of trees was not so hard.

Indo-China, we have noted, shares the indentured labor stigma with Sumatra. In this land dominated by Saigon, largest French port outside France, the indentures are for three years with one, two or three year renewals and there are penal clauses providing the usual fines and imprisonment for bolters or 'malingerers.' Housing, hospital and medical requirements are far below those enforced in Malaya and Sumatra. Considering, however, that the alternative for the native of Tongking is starvation in the overcrowded mud lands of the Red River or service with the French Army in Syria, something can be said for rubber work.

In the British and Dutch lands in the old days the white man socked a serf whenever he felt like it and there were terrific threshings dished out by the *mandurs*. Never were there atrocities of Congo or Putumayo varieties, however. A far higher class of master was involved, the coolies were under more or less observation by more or less civilised governments, and always the idea was establishing something permanent — preserving both the valuable trees and the valuable serfs who represented investments in recruiting and transportation charges. Today the white man lands no blows for a variety of reasons. There are laws against it; estate manager and assistant standards would seem to be higher than in the pioneer era; and in the case of the Javanese there always have been and always will be more than legal reasons for staying the hand of the unbalanced or cowardly exceptions among the masters. A Javanese is a Malay, which means that ordinarily he is a tractable soul, calmly accepting whatever befalls as the will of Allah. It also means, however, that when he decides the insult has been passed, blood will be spilled. Exactly what things constitute an insult only Allah and the Malays know. And because white men never can hope to learn all the answers, their restraint in the matter of blows has been insufficient to save dozens of them from the knives of underlings. In the history of the plantations, then, the labor bosses have been not killers but victims, and this situation continues to prevail.

The world demands rubber. To produce that rubber, jungle must be cleared. And when jungle is opened, malaria is loosed. There in bald outline is the one great atrocity story of Malaya, Sumatra and all the other Middle East rubber lands. There are exceptions, of course. Sometimes the land covered by malarial jungle has become healthy after being cleared. More often land covered with non-malarial jungle has become intensely malarial on being opened. And almost always it is a safe bet that a certain amount of the disease can be expected anywhere when opening jungle — especially with imported labor. How bad it was at its worst can be speedily established by a glance at the record of Negri Sembilan, one of the Federated Malay States. Its death rate per thousand among imported estate laborers was 85 in 1909. With the opening of jungle for new and expanding rubber plantations at its peak in 1911 the rate was 195! In 1912 it was 114. After 1914 it never again passed 50 save in the influenza year of 1918.

One out of every five on all the Negri Sembilan estates was struck down in 1911. Imagine, then, the mortality among crews of assorted natives not just on estates but actually engaged in the clearing work. And remember that deaths of a third and a fourth of the whites, browns and yellows so engaged was no notable thing. Nor was it a case of merely opening jungle at terrific sacrifice of life and thereafter watching the plague subside. In many spots it hung on unabated, yet into these death traps, whites, yellows and browns continued to be poured by the overlords in London and the export company agents. And think not that the absent owners and directors did not know what conditions were. These were the gentlemen who hung up wage offers 75% higher than Ceylon's, expressly to lure the Tamils frightened off by the appalling mortality. These were the gentlemen who fought the government's 1907 order that each estate erect its own hospital, one of their arguments being that *some estates possessed no healthy site* and that, anyhow, the supply of dispensers was quite inadequate.

Since then a great deal has been learned about Malaya's malaria. On one of the peninsula's best known estates, for instance, subsoil drainage reduced the death rate from 144 per thousand to 17 per thousand in a dozen years. Yet many estates continue to be malaria traps and are operating today as always. And it is on them that the

Chinese have been able to demand and receive wages of 2½ to 3 times the standard figure, for save in periods of the direst depression Tamils can't be had at any price.

In connection with the *Hevea*'s effect on general health conditions in the Middle East it might as well be noted that no forests currently are being opened on its account. Whenever the demand materialises, however, the work will start up again no matter what the skyrocketing of malaria statistics. Often it is done on contract and always it is a melodramatic spectacle as well as a malarial tragedy. Earth taken by force of ax and fire from a jungle that would reclaim any portion left untended, the white man working himself weary along with his natives — that is the general picture. In Ceylon the trained elephants go in and with their aid the spindling Tamils can get by. In Malaya there are no baggy pachyderms but sweat-smooth yellow and brown men — Chinese, Javanese and even Malays of the peninsula — for these bold swaggerers will take contracts for clearing jobs although they will work on no man's plantation. In Sumatra, Javanese mainly and a few Chinese. Felling the towering trees, these scurrying mortals expose dark glades which for centuries had remained untouched by sun. Up into the light shoot magical vines, ferns, orchids and flowering creepers in these days. Then the fire is applied, destroying vegetable and animal—ant, scorpion, serpent, blossom and leaf. Nothing remains to be seen save the smoldering ashes. Yellow men are sent out with hoe-like *chankols* to turn the top soil under and bite deep to destroy the weed roots. Brown men labor thigh-deep in the swampy land at the forest edge, digging ditches. Leeches cling to their naked bodies, mosquitoes and flies halo their heads, their sweat mingles with the muddy water in which they stand. As the estates shape up no dead trunks, no rotting stumps, no stubborn, killing lallang weeds are permitted to remain. Weeks of toil go into this wiping out of these remnants of the rank jungle growth. And weeks of back breaking labor in the drainage ditches. During this period it is that disease strikes hardest. The slimy mud of swampy low spots harbors all manner of vicious insect pests. And through it the coolies go, their bare skin burned darker brown, darker yellow by the flame of the sun. At eleven the tom-tom sounds. Food is brought to them — tea, rice, dried fish. Back to work they then go.

Roads must be constructed, of course, and buildings erected. And on the sloping lands there is terracing work. Planting itself is done by digging a hole for each seedling and setting out mathematical rows. The thinning is gradual and usually is not completed until the seventh year. Weeding, pruning, drain upkeep, road upkeep — these are the tasks during the five years until tapping begins.

Routine indeed is the real plantation life that now commences. Shivering into the half-dark and down the orderly rows of pale *Heveas* go the natives, who are routed out at 5 AM by the booming of the tom-tom and checked by sleepy whites. Removing a thin shaving of bark from the cut on each tree, turning the porcelain cup upright in its wire holder beneath the latex-directing spout driven lightly into the bark — this is the first round. The scrapings from cup and bark and any drops that have fallen to the ground are gathered up, to be made into crude of second or third quality. The cup is then replaced sideways so that it will remain clean and ready for the next day's tapping.

To the uninitiated, the plantation is a confusing place — unwavering lines of trees stretch endlessly into the distance. Corridors of matching trees are so much alike that the casual observer sees no difference in them. Only to the native worker and the experienced white manager is the path distinct, are the trees separate entities, each having its own personality. The coolie, going methodically about his business of tapping, is as much a part of the crude rubber machine as the latex shed or the plantation factory where the latex is coagulated and made ready for shipment. To him there is no interest in what happens after he empties his pails into the large cans at the shed. His work ends in this cement-floored shelter with its corrugated iron roof resting on pillars. Usually he is there by eleven o'clock with his burden of sticky liquid. In boom times of specialists for every task, he has no other duties. More often his own living quarters must be kept clean, his section of the white man's forest must be tended — trees pruned, nursed, ground kept clear, drainage ditches kept free from refuse.

Other coolies spend their working hours in the preparing of the crude. The latex sheds, placed at regular intervals in the sections, smell indescribably foul. Here the liquid is poured into cans to be sent to the plantation factory and the scrapings from cup, bark and earth are placed in troughs for washing and coagulation. And here the almost naked workers are covered with the slippery stuff. Into the face it gets, the hair, the hands — their loin-cloths are stiff with it. The smell of raw rubber, always a disgusting odor, penetrates the skin so that it can't be washed off. And because of this, the martyrs of the sheds pay double for their women. On some of the larger estates, narrow-gauge railway cars run through the prim, park-like manufactured forest, taking the loads of liquid-filled cans and the spongy white masses of wet rubber from sheds to central factory. On others pneumatic-tired bullock carts are used. At the factory the latex goes to the straining and coagulating room. This is one spot on the plantation where cleanliness is a necessity if first-class crude is to result. The collecting buckets, the tile floors, the vats and particularly the screens for straining the latex and catching the particles of dirt and bark must be rigidly inspected. Strained, the milk is run into aluminum-lined troughs and the coagulant, usually a solution of acetic acid in water, is added. If the product is not to be smoked, sodium bisulphite also goes in at this point. Partitions are inserted in slots in the troughs and the latex then is left over night. The next morning the slabs of wet doughy coagulum are put through a series of rollers of varying sizes and the resulting long white strips are taken to the drying room. So is the 'pale crepe' of commerce produced. 'Ribbed smoked sheets' are ribbed during the milling process, cut to the required length, put through a hot water bath, dried on racks and taken to the smokehouse for 10 days at 120 °F.

Preparation of the rubber varies, of course, with estates small or large, stand-pat or progressive. Especially in connection with the plantations owned by the tire corporations, industrial ways push themselves into the jungle and huge buildings rise. The old setup of factory—mills, vats and Diesel engine — smokehouse and warehouse, gives way to elaborate gadgetry or bulk handling in a simplified fashion.

On the face of Sumatra, pitted with latex sheds, breaks out a carbuncle — a spray 'shed' that looks like something picked up in an American industrial city. In this queer building, the latex is poured in a steady stream on a revolving disc at the top of a heated chamber, the resulting fine spray being dehydrated by the time it reaches the floor where it falls in small sticky granules and coheres into a mass. Pressed into blocks it is whisked by rail and ship and rail from the United States Rubber Co.'s Mid-East plantations to its covey of American rubber goods factories.

Ready to elbow in on sprayed rubber is rubber powder, patented in 1935. Here the process begins with spraying the latex on moving, heated, endless steel belts where it forms large globules. Dehydrated in passage through a heated chamber, they become small globules of dry rubber. Sprayed with zinc stearate to prevent their adhering to each other, they are detached by a revolving roller and rolled to resemble very small tubular beads. In this powder form or pressed into blocks, which may be readily crumbled into powder, the rubber is ready for shipment.

By tank-truck, tank-car, tank-lighter and tank-steamer, latex itself goes out of Malaya and Sumatra in liquid (2 parts of rubber to 1 of water) or concentrated form. Or it is shipped as a paste. Extensive plant equipment the latex requires, and stationary dock storage tanks as well as all the special paraphernalia for moving it. Only the industrially owned estates can afford such investments, but Dunlop Plantations, Ltd., owner of the concentration patent, is getting the small fry into the game by reaching out to collect their latex. Expansion of this idea, cooperative investment or the coming of special handling companies into the business, looms for the future.

In opening the jungle and in upkeep operations, modern efficiency has also been tried. Here, however, the coolie has defeated the cabletype steam plows, the gasoline tractors, mechanical saws and steam logging machinery. Machinery is expensive and little of it can be used' once the clearing work is completed. Efficient operators are almost impossible to find. In removing roots from forest soil or eradicating lalang grass, hands can do the work more efficiently than plows — and brown hands can do it much cheaper as well.

In all these matters mechanism eventually might be able to triumph.

But in actually getting the milk, it would seem that industrialisation can come no closer to the tree than the estate railroad line running to the receiving depots where the coolies make their deliveries. You can't pipe liquid from a tree yielding but a few drops of fluid a day and that only when cut. The alternative is an army of robots complete with tapping knives and pails and even such tin woodmen probably would refuse to work for coolie wages.

The one thing you can do is to install the factory speed-up system among the black and brown boys. This has been done. Prior to 1927 yield of rubber per estate employee never averaged 1,000 pounds a year in any rubber region. By 1929 it had climbed to 1,576 pounds in the Federated Malay States, by 1933 to 2,342 (more than a long ton). Cessation of planting partially accounts for the last high total, but most of it represents a figurative turpentining of the breech clouts of natives who in depression days fear unemployment nearly as much as do the tire builders of Akron, OH. Thus, by the same system of pay cuts, speedup and staff reductions as forced upon white employees, there has been achieved that cutting of costs to 3½ to 8 cents a pound.

On which note we pass from consideration of the Asiatic on the estates to the Asiatic on his own small holding. In British Malaya, the Malay owner assumes the role of manager, and his family and friends are relied on for the labor. Those not in the immediate home circle are paid a little less than seven cents Straits a pound on the rubber content of the latex collected in kerosene cans. Chinese and Indians pay on the same basis save when their milk forests begin to approximate the size of small estates. Then they run up palm-leaf huts and put on the Tamils at 35 cents Straits a day and housing. Even on the Chinese and Japanese estates that actually are estates, these Tamils are the favored workers. Japs have been tried by the Japs and found wanting, which is an interesting commentary on the cheapness of Indian labor, considering the world-wide reputation of the Nipponese as one who lives on rice and wind and toils tempestuously 24 hours a day. On the larger Tom Thumb layouts the latex generally is prepared in the holding's own 'factory,' a minute, ramshackle, open-faced hut. The process is a sort of homely parody of that employed by the Europeans. Pans discarded by the white estates or beaten out of kerosene cans are used instead of aluminum-lined troughs. Substituted for the four pairs of power rollers are two miniature hand pairs or sometimes only one. Sometimes coagulation is by evaporation and again by adding alum to hasten the process. Efficient is this last method, but it results in a brittle sheet. Off then go the sheets to the nearest public smokehouse, an enterprise which also maintains a public 'factory' for the smallest of small fry use. Some attempt is being made by the British government to teach the potential Harvey Firestones how to improve their methods of cultivation and preparation. For the benefit of browns unable to read the English publications, lantern slides and lecturers resembling Dr. Roosevelt's WPA puppet jerkers and news dramatisers are available, and this is expected to bring about the more abundant life any day now.

The educators are making more progress in Sumatra but the methods there were far more primitive to begin with. In Palembang and Djambi, states in the South-East of the island, most of the Malay producers are located. The usual Palembang 'factory' is a grass shed which represents a capital investment of from one to two days' work. Alum is the coagulant, a bottle or a round piece of wood serves as the roller and the product is 'dried' by hanging it in the sun for a day or so and then in the shade. When the capitalist has a picul (133.3 pounds) or so on hand he totes it to the nearest buyer, always a well-heeled Chinaman or Malay. Price paid the native chiefly depends upon moisture content and percentage of foreign matter. This varies from 2% to 60% with the Chinese doing the judging. Governmental advisers are hot for encouraging the natives to form cooperative societies and buy hand rollers. Their chief mission, however, is trying to talk them into conservative tapping. To these last instructions Allah's boys listen dutifully. Then they go out and put the knife to three year old trees if the price is right. When this last happens to be the case they raise rubber only and buy all their foodstuffs. Otherwise they call on the good earth to yield whatever is eatable and can be obtained with the least effort.

In Djambi, however, the Malay milkmen are come to regard rubber as their only source of livelihood. One village near the mouth of the Djambi River exists only to supply the *Hevea* abusers with fish, coconuts and chickens while the boats bring them cigarettes from Singapore. Not so long ago today's tappers exported rice to this last city. Now they import it for their own use. Every village up the Djambi has a few *Hevea* nurseries. Seeds are germinated in muddy water on the river bank and seedlings kept there until ready for planting. Aforetime the tree owners paid straight wages which finally got up around two guilders a day. Tappers whooped for half of the yield in addition and wages ceased. Now it's done on a straight 50-50 basis. Coagulation is by alum (although the Government has a law against it) and the six inch thick sheets are kept under water. When there is an accumulation it is suspended from rafts and floated down to the town of Djambi where it is bought by the Chinese and loaded direct from the water onto the Singapore boats.

For the whole of Sumatra wet rubber exports to Singapore 82% of the *caoutchouc* total in 1934 and but 56% in the first half of 1935 — the best of indications that the government-encouraged program of building smoke sheds and importing made-in-Singapore hand mangles now really is catching on. For the rubber mills of Singapore, it naturally means a decline.

More on small *caoutchouc* holdings will be found in the succeeding chapter, which deals with the white man's control attempts. To the estates we return to take up the subject of brown women and white men. Naturally, not all of the rubber land feminine minority is composed of tree and coolie tappers. Some of them become nurses and housekeepers (*bona fide*) to the white families long since become fairly numerous in the plantation houses. As conditions improved, European wives came East to share their husbands' lives. And according to the social code, white women could do no work of any sort without losing caste. So that positions as house servants multiplied as the European groups increased, a condition which prevailed until the last drastic depression. More fortunate, possibly, are the contract, indentured or free females who go or have gone to share the white man's bed and board, boss his servants and manage his household. In the early days some estate owners deliberately discouraged real marriages of their white employees. And encouraged the 'housekeeper' plan. Certainly it was a quick way for the new assistant to pick up the Malay dialect which is generally used in all parts of the Middle East. It also had the advantage of being cheap. And even a bad housekeeper was better than keeping bachelor's house in a land where everything conspires to make a white man uncomfortable. If she could do no more than keep his beer cool she might be considered a jewel. Women of all the tinted lands — Javanese, Japanese, Sundanese, Malays, Chinese — became adept in the business of serving as 'wives' of the rubber land whites. Sometimes they were lucky enough to encounter not one kind master but a series of them. Handed on from one manager or assistant to another, they had opportunities of saving for a trip back to their homeland, wherever that might be, or they might set themselves up in shops. The advantages that a clever woman had were numberless. Children were sometimes a problem. The tuan didn't want to be bothered with them usually, and abortions were common. When, however, the child was born, its future depended entire1y on the father. The Dutchman often claimed his offspring and sent them back to Holland to be educated. And still does. The Briton's unacknowledged son went into the native compound and grew up under the dreadful handicap of being accepted by neither brown nor white.

Less tolerant and less generous than the Dutch, the English certainly were, and perhaps less fortunate with their women. How today's new breed of planters from Albion feels about the business of brown mistresses is shown by the vehemence with which they are arguing that their London exploiters should pay sufficient to warrant marriage after 10 years in Malaya — which means at about 30 to 33. We quote from an official statement of the Incorporated Society of Planters (estate managers and assistants) issued at Kuala Lumpur, Federated Malay States, in September 1934:

'If the terms and conditions of service on estates make marriage impossible at a reasonable age, many disadvantages and embarrassments will be suffered by planters by liaisons with native women; by all the sordid adjuncts which attach themselves to prostitution, and by the unfortunately only too common concomitant of venereal diseases.'

So does the glamorous East now appear at first hand. And so it was in the old days when the first rubber estate assistants were going out. Veterans of some sort of service in the Middle East were the domineering managers. Often deeply interested financially in the *caoutchouc* ventures, they were autocrats in their own little worlds and they liked them. But to the boys from home the acres newly taken from the jungle seemed raw and dismal indeed. Rough roads led to them through dense green forests, blank walls of foliage, creeper-hung, monkey-haunted. Through burning air that became an exhausting burden to the unaccustomed Northerner. Through glaring sunlight that made the eyes ache. To come at last to the scarred brown plains where brown men were marking the planting spots for the young trees.

From sunrise to sunset the new man toiled at monotonous, exacting work in tropical heat, for white supervision was necessary every step of the tedious way to a finished acreage of growing *Heveas*. When the evening tom-tom sounded and the natives filed slowly back to their quarters, the white man was not always through with his day's work. Costs to compute, wages to count - work with pen and paper occupied part of his evening. After a bath and change of clothes, he might eat or try to eat an unappetising meal — butter melted, bread doughy, condensed milk spotted with ants or flies, half-cooked pancakes, questionable steak. Then, with his head buzzing from quinine, he would sit under a smoky lamp and try to work or read, being too tired, often, to do either. His house, a rude box-like affair on piles, was, of course, only temporary. New houses, usually of cement, were built for all white employees as soon as the sections were completed. But in the meantime the pioneers lived in their ramshackle wooden bungalows, tended by slow, inefficient servants, fighting malarial fevers with daily doses of bitter drug. Importations they were, all of them — man, *Hevea*, *cinchona* — set down in the steamy little islands and peninsulas of the Indies.

As the plantation gradually took shape and the trees grew ready for tapping, living conditions improved. Good roads were built, railway lines were pushed through, better offices were constructed and a regular life took form out of the chaos. Yet in that regular life, there are many drawbacks. Nothing can change the enervating effect of the weather. The working hours are still long and tiresome. Up at 5 AM checking the gangs of coolies, watching them trudge out to begin their operations. Eternally checking them. Walking through the never-ending lines of *Heveas*, inspecting each tree to see whether the tapping has been properly done, keeping records on the yield and health of each one, noting the need of pruning and weeding. Eternally inspecting — cups, pails, factory equipment, coolie quarters, buildings, yards and latrines. Always the necessity of keeping everything clean, not only to achieve good crude, but to keep the workmen in good physical shape. Disease, both human and plant, has to be guarded against. Bark, stem, root and leaf of the rubber tree may be attacked. And the assistants must learn to recognise brown bast which appears in the tapping cut, pink fungus on the forks of branches or at the junction of limb with trunk. These two are bark plagues. The rotting of leaves and fruit indicate leaf disease. Such pests as white ants and boring beetles also inflict themselves on the *Heveas*. But of them all, some planters think the root troubles are the worst. Spreading underground as they do, wet rot and the other forms may destroy a whole group of trees almost before the workers realise it. Speed is the essential point in curing sick trees. Therefore daily inspection is important.

Less important than a sick *Hevea* section are sick men white or brown, of course. Among the planters, no Englishman in Malaya, no Dutchman or Englishman in Sumatra, escapes without contracting malaria. At best the victim is liable to recurrent attacks for the balance of his life. At worst he helps swell such death statistics as we have cited.

Once there were compensations, however. Beginners' salaries of $135 to $180 a month with yearly raises bringing the amount to double in six years are hardly to be included, however. Very sweet these figures sounded back home, but up against the high cost of living in Malaya the low purchasing power of money, the fancy prices on all the things that had to be imported from the homelands, the necessity of maintaining a swanky front both to keep up with the other whites and to impress the brown man with all the outward marks of superiority — this took the salaries. It was on the bonuses that the now non-owning managers and their assistants relied. Into important money these ran in the boom years, in Sumatra at least. Helpings of $8,000, $12,000, $16,000 a year were handed out and a frenzy of spending went on. Expensive automobiles, riotous entertaining, visits home in the lavish manner — so went the bonuses. In Malaya there were very few like these, according to the aforementioned Incorporated Society of Planters. 'As opposed to statements as to the large commission earned and bonuses drawn by planters in the past, the average amount of such works out at $396 per capita per year of service,' it contends.

At any rate, there was a crash. Hundreds of assistants, most of whom had never participated in any noticeable bonus distribution, were sacked — nine hundred in Malaya alone. Most of them stayed on in the East — using up their small savings in vain job hunts. Then they had to fall back on what amounts to poor law relief from the Federated Malay States Government. How those who remained worked their heads off is shown by statistics. In 1929 the average acreage for which each white was responsible was 688. By 1933 it was 1,395. Yet with this doubling of detail work (which is far worse than in the old days), with actual laboring staffs cut beyond all reason, with maintenance funds wholly shut off, these planters so drove themselves and browns that the total output of the plantations hardly changed.

The usual reward has been theirs. Most of the home boards have decided that it would be a good idea to perpetuate the depression pay cuts and economies in full although the estates are now paying juicy returns. And the white man who serves the London rubber millionaires in the tropics now is himself beginning to notice something which has been a verity for decades but which he had blindly refused to see. He is realising that he is and has been right where he has always known his natives to be — in the ranks of the exploited.

We quote again from that recent planter organisation 'statement upon questions relating to the salaries, emoluments and the general conditions and terms of service of planters on rubber estates and other plantations in British Malaya':

'It is a simple fact that very great discontent exists. There are some hundreds of European Planters in Malaya who are living from hand to mouth, gravely endangering their health by exposure to the climate, yet saving nothing — quite unable to save anything — against the day, which may come soon or late, when they will be told by a doctor to quit if they value their lives. Many of these planters have wives whose health is suffering most seriously because the husbands cannot afford to send them for a change of climate as often as is necessary. Not a few have wives and children in England, and the bulk of their meager pay is sent Home to support these; and the Planter of, perhaps, 10 to 20 years standing is worse off as a mere individual than he was during the first year of his service in Malaya ... There are scores of men who have learned the local languages, acquired the art of handling labour, and soundly informed themselves on all that pertains to the welfare of an estate. At 50 they will be getting elderly, at 55 their agents will hint that younger men would be cheaper and more active, and with a possible 20 years of life before them, they stand a fair chance of being requested to resign. What can they do when that happens? Thirty odd years in the tropics, with an average seasoning of malaria, or some of the other tropical diseases, has unfitted them for manual labor, and they had had no chance of saving more than a few hundred or more dollars. Let Directors of companies think of it, and they will not marvel that men are restless and dissatisfied. In point of fact it is the sanest, steadiest, most worth-keeping men in whom dissatisfaction is found most acute. The man who is happy with a native girl, who drowns dull care in a plenitude of strengths, and whose philosophy is summed up in the phrase 'sufficient unto the day is the evil thereof,' may be satisfied. Steady, thoughtful men are not.'

3.4 Caoutchouc Corners

With rubber entering into every activity of daily life it would seem that any successful attempt to hoist prices by limiting production of this extremely necessary commodity would attract plenty of attention from the citizenry at whose pocketbook it was aimed. Actually, however, the populace of this rubber world for several years has been paying tribute to *caoutchouc* restriction without thinking about it and probably without even knowing about it.

In the United States such restriction is something the average citizen vaguely recalls having heard about a decade or so ago. He remembers that with it went a deal of tumult and shouting; the bellowing of international insults, accusations, recriminations; the spectacle of an undersized American tire manufacturer and an overstuffed United States Secretary of Commerce (often accused of pro-British leanings and sometimes of British citizenship) thumbing defiant noses at perfidious Albion.

The new restriction, arrived at without political and publicity uproar, is more efficient in its operations and is far more modest in its price raising aims. In that lies the secret of the lack of general attention accorded it. And since the raises it has imposed are by way of being reasonable to date, the oblivious citizen is not being unduly penalised as things stand. This moderation is a wisdom that cornerers and would-be cornerers of the *caoutchouc* supply have learned with the years. Tracing past efforts as well as present operations may, then, offer something to instruct or to entertain.

To speculators rubber early on seemed to be a commodity susceptible to easy control, this deceptive appearance arising from the fact that production always has centered in odd ends of the world. Back to the 1870s we must go then to find the first man to visualise the huge profits that would accrue from cornering the world's gum elastic and to act upon that vision. Naturally he was one of the Portuguese merchants of Brazil —specifically Joao Goncalves Vianna, so-called Baron de Gondoriz, who had gone to South America and entered the rubber business at an early age. With fine para selling between 40 and 50 cents a pound in 1878, Vianna persuaded a group of Brazilian bankers that there was money to be made by gaining control of Brazilian rubber. With the bankers' aid he bought heavily in home and foreign markets, forced the price to a dollar a pound and the ring profited enormously. Encouraged by his success, Vianna began to manipulate again in 1881, quit abruptly, but was back in the market supported by European as well as Brazilian speculators. This time he ran prices from 40 cents up to $1.25, at which figure the syndicate controlled $5,000,000 worth of rubber. British manufacturers paid tribute to the wily Brazilian. Americans were more stubborn. Heavy business increases which they had anticipated early in the year had not materialised and they had on hand 40% more finished goods than they usually carried. Well situated to make good their threat of closing their factories unless rubber prices dropped, the Americans forced the speculators to capitulate with heavy losses. This painful lesson served only to teach Vianna that he needed more capital to carry out his operations.

While he was raising it, a group of American rubber goods manufacturers working with greater secrecy towards a more reasonable goal maintained a highly successful pool for some five years. Organised in 1880 to buy and sell crude, this New York Trading Company operated so quietly that hardly anyone outside of the group knew of its existence. Each of the member firms bought and sold rubber, ostensibly on its own account, but in reality for the account of the secret corporation; in this way considerable control was exercised over market prices. On capital of $100,000 the furtive company paid dividends of $1,000,000 in its half-decade of existence and it has been called (India Rubber World) 'the closest approach to a monopoly ever experienced by the rubber trade' since it exercised a longer control than anyone else ever was able to effect. Operations of the combination were carried on through the importing firm of W. R. Grace & Co., partners in which included William Russell Grace, 'Pirate of Peru' then serving New York City as its first Roman Catholic mayor, and Charles Ranlett Flint who one day would proudly describe himself in Who's Who as 'widely known as the 'father of trusts."

It was under Grace that Flint, scion of a famous Maine clipper-ship family, evolved the interests that were to become his professions: company organising, Latin American exploitation and armament vending. He will reappear in this rubber chronicle from time to time.

Right now it is the Baron who reappears. In 1891 the major rubber interests of Para were persuaded into a reunion with Vianna, and this combination succeeded in interesting London bankers in the scheme. The resulting syndicate controlled a capital of $25,000,000 and during the nine months it operated, succeeded in cornering 90% of the world's supply of rubber. Prices rose to 87 cents a pound. American manufacturers again cut their requirements to the utmost, some even closed their doors entirely. The capital controlled by the syndicate, huge as it was, proved inadequate for buying up the new crop of rubber, which was larger than usual because of the stimulation of high prices. British bankers who were carrying warehouse rubber certificates found sales increasingly difficult, knew that production during the fall and winter would be further increased, asked the syndicate to put up more margin on its certificates. One small firm called on the Baron to cover his margins, and, when he failed to do so, unloaded a 50 ton lot of rubber. Another small holder unloaded a similar quantity. The forced sale of the 100 tons started a stampede which sent the price tumbling to 63 cents and broke the corner. The break came opportunely for American manufacturers, who could not have held out against Vianna much longer. Contracts had been taken for goods which had to be filled before the middle of September. That the supplies of crude in the United States had been reduced almost to extinction was indicated by the fact that American manufacturers bought over four million pounds of rubber during the last three weeks of August. Had the syndicate carried its stocks a little longer instead of selling out, and had it been able to continue taking all of the rubber received at Para throughout August and September, it could have forced the manufacturers to have taken at least part of the supply at any figure it wished and the Baron and his associates would have made millions. Instead the Brazilian capitalists and the English banks suffered heavy losses.

That finished the Baron, and for the next decade it was Charlie alone who officiated as the bugaboo making the blood of non-allied merchants and manufacturers run cold. Actually this Flint never did get around to launching any real cornering operations, but there is reason to suspect that certain of his speculative enterprises profited by the dark hints and sinister hisses in re monopoly he was in the habit of loosing. The first such episode occurred in 1892. Flint now was one of the leading importers of crude, but it was the name of the newly incorporated rubber shoe manufacturing trust, United States Rubber Co., that was used to throw consternation into the market. Incorporated in March, this footwear octopus did not organise and elect officers until November; in the intervening months word was out that the chief promoters were Flint and William M. Ivins, newly elected president of the United States & Brazil Mail Steamship Co. and another former partner in W. R. Grace & Co.

In its first announcement United States Rubber bore down on the fact that it expected to effect great savings by obtaining its raw material to better advantage and hurled dire threats to build its own factories to produce litharge, whiting, buckles, dry goods and other necessities on which rubber manufacturers had been paying 'heavy tribute' as high as 'over 40%' in one case. In line with this attitude, it was also made known that the Trust expected to obtain rubber at the very source of supply — the tree itself — instead of buying it at Para. A committee was appointed to arrange for purchase of both steamship lines on the Amazon and the press hummed with tales of a monopoly affecting not only American rubber goods manufacture and the Amazon carrying trade, but the conveying of rubber to New York as well.

In New York the India Rubber World recorded:

> 'At a meeting on April 7th, the committee reported that a representative of the steamboat interests on the Amazon would at once leave for New York with power to conclude the sale of the steamers to the US Rubber Co... The franchises of the Amazon river steamboat lines, by the way, involve liberal subsidies from the governments of Amazonia and Grao Para, which, it is hinted, on account of the shortage usual in Brazilian public treasuries is much more apt to be paid in concessions of valuable rubber forests than in money.'

> In Britain the rubber journal editors nearly blew out blood vessels over this contemptible Yankee plan 'to tamper with the rubber supply — to buy up all the rubber transportation steamers of the Amazon river, and, instead of buying rubber at Para, to procure it at the very source of supply, and send it off at once to New York.' In a May 9th editorial the India Rubber Journal of London literally jumped up and down and screamed:

'Are British merchants and manufacturers willing to sit still and permit this scheme to be carried out? Will it suit them to find the supply of rubber from the Amazon valley carried off to the States? Do they like the idea of going, cap in hand, to New York for their best Para and contenting themselves with the leavings of the Yankee mills? Surely British capital and British enterprise can find out the way to frustrate their knavish tricks.'

Now permit two months to elapse with the steamship monopoly failing to materialise. And here we have the Indian Rubber Journal's editor telling the world that all along he knew there was nothing to it:

'From Para to the forest defiles of the Peruvian frontier a mighty peal of laughter has resounded as the dwellers in the Amazon valley have become aware of the ridiculous pretensions of the United States Rubber Co. The latest spread eagle prospectus deceives nobody, and amuses everyone... Look at the absurdity of the thing on the very face of it. The rubber crop is worth about half a million a month and an entire population are dependent on it. Suppose the company do buy up the 31 steamers purchased by a Brazilian syndicate about a year ago from the Amazon Steam Navigation Co., Ltd., for £860,000. A fleet of new steamers could easily be put into competition at half the cost.

'American speculators may please themselves so long as American rubber merchants and manufacturers are fools enough to put up with it, by manipulating prices in their own markets; but any attempt to monopolise the world's rubber supply must be made to meet with well merited disaster. No means and no opportunity must be neglected to give this wretched abortion a kick or a shove on its downward path to perdition.'

It is interesting to note that this publication had no criticism of the knavish tricks and ridiculous pretensions of a British government which some years later launched a monopoly attempt while fondly dreaming of Yankee manufacturers coming around, cap in hand, to pay any price asked for best para.

As to Flint, a good guess is that he was spoofing the boys a bit and simultaneously doing himself a spot of good on the market. In his only ship deal with the Brazilian Government he was not a buyer but a seller who in a month's time fitted out an armed fleet to suppress the revolting navy of that new republic in 1893. In December, 1900, however, he was able to throw a still greater scare into the rubber world by incorporating the $30,000,000 International Crude Rubber Company. By now Flint was mainspring of both United States Rubber and another giant combination, Rubber Goods Manufacturing Co. Also he let it be known that Standard Oil was in on the new cornering venture to the amount of $7,500,000 and he could point to the presence on his International Crude board of two Standard directors, Henry H. Rogers and John D. Archbold. The board also included Albert Burrage, president of Amalgamated Copper, Flint, and additional representatives of United States Rubber. This lineup carried conviction, indeed. Early in 1901, however, it became known that Standard was out of the venture; Flint ceased to be treasurer of a United States Rubber wallowing in financial difficulties; and the scheme folded up.

In his sensational Frenzied Finance series in Everybody's Magazine some years later (1905), Thomas W. Lawson, speculator notorious for his career as stock-market errand boy for Standard Oil interests, affirmed that the oil corporation backed or agreed to back Flint and then pulled out in order to get even with James R. Keene. Active in promotion of International Crude, this last gentleman was said to have bribed William Rockefeller's secretary for confidential information which enabled him to clean up in Amalgamated Copper. Far-fetched to say the least, this 'explanation' is offered for what it may be worth.

With 1905-1906 began the series of 'valorisation' attempts by which the Brazilian Government sought to sustain rubber prices at a larceny level. In the realm of coffee this form of government aid to producers and profiteers was made to work, but *caoutchouc*, as usual, proved impossible to hold down. The plan involved payment of an official daily market price on all crude deposited at government depots and flotation of a $50,000,000 home or foreign loan which would include the purchasing fund and would provide for loans being made to producers to enable them to hold their products for higher prices. Profits from the sale of rubber through the Government agency were to pay interest and provide a sinking fund. This plan was never carried through to the proposed elaborate extent and though a squeeze by the cooperating traders of Para had carried prices to $1.50 a pound and held them above $1.25, the inevitable drop came in less than a year, hastened in this case by the 1907-1908 panic in the United States. Coinciding with the beginning of the auto demand this run-up made fortunes for some, lost fortunes for others, and in the final analysis was a failure in more respects than one, for it was the $1.25 price that first caused the establishment of *Estrada* plantations to any considerable extent.

In late 1908 groundwork was laid for another squeeze by the launching of a valorisation scheme which authorised formation of a privileged syndicate headed by the Bank of Brazil to act as middleman and presumably to 'stabilise' prices for both producer and consumer. If production was considered 'excessive,' the syndicate was empowered to discourage it by restricting exports until world prices were forced to a more profitable level. Over protests of political opponents who declared that the scheme would only result in driving out foreign exporting concerns and making the syndicate a monopoly able to manipulate prices at will, the Government went ahead. The syndicate bought practically all the rubber offered at prices profitable to the sellers and held the bulk for maximum prices. From 72 cents to a brief high of $3.06 in 1910, prices were driven up with the aid of the now really immense automobile demand. Now, however, some of the plantation trees set out as a result of that earlier squeeze were just beginning to bear and nemesis arrived in the form of a swelling *Estrada* output. With this added to increased supplies of wild from sources other than Brazil, and with buyers taking no more from Para than they could help, prices were hammered down to less than $2, to less than $1.50, to a point where the Government realised it had bitten off something too elastic to be chewed. Faced with the prospect of certain loss, it finally arranged with the syndicate to unload even at a sacrifice. Altogether the Bank of Brazil is reported to have lost $10,000,000 in its effort to stem the tide of falling prices, while the Government was brought close to bankruptcy.

In 1913 plantation output passed Brazilian, in 1914 it went far past all wild, but with rubber down to 65 cents in 1915 the Government again was persuaded to attempt valorisation. Now realising the futility of trying to purchase all the *caoutchouc* offered, the Government and its official bank were, however, to carry rubber on margin and do their best to sustain prices. Assurance was given that no large amount of rubber would be withheld from market. The whole plan was knocked in the head when the Senate showed more sense than the executive officials of the land by refusing to approve it.

For the plantation industry, smooth sailing depended on that torrentially increasing demand of the motor industry. In 1920 the American rubber industry sailed head on into the rocky coast of post-war deflation. American demand for crude rubber ceased to exist. Since the American factories consumed 70% of the world's rubber and their English competitors, who were similarly hit, used much of the remainder, it is not difficult to see what sort of a wallop crude rubber producers sustained. For the last quarter of 1920 prices were down to 19 cents a pound. Towards the end of 1921, world stocks of crude totaled 300,000 tons. Production for 1922 was expected to hit 360,000 tons of plantation rubber and 30,000 tons of wild. Just 690,000 tons - this was supply. Even the most optimistic could not foresee a consumption for 1922 of anything more than 250,000 tons - this was demand.

Since British planters controlled 75% of the world's production, 69% in British colonies and the remainder in the Dutch East Indies, it was natural that they should take the first action. In the fall of 1920, the British Rubber Growers' Association announced that 70% of the producers in Malaya and Ceylon had voluntarily agreed to restrict production by 25% from 1st November 1920 to 1st January 1922. This was the first serious attempt at control of rubber production since the plantations had really entered the rubber picture. The policy was adhered to fairly consistently and undoubtedly kept prices from going any lower than they did. In 1921 it was hoped that the 14 months plan could be extended. Only 55% of the planters would sign, however. Those in the Dutch possessions and in Ceylon refused, saying they had less need of the scheme because of their lower production costs. The idea was dropped.

The third quarter of 1921 had seen rubber touch a new low of 11½ cents. The moans of the planters swelled in volume. Accustomed to profits lush almost beyond the dreams of avarice and too prodigal to have set up reserves, they found adversity most unpleasant. To scream for restriction backed by the full authority of the government was the first thought that occurred. The rubber producers were not, of course, the only supplicants; they were, however, the only ones to whom the government paid attention. Unquestionably this was because any burden imposed upon rubber consumers in order to help rubber producers would rest easily upon England and repose 70% of its weight upon the United States. Mr. Winston Churchill, Secretary of State for the Colonies, once said:

...the whole industry in which quite £100,000,000 of British capital had been sunk, was falling into ruin... It was impossible for the Colonial office to witness the financial ruin of the rubber-producing colonies owing to the continual sale of their products below the cost of production. The exhaustion of the rubber plantations would be fatal for many years to the prosperity of the Malay states and would greatly injure Ceylon.'

On another occasion during the run of restriction, he blundered into frankness, however: 'One of our principal means of paying our debt to the United States is in the provision of rubber.'

It was in October 1921, that the British Rubber Growers' Association asked Parliament to appoint a committee to investigate the condition of the industry. The request was turned over to Churchill, who gave the job to a specially appointed committee, headed by Sir (since 1917) James Stevenson, Scottish traveling salesman in Scotch and Churchill's personal financial adviser. This gentleman, whose name is immortalised in the 'Stevenson Rubber Restriction Plan,' began his career as plain James Stevenson, knight of the grip on the road with Johnnie Walker, and ended as Lord Stevenson, first baron of Holmbury and managing director of John Walker and Sons, still going strong. He had nothing to do with rubber before his appointment and, dying in 1924, never knew how close he came to destroying British rubber supremacy. Aside from the whisky baronet and two subordinate Cabinet officials, the committee was made up of members of the British Rubber Growers' Association: Sir Stanley Bois, E. J. Byrne, William Duncan, H. Eric Miller, and Sir Edward Rosling. British manufacturers were not represented.

The group made its first report on May 19th 1922. Because of 'the grave objections to Government interference with industry, especially when it takes the form of restricting the output of an important raw material... it was only with reluctance and with a lively apprehension of the dangers which threaten both the industry and the countries in which it is so largely carried on that the committee agreed to consider a measure of compulsory restriction as an alternative to what seemed to be worse evils.' The committee looked with favor on two schemes, the first, a restriction of output; the second (Stevenson's own) a restriction of exports by duties on a sliding scale, but did not wish to recommend either until the attitude of the Netherlands government had been determined.

Before submission of this report, Edgar B. Davis, father of United States Rubber Co.'s plantations in Sumatra, had bobbed up in London with one of his usual epic proposals. No longer a director of US Rubber and no longer actively connected with the *caoutchouc* business, this plantation enthusiast acted solely as an individual in proposing to the leaders of the British plantation industry a plan for cooperation between said industry and American financial interests. The idea was a $250,000,000 international stock company to acquire estates or interests in estates in each of the rubber-growing countries direct or through holding companies of British, Dutch, Belgian, American or other nationality. A syndicate would be formed to provide funds to meet the requirements of the industry and participation in this financing would be open to capital of all nationalities interested. Points 12 and 13 of the Davis 13-point program were: appointment of a plantation rubber industry committee to meet with American interests, including rubber manufacturers, for discussions of the present and future potential surplus of rubber; channels for distribution of product to remain undisturbed, subject to such measures as might be mutually agreed upon for the disposal of surplus stocks. Among the advantages beyond the dreams of a Flint which would be secured through this Trust, Edgar enumerated: international market for securities, formulation of policies affecting output and surplus stocks; scattering of risks from diseases and pests, reduction in costs through economies of large scale management, centralised administration of labor, benefits to manufacturers and planters through large-scale production of a uniform product under new processes as developed.

Edgar had three sessions with the British Rubber Growers' Association which finally voted in September 'That we are unable to proceed further with the consideration of this scheme.' Needless to say, American manufacturers were having all the trouble they could handle at home and had less than no interest in helping to finance a solution of the British and Dutch surplus crude problem. The Davis dream became nonetheless a political argument for government intervention as the only means of keeping America from seizing control of the distressed British plantation industry. So late as 1925 J. H. Thomas sounded off in Parliament:

'In 1921 and 1922 there was an effort to bankrupt our rubber producing concerns, so that people in America could have bought up the whole lot, wiping out the investment of British small shareholders, and exploiting the situation.'

So fared Edgar, the internationally-minded volunteer Samaritan. The Dutch having, in the meantime, indicated their complete lack of interest in restriction plans, the Stevenson Committee, on October 2nd 1922, issued a supplementary report recommending that England go ahead alone on a plan similar to scheme two of the original report. The Stevenson plan became law on November 1st 1922. Its operation was not complicated. The actual output of each producer during the twelve month period ending October 31st 1920 was adopted as standard output. Each plantation owner could export 60% of his standard production and pay a duty of only one penny (English) a pound. If he exceeded this figure but did not export more than 65% he paid a duty of four pence not merely upon the excess but on his entire production. Each 5% increase over 65% added another penny to his tax; thus he paid five pence per pound if his production was between 65% and 70% of standard; six pence if it were between 70% and 75%; and so on up to an even shilling per pound for production exceeding 100% of 1919-1920 production. The scheme further provided that when the price for any quarter averaged one shilling and three pence (30 cents), the percentage of standard production which could be marketed at the one penny minimum duty would automatically be increased by five and if the price averaged 1s.6d., it would increase by 10. Conversely, if prices averaged less than a shilling, the percentage of standard production exportable at the minimum would drop to 55 and would continue to drop until an average of 1S.3d. was again reached.

Rubber prices, which had averaged under 24 cents for 1922, were carried up to 36 cents by the end of January 1923, in the first wave of enthusiasm for the Stevenson scheme. This had caused some concern in the United States and the Rubber Association of America, organisation of leading manufacturers, had invited the Rubber Growers' Association to send a deputation over to explain and discuss the situation. Sir Stanley Bois and two others made the trip in January and succeeded in convincing American manufacturers that there was no sinister purpose behind the act; that it was designed solely to end the violent fluctuations which were bad for everyone concerned.

The January rise was soon shown to have been premature as world stocks were large in spite of the increased business enjoyed by American manufacturers in the winter of 1922-1923. American rubber buyers overestimated this recovery and by June 1923, stocks of crude in American factories were at the highest point in history. Consequently stocks had to be reduced and this caused the price of crude to drop month by month until it fell as low as 19 cents in May of 1924. During 1923 and 1924 there was a steadily increasing amount of native rubber coming out of the Netherland East Indies and a substantial amount of rubber smuggled from Malaya to Dutch ports for shipment. This native rubber was much in excess of calculations. As a consequence, market rumors spread that restriction was bound to fail. It has been the British contention ever since that manufacturers who should have supported the market in the May-July quarter of 1924 when prices averaged under 22 cents failed to do so. The percentage of release dropped to 55 and prices came back somewhat, but for the next quarter were still under the 30 cent minimum and the export percentage dropped to 50.

Here was the great fault of the Stevenson scheme — its inelasticity. For at the same time that the export rate was being reduced, the demand for rubber again started to increase. In addition to the general improvement in the rubber industry, the balloon tire at this time began rapidly to replace the old high pressure tire and this had a considerable effect on consumption. The balloon, because of its greater area and smaller amount of fabric, used about 30% more rubber than the tire it replaced. The cut in production on the one hand and the increase in consumption on the other had the combined effect of reducing world stocks to the lowest point in years. By the end of December prices had climbed above 36 cents. Manufacturers either failed to realise what their 1925 requirements were going to be or else they remembered the 22 cent prices of the previous summer and were hesitant, in the early months of 1925, of covering at 40 cents. Possibly both of these conditions existed. At any rate, by June of 1925 nearly every rubber manufacturer of any consequence in the United States was in the market for rubber and the result was a sensational price rise. Average price for the month of April was a fraction over 41 cents. In July, spot prices hit $1.21, the highest they have ever been since 1916. Dollar prices continued during the balance of the year.

Until this time American rubber manufacturers, with one exception, had shown little alarm over the workings of the Stevenson scheme. One, in fact, was quite happy about it — the United States Rubber Co., aforetime *caoutchouc* manufacturing Trust and still (this was true until 1926) the world's largest rubber manufacturer. The only manufacturer owning rubber plantations to any noticeable extent, this corporation was drawing 20-25% of its requirements from its own properties. Since most of them were located in Dutch Sumatra and were not subject to restriction, the United States stood to clean up by averaging down the cost of its market purchases with its own product and holding a tremendous competitive advantage over the rival manufacturers.

The extremely able and vociferous foe of restriction was Harvey S. Firestone, Sr., himself the Firestone Tire & Rubber Co. of Akron, OH, USA. As the great slasher of tire prices, Firestone was for cheap rubber at all times. As one always hot for publicity, he saw a beautiful nationalistic band wagon to grab for himself. At every opportunity he referred to Winston Churchill's inept statement requesting the Bonar Law cabinet succeeding the Lloyd George cabinet to preserve rubber restriction and so pay the war debt. From all of his company's billboards and magazine advertisements shouted the slogans 'America Should Produce Its Own Rubber' and 'Americans Should Produce Their Own Rubber.' In the White House was a friend and camping companion from Ohio and Harvey made good use of access to the Harding ear.

How far apart United States Rubber Co. and Harvey stood is best demonstrated in a letter to Frank G. Smith, secretary of the British Rubber Growers' Association, from Henry Stuart Hotchkiss, vice president of United States Rubber. Published in the August, 1926, issue of the Association's bulletin, it was referred to as having been received by Smith prior to his visit to the East (on which he left in October 1925):

'As you know there has been a great deal of nonsense published regarding the American manufacturers' attitude toward restriction etc. This is largely based on Mr. Firestone's sensationalism and is not at all representative although certain papers like to make it appear so. You are familiar with the position that the General Rubber Co. (V. S. Rubber's crude rubber purchasing subsidiary) and the United States Rubber Plantations, Inc. (V. S. Rubber's plantation subsidiary) have taken and it would be most gratifying to me if you could take occasion to point out the fact that we, together with practically all of the other large manufacturers in the United States, are entirely out of sympathy with Mr. Firestone's campaign and that he is in a very small minority, although his opinions seem to be given all the prominence.'

The Hotchkiss letter did not by any means represent the opinion of the majority of American manufacturers at the time of publication. That it fairly represented their opinion at the beginning of restriction is certain, however. Disregarding the case of the gone and forgotten executive reputed to be in the sideline business of buying rubber on his own account when it was low and selling it to his company when it advanced, there seems no reason to doubt that the most of them honestly believed that the British were out not to kite rubber prices but to stabilise them at a reasonable figure, such stabilisation being something that would have been eminently beneficial to manufacturers buying in the world's most violently and unpredictably fluctuating market.

Add to this that none of them had any least taste for tagging after the Firestone calliope at any time and in any event and their attitude is understandable.

To the Rubber Association's January 1923, jolly good fellow parley with the British rubber grower representatives we have referred. Loudly hooting that session and refusing to attend, the Akronite had fared to Washington instead and attempted to stir Secretary of Commerce, Hoover to action. Also he induced Senator Medill McCormick to champion a bill appropriating $500,000 for a survey of possible rubber producing areas in the Philippines and Latin America. Deciding, then, that it was necessary to put on a show for Washington, he sent out a call for a February conference of rubber and auto manufacturers. The Rubber Association wired all its members, urging them to stay away from this meeting. Representatives of the Department of Commerce, auto manufacturers and a couple of minor rubber manufacturers' associations attended. While they were in session the rubber survey legislation backed by the President and McCormick was unanimously passed by Congress. In May, Firestone chased his own rubber land hunters into the field and forwarded his resignation to the Rubber Association, of which he was a past president.

When prices skyrocketed in 1925, Harvey was no longer alone in his opposition. Following conferences with rubber and auto industry representatives, the leisurely Secretary Hoover in December finally cut loose with public appeals for the cooperation of manufacturers and consumers in a drive against the squeeze — play prices by conservation and provision of independent American supplies. In a letter to Senator Arthur Capper on the 10th, he pointed out that if rubber prices remained at the then existing levels throughout 1926 it would cost Americans $666,000,000 in excess of what they would pay at the 1s.3d. figure. As a remedy he suggested the withholding of loans by American bankers, the elimination of waste, use of substitutes and stimulation of production in other countries. A week later, Hoover, with the backing of the National Automobile Chamber of Commerce, was encouraging auto owners 'to cut down sharply their consumption of rubber in tires' as much as to intimate that they had been deliberately using them up faster than they should. He was also expressing hopes that the Congressional investigation of the subject (December, January, February) would make it possible for American rubber buyers to pool purchasing (which would be flirting with Clayton Act Violation). Spot price at this time was $1.09 a pound, price three months forward was $1.08 and American manufacturers were contracted six months ahead at an average price of 79 cents. Some of them, incidentally, had favored working off all their 79 cent commitments on the public while making no further purchases and chancing what might happen to prices with all of the big buyers out of the market.

Or as Mr. Hoover later phrased it in a letter to one of the present writers:

> 'Some of the rubber companies did not support the campaign by which we broke down the raid upon us, because of their desire to profit in stocks of raw rubber.'

Majority, however, were now for government aid and so that call upon consumers to refrain from using rubber was arranged. The British at first professed to be amused by Hoover's roars. The London correspondent of the New York Times, on December 23rd, described the English attitude thus:

> 'The man who has a smattering knowledge of American political institutions affects to laugh. 'Don't worry,' he says, 'Hoover is just pulling off what they call a political stunt. He is grooming himself for the Presidential fight in 1928. He is twisting the British Lion's tail and there is no more in it than that."

In January representatives of the now chastened Rubber Association of America belatedly met with Hoover to consider 'proposals for the expenditure of $50,000,000 in annual appropriations of $10,000,000 for five years to plant rubber trees in areas under American control.' It was asserted that definite plans were to be announced within a short time and that Sumatra and the Philippines were under consideration.

Almost simultaneously, President Charles Clifton, of the National Automobile Chamber of Commerce, announced that auto manufacturers were going to back a $10,000,000 organisation to produce, purchase and deal in crude rubber. John J. Raskob was a member of the committee which was to complete the organisation of the corporation. Other committee members were C. W. Nash, of Nash Motors; A. J. Brosseau, of Mack Truck; Alvan Macauley, of Packard; A. T. Waterfall, of Dodge; and A. R. Erskine, of Studebaker. A short time later it was announced that 102 auto firms would be asked to buy stock in the American Motor Rubber Corporation, the name given the company which was supposed to acquire plantations as well as buy from small producers. Nothing ever was done in connection with any of these grandiose plans.

The widely heralded Congressional investigation of 'The Means and Methods of Control of Production and Export of Crude Rubber' also was pretty much of a dud. Opening witness H. Hoover demanded first of all a secret hearing, at which he read the notes exchanged by the State Department and the British Foreign Office on the subject of rubber. In his public testimony he expressed the belief that it was not the intention of the British Foreign Office that prices should rise the way they did; speculation, not the planters and officials, was responsible. He referred also to other monopolies which he felt endangered the 'sane progress' of the world and the future of international relations. Steps should be taken to secure relief, but Hoover was not prepared to indicate what form the measures should take.

Democratic newspapers leaped at the chance to contrast Republican eagerness to harry foreign monopolies and ignore domestic ones, taking great pains to point at Andy Mellon's Aluminum Trust as a shining example. The World in a cartoon pictured the British rubber monopoly as a horned devil and the Aluminum Trust as an angel with wings and editorially commented: 'Mr. Hoover attacks the British trust but seems to forget that he belongs to a government that believes in protecting American Trusts by the highest possible tariff.' Other publications reminded Hoover that he had loaned his assistance to the attempt of American beet sugar interests to force Cubans to reduce the production of cane sugar as a measure for obtaining higher prices; pointed out that American farmers were at that very moment clamoring in the House for the creation of an export corporation financed with public money to operate for the purpose of raising prices of farm products; suggested that 'it would be a whole lot more honest to admit that in this game the British have taken a leaf out of the American notebook and are doing the thing that we did for many years and are doing still' (Baltimore Sun).

The London Morning Post found occasion to remind the badgered Secretary of Commerce that in 1919 the United States government had issued a public recommendation to farmers to reduce their spring wheat sowings in 1920 by 12%.

Other English journals began to grow indignant over prices America had charged for commodities in the past, especially cotton, of which England is one of the chief buyers. A statistician in the India Rubber Journal figured out how much more England had paid for cotton — and tobacco than she would have paid at pre-war rates, demonstrated how much less the United States had paid for rubber than it would have paid for it at pre-war rates and pointed to a tremendous balance in Uncle Sam's favor.

The presidentially aspiring Secretary of Commerce could take all this and like it, however, for by February 6th 1926, spot rubber prices had dropped to 66 cents and he was able to issue a statement pointing out that he had saved America $350,000,000. Meat for Herbert this was, but poison for tire manufacturers who had wanted lower prices but no such precipitate tumble as this. Bought far ahead on high-priced rubber and under the politic necessity of cutting tire prices as a result of the widely publicised Congressional hearings which had emphasised that casing costs were controlled by *caoutchouc* prices, the manufacturers were clipped for some $115,000,000.

Despite Mr. Hoover's belief that his proclamations 'broke down the raid upon us,' it is patent, however, that all the sound and fury of the campaign against restriction accomplished little if anything. And in the sound and fury we include the Congressional hearings, the lofty plans of the Rubber Association and the Automobile Chamber of Commerce which never got beyond the committee room, the long series of earnest and ineffective State Department representations to the British Government, the Firestone advertising and the Firestone search for new rubber lands, the sadly belated move by which a rubber buying pool in 1927 undertook to eliminate manufacturer bidding against one another for crude. The increased usage of reclaimed rubber in this country (54,488 tons in 1922 contrasting with 189,500 in 1927) did help to sink the Stevenson plan, but the Hoover 'campaign' hardly can be credited with bringing that about.

Also helpful was the smuggling of rubber out of Malaya, often with the connivance of corrupt officials. Pursuit of the rubber runners apparently was as dangerous at times as was the chasing of rum ships off the American coast during the same period. Such is the impression gleaned from a story appearing in the Straits Echo in 1926:

'With her searchlight smashed and the well scrubbed decks stained with patches of blood, the rubber restriction launch, Sinbang, cruised into harbor ... Those who boarded the chaser were presented with the extraordinary scene of four dead bodies ... The weary crew appeared with bandages over wounds which were serious ... Police officials ... took statements from the crew. They were told of a surprise attack by some Bugis smugglers. The searchlight had been used and although it was a very dark night, those on the chaser were able to make out a fair amount of rubber on board the smuggler's kotah. The launch approached the kotah and went alongside, when the smugglers scrambled on the decks of the Sinbang and attacked the crew with knives ...'

A writer in the Straits Times in 1927 charged that at least 30,000 tons of rubber had been smuggled out during each of the three preceding years. This, he claimed, was the result of the policy of employing temporary officers, usually unemployed rubber planters, instead of permanent officers of the civil service. The temporary men, it was claimed, were only too willing to receive bribes, and it was not unusual for a rubber dealer to offer as much as $10,000 if accumulations of uncouponed rubber in his stores were overlooked.

The one factor which really defeated all British calculations was neither reclaim nor rubber runners, however. Before proceeding to it, it would be well, however, to note exactly what did happen to the Stevenson plan in 1926-1927. With demand running away with production towards the middle of 1925, Leopold C.M.S. Amery, Colonial Secretary in the Baldwin cabinet, saw that production could not adjust itself because of the inelasticity of the scheme which permitted increases of only 5% and these only every three months. In November, as an emergency measure, he announced that export percentages would be raised to 100% on February 1st. Prices began to fall in December, as we have seen, and by the end of the February quarter were a shade under two shillings. In an effort to remedy in 1926 the failure to secure stability in 1925, the restriction scheme was modified. The pivotal price was changed to 1S.9d. and it was announced that releases would be cut from 100% to 80% if the average price fell below this figure for the quarter beginning August 1926. It was further provided that the exportable percentage should not go below 60%. Rubber values continued to slide. Price for the quarter which started May, 1936, just met the pivotal price. In the August quarter it fell below and production was forthwith cut to 80%. For the November quarter it was a shade lower and the percentage of release was cut to 70. The same thing happened in the February, 1927, quarter and releases dropped to the 60% minimum. During 1926 and 1927 the British were increasingly discovering that their power to fix prices was growing less every day. Daily it became more apparent that the scheme no longer had the effect it once had in influencing prices.

The factors, unforeseen in 1922, which defeated the British plans were the Dutch bent for scientific tropical agriculture and their new policy of encouraging their subjects to develop territory themselves. In 1921 the Dutch East Indies had contributed 72,245 tons to the world's rubber total. In 1922 it was 102,548 or 16% of the world total. By 1926 it was 203,634 and by 1927 was 229,000 or 34% of the world total. To this rise the white-directed plantations had contributed by hoisting their production from 68,245 tons in 1921 to 136,000 in 1927, a gain of 67,755 tons. More astoundingly the natives had upped their 4,000 tons of 1921 to 93,000 in 1927, a gain of 89,000 tons! And in that total gain of 156,755 is the answer to the question of why British restriction failed.

Dutch planters had opposed Netherlands participation in restriction from the start because they knew they could produce more cheaply than the English. They believed the thing to do if rubber prices were low was to improve methods until production costs were still lower. Bringing new acreage into bearing and utilising bud-grafting tells the story of their advance.

To bud-grafting the British had paid little attention. And it was not a matter of general knowledge among them that Holland's subjects were getting into the rubber game on a large scale.

In addition to receiving free land the natives in the Indies had other advantages over the English planters in Malaya. They did not have to stand the cost of roads, permanent buildings, machinery; had no expenditures for managing directors, superintendents and huge staffs; no payrolls to meet, no Jabor to recruit at distant points. They could in fact make a fine profit at prices below the Englishmen's cost.

When rubber prices boomed in 1925, the natives found themselves rich beyond their wildest Oriental dreams. Bazaars in the Sumatran towns were jammed with Javanese buying bright clothes, jewelry, musical instruments. All of them flashed 50 and 100 guilder notes where formerly 50 guilders represented three months' salary. Other luxuries which were in great demand with the newly rich were bicycles, oil lamps, perfumed soap, flashlights, sewing machines, phonographs, watches, safety razors, canned fruit and biscuits.

In 1922 Malaya had exported two and a quarter times as much rubber as the Dutch East Indies. In 1927, Malaya was still the world's leading exporting country but its total of 240,000 tons was only 10,000 tons ahead of the Dutch. Such comparisons finally convinced the English that the longer restriction continued the smaller would become their share in world exports and the more marked would be the increase in that of the Dutch. British planters in 1926 had begun to awaken to what was happening and their protests became increasingly audible. In November 1927, Secretary Amery recommended to the Prime Minister that the Cabinet thoroughly review the restriction program. On February 8th, 1928, the Hambling Committee was appointed to inquire into the operation and effects of rubber restriction. Its confidential report was never published but on April 4th, Prime Minister Baldwin announced 'The Government has . . . decided that all restrictions on the export of rubber from British Malaya and Ceylon will be removed on November 1st 1928.'

When restriction of production ended, the British producers tried to regain their old place in the sun and turned out rubber to the full extent of their capacity. Dutch and native production continued to increase steadily as more and more trees, planted during restriction, came into bearing. As a result, surplus world stocks amounted to 300,000 tons in 1929. As production continued to outrun consumption, this figure rose to 490,000 tons by the end of 1930 and 620,000 tons at the end of 1931. These supplies, mostly accumulated in the United States, depressed prices from 16 cents a pound in December, 1929, to 9 cents at the end of 1930 and 4.6 cents in August, 1931. The depression which struck in 1930 cut the demand in this country. At the lowest level of the depression in 1932 rubber stocks were fairly stable, but prices continued to fall, touching an all time low of 2½ cents.

The natives quit tapping as prices dropped to unattractive levels. During the whole of 1932 their exports were only 61,000 tons. As prices gradually rose, more and more started tapping again and in June of 1934 they were producing at the rate of 240,000 pounds a year. This ability of the native to get in and out of the market at will — producing when he can make a profit and turning to other crops for his living when he cannot — made it seem likely that eventually, in the absence of some form of restriction, rubber production would fall more and more under native control. The white planter, unable to operate so flexibly, but instead forced to produce all of the time, regardless of price, was in a bad spot.

Until 1932, the British continued to try to persuade the Dutch to cooperate on a restriction policy. Then they gave up, indicating that the Dutch would have to open any future negotiations. By 1933 the Dutch began to change their views. World stocks had risen to such a point that they saw that even their own plantations must continue to run along uneconomic lines unless price control were established. Tax collections had fallen as a result of low prices, and the East Indies budget showed a deficit of 120,000,000 guilders for 1933-1934. The Dutch had also gained experience in the control of tin and of sugar. Whatever the reason, the Indies government changed its mind about its inability to control native production and decided to see what could be done with restriction. In May, Prime Minister Dr. Colijin announced that he thought restriction might be desirable 'if a practical and useful scheme applicable to native production and acceptable to the British government' could be devised. He also intimated that if the producers could reach an agreement, the government would then take a more active part. On June 21st, a committee of the Dutch International Association for Rubber Cultivation met with representatives of British, French and other producers in the Middle East. Negotiations were carried on throughout the year with the utmost secrecy in order that the market might not be disturbed.

Not until April 29th 1934, was announcement made of the Producers' Agreement, signed by representatives of the British, Dutch and French rubber growers and delegates from Siam and the white rajah of Britain's protectorate of Sarawak in Borneo. The plan was confirmed on May 7th by the British government draft of an

Intergovernmental Agreement, relating to British Malaya, Netherlands India, Ceylon, British India, Burma, French Indo-China, the State of North Borneo, Sarawak and Siam. The plan was to take effect June 1st 1934, and to continue until December 31st 1938 with a provision for renewal thereafter.

The purpose of the Agreement as expressed in a vague and general way in the preamble is 'to regulate the production and export of rubber in and from producing countries with the object of reducing world stocks to a normal figure, and adjusting in an orderly manner supply to demand, and maintaining a fair and equitable price level which will be remunerative to efficient producers.'

According to the Agreement, basic production quotas were established for the various territories, founded on average production for the years 1929-1932 inclusive. These quotas ranged from Malaya's high of 504,000 tons in 1934 increasing to 602,000 in 1938 to Burma's low of 5,150 tons in 1934 and 9,250 in 1938. The Netherland Indies were granted a quota of 352,000 increasing to 485,000. Percentage for export was based on these quotas, the percentage rate being fixed from time to time by an International Rubber Regulation Committee. Each country had representation on the committee according to its importance as a producer and its delegates had one vote for every 1,000 tons of the basic quota. Besides authority to establish and change the restriction percentage, the committee was charged with the responsibility for doing whatever was required at any particular moment to insure the proper working of the plan. The new plan also provided for the suppression and elimination of smuggling; prohibited new planting except in Siam; prohibited replanting without special permission; prohibited exportation of rubber seeds, flowers, buds, twigs, branches and roots in order to discourage new planting outside of the restricted areas. The agreement contained no mention of what was considered a fair price nor did it state how restriction was to be accomplished, leaving that matter up to the individual governments involved.

The scheme was a great improvement over the Stevenson Act. From the planters' viewpoint it was better because it was comprehensive, only 13,000 tons of rubber having been produced in 1933 outside of the signatory countries. It was better for consumers because it was much more flexible than the Stevenson plan. The plan was not hailed everywhere with delight, however. Perhaps the most distinguished of the skeptics was Sir Cecil Clementi, Governor General of Malaya, who declared it his opinion that only the plantations would genuinely restrict their output, while native grown rubber might even increase in tonnage. Sir Cecil got the sack for his outspokenness. Before June 30th his resignation had been tendered to the Colonial Office on the grounds of ill health and Sir Thomas Shanton Whitelegge Thomas took his place. The plan hit a temporary snag in September when the Siamese legislature refused to ratify, demanding a 40,000 instead of a 15,000 ton quota. The Cabinet immediately resigned and matters were further complicated by the absence and then the abdication of the King. After considerable negotiating the Siamese request was granted. Its situation with respect to Malaya was ideal for a smuggling base, and it was felt that it would be better to have Siam in the Plan than out, even at the cost of a few concessions.

Greatest task in getting the plan into effective operation was the division of the basic quota among the producers in each territory. The problem was more or less a familiar one in Malaya because of previous restriction machinery. The Dutch, in splitting up their quota, were faced, however, with a vastly troublesome factor of native production. At last they decided that it should be permitted to bear the same relation to the estate quota that it had in 1929, when the Orientals exported 71.5% as much as the estates. The Volksraad, or People's Council in the East Indies, fought determinedly for a 50-50 split for the brown men and refused to pass the legislation as it was offered. The Volksraad then found out just how much power it had because the law went into effect by government decree. The natives have since been placed more nearly on a parity with the estates in the division of the Dutch quota. An increased quota granted the Netherlands Indies on December 3rd 1935, effective at the first of this year, was applied entirely to them and brought the ratio of their production to estate production up from 71.5-100 to 93-100. The increase raises the basic quota of the Dutch to 500,000 tons for 1936, 520,000 tons for 1937 and 540,000 tons for 1938.

When restriction first went into effect, the East Indies government decided to control estate production by individual restriction under license, much the same as is done in Malaya. It was obviously impossible to apply this system immediately to the hundreds of thousands of brown producers and these, it was hoped, could be held in check by means of a high export tax on their rubber. Proceeds of the tax were supposed to be used for financing improvements in native districts, but frequent complaints are voiced that this has not been done. The use of the export tax device is based on the connection between rubber prices and the export of native rubber, so that when an artificial depression of rubber prices is caused in the interior by means of a high export duty the result is an almost automatic limitation of native production. The objection to this system is that it hits the native not only by reducing his production but also because he receives a lower net price for his rubber.

Towards the end of 1935 it was becoming evident that the Dutch were having more and more trouble controlling the indigene producers.

Export duties have been increased steadily ever since restriction went into effect, jumping from 4½ cents (United States) a pound at the end of 1934 to 10½ on April 1st 1936. In spite of this, shipments by the Orientals exceeded allowances by 12,000 tons in 1934 and by 34,000 tons in the first nine months of 1935. As can be seen, the export tax was mounting toward its maximum which is, of course, the market price of rubber.

In order to live up to the restriction agreement and not exceed its international commitments, the Government in October 1935, announced that it would use surplus funds accumulated from the tax on native rubber to buy from the estates licenses for 20,000 tons.

In other words, estate owners were paid for not producing rubber in much the same fashion that farmers of the middle Western United States not long ago were paid for going into the lucrative not-raising hogs and not-raising-corn businesses. This drastic action held the Dutch exports for 1935 to an excess of 15,000 tons over the quota. The very evident fact that the original Netherlands Indies quota was not sufficient to permit them to handle effectively the indigene problem was the reason for the granting of the increased quotas mentioned above. According to the Dutch this increase was necessary in providing an irreducible monthly export large enough to permit those totally dependent on rubber to earn a living.

With the granting of the December 1935, increase the Dutch gave assurances that they would positively be able to regulate production. Whether they can do so remains to be seen. The Government realises that the export tax method is not as satisfactory as individual restriction. Beginning in the fall of 1935 it began making intensive efforts to register every native rubber tree in order that the latter plan might be put into effect for the brown men as well as for the estates. It was believed in the early part of this year that individual native restriction would be announced by this fall, probably to take effect at the start of 1937.

Percentages of basic quotas permitted since restriction went into effect have been steadily decreased. In June and July 1934, the permitted percentage was 100; this was reduced to 90 for August-September, 80 for October-November, and 70 for December. Prices which had reached 15½ cents in August 1934, weakened because of the recession in American business and the Siamese trouble which indicated that restriction was not running smoothly. Immediately the British wanted to lower the percentage for the first quarter of 1935 but the Dutch wanted to raise the figure, thinking this would reduce the price and lower native production. The Dutch won this argument and the percentage figure went up to 75 for the first quarter of 1935 but dropped back to 70 for the second quarter, 65 for the third and 60 for the fourth. The percentage was also fixed at 60 for the first six months of 1936 but increased to 65 for the next half year.

At the time this is being written it is impossible to say whether restriction will be continued after 1938 or not. The system of control has been reasonably successful and so far it actually has been operated as a plan to control and not to extort. The

15-16 cent price of rubber in effect in 1936 is sufficiently high to give efficiently managed estates a good profit and it is not so high that an unfair burden is placed on the manufacturers. The crux of the whole situation is Dutch ability to control native production and export. The potential production of East Indies native rubber is so enormous that the last two authorities to survey this subject arrived at figures of 750,000 tons a year and a million tons a year. The latter figure is virtually the same as the world total production record set in 1934. Should prices soar in the old squeeze play fashion it would seem that no control measure short of armed force could hold in this reservoir of rubber. It is, then, the native of the tropics, exploited since the birth of the *caoutchouc* industry, who looms today as the controlling figure in the market, an act of poetic justice long delayed but now particularly welcome to the American consumer to whom the brown man represents his best guarantee against highway robbery prices in the future.

3.5 Industrial Imperialists

Not to be wondered at is the fact that the world's largest rubber consumers today are on the way out of the rubber market. For a tire manufacturing corporation to add the strange business of tropical agriculture to its industrial worries is for it to embark upon a troublesome and speculative venture not lightly to be tackled. Into this move, however, have been forced the first, second and fourth largest American rubber goods manufacturers, the one important British one and the one important French one. And to their number has been added the largest single automobile manufacturing enterprise in America and the world. Involved are a variety of minor motives, but the compelling one is the fantastic and inexcusable fluctuating of the rubber market that has caused crude prices to range from \$1.25 to 2½ cents a pound in the last decade, from \$3.06 to 2½ cents in the last quarter century. Those are the extremes, but even more telling, perhaps, is the fact that in but two or three years out of the last 30 has the fluctuation been less than 25% of the average price for each year. Costing the rubber manufacturers millions upon millions of dollars in inventory write downs, it is these market antics, combined with the senseless tire price-slashing of the manufacturers themselves, which for 15 years have made this the most profitless division of Big Business.

Whether rubber prices are consistently high or low makes little difference to the rubber goods manufacturer so long as they do not reach heights so forbidding that the use of *caoutchouc* articles has to be curtailed by consumers unable to afford them. Essential, however, if this manufacturing game is ever to be a sane one instead of an extravagant parody on sound business, is some fairly consistent raw material price, some reasonable approach to stability. This reasonable stability was the one thing denied the manufacturer by the Brazilian speculators and the Brazilian government and in the long run the denial cost South America its place in the rubber sun. It was the one thing the British allegedly were aiming at when they launched the Stevenson restriction plan. Instead, however, they framed the measure in as extortionate a fashion as possible and then permitted the London speculators to take over. The result was the greatest fluctuation since Brazilian days and Britain's reduction from rubber dominance to no more than a competitive first place. To date, as we have noted, the current British-Dutch-French-Siamese restriction scheme has trended towards stabilisation at a reasonable figure rather than extortion and fluctuation, but there is a big question as to whether this common-sense attitude has not arrived too late.

Last year just one Akron rubber corporation, and that not the largest, consumed 64,000 tons of rubber apart from reclaim and ground scrap — more than the world production of 1906, more than half the world production of 1914. Remove buyers of this caliber from the market and it well may mean the doom of the independent rubber plantation on European lines. We can visualise the possibility of chief consumers supplying their own needs and taking the surplus to market in competition with the tree milk from the small holdings of the Mid East natives. Not yet do the leading factories as a whole control anything like a major portion of their own requirements, but the lands for which they have contracted, the acreages which they are planting, make certain that they are going to do that or more.

That the manufacturer-planter is not confining himself to Sumatra, Malaya and Indo-China, that he is turning his attention to West Africa's one independent state and to South and Central America is, of course, the most interesting feature of the whole trend. In the Philippines he has been rebuffed (although there is one factory-owned estate there) but elsewhere the move away from plantations operated under European governmental control in the Middle East to vast plantation concessions directly negotiated with tropical governments, comedy and otherwise, is ushering in an era of industrial imperialism certain to affect powerfully the politics and the peoples of the old wild rubber lands and likely to make itself felt in the realm of international relations and world affairs. The Firestone Tire & Rubber Co.'s mortgage on Liberia, Henry Ford's personal duchy within a Brazilian republic, Goodyear Tire & Rubber Co.'s entrance into Panama and Costa Rica — these are developments of more significance than the possible undermining of Middle East rubber rule.

Rubber manufacturers with the greatest stake in Sumatra and Malaya are, naturally, those earliest in the 'vertical trust' field. First of the manufacturing giants to interest itself in the business of supplying its own rubber was United States Rubber Co., the *caoutchouc* shoe manufacturing trust which enunciated control of its own raw materials as a founding aim in 1892. Charles Ranlett Flint, promoter of the trust, was the originator of its belief that the nearer the rubber tree the sweeter the profit — a belief that has been a guiding principle in United States Rubber's operations down to the present day. Rubber-broker Flint never seems himself to have moved the trust one bit closer to the tree, however. His chief use of United States in his rubber corner schemes, *bona fide* or otherwise, appears to have consisted in adroitly connecting the name of this fearsome trust with such speculations. Not until late 1901 did the corporation begin saving commissions by purchasing crude directly under its own letters of credit instead of through importing houses, and this was after Flint's departure and the elevation of Samuel P. Colt to control.

Of imperialistic negotiations in South America we gain a glimpse in President Colt's 1903 statement to the stockholders, 'We were interested in the 'Acre concession,' so called, granted by the Bolivian Government to F. W. Whitridge, Esq., in association with Sir Martin Conway, which, however, owing to complications with Brazil has been abandoned, Brazil paying an indemnity.' A year later, however, the trust did move to gain another advantage over its smaller rivals by setting up its own purchasing houses at both Manaos and Para. This made it the only consumer in the world so fortified, but it continued to fish for forest concessions as well. In 1905, for instance, Colt turned on the applause for one of the directors, Commodore E. C. Benedict, who 'in consequence of the continued advancing prices in crude rubber from the beginning of the year, in November last organised and conducted at his own expense a trip to the Amazon in the interest and for the benefit of the company ... We look for great benefits in the future as the result of this expedition.'

While awaiting these benefits United States had set up a $1,000,000 subsidiary, General Rubber Co., to handle the Para and Manaos business and with the 1905 absorption of Rubber Goods Manufacturing Co., it expanded this purchasing agency in keeping with its new importance as the largest consumer of crude rubber in the world. Capitalisation of General was increased to $3,000,000 and there was authorised a $9,000,000 issue of 10 year debentures to provide sufficient working capital. The next year a selling department was opened and through acquisition of the house of William Symington & Co., Ltd., at London and Liverpool, General added not only representation in the English crude market but in the other markets of Europe as well. This eliminated bankers' commissions as Para and Manaos now drew directly on the London house in payment for rubber purchases; not yet, however, was United States really getting right up next to the tree.

Still riding the corporation was the idea of nabbing wild rubber forest concessions as in 1905 when we find Colt scoffing at the idea that plantations in the Middle East or anywhere else could be an important factor in *caoutchouc* supply. How long he continued to have Brazil on the brain is shown by the May, 1910, report to stockholders in which he darkly noted, '... special attention has been given to the initiation of plans for ourselves producing both in the 'Far-East' and in Brazil the supply of crude rubber needed by the Company, which we confidently expect will result in our obtaining in the not distant future from our own rubber properties a substantial portion of our requirements.'

That was the first the stockholders heard of the 'Far-East' and the last they heard of Brazil. Right along with the great plantation boom rode the United States, making its first actual estate investments in the East in 1910, bringing its plantation acreage to 32,500 (more than double the area of Manhattan Island) in 1912. By 1913 it had 10,000 coolies toiling for it in Sumatra; 20,000 acres planted of 80,000 owned; 2,500,000 *Hevea* trees. One factor making for rapid progress was that much of the 80,000 acres had been planted in tobacco and the task of conquering jungle before setting out the *Heveas* was avoided. Already in 1913 this development was the world's largest rubber plantation and it has continued expanding and retains that distinction to this day. By 1917 United States had its own rubber purchasing houses at Singapore and at Colombo in Ceylon as well as in Liverpool, London, Para, Manaos; had upped its area in Sumatra to over 90,000 acres; had nailed other important holdings and particularly on the Malay Peninsula. In 1918 United States Rubber Plantations, Inc., capitalised at $30,000,000, was organised as a General Rubber subsidiary to take over direct control and management of the estates. In the complicated United States setup its subsidiaries in turn are Malayan American Plantations, Ltd., incorporated in the Federated Malay States and operating five plantations, and these Sumatran jawbreakers: Holland American Plantage Maatschappij, Si Pare Pare Rubber Maatschappij, and Nederland Langkat Rubber Maatschappij. Very few Americans figure in the on-the-spot operations. In Sumatra, for instance, the 15,000 United States Rubber coolies are bossed by some three-score whites of whom but half a dozen are Yankees. Over two-score of these dignitaries are Dutch.

As of today the corporation has 70,000 acres of its Sumatran holdings planted and has another 30,000 planted acres in Malaya. Of this 100,000 total, 90,000 is in mature trees and the 75,180 acres tapped in 1935 yielded an average of 559 pounds an acre. For 1934, shipments were 22,717 tons about a third of United States Rubber's own requirements. For 1935 they were 17,850, the decrease being wholly due to governmental restriction. On its books United States carries the plantations at $18,000,000, from them has collected profits of $40,000,000 including the $6,000,000 of 1926. It is, then, the best investment the one-time rubber trust ever made and by far its most important single asset.

Not to its Flints, its Colts or any of its practical rubber men does United States owe this development, however, but to a singular character who breezed into and through its councils on the way from shoes to oil. Reader, we give you Edgar B. Davis, the most likable, the most unexpected and withal one of the ablest gentlemen in the whole centuries-long history of *caoutchouc* from Hernando Cortez to the brothers du Pont. Already Edgar has bobbed up in this chronicle to hand a $25,000 tip to Henry

Wickham and to startle whimpering British rubber growers with that proposal for a $250,000,000 international rubber estate trust. He will appear again in an after-Waterloo episode in the career of Frank A. Seiberling, the Little Napoleon of the tire industry, but since his fathering of the factory-owned plantation idea constitutes his chief claim to a place in the rubber records, it is best to complete his identification right here.

Spectacled, benevolently round faced, bald in front, over six feet in height and tipping the scales at more than 250 pounds — that is the picture of the physical man in his prime. Today the poundage figure no longer applies, for some of the weight has slipped away even as the fortunes of this *caoutchouc* caliph. Edgar B. Davis is 63 now and reputedly involved in financial reverses, but behind him is such a glowing, memorable and generous career as no other figure in American industry ever has boasted. Born in Brockton, Massachusetts on February 2nd 1873, Edgar, son of a small scale shoe manufacturer, had the regulation whirl at reporting on the hometown Enterprise and settled down then to the predestined career in footwear. Beginning as a clerk in one of the city's three-score plants, the Davis rise was so rapid that by 1904 he was treasurer and sales manager. By 1908 the breakdown had arrived and he was on an ocean voyage for his health.

In Malaya and Sumatra, the tourist saw some of the early rubber plantations and then and there the United States Rubber Co.'s Mid East estates were born. Bursting with his subject, the super-salesman checked in with Colt's corporation, sold Samuel on the idea and went bouncing back to Malaya with the necessary financing behind him. Enlisting the services of W.J. Gallagher, director of agriculture for the Federated Malay States, and H. Ketner, a veteran planter, he spent months in investigating and finally picked North East Sumatra. The first purchase was the Soengei Sikassim estate of the New Assahan Tobacco Co. and nine subsequent purchases of adjoining pieces brought United States Rubber's one gigantic plot to 90,000 acres. The first purchase was made in May 1910, the first planting done in June 1910, which gives you some idea of how Edgar works. And it was Davis who, as managing director of plantations from the beginning through 1918, brought the estates along from the idea stage to the greatest *caoutchouc* development in the world.

In February of 1919 (same year in which Colt relinquished the presidency of United States) Edgar dropped from the rubber corporation board of directors. Up to his ears now in oil dreams, he sunk the whole of his $2,000,000 rubber fortune in a series of 'dusters' in Texas. Undismayed, he borrowed $57,000 from a friend among the *caoutchouc* barons, brought in oil on the next try and a few years later sold out a part of his holdings in the Luling fields for $12,000,000. On record as an advocate of the distribution of wealth was Edgar by this time. Published in 1920 his 8-page pamphlet, The Dawn of a New Day: Containing suggestions for the introduction of a new industrial order, anticipated the Hueys and Townsends by nearly a decade and a half. Point one of Edgar's pre-Roosevelt and beyond-Roosevelt new deal called only for giving workers an opportunity to buy stock. With point two, however, he really cut loose and demanded formation of a corporation to be capitalised for at least $100,000,000, christened the American Cooperative Association, and dedicated to promoting the interests of 'justice to labor.' The sum could 'easily' be raised, he thought, through 'a campaign organised on the order of the Liberty loan drives.' As point third and last he called on the President and Congress to appoint an Industrial Relations Commission 'which will hear and evaluate all facts bearing on the division of surplus profits between Capital, Labor and Management and recommend fair bases for distribution.' In pre-Roosevelt 1932 he returned to the charge and through the pages of the North American Review announced 'this country cannot remain half capitalistic and half destitute.' Here he summed up his philosophy: 'Workers on farm, in mine and factory must share in the profits of business. The permanent prosperity of the manufacturer, shareholder, merchant, capitalist and general public is dependent upon the economic security of the workers and the safety of their investments, and the urgent need today is that business should be done at an ample profit with generous distribution to workers, management and shareholders.'

The thing that makes all this important is that here is one share the-wealth Messiah who actually has practiced what he preaches. When he pulled out of Luling, Texas (population 1,500), it was to leave behind him a $2,000,000 bonus. Employees who had been with him three years received a sum equal to all they had drawn in salaries or paychecks during that period. From $200,000 each for five employees, the gifts ranged on down to a few thousands or a few hundreds. Having taken care of his immediate family of workers, Davis thereupon handed Luling $1,000,000 for a public park and the State of Texas $1,000,000 for agricultural research. And when he heard that the blacks were barred from the park he forked out another $1,000,000 for a negro duplicate. He set up a fund for the benefit of Luling widows and orphans, paid $10,000 for the 'best' painting of a Texas wild flower, went into the business of hauling a trainload of singers from New York to San Antonio each spring to provide the *Estrada* music. Nor was his Massachusetts birthplace overlooked while he was taking care of his adopted home to the South. Brockton drew a Plymouth County Development Company endowed with $1,000,000 which transformed the Davis homestead into a public park and presented the community with a pair of super-swimming pools. It drew also a widows' and orphans' fund. And the oil millionaire bought a Brockton bank just so he could make an old schoolmate, J. Joseph Cooper, the president. In lending a hand to another hometown schoolfellow, Davis was able to play both Massachusetts and Texas angles. This beneficiary was J. Frank Davis (no relation), whom he encountered in a San Antonio hotel 35 years after they had last seen each other in Brockton public schools. J. Frank, now a Texas journalist, had emerged from Boston newspaper work permanently crippled as a result of a fall on the ice while on an assignment to greet the pole-conquering Peary. Edgar thought it would be a fine idea for J. Frank to write a play and gave him a theme — the transmigration of souls. The resulting drama was that reincarnation epic, The Ladder, which forever will remain notable in the annals of the American theatre as the greatest flop of all time. Five and ten thousand dollar checks Edgar tossed around to the play doctors before the opening. That was in October 1926. For the next year he kept the drama going before average audiences of five to six persons. Then he decided that the only way to bring in the customers would be to make admissions free. This was done and, once the queues started forming for every performance, there was a 75% market for pasteboards obtained by speculators who packed the waiting lines with their own agents. Altogether The Ladder's two-year run cost its astounding backer more than $1,500,000.

Quite evidently more oil money was going to be needed to keep this sort of thing going. Davis, with a theory that immense amounts of oil are to be found at depths below which the commercial product usually is obtained, had reserved such deeps in selling his Texas properties. In September, 1932, came word that he had brought in a wildcat well for 2,500 barrels a day. This was at Buckeye, Texas, 150 miles South-East of Luling and he had gone down 7,800 feet to get his 'black gold.' Late in 1935 a different sort of story came out of Texas. Edgar B. Davis' oil concern, United North and South Development Co., was declared insolvent by a Federal court judge when creditors and stockholders opposed a reorganisation plan. How long it will take Edgar to bob up with another fortune now remains to be seen. When it does happen it will only be well after the event that it is heard about, for all news concerning Davis is second hand. He refuses to be interviewed and dislikes being photographed, which may account for the fact that one or the other of the great American political parties has overlooked the best bet in its history by failing to nominate him for the Presidency. And may also account for the fact that Who's Who In America does not recognise him.

Potentially as important as the Davis-fathered plantations on Sumatra are those of Goodyear Tire & Rubber Co., only American manufacturer to imitate United States Rubber and take the plunge into Mid East rubber growing. Goodyear began by acquiring 17,000 acres of jungle on the Dutch island in 1917. Large-scale clearing was launched at once and planting was well underway when the 1920 crackup halted operations. One result was the adoption of bud-grafting for all planting of 1923 and later. For the whole estate Goodyear claims an average yield of 620 pounds an acre. This plantation yielded its first rubber in 1923, turned in a profit as Britain ran up rubber prices in 1925-1926 and is all in bearing today. A second tract of 40,000 acres was acquired in 1927 as a direct result of the Stevenson restriction plan and its fluctuating prices. Planting of this domain was started in 1928 and completed in 1933. Producing *caoutchouc* now, it will be entirely in bearing by 1938 and reach full production in 1944. Bud grafting was employed throughout and the yield is expected to reach 1,000 to 1,200 pounds an acre. Acquired in 1931 is a third tract of 32,000 acres which brings Goodyear's Sumatran acreage virtually to the same figure as United States Rubber's. No development work has yet been done on this area, which is being held in reserve while the Akron company ponders rubber prices, rubber over-production and the workings of the current Mid East restriction measure. From the United States, Goodyear differs by interesting itself in a secondary line of defense in case British-Dutch-French-Siamese restriction starts squeezing too tightly the production which manufacturers are allowed to take from their own acreages.

Figuring in this plan of action are the Philippines, Panama and Costa Rica. In the Philippines experimental plantings of *Hevea*, *Ceara* and *Castilloa* were made by several individuals in the 1905-1907 period and by 1915 the Basilan Rubber Plantation Co. (Swiss, German and American capital) was making its first commercial tapping of 15,000 *Heveas*. North East of Borneo and geographically a part of the new world of rubber, the Philippines have some extremely suitable *Hevea* territory and especially on huge Mindanao (which has a total area of 36,900 square miles), Basilan (total area 600 square miles) and a number of smaller islands in the Sulu Archipelago. Altogether, it has been estimated, these offer at least 1,500,000 acres well adapted to the *Hevea brasiliensis*. By 1922, however, there were but four small rubber estates on the islands - capital having been very effectively steered off by a land law limiting holdings of anyone corporation or individual to a maximum of 2,500 acres and by a ban on the importation of Asiatic laborers. Then came the Stevenson restriction squeeze, the Harvey Firestone-fostered American Government expenditure of a half a million dollars on surveying possible rubber-producing areas and Harvey's own personal survey. It was to the Philippines that Firestone turned first of all in 1923 and despite that troublesome land law he was unable to get them out of his mind. So it is that in early 1926 we find Harvey Firestone, Jr., being welcomed by Governor-General Leonard Wood and the Supreme National Council as he arrived at Manila on a Prince of Wales tour of the rubber world. Into Mindanao went the junior Firestone with his rubber experts and engineers to confirm what he already knew. The result of his visit was a Firestone-drafted 'Rubber Land Act' hoisting the holding limit from 2,500 to 500,000 acres under 75 year lease and with no lessee being allowed more than 50,000 acres in any single tract. In August 1926 this was introduced in the Philippine legislature to the tune of a terrific ballyhoo about big pay and happy days ahead for the Moros of Mindanao and a commercial cleanup for the whole island world as it forged around the other rubber lands of the globe.

Right there is where Filipino independence hung in the balance. Whether such independence is a desirable thing for the islanders always is open to argument and never more so than now when emancipation day has been officially set. That centering of American rubber production on the islands would have barred the way to freedom hardly admits of controversy, however. And so the fighters for Filipino independence saw it from the beginning. Said Manuel Quezon, president of the Philippine Senate:

> 'We are unalterably opposed to any amendment of our land laws by the American Congress. In the Jones Law of 1916 Congress placed the disposition of the public domain of the Philippines in the hands of the Philippine legislature. We think there would be no justification for Congress to resume control of our public domain.
>
> 'Since the establishment of civil government in the Philippines, Congress has never assumed legislative control of our public domain except to the extent in limiting to 2,500 acres the area of public land that could be owned by a corporation. In other words the limitation originally was imposed by Congress itself.
>
> 'The reasons for our attitude are both political and social. Politically we fear that the ownership of large tracts of land by American corporations would increase the opposition to our future independence and perhaps actually prevent Congress from granting it.'

So the 'Rubber Land Act' of the Firestones was turned over to a committee formed to 'study' it. The committee efficiently buried the proposal. So ended the dream of a *caoutchouc* kingdom under the American flag. Giving up the Philippine idea doesn't mean forgetting it, however, and the Firestones still are a bit huffy about the whole matter. 'The Philippines' lost chance for a rubber empire' is the way they still refer to the episode, the inference being that the islanders are plenty sorry by this time for having kicked that magnificent opportunity away. That the bustling businessmen of the archipelago regretted the loss of the development is not to be questioned, but just how the average Filipino would have been benefited by it is a bit hard to figure out. Nor is there record of any ordinary citizen's having then or since mourned the withdrawal of the Harveys, although total island rubber production for 1934 was but 450 tons.

As to Goodyear, it arrived in 1928 after the Firestone retreat and snapped up the 2,500 acre unit allowed by law. Annual yield from such a patch means little more than a factory supply of a day or so. Why then is Goodyear bothering with this backyard garden in the Philippines? The tire company's 1936 answer: 'It is being developed against the possibility that political conditions might change later, which would permit the Philippines to become an important source of rubber supply.'

Meanwhile, and as a result of the Mid East restriction plan now in effect, these hopeful ones are looking around in Latin America. First business was done in 1935 in Panama, a country investigated by United States Government and Firestone Tire a decade before. Here the globe-girdling Akron corporation acquired a 2,500 acre experimental tract after looking over the sample *Heveas* brought up in the isthmian Experimental Gardens. The concession is in the Province of Colon adjoining Gatun Lake and just outside the Canal Zone. The lands devoted to rubber are exempted from taxation in ambitious Panama, incidentally, and large tracts of public lands can be acquired by negotiation of special contracts with the approval of Panama's Congress.

Goodyear's thousand acre experimental deal with Costa Rica, just North of Panama, was made in 1936. In this 'hot country', important tracts of land once granted to concessionaires, who have allowed their titles to revert, are held by the Government in fee. Legislative approval is all that would be required should Goodyear want to close a deal for any of these. Under old private titles, individuals have areas of up to 50,000 acres and there is even reported to be one 400,000-acre holding. Should *Hevea* and labor conditions prove to be right, then, it is quite probable that the future will see the Akron tire colossus heavily interesting itself in both Costa Rica and Panama.

A trifle ahead of Goodyear in extent of Middle East planted holdings and second only to United States is England's huge Dunlop Rubber Company, an industrial concern in every way resembling the American giants. Dunlop, after messing around with an interest in a Liberian exploitation hereinafter to be detailed, settled down to building estates in Malaya and today has 60,000 planted acres on the peninsula. As antagonistic to restriction schemes as are the American companies, it is yet not likely to look beyond the British Middle East in extending its rubber holdings. So with Michelin, dominant French manufacturer, and French Indo-China. With only two of its three plantations old enough to be giving beginning yields, Michelin additionally will have no important restriction worries for some years yet. Its two older plantations have 23,000 acres of bud-grafted trees.

Firestone Tire & Rubber Co. and more particularly the Firestone company of 1923-1926 has, of course, been the real globe-comber in the quest of rubber lands. The Philippines it checked first, Mexico second, Central America third, Liberia fourth and Sarawak on Borneo fifth. Harvey Firestone's current concentration on Liberia is not due, then, to his conceiving it as the finest *Hevea* land in the world, as current publicity would have one believe, but to this country's having been almost the only non-British area in which he was able to set up on the large scale. Philippine, Mexican and Sarawak negotiations went forward all during the time he was ironing out details with the Liberians and save for such little matters as Filipino patriots, Mexican revolutionists, a changeable white rajah and, perhaps, experts with no stomach for the Central American jungle, he would be carrying on in several of these spots instead of shooting the works on West Africa.

Harvey led off in May, 1923, by cabling his Singapore manager to send a couple of experts into the Philippines. Their report on soil and climatic conditions was extremely favorable, but we already know what happened to that project in the long run. While starting the wirepulling on this one, Firestone made his second 1923 move by sending two of the Philippine explorers into the State of Chiapas in Mexico's extreme South. While they were duly probing a couple of the old abandoned Castilloa farms, the usual revolution broke out and the pair took it on the lam by way of Guatemala. In September, 1925, however, the undeterred Harvey from his vantage point in Akron concluded a lease for one of the Chiapas properties — a 35,000 acre empire with 350 acres in *Heveas* and a couple of thousand in Castilloas—and sent a veteran Mid East planter to the scene. Since the flight of his predecessors, he found, all buildings had been burned to the ground by Mexican playboys and the other improvements had been kicked apart. Convinced that a Firestone Park in Mexico would be liable to raids at any time, Harvey regretfully abandoned America's neighbor nation.

On Panama's Southern province of Darien, meanwhile, he had received a glowing report from Richard Oglesby Marsh, earlier a Washington partner of General G. W. Goethals of Panama Canal immortality and more recently a junior Brain Truster in F. D. R.'s Washington. Marsh, long interested in rubber, whooped it up for Panama from the moment the United States launched its Firestone-inspired quest. Goethals sent him to Secretary of Commerce Hoover, and Hoover sent him to Firestone and Ford. On commission from Harvey and with two Harvey-designated companions, then, Marsh surveyed the better known parts of Panama and thereafter turned his attention to unknown Darien. One of his partners, he says in his White Indians of Darien, was 'a Firestone rubber-buyer, whose experience had consisted of sitting in the bar of the elaborate Raffles Hotel at Singapore, changing into his 35 white linen suits (so he said) and examining sheets of crude rubber brought in for his approbation ... The very sight of the jungle made him ill.' The other was a lawyer delegated to pass on title claims and 'law and order.' Dislike for jungle going, Marsh very strongly intimates, caused these gentlemen to turn thumbs down on Panama as a possible site for rubber growing. Marsh himself tried to sell Harvey and Henry on a thorough investigation of Darien, but the report of his companions prevailed and it was with the backing of a non-rubber industrialist and accompanied by scientists of the University of Rochester, American Museum of Natural History and Smithsonian Institution that he returned to Darien. There he scooped up white Indians and carried them off to Washington — one of the oddest of all the results directly or indirectly stemming from the Firestone hunt for non-British *caoutchouc*. Another was the hell raised on Marsh's second return to Darien when he led the San Blas Indians in an armed revolt against their oppressors — negro hunters of wild rubber and Panamanian officials and 'police' troops. The rebelling reds drove out blacks and Panamanians in a series of battles, set up the Republic of Tulc and only rejoined Panama when highly favorable terms were arranged. Or so Marsh tells it in recounting his own expulsion from the isthmian country, protected in his departure by Uncle Sam but under penalty of arrest and punishment if he ever returns to the land between Costa Rica and Colombia. To the junior Harvey Firestone's 1926 Prince of Wales swing around the circuit we have referred. Leaving the Philippines early in that year he received a cable from the third Rajah Brooke of Sarawak inviting him to move right into that part of Borneo and start cultivating *Heveas*. This was a welcome bid, for right then it seemed that none save Liberia's rulers wanted Harvey plantations and even the Liberians weren't quite so eager as was deemed desirable. So over to the land of the white rajah paddled young Harvey and inspected the outer fringes of the 500,000 acre concession eagerly offered by a ruler whose father and predecessor (the nephew of the original pirate-battling Brooke) had devoted a 49 year reign to resisting any attempt of outside capitalists to muscle in. The setup looked very pleasing to the Crown Prince from Akron and on May 1st 1926, conditions of an exploitation agreement were officially confirmed on behalf of the Rajah by the Chief Secretary of Sarawak. Harvey went on to Liberia, but, before he reached London on his way home, the Borneo agreement was all off. Sarawak, incidentally, is unique in the British empire — an independent state with Great Britain controlling foreign relations under an 1888 protection agreement, but having no right to interfere with internal administration by the Rajah. The Firestone deal apparently fell under the head of foreign relations. Or, as a Harvey laureate once phrased it, cc••• the long arm of the British Empire had reached to Sarawak, and the independent Rajah had been tapped on the shoulder.'

Which brings us to Liberia. Here also the British had been ahead of the eminent Akron Anglophobe. They had, however, been gone long enough for the Firestone forces to feel safe in taking over their abandoned plantation without fear of contamination. Only republic on the African continent and one of the world's two black republics, Liberia is a Noble Experiment dating back to the early nineteenth century activities of the American Colonization Society. By purchase and treaty it acquired land along the West Coast and considered that in sending black freedmen there it was returning them to their 'home.' Africans were Africans to these philanthropists and it never seems to have occurred to them that the intrusion of strange blacks with ideas of ruling was more to be resented by the native Krus and Buzis and Greboes than if so many white men had come among them. Out of this clash between ebony and ebony have sprung more wars and skirmishes than almost any other State on the continent can show. Rapid, however, was the economic progress of the strange blacks for a time. But even speedier their fall into decay. Starting out as industrious farmers, planters and small shippers, they received an ill windfall when the United States funneled in 13,000 Congolese taken from captured slave ships. For their initial support $150 apiece had been appropriated by Uncle Sam and the first settlers lost no time in dividing up money and Congolese and graduating into a slaveholder life of ease in excusable imitation of the ways they had observed across the Atlantic. The American Civil War was another bad break. That definitely shut off emigration from America, for the emancipated black preferred to remain right there as he does today despite all his sentimental attachment to the little republic in the African bush.

Nor has he been wanted in Liberia for many a decade. Descendants of the original Americo-Liberians and their intermarriages with the aborigines have degenerated into a papsucker political class of 10,000 or so individuals who live not by agriculture, not by trade but by 'governing' the 40-50, 000 small farmers, sailors and other 'civilised' Africans along the coast and 'governing' and exploiting the 1,500,000 to 2,000,000 tribesmen of the interior. A sweet racket it is and no reforming black intruders from the United States are desired.

It was only with the present century and notably with the organisation of a Frontier Guard army in 1907 that any real extension of authority over the interior began, incidentally, but today the politicians at the capital, Monrovia (population 10,000), have all the constituents well under control save for the seafarers and battlers of the Kru country. There so recently as the end of 1931 the Frontier Force knocked off a few hundred revolters, burned some 40 towns and chased 10,000 tribesmen into the bush. There so recently as 1934 occurred disturbances reported as Frontier Force attacks on disarmed native villages and characterised by a denying Liberian Government as an attack on a Frontier Force party by rebels deluded by belief in a coming white administrator. Along with use of terrorist troops the politicians retain their control over the tribesmen by appointing the paramount chiefs whose loyalty now is to the President at the capital instead of to the people who themselves did the selecting in the old days. Only black landowners are eligible to vote in Liberia and since there are but 6-15,000 of them, it is not to be wondered at, that the True Whig party has been able to remain in power for over 30 years with little opposition, that an oligarchy of six or seven families has been able to dominate this party, and that the office of President is virtually that of dictator. For the rest Liberia has been a stronghold of indigenous slavery and a center for forced labor exports; it is without a harbor, without sanitation, without public works worthy of the name; is, in short, the least developed, worst governed, financially shakiest and, perhaps, internally unhappiest State on the map of Africa. It is, however, a charter member of the League of Nations.

Liberia's closed door policy has not meant protection of the indigenes, but merely protection of the black interlopers in their exploitation of the indigenes. Originally that policy applied to trade on the coast as well as interior penetration, but for half a century the Liberians have been content to have the white Europeans and Syrians look after all the trade while they busy themselves with loafing on the government payroll and battening on the tribesmen as tax collectors, slave exporters and owners of private plantations maintained by labor levies assessed in the name of the government. Peculiarly, however, they always have been willing enough to unlock the interior for white rubber men. So it is that today we find Harvey Firestone owning the only large private enterprise in the country; possessing the only white interest or holding of any sort save coastal commercial and trading houses. And examining the history of his predecessors we find that they too, almost without exception, were rubber men. It was in 1898 that the first real concession in Liberian history and the first exploitation of rubber on any noticeable scale simultaneously arrived. To a London outfit, the Liberian Rubber Syndicate, went a 26 year grant to collect and export the milk of the indigenous tree and vine. This concern began the organising of collecting in the interior but fared none too well, and, after passing through various French, Dutch and German hands, it wound up in 1904 as part of a new British corporation organised to exploit anything and everything in Liberia. This was the Liberian Development Company which never operated save through the Dunlop-financed subsidiary organised to handle the wild rubber concession. The other grants it had schemed to obtain all were knocked out by the Liberian legislature in 1906.

So early as 1871 the free blacks had made their first bid for European banker aid in their financial troubles by floating a $500,000, 7% bond issue guaranteed by the customs of the nation. Of this amount $150,000 went to the bankers. An amount never exactly determined was expended by the then - President of Liberia on self and friends and about $100,000 actually reached the Government treasury. With a new President on the job in 1874 Liberia declared a moratorium on debt and interest payments and it was not until 1898 that it agreed to payoff a principal of $350,000 to $400,000 at a progressive interest rate of 3-5%. The country was again sorely pinched for money in 1904 and so we have the Liberian Development concessions and Sir Harry Johnston negotiating a $500,000, 6% loan to the company through the Erlanger Company of London. Of this sum, $150,000 was to go to redeeming the floating indebtedness and Treasury Notes of the country and the balance to the rubber company for use in constructing roads, paying off company indebtedness and financing a bank. The *caoutchouc* company received about $200,000, built a 15 mile motor road (promptly destroyed by tropical rains), stuck a launch on one of the rivers and bought two automobiles. About then it announced its funds were exhausted. It never did account for its expenditures. In January, 1908, Barbadian-born President Arthur Barclay indignantly reported to his legislature that Sir Harry was proposing that the government buy out the Development company for half a million dollars. In the windup the unused balance of $150,000 was returned to the Government, which made itself directly responsible to the Erlanger group.

Rubber exploitation by the British continued, however, and in addition to gathering wild (Liberian rubber exports were 101 tons in 1906 and 46 in 1911), the Liberian Rubber Corporation set out a plantation at Mount Barclay ten miles from Monrovia. Planted first in *Funtumias*, this was replanted with *Heveas* in 1909-1910 and by 1912 the company had reduced capital and was continuing as a plantation proposition exclusively. By 1914, the year in which some of the trees first were tapped, some 1,400 acres had been planted and in 1916 the entire plantation came into bearing. Liberia's export tax of six and twelve cents a pound on rubber (six cents when exported directly by the plantation) was one of numerous factors making it virtually impossible to compete with the East and in the War year of 1918 the British gave up. The 1,400 to 2,000 acre plantation reverted to the Government, which tapped it in 1919 but was unable to maintain interest in this activity for any longer time. The chart of Liberian rubber production thus runs: 1917: 31 tons; 1919: 127; 1920: 57; 1921: 6; 1922: 0; 1923: 3.

So much of detail we give on this planting because, experimentally taken over by Firestone in 1924, it sold him on the Liberian idea and caused his optioning of a million acres more. As to the early loan history that, too, is important, for the matter of Liberian borrowings continues to be inextricably bound up with the history of rubber collecting in the Republic. Just as the Mount Barclay experiment by the British was the nucleus of Harvey's plantation, so has the Republic's original borrowing gathered accretions to become the Firestone mortgage of today.

Harvey's first agreement with the Liberian Government was the leasing of the Mount Barclay plantation for 99 years at $6,000 a year. Another, in connection with his million acre lease, obligated him to construct a harbor for Monrovia for which he would be reimbursed in a sum not to exceed $300,000. This proved to be an impossible job and the Akronite gave it up and absorbed an alleged loss of $115,000. The 99 year main lease, executed in tentative form in 1925 and in final form in 1926, permits the rubber baron to select his maximum million acres at will and with only the 'tribal reserves of lands set aside for the communal use of any tribe' being exempted from its operation. It gives him the right to engage in any operation upon his land and exempts all plantation products and machinery and supplies from internal tax or from customs duties save for a 1% tax on rubber exports. For land actually under development Harvey pays an annual rent of six cents an acre.

One thing that Firestone insisted upon before closing the deal was that Liberia obtain a $5,000,000 American loan. Such was done after considerable bickering. Designed to protect Harvey's investments in Africa, this 40 year, 7% loan of 1926 is in the ultimate $5,000,000 amount and $2,500,000 already has been advanced. It was extended to Liberia by the Finance Corporation of America with the National City Bank of New York acting as fiscal agent. Finance Corporation is, of course, Harvey under a more sonorous name. The loan not only bans Liberia's obtaining any additional loan or creating any floating debt until Firestone has been repaid, but also bars it from negotiating any refunding loan for a period of 20 years. In other words it gives the Akronite a financial monopoly in Liberia for a fifth of a century, at least. At the time it was made, Liberia's chief indebtedness was a 1912 loan with a face value of $1,558,000 extended by British, French, Dutch, German and American bankers at the suggestion of the American State Department. It had gone into redeeming previous bond issues and internal indebtedness notes, the greater part of the new bonds being issued directly to creditors in payment of claims and nearly half of them going to London. To secure charges on this loan, collection of customs, rubber taxes and head money were placed under an international receivership. Announced idea of the Firestone loan was to rid Liberia of this receivership and to enable it to payoff internal and floating debts, which had risen to over $600,000. Finance Corporation immediately purchased bonds to the extent of $2,500,000 at 90 and it has an option on the second $2,500,000 which can be issued only when customs and head monies reach $800,000 annually for two consecutive years. On the first $2,500,000 Liberia netted $2,027,700 of which 90% went to retiring the existing obligations, the international customs receivership thereupon giving way to exclusive American supervision of both internal revenues and customs. According to the Firestone apologists, Harvey's insistence on making the plantation deal dependent upon Liberian acceptance of such an American loan, was motivated by belief in the necessity of extinguishing European (and more particularly British) rights in Liberia and preventing the Government from granting competing concessions to foreign concerns which would impair his own. The 20 year provision against a refunding was, of course, to prevent the British from again coming in.

That Firestone 'forced' this loan upon the Monrovian ruling clique which is Liberia is often charged. The charge does not stand up. Nor does the claim that he schemed from the first personally to obtain financial control of the country. The records show that he originally wanted the United States Government to make such a loan. This would have saved his hazarding the loan millions and at the same time would have given him better protection for his rubber investments. As it is the loan has been a source of continuous grief to him.

On paper it would seem that the conditions are hard upon Liberia, but in actual practice it must be remembered that there is nothing to make Liberia live up to them. Ignoring the loan agreement, it almost immediately began the accumulating of a new floating debt (which passed $600,000 by the end of 1933) and in December, 1931, it calmly suspended payment of interest and amortisation on the Firestone debt. Thereafter Harvey, the State Department of the United States and the League of Nations, to which Liberia had appealed for assistance, did all the worrying. A veteran problem child so far as the League was concerned, Liberia was supposed to be cleaning up slavery, disease and other disgraceful conditions — a situation which made the League interested in its financial rehabilitation. The slavery probe had forced the resignation of President Charles Burgess Dunbar King, negotiator of the Firestone concession, in early 1931 and his successor, Edwin Barclay, never had been recognised by the United States, which was insisting on 'sincere' and 'comprehensive' reforms in the forced labor matter. Not until the summer of 1935 was anything ironed out. The League's original plan of assistance included such recommendations as Firestone's reducing the size of his concession, raising the annual acreage rental from six to 50 cents and consenting to suspension of interest and sinking fund payments on the Finance Corporation loan until revenue and receipts of the Republic had reached $650,000. It also suggested a lower interest rate on the loan and the Finance Corporation's advancing at par the $247,000 still outstanding from the $2,500,000. Firestone and the State Department of the United States were as one in clamorously refusing to go along on this. Eventually, with the Liberian moratorium still in effect, there was worked out a proposal satisfactory to Harvey. He conceded permanent reduction of the interest rate from 7% to 5%; agreed to a five year renunciation of interest accruing during anyone year if the total revenue of Liberia should be insufficient to pay running costs; and guaranteed an immediate loan of $150,000 as initial working capital to start the plan of assistance. This arrangement Liberia refused to accept and the League gave up.

Harvey thought that the United States should play a stronger hand. He said, following a 1934 conference with Secretary of State Hull: 'We believe that the American government should accept the leadership in assisting Liberia to rehabilitation through the league plan, as the American people are responsible for the establishment of that country.'

That leadership America now assumed. In June, 1935, negotiations brought diplomatic recognition from the United States after the five year break as Liberia accepted a supervised rehabilitation plan much the same as the compromise previously rejected by the blacks. Temporarily peaceable, then, is the turbulent Liberian financial situation. That it will long remain so, only a thoroughgoing optimist would affirm.

The fact that the American State Department originally encouraged the Firestone plantation leases and Firestone loan, the fact that it went to bat for Harvey in breaking the moratorium deadlock, raises the question of how far it might go in the future to protect the Akronite's interests. American intervention in Liberia certainly would be in direct conflict with the policy under which President Roosevelt pulled the marines out of Haiti, only other black republic in the world, and it is safe to say that the current administration would do no more than lodge representations no matter what happened to Harvey's investment. But since Firestone is a consistently heavy contributor to the GOP campaign chest, a Republican administration friendly to the tire builder might conceivably go all the way to help him out should his stake in West Africa be placed in jeopardy.

And aside from the matter of Firestone investments there has long been a strong sentiment for American intervention based on honest belief that Liberia and Liberia's tribal millions would be better off under Uncle Sam than under the Monrovian politicians. Whatever formal stand they might have to take in League of Nations councils, Britain and possibly France would welcome this, for they look on their West African neighbor as a source of disease and disorder spreading beyond its own boundaries. In Liberia itself the Krus long have wanted to be under the British and in the interior many a tribe has been eager for a white administrator. That many American blacks would favor intervention is unlikely, yet George S. Schuyler, well known black journalist, affirms that 'only the lunatic fringe of Garveyite Aframaniacs' still sees the Republic as a glorious vindication of the black race's capacity for self-government. Schuyler, on returning from Liberia in 1933, reported, 'The Aframerican who goes there a resolute advocate of Liberian independence is more than likely to come away convinced of the necessity for American intervention.' He also recorded, 'It is the consensus of opinion among the Americans in Liberia that the thing most necessary there now is American intervention and benevolent supervision for a few years, with perhaps intelligently directed immigration of Aframerican farmers and artisans later on to infuse some new life into our African step-child.'

This is not immediately on the books, as we have said, and actually there is a strong question as to whether the current Roosevelt administration would even perpetuate the pre-Firestone tradition of sending in a cruiser or so at threat of foreign aggression. In Liberia, the Government-controlled press is, at the moment, steaming up such an aggression scare. For long Monrovia had seen Britain and France as the monsters threatening to devour. Then it was France alone. Now it is Germany. The report around the black capital is that Hitler was only awaiting the outcome of the now completed Italian conquest of Ethiopia before gathering Liberia in. Referring to the weakness and ineffectiveness of the League of Nations in the Ethiopia matter, Monrovia's Weekly Mirror whoops:

> 'It is evident that Liberia is destined to be the next objective of European imperialism, and unless Liberians refuse to be a nation of orators and cease from idealising, and look to machine guns and bullets and gases and explosives as the god of the ark of the covenant, and not the covenant of the League of Nations, they are doomed because the republic is unarmed and is incapable of any resistance to organised attack from without.'

That Germany, stripped of its African rubber colonies in the World war, would be inclined to pay any attention to Harvey's claims in event of its actually pocketing Liberia is extremely unlikely. Nor would it be restrained by fear of giving offense to the United States, for the ultimate offense would be given by the mere act of invasion. This it was that kept Britain, France and Germany from occupying the Republic in pre-World war days, but currently Naziland doesn't seem to mind offending the world or any part of it. The odds are, of course, that the invasion is a figment of Monrovian imagination; nevertheless it provides matter for interesting speculation.

That Firestone so far has been a loser by his African loan we already have shown. Also, as we have seen, he dropped $115,000 in that ill-fated harbor experiment. Additionally he claims to have spent $200,000 on medical treatment, $56,000 on a hospital, $275,000 on 125 miles of roads within his plantation, $30,000 on his radio station (which offers a public service as well), and $10,000 on a trade school and farm 'to help the natives of Liberia become more useful citizens.' And to have contributed $650,000 to extending Government roads and $36,000 to various medical, biological, anthropological and forest surveys by Harvard and Yale. The total he actually has sunk into these trimmings and his main business of *Hevea* cultivation remains matter for speculation. His company's 1936 report is 'approximately $8,000,000,' the same figure it has been using since the beginning of 1933.

Firestone occupied the Mount Barclay plantation from the beginning, of course, and of the million acres allowable at his discretion he has so far helped himself to exactly 110,000, of which 55-60,000 are planted (the seeds came out of the Middle East) and another 50,000 marked out. Of the 60,000 acres, 12-15,000 are in bearing this year, according to assorted announcements. Since a planted area of 50,000 acres had been announced by so early as September 1931, it can be seen that there has been very little expansion in the last five years. Record low prices for rubber which prevailed until just recently are chiefly responsible for this halt, and the Liberian moratorium on debt payments to the rubber baron may have figured in this also. Whether the land is showing itself as suitable to *Hevea* cultivation as had been hoped is one question that the present writers do not now profess to be able to answer. Naturally, the Firestone company says that it is and the British say that it isn't. United States Department of Commerce statistics show only that little rubber has come out of Liberia to date, that Firestone production seems to have been much slower in coming along than had been expected, but that 1935 seems to have marked the arrival of regular production on a considerable scale.

For the whole of Liberia exports of crude rubber amounted to 170 tons as Firestone began his major clearing operations in 1927. In 1928 they were 108 and in 1929 a total of 187, of which three tons represented latex. This last was the year in which Harvey announced that his first harvest from other than the Mount Barclay inheritance, his first imports on a commercial scale, would take place in 1931. In 1930 total Liberian exports dropped to 100 tons of which 12 were latex; in 1931 the figure was 102 of which less than a ton was latex. In the 1932 of record low prices there were no exports of rubber whatever and for 1933 the figure was but 27 tons. In 1934 the United States imported 63 tons of dry rubber and 278 pounds in the form of latex in three shipments from Liberia. During the first 11 months of 1935 there were imports in every month save April and the total leaped to 649 tons. From now on that figure should multiply at a rapid rate and a check of annual import statistics is all that will be necessary in determining how well Firestone is going to make out. Since the present planted area is largely bud-grafted (the buds came out of the Middle East), the Department of Commerce estimates that at maturity it might yield 15-20,000 tons annually.

Naturally the Liberian Government hasn't been getting rich on rents and taxes from such development. On the 50,000 acres, rental was but $3,000 a year; on the 110,000 it is but $6,600. Export taxes on rubber has been too microscopic to be discernible, of course, and if 16 cent *caoutchouc* continues, the possible 20,000 tons to come in the future from the present planted area would yield but $71,680. You can write it down that Firestone never will plant a million acres but trying to figure out how much he will plant is sheer guess work. He has no idea, himself. Originally he talked spaciously of spending $100,000,000, developing the whole million acres, employing 300,000 to 350,000 men, producing 200,000 to 250,000 tons of rubber annually. This was before he knew anything about bud-grafting and, if Liberia gives any kind of rubber results at all, a million planted acres would produce 400,000 tons or almost half the world production of 1935 (865,000 tons).

For such amount of rubber Firestone could, of course, have no possible use as consumer or seller unless he can arrange to have some plant blight desolate most of the Middle East. Fully and successfully planted in high-yielding bud-grafted trees the 110,000 acres on which some work has been done would give him a good two-thirds of the rubber he uses in his present operations. Imagine his business expanding sensationally, even imagine him getting into the game of peddling rubber to manufacturing competitors, and still the crystal doesn't show him approaching any half-million-acre planting save in an excessively remote future. The million-acre option is an extremely nice thing to have, however. It keeps out competitors and it doesn't cost anything to hold it as insurance against the desolating of the East or the ascent of China to a two-cars-in-every-garage scale of living.

It is well for Liberia that the million acres never will be developed, that Firestone never is going to want any 350,000 laborers. Full of misinformation and talking through his hat he was in those first days when he described the indigenous labor supply as 'practically inexhaustible.' Actually the total able-bodied male population of Liberia is no more than 3-400,000, so that if his original spread-eagle plans were carried out either the whole male population, willing and unwilling, would have to be put to work for Harvey, or tens of thousands of coolies would have to be brought in over Liberian protest. In the Belgian Congo, partnership recruiting by the government-industry twin has succeeded in turning up a worker supply of but 170,000, including forced labor. And the Belgian Congo has a population of 8,000,000 to 10,500,000 as against Liberia's 1,500,000 to 2,000,000.

Firestone has benefited the tribesmen by making employment available to thousands among them who want it. His wages are not the native-astounding ones of the rapt publicity yarns, but they are the shilling to a shilling-and-a-half a day standard throughout West Africa. And for Liberian natives to obtain this or any sort of wage before, it had been necessary for them to leave their country, since within its borders virtually no gainful occupation was open to them. Firestone has had about 10,000 men (under about 75 whites, chiefly Americans) working most of the time as against 18,000 at the peak of original clearing activities. And with increased activities this year he is putting the number back up to around 15,000. On the treatment of labor on the Firestone plantations no complaints have ever been made save in connection with pay matters. And it is well established that since 1930, and probably several years earlier, the company has employed none but voluntary workers. Not only are the blacks free to leave at any time, but they are under no compulsion regarding the number of days they shall work in any given month. In 1930, for instance, the 1930 League of Nations - United States - Liberia commission to investigate forced labor found that the average daily turnout was only about 60% of those on the payroll.

In the beginning years, however, there is no question that the Akron outpost used some labor recruited by force and guile, workers who were unwilling to go to the rubber farm, workers who were taken there to serve three-month tricks without their having any notion of where they were going or what the company and its activities were all about, and workers who had no other idea than that they had been nabbed for Government service. In its report the committee of inquiry recognised this:

> 'The Commission finds that labour employed for private purposes on privately owned plantations has been impressed for this service on the authority of high Government officials. That there is no evidence that the Firestone Plantations Company consciously employs any but voluntary labour on its leased rubber plantations; but this, however, was not always the case when recruiting was subject to Government Regulations, over which the Company had little contro1.'

Firestone's manager, D. Ross, told the commission that about 10% of his 1927 labor had been supplied by the Government. Whatever was supplied by the Government was supplied by forced labour recruiting methods, it is safe to say.

It was the capturing, kidnapping or enticing of bucks to be exported to Spain's fever-ridden African cocoa island of Fernando Po that belatedly touched off the Liberian forced-labor investigation giving the Firestone plantation almost a clean bill of health. Not for any other part of Liberia could the commission say anything at all of the kind. Indiscriminate forced labor on state roads and the private rubber plantations and farms of Government officials presented to it as appalling a picture as the Fernando Po racket. And also contributing to the picture of sweetness and light in Liberia, it found, was the indigenes' own system of seizing, buying and selling slaves in a considerable system of inter-tribal domestic slavery violating the Constitution of Liberia.

Not only did the investigation result in the departure from office of the Hon. Charles Dunbar Burgess King but in the similar resignation under fire of his vice president, Hon. Alien N. Yankee. Additionally involved was a posse of King relatives, Cabinet officers, legislators, representatives of all the upper crust of Monrovia. How much cleaning up Liberia has done since the King-Yankee ousting it is impossible to say although we do know that the American refusal to recognise President Edwin Barclay (Secretary of State under King) was based on denial of the effectiveness of his reform measures. Presumably the recognition last year granted in exchange for Liberia's doing right by Firestone is founded on the belief that under the plan of assistance the demanded 'comprehensive system of reforms' can be effected.

In fairness to Firestone (who never has seen Liberia) we must emphasise that his labor record is about as good as any company entering the Liberia of 1926 could have achieved. Additional credit accrues to the rubber company when we remember (quoting the investigators) 'the precipitous character of the Company's pioneering operations in Liberia, the unfamiliarity of its employees with African conditions, the unsuitableness of equipment and outfit, and the inexperience in the control of primitive labour.' And that while Harvey's entrance on the Liberian scene undoubtedly gave strong impetus to Government road building he could not have foreseen what that would mean to the natives drafted for the work. Unfortunately the Firestone development can't save luckless tribesmen stuck for public works toil in the future, but it is pleasant to reflect that the native voluntarily seeking employment can turn to the rubber estate instead of having to sign for Fernando Po, which always has recruited some free labor along with the victims of force.

The summing-up on Firestone, then, is that whether he has exploited the Monrovian politicians or been exploited by them, he probably has benefited the natives of the interior more than he has harmed them. That by pouring millions in money and a decade of work into Liberia in exchange for a monumental pile of worries and a trickle of rubber now just beginning, he has learned that one unable to tolerate British or Dutch or French jurisdiction over his rubber-growing must reconcile himself to paying a price for his independence. That if he had deliberately sought to pick from all the lands with climate and soil suitable to *Hevea* the one that would require the most toil and grief and soul-searching to develop, he couldn't have done much better than Liberia. Considering the expense he has been to, the cracks about 'Liberian slaves' emanating from labor leaders of America, the future uncertainties as to labor supply, loan payments and foreign aggression, his bitterness about Filipino 'ingratitude' becomes understandable.

Henry Ford, friend of Firestone's, feels the same about Brazilians, it may be, but for a different reason. The Brazilian politicians (those of Para state, at least) were as anxious to have Henry come in as the Filipino statesmen were to keep Harvey out, but Ford has found no happiness in being welcomed. On expenditures the Ford company is putting out no information whatsoever, but the figure generally used is the same $8,000,000 the Firestone company sticks to in connection with Liberia. As against

Harvey's 55 or 60,000 planted acres, however, Henry has but 10,000 to show and not a single pound of rubber has he taken out. He can, however, point to a record of political interference, labor disputes and planting failures more than overmatching his friend's troubles with harbors, loan moratoriums and a slaving officialdom. Most interesting rubber plantation development anywhere in the world is the Firestone one today. The fact that it marks the first large-scale, modern attempt to grow cultivated Para outside the Middle East is one reason for this. Chiefly, however, it is the acreage to which that estate might be expanded which attracts our attention. Basically the spectacle of returning *Hevea* milking to Brazil is at least as interesting as trying it out in Liberia, and on the potential acreage count Ford has Firestone topped three to one. Less interesting to the rubber world, however, and rating less space in this book, is the Ford venture for one very good reason - the fact that the extent and depth of the Detroiter's interest in *caoutchouc* is more than open to question today.

Into Ford automobiles, it has been estimated, go some 23,000 tons of rubber a year, with tires accounting for the largest part, and with Ford himself manufacturing no more than *caoutchouc* steering wheels and other such relatively unimportant gadgets. Assume that Henry could build his own casings cheaper than he can buy them (the odds are against this) or assume that he would deliver his own *caoutchouc* to a casing plant and pay a fee for the building work, we can easily see his using 23,000 tons or more out of Brazil providing he can raise this rubber for less than the Middle East selling price (which remains to be determined). There remains, however, the question of what the man can do with three million acres of potential rubber land when, if it produces up to Mid East bud-grafted standards, some 60,000 acres would give him all he can use. Picture his going into the crude rubber peddling business, even, and there's still nothing in the cards that shows three million or two million or one million or a half million or a quarter million acres being planted. For some years, patriotic Detroit organs and bombastic Brazilian officials have talked loosely of Ford's 'controlling' the world *caoutchouc* output 'within a decade.' This, of course, is utter nonsensical rot. Drop the 'within a decade' and it still remains comic page stuff, for it has to be based on the naive notion that everyone in the rubber-growing business is going to drop out just because the Detroiter has bowed in. The most he can do, if successful, is set out sufficient rubber to become an important factor in determining rubber prices, and it is extremely unlikely that he will do this much. There is, in other words, no margin in it for him.

That, of course, brings up the question of what got him into the rubber business in the first place. It was the $1.21 a pound rubber of 1925 plus the 'America Should Produce Its Own Rubber' ballyhoo of his friend, Harvey. Today, it's a safe guess, he often wonders just why he is in the business. How far he will go with it, not even Henry himself knows. As to his Detroit lieutenants, they have no idea of what it is all about. Except that it's costing a lot of money. That Henry could make use of a limited rubber tonnage, we have seen. On this matter of use the only statement direct from the old gentleman himself is a 1928 declaration that tires would be manufactured in Brazil in connection with the plantation. 'Manufacture of finished products from Brazilian rubber should be located in Brazil,' he was quoted as adding. Unless he expects to get out only a very minute quantity of rubber, this is the old tot for Latin American consumption. Brazil, which already has tire and other rubber factories, is far from being an important consumer of rubber articles, and it certainly isn't going to be able to ship them into the United States in competition with the American product. Also included in the 1928 statement was a Henry declaration that he might not only cultivate rubber but nuts, cotton and whatever else had a chance of being profitable. Along this same line, J. R. Weir, his 'director of agricultural research,' who commutes between Para and Detroit, recently spoke of experimenting with fibrous plants, oil-yielding and food nuts and soy beans. Actually, Ford doesn't know just what he is going to grow and what acreage he might use. Rubber he regarded and still regards as the principal crop, but the puling 10,000 acres planted in 10 years is the best indication that he's far from making up his mind as to whether it's really going to be important. And the popular picture of some years ago showing our Henry steaming right into Brazil and slap-bang putting rubber growing on a Dearborn-style efficiency basis, shaming all the planters of the world, has become a very comical thing, indeed. He is, already, a failure as a tropical agriculturist to this extent: any company formed to make money out of rubber raising would long ago have been broke after spending so much and obtaining nothing. The thing that the chortling British forget in considering Ford, however, is that if this man wants Brazilian plantation rubber strongly enough, he's going to get it and in such quantity as he desires. That the *Hevea*, properly cultivated, will do as well in Brazil as in the East is not to be doubted, and if there's no limit to one's purse that cultivation and the unwilling necessary labor can be obtained. Funny some of the Ford blunders may have been, but no funnier than the original British ones in Ceylon and Malaya. Short, then, of a revolution chasing Henry and all his works clear out of Brazil, the whole question of Ford plantation output ultimately hinges on just how badly the Detroiter wants rubber. Our guess is that he's in deep enough still to want a quantity of it, but of what that quantity ultimately will be we're as ignorant as is Henry.

Ford began his Brazilian investigations by sending in his experts in 1926 in the wake of a 1925 United States government survey group financed out of that $500,000 appropriation sponsored by Firestone and his friend, Harding. In 1927 the Detroiter obtained his concession from the State of Para, the deal involving purchase of a concession granted a politician, Dummont Villares, by Governor Dyonisio Bentes, familiarly known to his political foes as 'Concrete Cement.' Over the disposition of the Ford money — supposedly a half million dollars — there was a good deal of scandalous uproar in the Para press and a number of government officials were involved. At its height Villares took off for Paris and a Bentes newspaper published letters from this promoter and a partner affirming that not a cent had been paid to Governor or Government. Adding mightily to the wrath of the press was the fact that all terms of the grant to Ford were concealed until after the deal had been completed, and that when they were brought to light (May 1st 1928) they proved to be sweeping indeed. This original Ford rubber empire, in which Ford never has set foot, consisted of some three million acres back from the East bank of the Tapajos river and about 150 miles South of its junction with the Amazon at Santarem. It will be remembered that it was up this tributary that Wickham went for 50 miles before cutting over into the plateau forests of its West bank for his seed gathering. Ford's grant includes unlimited permission to engage in commerce, industry, banking and navigation, with none of his activities subject to government inspection and with all of his subsidiaries to be free of all municipal or provincial taxes. All his imports and exports are free for 50 years, and he has power of expropriation over any lands deemed needful for highways or railroads from Santarem to the limits of his concession. Also he polices his own domain, and there he has put Prohibition in effect and infractions of the moral code mean banishment for the sinner. All in all somewhat of an independent duchy to which the Duke of Detroit well might retire if the Communists ever take over Detroit and Dearborn. So far his possession of a part of it and of a new parcel has been productive of good to the Amazonian who wants regular work at regular pay and can endure sanitation and eatable food. And it has been a godsend to long-expiring Santarem and to Para city, for in these communities, especially Para, he has been spending vast sums of money in purchases of supplies.

Ford's first move towards developing the duchy came in July, 1928, when the Lake Ormoc, former United States Shipping Board vessel, took off from Detroit with a load of saw mills, structural steel, cement mixers, machinery, boilers and other American exotics for the Tapajos. Since then Henry has had a great success in building a model South American town. Complete with school, hospital, power house, ice plant, radio station and Boy Scout troop is his Boa Vista and 90% of the workers now wear shoes as against the 1% of first days. Down river and just 30 miles from the Santarem which this year paid toll of hundreds of lives to the worst malaria epidemic in its history another development, Belleterra, is coming along. For 687,000 acres here Ford in 1934 traded the government 687,000 acres of the Boa Vista domain — which even yet has hardly been explored. Whether the switch was dictated by dissatisfaction with planting results or by his belated discovery that Boa Vista is accessible to boats of any size in but few months of the year, never has been disclosed. It is very evident, however, that for all his road building and running estate railroads back from the river, Ford has but little rubber planted. At the original site he had cleared only 8,500 acres by the end of last year. Of these, 8,100 were planted. At Belleterra, 3,700 acres were cleared and 2,350 planted. The fact that the whole 1929 planting of 1,000 and some acres died is one answer to why no rubber has yet come out of the Ford estate. These were native Tapajos seedlings apparently set out helter-skelter in order to live up to concession terms which called for Henry's planting 1,000 acres a year for each of the first four years. To obtain seeds from the upriver trees Ford sent collectors into *Amazonas*, Matto Grosso and the Acre. They gathered a million which were promptly seized at Manaos by *Amazonas* officials jealous of Para and held for payment of heavy duties. Ford went to the supreme court of Brazil and was thrown out on technical grounds, the petition not being properly drawn. Eventually he won his case — but the seeds, of course, were long dead. More seeds were obtained. Then there was trouble with grasshoppers, trouble with hilly land, trouble with everything. The art of terracing finally was learned; seed beds were established; eventually Henry's experts heard of budgrafting. With this last news, Ford had to give up the idea of a self-sufficient Brazil and over 2,000 clone stumps were brought in from the Middle East, budwood from these being budded to seedlings in the nursery and planted area. Future plans call for planting budded rubber only. In the meantime a large part of Ford's planted area is of tappable age, but commercial tapping, for some reason, has not yet been started.

Considering Ford's planting troubles, it is well, perhaps, that he failed in plans to get ahead with large-scale rubber development to begin with and that expansion has been small to date. Perhaps the principal factor that held him down was inability to get labor, and while this may have amounted to an unintentional benefit then, it looms as a serious problem in case he actually has passed through his rubber growing apprenticeship. Competent to raise rubber on a large scale he now may be. Heeled as he is, he certainly can be expected to carry his own rubber and turn a transportation handicap into a transportation advantage over a Middle East immensely farther from New York than is the Tapajos. But confronting him is this labor problem that well may prevent his ever competing successfully with the East and its coolies on a cost basis. Ford as a Lady Bountiful awing and delighting the Brazilians with such wages as they never had dreamed of is one of those nonsense creations of the North American feature writer. He does, however, offer just about the only cash employment available in the Amazon valley; his wages of 60 cents a day and up compare favorably with those paid in Brazil's cities; his workers are well treated and general living conditions on his properties are incomparably superior to anything ever before offered up the Amazon. Nevertheless, he can't get the men he wants. The liquor ban, the prying into the immoralities and the general 'efficiency' system have considerable to do with this, doubtless; but the chief reason would seem to be that the Amazonian just doesn't care for steady, routine employment. By force he can be made to sweat under such conditions, but left to himself he'd rather flirt lightly with a little independent fishing and farming on a semi-starvation basis. To work a few months for Ford and bounce away feeling wealthy is one thing, but in nearly 10 years in the Amazon the Detroiter has caught but a very few thousand willing to settle down as steady workers. Nor can he go outside for them. Brazilian feeling runs so strongly against this that a mere rumor of his importing American negroes brought official protests, while the actual presence at Boa Vista of 50 West Indian strays; combined with newspaper reports that labor was to be brought wholesale from Barbados, touched off a genuine riot.

When the first call for Ford workers was sent out in 1928, 5,000 were wanted and 1,600 responded. Many of these cleared out at the end of the first three months, more or less satisfied with their takings. The first strike came shortly after work was started and was for a pay raise, which was granted. Shortly there was another one and the state military police had to be sent in, for the Ford force was yet in process of organisation. Eventually about 90% of the staff went back to work and the usual hooey about 'Bolshevik' money having been used to promote the strike was loosed. Early in 1929 came the big race riot with a West Indian stabbed, a couple of Brazilian heads cracked and the Ford management bowing speedily to the mob demand for expulsion of all the British blacks from the development. In August of the same year came another two-day strike with 600 of the 2,000 workmen parading in from the field and c1amoring for higher pay. According to a chronicling Briton in Brazil too prejudiced against Ford to be an entirely trustworthy witness, 'This last strike was perfectly peaceable and well conducted, but on hearing of it beforehand the American staff took prompt and heroic measures. They immediately dispatched a swift launch with their women folk down the river out of the alleged danger zone. This action, known of course to everyone, has done more than any other yet to hold up the North American to ridicule in Brazil.' (India-Rubber Journal, October 31st 1930.)

In December, 1930, with 3,000 men on the job, there was another big strike. This one has variously been attributed to 'communistic agitation,' to eviction of former occupants of the land as clearing proceeded, and to worker objections to punching time clocks and standing in line with trays of food at the mess hall. It climaxed, anyhow, with some hundreds of Brazilians marching on Henry's headquarters with pitchforks and guns as Para city police flocked in by airplane. The plantation 'discharged' all 3,000 workers and shut up shop, and in early 1931 it was reported that Ford was through with Brazil forever. This was entirely premature, however, for in a few months he was going again with a new staff of 1,500, the approximate figure he has since maintained. According to 1935 word from Brazil he had about 2,500 people (not workers) at Boa Vista or 'Fordlandia' and about 1,000 at Belleterra.

In addition to his labor problem he, of course, faces another one alien to the Middle East — the question of governmental stability. There have been newspaper cries for terminating his lease in the past and in a land given to revolutions it well might be that an anti-Henry faction or one dedicated to putting the revenue-raising squeezer upon him some day will come into power. As this was being written all Brazil still was under martial law as a result of the November 1935, uprising blamed on 'the Communists,' and there is no question that trouble still is simmering. The Ford company's idea is that it is protected against possible contract violation and excessive taxing in the future by a clause which eventually makes the Government a partner in the enterprise. This is the concession provision that after 12 years of operation the company shall annually pay 7% of profits to the state and 2% each to the two municipalities in which the property is located. The future will tell whether that proves sufficient for the Brazilian politicians. As also whether there ever will be profits.

3.6 Temperate Zone

Of trees, vines and shrubs from which rubber has been obtained or in whose latex rubber has been found, no complete listing ever has been made. Methodical German botanists have reliably described some half a thousand, and the late rubber-researching Thomas A. Edison was variously quoted by interviewers as claiming to have found hundreds of wild and ornamental United States plants containing 'appreciable amounts of rubber' or 'enough rubber to make them potentially profitable as a rubber source.' The old idea was that rubber was native only to the tropics or sub-tropics. Yet Edison deposed that he had found it as far North as the Yukon, and Soviet Russia has extracted workable quantities from the roots of wild plants native to Kazakstan.

It must not be forgotten, however, that Edison's 'appreciable amounts' were negligible indeed (and usually almost microscopic) and that the Soviet bulletins all have been as exaggerated as most of the other tidings pro or anti out of that vast land. It remains generally true that plants producing rubber in any quantity even worth noting are confined to the region between thirty degrees North and thirty degrees South of the Equator. And among such of these as occur near the margins of that belt, only the guayule bush of Mexico and the South-Western United States has ever yielded rubber sufficient to figure in commerce.

Little known to the general public but well known in the rubber industry, almost wholly unpublicised while the publicity tomtoms were beating out the Edison golden rod chant, the extraction of rubber from the guayule shrub remains, then, the one even moderately successful American effort at developing *caoutchouc* collecting within the boundaries of these states. Older by far than the other serious efforts to assure the United States its own rubber, the guayule business has supplied important quantities in the past, and can yet reach heights of importance even beyond that it attained in very definitely contributing to the deflation of 1910's preposterous prices. Meanwhile the shrub continues to supply raw material in quantities which, if negligible compared to the *Hevea* figures, yet entitles it to commercial rather than experimental rating.

Insignificant in appearance compared to any kind of a rubber tree, the two-foot high, silvery gray guayule bush really is one of the champion rubber producers of the world when comparative size of the plants is taken into consideration. Analysis of the whole plant shows that the average wild guayule contains 10% of rubber, and selected cultivated specimens have yielded 22%. Old *Castilloa* trees have been felled and bled of a hundred pounds of rubber, but, considering that the weight of the whole tree was probably 20,000 pounds, this amounts to but a sorry fraction.

Wild, the guayule shrub, a relative of the back fence sunflower, grows in only one region of the world, a 600 by 250 mile oval of prairie in Southern Texas and North central Mexico. Known for nearly two centuries, it was scorned as a rubber source until almost the beginning of the present one. If used at all by the natives of its home desert, the method of obtaining rubber was, no doubt, the same as that employed by current Mexican children, who simply chew the bark of the plant. Today, however, guayule is cultivated as a farm crop in the South-western United States and is yielding at the rate of a million and a quarter pounds of rubber a year. Surprising, indeed, this figure will seem to those whose knowledge of the home-grown rubber problem is confined to accounts of Edison's much publicised but not very fruitful experiments.

In the intervening years, who or what changed the lowly state of this ignored vegetable into its present position of respectability and esteem? These staged the transformation: an old man with a coffee mill, a fruit jar and a handful of pebbles; the financiers Ryan, Guggenheim, Aldrich and Baruch; Francisco Madero, who became president of Mexico and was murdered; and a half dozen scientists who fought and struggled and toiled to civilise a plant which did not in the least wish to be civilised.

To begin with, guayule was reported about the middle of the eighteenth century by Father Negrete of that black-robed brotherhood whose members have always accepted the challenge of the out-of-the-way places of the earth — the Society of Jesus. The Indian youths were forced to chew in obscurity for another century before the shrub again attracted notice. This time it was discovered to botanical science by J.M. Bigelow, M.D. Serving in 1852 as a member of the Mexican Boundary Survey, Doctor Bigelow became acquainted with the plant in the region of Escondido Creek in Southern Texas. For another quarter century the shrub got no attention other than mastication by playfully inclined Indians. In 1876 the Society of Natural History of Mexico took up the study of the plant, and in the same year a sample of guayule rubber was included in an exhibit sent from Durango to the Philadelphia Centennial.

The first effort to exploit the bush commercially was made by John Cheever of the New York Belting & Packing Company in 1888. Cheever sent an agent into Mexico with orders to obtain a supply of 'rubber bark.' His orders were carried out, much too thoroughly in fact, for the agent proceeded to ship 100,000 pounds of the entire shrub. A satisfactory rubber was extracted, but the freight bills on the tons of useless wood discouraged the company from any further experiments. Almost until the end of the century little use was found for guayule except as a fuel for the smelting furnaces. Because of its high rubber and resin content it burns with a fierce flame, and it was eagerly sought by the peons for this purpose. Thousands of acres of the shrub were wasted. Oddly enough, a great deal of the shrub was consumed in the furnaces of the American Smelting and Refining Company, owned by the Guggenheims whose millions later helped finance the company which has done all of the development work on guayule.

Until 1928, when Dr. David Spence proved differently, it had always been thought that the rubber in guayule existed as such in the cells of the bark and wood. Consequently most of the early methods devised for getting the rubber out of the plant depended on dissolving it out. The first patent for guayule rubber preparation was issued in 1899 to an Italian, William Prampolini, for extraction by solvents of what he termed 'Twentieth Century Gum.' The first guayule factory in Mexico, established by Adolpho Marx at Jimulco in 1903, employed a solvent process. Using similar methods, a factory in Germany did a successful business on imported shrub until 1905, when the Mexican government found out that the shrub was valuable and imposed an export duty of 15 pesos per ton. Dissolving the rubber out of the shrub was unsatisfactory from the standpoints of both cheapness and completeness and elaborate recovery systems had to be used to preserve the solvent if the price was to be kept at all within reason. The guayule business never would have grown to its present importance if the use of solvents had continued to be the best method of separating the rubber from the woody fiber. That is why an old man with a coffee mill occupies such an important place in the guayule picture.

At 62, William Appleton Lawrence had been a manufacturing chemist for better than 40 years and was a partner in the firm of Whiting and Lawrence which extracted hops by means of gasoline and was the country's largest consumer of that liquid. When E. B. Aldrich, managing the affairs of a company which was planning to exploit Mexican guayule, asked the Standard Oil company to recommend a chemist who knew something about gasoline extraction processes, he was sent to Lawrence. Beginning work on the problem in July, 1901, it took the elderly chemist just six months to prove that Aldrich was wrong in assuming gasoline extraction was the answer. But if he could not get the rubber out of the wood, he would get the wood out of the rubber. With this in mind he tried an alkali process, in which he reported some merit, but which was not very satisfactory. While experimenting on this method, he got a clue to the puzzle which turned out to be an extremely simple one. In July, 1902, he found that by merely grinding the shrub in the presence of water, a fiber rubber was formed. Working in a temporary laboratory at his home in Jamaica, Long Island, with the assistance of his chemist daughter, Clara Louise, he ground up some dried shrub in the household coffee mill, and rolled it in a glass fruit jar with water and a handful of pebbles. 'Rubbing and pressure in the presence of water' described the process in the patent which he took out in 1903. In the same year, Aldrich and Lawrence went to Mexico to make sure of an adequate supply of the shrub and on their return set up a small factory with a regulation size pebble mill in New York. The first lot of rubber was extracted by the new process in 1904 and sold to the Manhattan Rubber Company. Then followed the building of a small factory at Torreon, Mexico, in 1905. Mention should be made of the discovery in 1904 by a French chemist, E. Delafond, that when pulverised guayule was heated under pressure, the woody fibre became waterlogged and separated from the rubber. A modification of this method is used today in conjunction with the Lawrence process. On January 6, 1906, the Continental Rubber Company was incorporated for $30,000,000 to exploit the Lawrence patents, purchased for $2,400,000, and the Delafond patent, acquired at a relatively insignificant figure. E. B. Aldrich was president of the company. The principal shareholders were Thomas Fortune Ryan, Senator Nelson W. Aldrich, Bernard M. Baruch, Meyer Guggenheim, and Senator Aldrich's son-in-law, a man named John D. Rockefeller, Jr.

Guayule could have been had outright by the rubber manufacturing Trust at this time but President Samuel P. Colt was inclined to caution when dealing with Ryan, Aldrich and Company. In reporting to shareholders on the progress of United States Rubber's raw rubber purchasing subsidiary, General Rubber Co., he said:

'A suggested consolidation with the Continental Rubber Co. was deemed by your directors to be non-advisable in the present development of the so-called mechanical process of obtaining crude rubber through the grinding up of shrubs producing the gum, which is done extensively by the Continental Rubber Co., but that company and the General Rubber Co. have now agreed upon the terms of an arrangement which insures complete harmony and co-operation hereafter between the United States Rubber Co. and the Continental Rubber Co. and between those connected with both companies.'

And while the Trust was making these special arrangements for a supply of the desert *caoutchouc*, Northern Mexico was going *guayule* mad. Many poor ranch owners were made wealthy by selling the shrub on their land for $100 an acre. Gathering in the wild desert lands was done under contract by natives, who pulled the plant and loaded it on their burros. Carrying it to a central station, they baled the shrub in hay balers and dispatched it across the desert by 14 mule teams. The price, which had been $7.50 a ton, rose within a short time to $50 per ton of gold.

The silvery shrub proved literally that to the Madero family. Already owners of vast lands, when it was first thought that guayule might become a valuable rubber producer Francisco Madero and his brother Gustavo Madero formed the Compania Explotadora Coahuilense and acquired two million acres of wild *guayule*. The Maderos were Continental's greatest rival and their eight factories added greatly to their fortune before they sold their interests to the Intercontinental Rubber Company (reorganisation of Continental) in 1911. Although Francisco was head of the company until 1910, he was not active in its affairs much after 1907. His story in these later years was: presidential candidacy, imprisonment, flight, rebellion, inauguration, and, in 1913, overthrow and assassination. During his political life, his *guayule* deal was more than once cited to discredit his claims of championing the peon and the cause of land division; given a British twist, these tales still bob up in the John Bull press today. A popular version runs like this: Madero, inheriting a vast tract of barren land on which grew quantities of an altogether worthless shrub called *guayule*, used it as a blind in obtaining American oil money supposedly for the purpose of extracting rubber but actually given him to finance a revolution since the Americans were trying to drive the British out of Mexico and Diaz was friendly to English oil interests. This yarn would seem to have little if any foundation in fact. Actually one of the principal forces against which Madero had to contend was the American big business element fearful of the passing of the concessionaire system and cheap labor and a general dislocation of its advantageous position.

Be that as it may, *guayule* factories, from the beginning, were faced with the serious problem of a constantly diminishing supply of raw material and the more successful their production the more serious became the problem. *Guayule* does not replace itself readily. Roots broken off where peons uprooted plants produced a few bushes and some grew from seed, but the replacement was so slight that in 1908 the United States Consul at Matamoras was dolefully reporting that the extermination of the plant was in sight. He was very nearly right. Production of *guayule*, which had amounted to only 425,000 pounds in all the years prior to 1905 leaped to nearly four million pounds in 1906, double that amount in 1908, and to 21,375,000 pounds in 1910. Between 1906 and 1912 the world's rubber supply was increased by some 128,000,000 pounds of *guayule*, only two million pounds less than the total output of plantation *Hevea* to the end of 1912. The year 1911 saw the closing of the factories, but for another reason. The Madero revolution shut them and in doing so saved the wild plant from certain extinction.

With millions of dollars invested in Mexico, it had not taken the Continental company long to see what the ultimate fate of the *guayule* industry, as then conducted, was going to be. As early as 1907 the company began its first experiments toward cultivation of the plant, on its Cedros ranch in Mexico. The results of these first experiments were negligible. In fact, when Dr. W. B. McCallum was hired by Intercontinental in 1911 to see if he could either bully or coax the shrub into accepting the ways of civilisation, about all the information he had to work on was the long prevailing conviction that the plant could not be grown from seed and that when cultivated, it would not produce rubber. Had it not been for McCallum the *guayule* industry would probably be a memory today. He finally succeeded in transferring the plant from its home on the range to cultivated fields, but it took him 17 years of unceasing, patient labor to do it. No sooner did the botanist overcome one phase of the plant's rebellion against domestication than he was faced by some other apparently insurmountable obstacle.

No mountain boy battling bitterly against the indignity of shoes ever put up a more obstinate struggle against civilisation than did *guayule.*

The first efforts at cultivation were made by cuttings. After thousands of cuttings were made from plants in all sorts of conditions and not one in a hundred would root, the idea was abandoned. The botanists then turned to the seed and found that the so-called seed that was obtained for them from wild plants was largely chaff. The seeds are extremely minute — a half million of them weigh little more than a pound — and the plant's seed vessels are often barren or half developed. Thus, their first job was to take the shrub, feed it, starve it, water it, plant it under all sorts of conditions until they found just the conditions that result in full seed vessels. About this time revolutionary disturbances in Mexico became so acute that factory operations and experimental work both had to be abandoned. McCallum was instructed to salvage as large a seed supply as possible and start again in California. In 1912, he became a second Henry Wickham, fleeing from one country to another with seeds that were to found a new agricultural industry.

Of the first bushel of seed sown, not one sprouted. The trouble was found to be due to the inactivity of certain ferments that incite germination and all progress was halted until means were found to prod the seeds out of their lethargy by treating them with chemicals. Getting the seedlings to grow was the next step. Used to the dry, well-aired soil of the desert, they died in great numbers if the soil was excessively moist or closely packed. The natural growth of the seedlings is slow, however, and the plants are no more than an inch or two high when a year old, and are much too small to be set out. The rich soil and abundant moisture requisite to the necessary forced growth were just the things inimical to the plant's life. This problem of speeding up the plant's growth was one of the most vexing faced, but extensive experiment finally developed the correct physical condition necessary to produce a transplantable seedling. As an outcome of the same experiments the scientists reduced the 20 year period which the plant required to grow to maturity in the desert to four years. Next it was found that the young plants were unable to take root again after being transplanted, and for three years this difficulty threatened to end everything. Eventually the transplanting time that would give strong root development was found.

Plenty of shrub was brought through to maturity. At last science had triumphed! And then the scientists learned that what they had wasn't worth anything. Shrubs which on the range produced 15% of rubber dropped off to 4% and less. This phase of the battle lasted for a full 10 years. First, the botanists learned that the plant did not produce rubber during growth periods. The wild plant grew only in the spring; the rest of the year it slept — and produced rubber; but it took 20 years to grow up. The tame plant, brought to adulthood in four years by forced growth, produced little rubber. The combination of fast growth and plenty of rubber seemed impossible of achievement. At last McCallum detected a seasonal rhythm between the functions of growth and secretion of rubber. He found that moderate growth, if it came at the right time and was not prolonged, could be followed by satisfactory rubber production. By irrigating at the proper time to make the plant grow, and by the withdrawal of irrigation at the proper time to throw it into a dormant state, he finally made the cultivated plant as good as the wild one. During all this time and for some years afterward, efforts were also being made to improve *guayule* by selection and breeding of better strains. This work was extremely slow at first because it took four years to check the most important feature — rubber content. Thousands of plants were analyzed for rubber and at the same time a careful study was made of their physical characteristics. The botanists finally became skilled in recognising the rubber possibilities of a plant from its appearance, and this speeded up the work enormously. In making selections, hundreds of thousands of plants were gone over by hand. Hundreds of different strains were isolated, grown and reduced to constant types. After painstaking examination of several million shrubs, the total number of plants obtained that served as the starting point for the commercially successful strains did not exceed 10.

On February 6th, 1931, a $150,000 guayule extraction factory was opened by American Rubber Producers, an Intercontinental subsidiary, at Salinas, California. It was the first factory in the United States built to process this new farm crop, and it was equipped to handle the output of 7,000 acres. The methods employed at Salinas exemplify highly developed machine agriculture and it is for this reason alone that *guayule*, produced by relatively costly labor, can compete with plantation rubber produced by coolies. The speeded-up native on the plantation can produce about 2,300 pounds of rubber a year. One man plus his machines can produce 25,000 pounds of *guayule*.

At the nursery, seeds of high yielding plants are picked by a machine featuring revolving brushes and a vacuum cleaner. The seeds are treated chemically to insure germination and are sown by a seeder in the nursery in spring. The tiny plants are watered by overhead sprayers. When they are 10 months old they are topped, dug by machine and transplanted in the fields by a machine similar to the common cabbage or tobacco planter. For four years it is only necessary to cut down competing weeds with mechanical cultivators. During this time there is a profitable increase in the size and rubber content of the plant. Growth continues after four years, but at a rate so slow as to be unprofitable. In the fourth years the plants are uprooted by a machine which beats them at the same time to remove dirt. They are left to dry and then picked up by an outfit that chops them into half inch pieces and blows them into trailers.

At the factory the shrub is crushed between rollers and fed with water to tube mills where it is rotated with flint pebbles which tear the fibre apart and cause the rubber to come together in little 'worms.' The worms float on the surface of the discharge liquid, but as they still contain a great deal of cork, the mass is heated under pressure and the cork becomes waterlogged and sinks. The rubber is then dried and sheeted.

At present, *guayule* is a more resinous rubber than *Hevea*, but it is used where this is a not undesirable property. The resinous material can be wholly removed when the demand arises by a simple process discovered by David Spence in 1928. Finding that most of the supposed resin is protein-like material which is not acted on by water, he submitted the shrub to a retting process such as is undergone by flax. The shrub is permitted to ferment in wet storage for several days and the objectionable material is changed to a soluble form which is removed in washing.

As seen by George Carnahan, president of Intercontinental, the future of *guayule* is this: 'As eventually to be developed in the United States, *guayule* growing will be in the hands of the individual farmer who will contract his crop and be guided and financed by the factory organisation in his vicinity much in the same fashion as the beet sugar business is now conducted. He will plant one-fourth or one-fifth of his total area each year and will thus iron out his labor peaks and give himself an annual income.'

Plantings of the shrub to date have shown that it can be successfully cultivated in Southern California, Texas and Arizona. At the Salinas plant, 1,254,600 pounds of rubber were extracted from the 1934 crop of shrub harvested by farmers of the vicinity on marginal land not well adapted to other crops. Some 56,000 pounds of this were shipped to Canada and the remainder went to American manufacturers.

On the future price ranges of rubber from the *Hevea*, on the question of whether prices of synthetic rubber can be reduced to a level comparable with that of the natural product, depends the extent to which this American rubber producing enterprise may be expected to grow.

Between 1921 and 1930 American imports of the wild *guayule* from Mexico amounted to over 21,000 tons of which 16,000 were brought in during the last four years of Britain's Stevenson restriction plan. In 1930 they ceased entirely, importation not being resumed until 1934 brought the second restriction attempt in the East. Shipments of 50 to 100 tons a month were reported through 1934 and 1935, and today one of the larger American rubber manufacturers has ideas about getting into the business. This is Goodrich of Akron, only one of the leading tire corporations not growing *Heveas* in some corner of the world. Its well-advanced paper plans call for a $200,000 processing plant located on a leased *guayule* region midway between the border and Mexico City. Production of 100 tons a month is contemplated. There is no thought of cultivation as in California. Because of the differences in climate, seed can be thrown on the ground in Mexico to sprout when conditions are propitious. Goodrich, therefore, expects to 'reforest' rather than cultivate.

Wholly mercenary was Tom Ryan's motive for getting into *guayule*, of course, and wholly patriotic and scientific were Thomas A. Edison's reasons for tackling the problem of a native rubber supply. Practically speaking, however, it cannot be said that the greatest of American inventors achieved success in his 1925-1931 philanthropic experiments on his Florida acreage. Increasing the rubber content of golden rod and extracting and vulcanising this *caoutchouc* represents an interesting stunt of the rabbit-out-of-the-hat variety, but not necessarily an achievement of any importance.

Edison's interest in the rubber plant experiments dated back not to his presidency of a Castilloa plantation company, not to his marriage to an Akron girl, but to those same camping trips on which Firestone steamed up Ford on the 'Americans Should Produce Their Own Rubber' subject in the Stevenson restriction days.

That so much was heard about the golden rod act was solely due to the fact that Edison was both a genuine titan of our time or any other time and usually 'good copy' as well. There was real drama in the spectacle of this hero of so many scientific triumphs invading a strange field of research at 78, and with all the zest of youth. There was legitimate color in the progress of the experiments by which 15,000 plant entries were narrowed down to 4,000 by testing, to 600 and then to 40 by planting experiments, finally to golden rod alone (although golden rod rubber experiments did not originate with Edison). But there was an immense amount of phony melodramatics about the vulcanising 'triumph' that 'crowned' Edison's efforts 'five days before his death' at 84.

Aside from successful cross-breeding, here is how the work progressed. In 1930 Edison patented a process for extracting the rubber from golden rod, the only such process he evolved. It is no more and no less than the *guayule* procedure and it does not work so well with golden rod as with the desert shrub. In a February 1931, birthday interview Edison allegedly said that a factory would be built and production on a small commercial basis launched within 2½ years. This more than over-optimistic declaration attracted little attention for, even with the Edison connection, the golden rod story had become considerable of a chestnut by then. Harvey Firestone, visiting his ailing friend in early August, 1931, told interviewers of having had several samples of vulcanised golden rod rubber exhibited to him. The press associations carried the story and the newspapers buried it anywhere between pages 17 and 46. On October 18th came the end of all for Edison, however, and a few days later a news-hawk remembered the golden rod. No sooner had he filed than every press service in the land started shoving ream after ream of stuff about the great vulcanising secret snatched from the very jaws of death five days or six days or a week before.

Prizewinner of the lot was one which Mr. Hearst's International News Service slugged with a special copyright line. For weird misinformation it has had few equals even among newspaper 'science' features. Excerpts:

> 'West Orange, NJ, October 21st - Thomas A. Edison beat death by five days to his last great invention, the manufacture of synthetic rubber from golden rod and other ingredients.
>
> 'Synthetic rubber, this correspondent learned from authoritative sources Wednesday, finally was proven practical in the wizard's laboratory last Tuesday, even as Edison sank downward through a stupor into the coma which ended in his death.
>
> '... Edison was almost certain he had discovered the process for its manufacture when he took to his bed. He already had worked out a method of manufacturing latex, the ingredient which fuses the other elements of the synthesis into a practical and durable rubber suitable to commercial use.
>
> 'One of the men associated with Edison admitted Wednesday that the puzzle was solved last Tuesday.
>
> 'It came all of a sudden,' he said. 'We were dead against a stone wall when it flashed upon us. We tested the method, retested it and then telephoned Mr. Edison's home. They told us he was unconscious. But I don't think he was. I think he fell asleep and sent the solution over to us by thought wave. Anyhow, synthetic rubber is commercially practicable."

That was the worst of the lot. The best reported that the big discovery meant a way had been found 'to extract sufficient latex to make its manufacture profitable.' Since then a half decade has passed and what is the news of golden rod? Well, in November 1931, the Edison Botanic Company was formed to continue the experiments on golden rod and other rubber-bearing plants. It was one-third Ford, one-third Firestone and one-third Edison's children. In 1934 it was decided to turn the whole thing over to the government. Ford and Firestone, contemplating tax blanks, attempted to write off as gifts the $20,000 each had contributed to Edison's work. But the government informed them that they couldn't write off the sums because an invention had been produced. So they decided to retain the big discovery. The rest is silence.

Meanwhile the Department of Agriculture is plugging quietly along with its plant experiments as it was doing before Edison bowed in. Edison's good job of cross-breeding had resulted in a golden rod 12 feet high and containing 6-7% of rubber. This content the Department scientists reputedly have upped to 12%, but this has not been confirmed.

As to golden rod's ever qualifying as source of an emergency wartime supply of expensive rubber, *guayule* is so far out in front of it that there is no comparison. More importantly, the synthetic rubbers have eliminated any need for a native plant source in event of military emergency. And this despite the Edison 1930 pronouncement that 'The chemical development of synthetic rubber has no future whatever when plant rubber is quoted at about 23 cents a pound.' Stories of the golden rod type will continue to bloom, however, as was evidenced not long ago when one, Herman E. Pitkin of St. Louis, bounced into brief notice with a patent application based on his contention that the poinsettia is all ready to supply America with 60% of its rubber requirements.

Of native plants additional to *guayule* and golden rod, a desert milkweed *(Asclepias subulata)* has been the chief one with which the Department of Agriculture has experimented. Of indigenous and exotic plants interesting not because of their potentialities as producers but because of their very commonness are such familiar plants as sow thistle, Indian hemp, wild lettuce, greasewood, *estrada*, dandelion, rabbit brush, dogbane, bananas, various cacti and some forms of mistletoe. Intisy received respectful consideration for a time. A shrub that yields a free flow of latex from any cut, it reputedly had been slashed to extinction in its native Madagascar. D r. Charles F. Swingle of the Department was able to locate specimens after weeks of searching, however, and the shy stranger was introduced into Florida and California. Madagascar is also the home of the *cryptostegia* vine, long grown in Florida for ornamental purposes. From its two varieties the Department has produced a hybrid with a higher rubber content than either ancestor.

Among major rubber trees imported into the United States, both *Hevea* and *Ficus elastica* must be mentioned. *Ficus* is familiar not only as the parlor 'rubber plant' but as a huge tree in California coastal regions and in Southern Florida. It is planted as a shade tree only, however, rubber yield being negligible in this climate. Some 30,000 *Heveas* the Department of Agriculture now has growing in Florida and it has record of one stray *brasiliensis* which survived out of those planted there in the early years of the century, the years of the Mexican boom. The *Hevea* has shown no signs of yielding in this country as it does in the East. And the difference between the American wage scale, even in these depression days, and the *Estrada* pittances is such that were Akron ringed about with rubber-yielding *Heveas* it still would be far cheaper to import crude from Malaya and Sumatra than to pay local tappers. Mechanical methods of extracting rubber must be used in any practical production of *caoutchouc* in this country and here it is that *guayule* scores, as we have noted.

Imported into the Soviet Union, incidentally, *guayule* figured with the indigenous Soviet shrubs tau-sagyz, kok-sagyz and krim-sagyz in the prodigious 'Russians Should Produce Their Own Rubber' ballyhoo which in 1931 and 1932 waxed as ridiculous in its own way as the Hearst screams are in theirs. The Government promised 10,000 tons of mechanically harvested home-grown plantation rubber by 1937, some 40,000 by 1942, over 200,000 in short time thereafter. The Red press hoisted the guaranteed production figure to 20,000 tons or so for the end of 1935 [1] and the foreign correspondents ran wild with pictures of Russia rivaling Ceylon and Sumatra in a decade [2]. Figures on tau-sagyz shrub plantations running into millions upon millions of acres were nonchalantly tossed around and existence of the highly capitalistic Intercontinental was resolutely ignored by patriots picturing the proposed mechanisation of rubber gathering as a new and startling idea. Then came Russian semi-success with synthetic rubber and today very little is heard about the miracle shrubs.

Actually the Soviet scientists have accomplished tangible work in the extraction of rubber from plants native and domesticated and there is no insurmountable reason why Russia should not some day harvest its own planted rubber in commercially important amounts. Russia competing with the Middle East plantations is, however, strictly a fantasy for the Jules Verne - H.G. Wells shelf.

Tau-sagyz was discovered in 1929 and kok-sagyz and krim-sagyz in 1931; some amount of rubber has been extracted from these wild plants; experimental plantings have been made and immense plantations have been brought into bearing on paper: this much is agreed on. For the rest your guess is probably as good as ours and perhaps as good as Walter Duranty's.

[1] See Economic Review of the Soviet Union, October 1st, 1931. Also compare article by V. Makagon in *Sovietski Kauchuk* (*Soviet Rubber*), November-December, 1933.

See Walter Duranty's wireless to the New York Times, January 25th, 1932.

4 Laboratory and Mill

4.1 The Problem

Carl Sandburg has written of the herring peddler with the face of a man terribly glad to be selling fish, Arnold Bennett of the chiropodist who ecstatically sighed, 'It's a grand thing, a nail is!' Of the company of this fish crier and this toe tinker was Charles Goodyear, perhaps the only man ever on this earth who had a passionate love for rubber. Of a certainty it was a romance made in heaven, for without the affair between the Connecticut Yankee and his *caoutchouc* mistress there might have been no rubber industry of importance, no book for your present chroniclers to write. Had there been no Goodyear, would someone else have come along to make the essential discovery of vulcanisation and give birth to the *caoutchouc* industry as we know it? Perhaps — and perhaps not. There is evidence that vulcanisation had been achieved before Goodyear without the massive-browed experimenters having sense enough to realise what had happened. There is complete proof that long before Goodyear the best minds of the world had available all the facts necessary to achieve vulcanisation and to recognise what had happened. Their complete failure to do so is a persuasive argument in favor of the theory that, without the Yankee genius, rubber might well have been given up as a major raw material of industry and permanently classified as a semi-satisfactory substance for erasers, overshoes, suspenders, hose and raincoats. At best, discovery of the method for utilising its peculiar properties certainly would have been postponed for many years, perhaps long enough to have handicapped terrifically the growth of a motor industry which found processed caoutchouc waiting for it when it came into being.

From its first introduction into Europe, raw rubber appeared as a question mark to the scientists. Not immediately, however, were they aware of the major problem that Goodyear would have to solve — the problem of treating this substance to prevent articles of *caoutchouc* from hardening like iron in winter, turning sticky or even liquid in summer. To some extent the objects of Indian manufacture presented this difficulty, but it did not become a pressing one until civilisation itself promoted the substance to importance by attempting to make it the basis of important manufacturing industries. Before the major problem arose, then, there were other problems to conquer. First of them all, it seemed to the researchers, was the discovery of some method of restoring the hardened crude to a liquid form in order to duplicate the manufacturing methods of the aborigines, the Fresneaus and the Condamines, who used to produce their articles of rubber directly from the tree fluid by dipping molds into it or spreading it on fabrics. From the time of Condamine's 1745 appearance before the French Academy, then, communications concerning experiments designed to find a solvent for solid *caoutchouc* occasionally appeared in the scientific journals. First success was reported by two French investigators, Herissant, a physician, and Macquer, a chemist. In 1763 this pair suggested the use of turpentine as a solvent and five years later Macquer found purified ether even better. Rubber prepared by the latter method, when used to coat wax molds, gave a film similar to the original material and was free of the adhesiveness which resulted when turpentine solution was used. Rubber tubes were the first articles Macquer made and of them he wrote: 'The solidity of this material, its elasticity, its property of resisting water, salt, spirits of wine and many other solvents make it suitable for the manufacture of flexible and elastic tubes, which will render important service in many mechanical arts.' Ether was little better than a chemical curiosity, however, and incapable of becoming a solvent for rubber on a commercial scale. According to *Caoutchouc and Gutta Percha*, an eighteenth century pamphlet issued by the Society for Promoting Christian Knowledge, Frederick the Great was the first to apply the ether-*caoutchouc* solution to an article of dress: 'He procured a pair of lasts, in the form of riding boots, and caused an ethereal solution of *caoutchouc* to be applied to these lasts, until a coating of *caoutchouc* of sufficient thickness was deposited on them. The lasts were then removed and the monarch wore the first pair of riding boots made from this process, and probably also the last, for ether is too expensive a solvent for *caoutchouc*.'

Almost until the end of the eighteenth century unimaginative England could find no use for Condamine's 'resin' other than the rubbing out of pencil marks. First mention of this use is found in a naive stop press bulletin appended to the preface of Dr. Joseph Priestley's, *Familiar Introduction to the Theory and Practice of Perspective*, published in 1770: 'Since this work was printed off, I have seen a substance excellently adapted to the purpose of wiping from paper the marks of a black lead pencil. It must, therefore, be of singular use to those who practice drawing. It is sold by Mr. Nairne, Mathematical Instrument Maker, opposite the Royal Exchange. He sells a cubical piece of about half an inch for three shillings; he says it will last several years.'

It was because *caoutchouc* supposedly hailed from the West Indies and because it would rub out pencil marks that 'India rubber' it became; and 'rubber' it has remained ever since throughout the English-speaking world. The more colloquial 'lead eater' name which *caoutchouc* bore in the early days dropped out of the running, even in the country districts, by 1860. In Condamine's country according to Valmont de Bosnare's *Dictionary of Natural History*, the practice of rubbing out pencil marks with this substance instead of breadcrumbs was introduced by a Magellan, descendant of the great navigator. Officially, France called the substance *caoutchouc*, colloquially it became *peau de nègre*.

In 1790, Antoine François de Fourcroy published the results of his *caoutchouc* experiments and reported that 'by the mixture of fixed and volatile oils it is dissolved in order to form that fat and flexible varnish which can so advantageously be laid upon silk.' The year 1791 was notable for three events. In England, Samuel Peal, on May 2nd, was granted the first English rubber patent, covering means of 'rendering perfectly waterproof all kinds of leather, cotton, linen and woolen cloths, silk stuffs, paper, wood ... and other substances for the purpose of being worked up into shoes, boots, and other wearing apparel and to be used on all occasions where dryness or a power of repelling wet or moisture may be required.' For this purpose Peal favored the use of a turpentine solution, but also noted that 'the gum may be used with equal advantage in its native fluid state.' In Italy, Fabroni discovered that rectified petroleum dissolved rubber. His discovery, though important in itself, was not particularly applicable because petroleum, known from earliest times, did not come into general use until the exploration of Pennsylvania's oil fields began in 1860. France's M. Grossart contributed to the initial volume of *Transactions of the Philosophical Societies of All Nations* a very clear exposition of a method of making rubber tubes for use in surgery. Said Grossart:

'Its singular elasticity, its flexibility, and the little action most substances have upon it, have caused it to be considered as very valuable ... in the art of surgery. It has hitherto been impossible to procure instruments of this substance, inasmuch as almost the whole quantity of *caoutchouc* brought into Europe comes from Brazil already fashioned into bottles, birds or other figures; this has rendered its use extremely circumscribed ... It would be easy could we procure it in its original state of fluidity to give it all the forms under which it might be useful to us, but the communications with Brazil are very difficult for the greatest part of Europe. The necessity of passing the line, in bringing it to our parts, is another obstacle to the juice arriving in a proper state ... It is decomposed by heat in the same manner as milk, exhaling at that time an extremely foetid odour and having no longer its original properties.'

Deciding that it was unnecessary to dissolve the gum completely, Grossart sought a method 'of soldering it and of not acting upon it more than might be necessary to cause its softened parts to reunite.' To this end, he cut long spiral strips from the rubber bottles, softened these in ether, and wrapped them about a rod, pressing the edges firmly together. Binding the whole tightly with tape until it was dry, he produced a tube of one solid piece of rubber. 'It is, therefore, at present easy to make of *caoutchouc*, whatever instruments it may be advantageous to have of a flexible, supple and elastic substance which is impermeable to water ... and resists the action of acids, as well as that of most other solvents. As to the durability of these instruments, few substances promise more than this, because it may be soldered afresh in a damaged part.'

The 1791 article in the *Edinburgh Bee*, to which we referred in a previous chapter summarises nearly all that was known about rubber at this time and forecasts with remarkable vision its future uses.

'... It is easy to see,' wrote James Anderson, 'that the uses to which this substance might be applied in arts and manufactures are innumerable, and as such can be effected by no other known substance in nature. Yet so blind have mankind hitherto been to these advantages ... all that has been done is, to suffer the natives to mould it into the form of a small kind of bottle ... and these, when brought to Europe, are applied to scarcely any other use than being cut to pieces for the purpose of effacing marks made upon paper by a black lead pencil, or that of idly amusing children by stretching it out and observing how perfectly it again recovers its pristine form ...

'It is now time that we should begin to make some use of this very valuable substance, which, probably a hundred years hence, will administer in a variety of ways to the accommodation of our descendants.'

The Scotsman pointed out a few of the useful purposes which occurred to him; these included the use of rubber in place of leather in boots and gloves; 'caps to defend the head from wet'; umbrellas; 'buckets of canvas or other cheap substance ... made water-tight and incorruptible by ... covering with this matter'; waterproof tents, sails, and other army and navy uses 'so numerous as not to admit of being here specified'; balloons of rubber instead of silk; 'it might also be moulded into the form of riding whips ... and after they were worn out, they might be employed as torches ... As a matter for surgical purposes, it might be employed on many occasions.' Geographical globes which 'are at present an article of great expense' could be made of '*caoutchouc*' at moderate price. Anderson listed numerous other uses, but immediate realisation of his ideas was prevented by lack of suitable solvents or proper machinery.

An event of 1802 which was eventually to have a profound effect on the rubber industry was the first practical use of coal gas for illumination at the Boulton and Watts Foundry in Soho. The revolutionary idea was developed by foundry engineer William Murdock, who had successfully lighted his house with gas when at Redruth installing pumps in Cornish tin mines. At this time, lighting of towns and factories was accomplished by dim and sooty lamps which burned vegetable oils. Consequently, the use of gas, which was clean, convenient and gave far better light, achieved immediate popularity. Gas companies began to spring up all over England and Scotland. Coal, robbed of its content of gas, gives tar and ammonia also, and the new industry soon found a large supply of these by-products piling up with no way of disposing of them. In solving this problem, Charles Macintosh brought into being the rubber-proofing trade in much the form it retains today. The name 'Macintosh' is a familiar one, its owner, along with Derby, Chesterfield and Albert, having been one of the few men who has had an article of clothing named for him. Macintosh, born in Glasgow on December 29th 1766, became prominent as a chemist while still a young man. In 1819 he was producing dyes and mordants for the textile trade and was a consumer of ammonia which was used in manufacturing 'cudbear,' a quaintly named dyestuff made from lichens. With this in mind, he entered into a long-term contract with the Glasgow Gas Works to receive all of the tar and ammoniacal water produced at its plant. After separating ammonia in the process of converting the tar into pitch, naphtha is produced. The thrifty Scotchman wanted to save all of the waste products and it occurred to him that the new oil might be a suitable rubber solvent. From the very start, Macintosh purified his naphtha by distillation and it is more than probable that he was the first man to isolate benzene, although the annals of chemistry unanimously accord this distinction to Michael Faraday. By dissolving parings of caoutchouc in the naphtha, Macintosh produced a thick solution 'resembling transparent honey' with which he coated all sorts of fabrics. The new solvent was what had long been sought. It evaporated rapidly and left the rubber in its original state, and not sticky as turpentine did; it was much cheaper than ether. Macintosh pressed two of the coated fabrics together and had a material which was really waterproof. Since the rubber was incased between two layers of cloth, the stickiness caused by warm weather, which had been an objection to every previous rubberised fabric, was overcome. A patent was issued for the new process on June 17th 1823. Garments made from the new fabric became very popular. First known as 'Water-proof Double Textures,' they soon came to be universally called 'Macintoshes.' The popularity of the macintosh was not achieved without difficulty, however. Tailors, who made up the waterproofed cloth into coats, insisted on using needle and thread to sew the parts together instead of cementing them and the resulting products leaked at the seams. Quite often the wool from which cloth for the waterproofs was made had not been freed of its natural grease and this hastened the deterioration of the rubber. Apparently, too, Macintosh had to combat the physicians of the period. His biographical memoirs include a bitter note to the effect that: 'The medical faculty having lent their aid to run down the use of waterproof (apparently from having found it a decided enemy to their best friends—colds and catarrhs), the use of the article in the form of cloaks, etc., has of late become comparatively extinct.' The manufacturer had little reason to display this pessimism, however. Within a year after taking out his patent he began receiving substantial government orders for waterproofs and he came to the conclusion that his chemical works was not going to be big enough to handle the prospective business. In 1824 he entered into partnership with Hugh Hornby Birley and Joseph Birley, cotton spinners and manufacturers in Manchester, and R. W. Barton to form Charles Macintosh and Company.

It often happens that the utility of an invention is no sooner demonstrated, than claimants other than the inventor begin proving they had the idea first. Macintosh's discovery of the value of naphtha as a rubber solvent was no exception. Claims of Professor James Syme to priority differ from most contentions of the sort because they apparently have merit. As the famed Scottish surgeon told the story, he early became impressed with the need of a cheap solvent for caoutchouc, from which it could be recovered unchanged. He finally found what he was looking for and employed the solution for making tubes and for waterproofing textiles. 'A silk cloak, which afforded complete protection from the heaviest rain, and could be employed as a pitcher by turning up its skirt, was an object of wonder to all who saw it. My friends talked of a patent, but being then about to commence the study of a profession with which consideration of trade in those days did not seem to be consistent, I addressed the following letter to the editor of a scientific journal resident in Glasgow.' The letter, published by Dr. Thomson, editor of the *Annals of Philosophy* in August, 1818, relates how Syme distilled coal tar and obtained a dark amber liquid which he re-distilled to obtain a light straw-coloured oil which was very inflammable and extremely volatile. This proved to be the solvent he had sought for caoutchouc.

Leaving Edinburgh's Syme and Glasgow's Macintosh to their credit quarrel, we come now to a consideration of the dominant figure in the history of Britain's rubber manufacturing industry — Thomas Hancock. Hancock is accorded this distinction for his many achievements, which include the discovery of the processes of masticating, compounding and moulding rubber. He is best known for his appropriation of Goodyear's vulcanisation discovery, a shabby trick which has led the more rabid of Britain's patriots to claim the honor of the invention for him. Hancock himself never was brazen enough to lodge this claim, but he was small enough to omit Goodyear's name from the Hancock narrative of the origin and progress of rubber manufacture, referring to him only as 'the inventor.' Hancock was born at Marlborough in Wiltshire in 1786, son of a cabinet maker and timber merchant and third of a family of twelve children. About 1815 he entered into partnership with his brother John in a coach building enterprise in London, with a factory in Pulteney Street, Golden Square, and a shop in St. James Street. Hancock himself states that he does not remember when he first began to notice the peculiar qualities of rubber, but several biographers, apparently better informed than their subject, insist that he was attracted by its possibilities as a waterproofer for the garments of stage coach passengers, who fared badly on the muddy English roads in wet weather. By 1819 he was experimenting with turpentine solutions, which he found to dry so badly as to be almost useless. He then thought of applying the material as an elastic to articles of dress, began cutting strips of rubber from bottles imported from Para, and on April 29th, 1820, patented the fastening of these to the wrists of gloves, to waistcoat backs, to 'pockets to prevent their being picked,' to trouser and gaiter straps, braces, stockings, garters, stays, to 'boots, shoes, clogs and pattens, when the object is to put them on and off without lacing.' It was for the manufacture of these articles that the first rubber factory in England came into existence in 1820. Waste cuttings of rubber began to pile up on Hancock's hands, a serious item. Having noticed that freshly cut pieces united readily, he tried mincing the scraps very small and pressing them together, but the operation was slow and not very successful. Then Hancock thought of tearing the rubber to shreds and began constructing what developed into the industry's first masticator. The apparatus consisted of a hollow wooden cylinder studded with teeth, within which a spiked core was rotated. Charging the machine with scrap, Hancock began turning the crank, expecting to reduce the rubber to shreds. He soon became aware that some unexpected action was going on, for the machine steadily became more, instead of less, difficult to operate. Opening it, he found a ball of rubber, quite uniform and very hot. The ball was plastic, could be shaped into a block by pressing it in a mould, and consequently could be cut more readily than the irregular pieces formerly available. This 1820 apparatus solved for Hancock the problem of utilising waste material; it also led him to investigate the effect of mixing other materials into rubber and in this way he became the first rubber compounder. The wooden experimental machine would at last no longer hold together and two larger, sturdier machines, still hand operated, were constructed. The demand for Hancock's rubber products increased so rapidly that in 1821 he was forced to move into larger premises on Goswell Road, London, site still occupied by the Hancock factory. Here he had a horse mill installed to drive larger masticators and was able to turn out solid blocks of 15 lbs. Hancock had not patented the machine, and, in order to keep it secret, his men were pledged to silence and the machine was referred to as a 'pickle.' The process was kept under cover until 1832. Hancock soon learned that his masticated rubber dissolved much more readily in naphtha than did the crude. Needing only the right to use the solvent in order to be able to make better macintoshes than could Macintosh and in half the time, Hancock was granted a license in 1825. Macintosh immediately found that his competitor had a decided advantage over him as he was using solutions which were twice as concentrated as any before known. By 1830, the Macintosh Company was convinced of the superiority of Hancock's solution, entered into a trade agreement and soon afterwards the two companies amalgamated.

Though it is a common belief that the use of rubber did not begin until the discovery of vulcanisation, it was widely employed as early as 1825, 14 years before Goodyear made his great discovery. In rollers and printer's blankets it was used, and in carding machines, driving belts, and billiard table cushions. Rubber fire hose was made by the Macintosh company and according to an 1827 issue of the *Mechanics' Magazine* was successfully employed during the fierce conflagration at Fresh Wharf, in London. Rubber had no sooner ousted leather from the fire engine than the brewers, who had long put up with a great loss of beer from leather hose, decided to try it out. At first, the rubber was condemned for imparting a vile taste to the brew, but after it was allowed to soak for some time in waste liquid, it became sweetened and was found satisfactory. By 1825, rubber was being used for a number of surgical purposes and shoes of cut sheet rubber were in some demand. In 1835, the world's largest rubber factory — Hancock's — was using three to four tons of caoutchouc per week. The French rubber industry got its start under Hancock's guidance in 1828 when Rattier and Guibal, who had formerly sold English-made goods in France, established their own factory at Saint Denis. Hancock supplied the machinery and instructed the French in manufacturing methods; the solution he supplied from his English factory so that he would not have to divulge the secret of the masticator. The Englishman's other important contribution to the pre-vulcanisation rubber industry was the spreading machine, which he invented in 1837 to replace the primitive methods then used in Britain for coating fabric with rubber in solution.

In the United States the rubber industry got away to a late start. First to import gum shoes from Brazil in quantities was Boston merchant Thomas C. Wales (1823) and first to advertise them in the press was J.W. Goodrich, Boston stationer (1824). Wales sold at prices ranging from $3 to $5 a pair and realised such a profit on the misshapen affairs that he had American lasts sent to the Amazon. One of the first to manufacture shoes in this country was S.C. Smith of Providence, R.I. whose original New York store was at Maiden Lane and William Street and later was moved to Chatham Street. Foremost of the early manufacturers, however, was Edwin M. Chaffee, foreman in a Boston patent leather factory. This Yankee thought that he might be able to make a rubber varnish which would render leather waterproof and at the same time give it a smooth surface. He began his experiments in 1831 and soon had a mixture of rubber and lampblack in turpentine which gave the finish he desired when spread on cloth or leather. Chaffee, John Haskins and Luke Baldwin continued their experiments at covering cloth with rubber with a spreading machine of Chaffee's invention. So enthusiastic did they become that 1832 saw the formation of the first rubber company in the United States. The charter, issued on February 11th 1833, states that 'Lemuel Blake, Luke Baldwin, Edwin M. Chaffee and Charles Davis, Jr., are created a body corporate, by the name of the Roxbury India Rubber Factory, for the purpose of manufacturing at Roxbury, in the county of Norfolk, India-rubber cloth and leather and other India-rubber goods.' The new company began with a capital of $30,000 which was increased to $240,000 in 1834 and to $300,000 the following year. It manufactured shoes, life preservers, coats, caps, wagon covers and carriage traces. During 1833-1834 Chaffee was granted United States patents for making hose and mailbags as well as for improvements in the manufacture of boots and shoes. President Jackson visited Boston in the winter of 1834 and the alert management of the Roxbury company presented him with a suit of rubber clothes, for the day was rainy. These Jackson wore as he rode through the streets and the fame of the company's products was notably increased.

Chaffee was the inventor of the two machines which are still common to every rubber shop in the world - the mill and the calendar. The Roxbury man has never been given the place he deserves among the founders of the rubber industry. Nothing resembling Hancock's masticator has been used for years, but the mills and calendars that roar in present day rubber factories are but little different from the machines Chaffee built over a century ago. They have grown larger and more powerful and have many refinements, but essentially they are the same. The Yankee's first mixing mill was constructed towards the end of 1835. Labelled 'preparing machine' in the patent, the device differed greatly from Hancock's, for instead of the rubber being mixed inside a chamber it was compressed between two steam-heated iron rolls. The rolls turned at different speeds, as they do to this day, and Chaffee found that the combined rolling and slipping action tore and softened the rubber until it formed a smooth sheet on the roll and easily 'partook of the coloring matter' and other ingredients he placed on top of the batch. Following his successful introduction of the mill, Chaffee received a very respectful hearing when he proposed to the directors of the Roxbury company that he could save the $50,000 they had been expending annually for rubber solvent. Until this time, all the waterproof fabric had been produced by coating the cloth with rubber in solution. Rubber and lampblack were mixed on the mill and then dissolved in turpentine. The solution was 'flowed' on to fabric from a box so arranged that a steady supply poured on to the cloth as it unwound from a roll. The coated cloth was then festooned on wires which extended the full length of a long building and left for several days to dry. Chaffee's new idea was to cut out the loss of expensive solvent and at the same time speed up operations by spreading the rubber dry. Though his proposed machine looked expensive he was told to go ahead. The coating machine, completed in 1836 at a cost of $30,000, not only saved 30 to 36 barrels of turpentine a week, but reduced the labour by half. Fabric prepared by the old flowing process had sometimes required eight or ten coats, each of which needed a long time to dry; now an equal thickness of rubber could be deposited after three or four coats which could be applied in rapid succession. The coating machine weighed 30 tons and was called the 'Monster' or 'Mammoth' because of its great bulk. It consisted of four hollow rolls placed one above the other, each six feet long and heated by steam. The roll next to the top moved much slower than the others, creating a slipping action between itself and its neighbours. The cloth to be coated passed into the machine between the two middle cylinders, then down and around the bottom one. The rubber compound was fed in between the two top rolls and as it came into contact with the fabric was wiped from the roll and into the cloth.

Establishment of the Roxbury company had been followed by the incorporation of a number of others capitalised at $50,000 to $500,000 and located in Boston, South Boston, Chelsea, Woburn, Framingham, New York City, Staten Island and Troy. With several million dollars invested in the industry, America's first rubber manufacturing boom was on. It had an extremely short run. Save for the Roxbury, not one of the corporations or individuals engaged in the business lasted for more than one to three years. Rubber that decomposed and lost its elasticity, shoes that stiffened in winter, suspenders that sometimes melted at body-temperature, raincoats that hardened like armour or melted and ran away were enough to disgust anyone, and, in the consumer reaction against caoutchouc that set in, it is no wonder that company after company failed with a loss of every dollar invested. Without exception these concerns had been using turpentine as the rubber solvent, the result being that the natural adhesiveness of the rubber was accentuated by this oil. Able to manufacture without using any solvent as a result of Chaffee's invention of the 'Mammoth,' the Roxbury had a somewhat better product but a far from satisfactory one and it too went under. Later it was reincorporated as the Goodyear Manufacturing Company and still later became the Boston Belting Company. In October, 1843, the 'Mammoth' was sold to John Haskins at public auction for $525. The patent on it which Chaffee had originally sold to the Roxbury company for $10,000 went to Haskins for $1.50. A year later Goodyear purchased the 'Mammoth' and patent from Haskins and subsequently transferred it to the Naugatuck India-Rubber Company.

Because of the more equable climate, the rubber manufacturers of England were not so beset with perplexities as were those in America. Their goods became sticky in summer but did not become a melted mass; became stiff in winter but did not become as hard as iron. Nevertheless, Hancock vainly sought to find methods of 'divesting rubber of its adhesiveness, particularly the surface of solutions of rubber spread upon cloth to render it waterproof' in order to avoid 'the necessity of employing a double texture for this purpose, according to Mr. Macintosh's principle.' And in the case of the macintosh, too, the wearer would find his coat oozing away if he was careless enough to stand near a fire. Even apart from the impossibility of making an auto tire from unvulcanised rubber, the English industry could have had no more than extremely small growth save for Goodyear, then. As to America, the industry would have died with the Roxbury company had there not been an answer to the question of controlling caoutchouc so that its consistency did not vary from rock to putty over a year.

That the problem was answered was due to the passionate pertinacity of Charles Goodyear, a man so fired with conviction that the sticky stuff had immense potentialities that he persisted in his wrestlings with it though reduced to beggary and imprisonment. No greed for wealth impelled this man who, by turning crude rubber into a new material of myriad uses, became a benefactor of nearly all mankind save the Amazon reds and the Congo blacks. A seeker who toiled and suffered out of an inner necessity to find his answer, he was great as greatness is measured in any field, great beyond comparison in the record of rubber. About this queer, humorless, fanatical fellow there beats most of the glory rubber's history can claim and the manner in which he was rooked and exploited is as nasty if not as terrible a story of greed as caoutchouc's chronicle reveals. Like John Brown of Akron, Ossawotomie and Harpers Ferry, he considered himself an 'instrument in the hands of his Maker.' Answering the question of how his attention was first turned to 'gum-elastic,' he wrote that he would prefer the whole to the Great Creator, who directs the operation of mind to the development of the properties of matter, in his own way, at the time when they are especially needed, influencing some mind for every work or calling. The creature may imagine he is only executing some plan of his own while he is the instrument in the hands of his Maker contributing to execute His purposes, which, though we cannot fathom, we may believe, involve, with the highest elevation of mind and morals, the highest improvement of things material.' Even better exemplifying his religious attitude towards rubber, are the ecstatic hosannas he loosed at reflecting upon this divine substance. Wrote Goodyear, the poet:

'The most remarkable quality of this gum, is its wonderful elasticity. In this consists the great difference between it and all other substances. It can be extended to eight times its ordinary length, without breaking, when it will again assume its original form. There is probably no other inert substance, the properties of which excite in the human mind, when first called to examine it, an equal amount of curiosity, surprise and admiration. Who can examine, and reflect upon this property of gumelastic, without adoring the wisdom of the Creator?'

For serving the god of caoutchouc in such reverent fashion he was to be penalised by afflictions that would have turned the heart of Job, adversity that would have crushed any but a fanatic or a fool. Almost it is probable that the whole history of invention cannot produce such another story as his. It is the tale of a man who starved and suffered; who made his experiments alternately in jail cell and sickbed; who was forced to sell his children's schoolbooks to buy food; who was an unwilling intimate of pawnbrokers as well as jailers; who was derided and scoffed at and considered demented until he made his great discovery; who thereafter was victimised by sharp bargainers and pickpockets and spent his last years still debt-ridden and shattered in health. Many men who stand by their convictions in spite of personal suffering will yield to spare their families. Not even this would Goodyear do. Testimony given by a New England neighbor in one of the court suits in which the inventor was involved pictures a family going into the woods and fields to glean fuel, a family with no money to buy bread from one day to the next, children digging half grown potatoes for the sake of having something to eat. The same witness told of furnishing milk and being urged to take furniture and bed-clothes in payment; of the inventor's three-year-old son dying and being hauled to his grave in a wagon with the poverty-stricken family trudging behind.

Like John Brown, that other humorless, fanatical and gigantic son of the new century, Charles Goodyear was born in Connecticut in 1800. He was the eldest of the six children of Amasa and Cynthia Bateman Goodyear, and New Haven, incubator of inventors, was the scene of his December 29th arrival. No upstart family were these Goodyears. The first of them in America, Stephen Goodyear, was an associate and later successor of Governor Theophilus Eaton, head of the company of London merchants who founded New Haven in 1638. At the time Amasa was first storked, he was a merchant engaged in trade with the West Indies, but while Charles was still young he bought an interest in a button-making patent and moved to Naugatuck. In 1807 he commenced manufacture of the first pearl buttons made in America. An inventor, his patents included several valuable improvements in farm implements. One of the most profitable of these was a spring-steel hay and manure fork, which came into wide popularity after Goodyear had overcome the initial reluctance of farmers to abandon the old style, heavy, awkward wrought-iron tools beaten out for them by local blacksmiths. In addition to buttons and manure forks, other items of Goodyear manufacture included spoons, scythes and clocks. The inventive trait which first cropped out in Amasa appeared in the next three generations of Goodyears, a record probably unequaled by any other American family. The second generation, besides the immortal Charles, included his younger brother Nelson, one of the inventors of hard rubber, and cousin Ellsworth, son of Bela, who contributed the process for making hollow rubber balls by placing water inside the blank prior to vulcanisation and letting the resultant steam swell the rubber into the shape of the mold. And there was also cousin Austin Goodyear Day, who evolved a cleaning method which gave the inferior grades of rubber some utility and patented a compound which came into almost universal use for insulation. Charles' firstborn, Charles, Jr., aided his father throughout his life and after his Father's death invented the 'Goodyear Welt,' epochal shoe improvement, and the machinery for making it. The invention provided the basis for the Goodyear Welt Shoe Machinery Company, which in 1899 joined with the McKay company to form the United Shoe Machinery combine. William Henry, brother of the second Charles, was an archaeologist with notable findings to his credit. The fourth generation furnished a second Nelson, son of the second Charles, who patented a variety of apparatus for generating and burning acetylene and other gases in connection with oxygen as a source of great heat.

According to his autobiographical *Gum-Elastic*, published in 1853, Charles Goodyear first encountered caoutchouc at an early age. 'While yet a schoolboy,' he wrote, 'the wonderful and mysterious properties of this substance attracted my attention and made a strong impression on my mind. On first obtaining a thin scale from a newly imported bottle of the Indian manufacture, it occurred to the writer that if any method could be discovered of preventing the surfaces from adhering together, when brought in contact, it would constitute a beautiful fabric for many purposes.' When Charles was sixteen, William DeForest became tutor in the Goodyear family. Years later, DeForest, remembering the promise his pupil, now become brother-in-law, had shown, financed the rubber inventor and aided him greatly in getting his discovery before the public. At 17, Charles hied himself to Philadelphia to serve a four years' apprenticeship in the hardware firm of Rogers and Brothers, one of the largest importing houses in the country. Followed six years in partnership with his father as A. Goodyear and Son in the manufactory in New Haven. The reputation of their steel forks and other implements grew so rapidly that in 1826 Charles and his family — he had married Clarissa Beecher in 1824 — returned to Philadelphia to open the first establishment for the sale of domestic hardware in the United States. Goodyear is authority for the statement that the idea was considered visionary as, up until this time, the whole trade had been in imported articles. At first the firm was very prosperous and 'a handsome fortune was accumulated.' In 1829 the younger Goodyear's troubles started with a breakdown in health, caused by severe attacks of dyspepsia, a disease that deviled him all the rest of the way to the grave. Then came business losses. The firm had extended its operations quite widely in different states, especially in the South, and had allowed customers too liberal credit. Many failed to meet their payments in 1830 and Goodyear found himself obliged to suspend payment. Wishing to avoid bankruptcy, he induced creditors to grant him a long period for the discharge of claims. This proved to be an unfortunate course. The claims were continually shifting into the hands of strangers and under the laws then existing, Goodyear was repeatedly thrown into prison during the ensuing 10 years because of his inability to pay. Part of his indebtedness he canceled by disposing of the good will and control of the steel fork manufacture. In jail or out, he continued working on various inventions — when in, through the kindness of a jailer who supplied bench and tools — and perfected one shortly before turning his mind to rubber. All of the creditors were eventually paid even though some of the claims had been outlawed by time and Goodyear released from them.

His serious connection with rubber began on the day when he stopped at the Roxbury store in New York to inquire about life preservers, an odd subject always much on his mind. He was shown one and immediately his inventive bent asserted itself, for he decided that he could certainly improve on the method by which it was inflated. Some months later he was again in Gotham and showed the agent an improved valve he had developed, hoping to sell it to the company. The Roxbury representative told him of the troubles besetting the factories which had started with such a flourish a few years before. He told him how the Roxbury company had turned out large numbers of shoes in the winter of 1833-1834 and then had $20,000 worth of goods returned the following summer, half melted and emitting so villainous a smell that they had to be buried. He told him how the rubber industry's difficulties were increasing steadily as more and more people became disgusted with a material that was soft and sticky in summer and rigid in winter. Finally he informed the inventor that the industry was on the verge of collapse unless some remedy was soon discovered and added, on behalf of his own company, that he would guarantee a huge reward to the person who solved the difficulty. This outburst astonished Goodyear who was 'not before aware that the manufacture was so imperfectly understood' and had supposed that before great sums had been invested in the business, its obstacles had been conquered. Here was just the sort of new field of enterprise he had been looking for.

Back in Philadelphia, he was arrested in connection with one of the suits he was constantly facing, and was compelled to reside within jail limits. Under this restraint he made his first experiment in trying to control rubber. The kitchen of his small cottage was his laboratory; his fingers were the only mechanical power of which he had command, with them he mixed the gum with turpentine, then spread it on a marble slab with a rolling pin. Deciding that it was useless to work with the pure gum, he tried the effect of other materials. One of his first trials was to mix magnesia into it. He was pleased with the result, for he had a white product and nothing but black had previously been mixed in this country. The surface seemed to dry thoroughly and Goodyear supposed that the magnesia had accomplished this in the manner that some substances dry paint. A book was bound with the compound; soon, however, this softened and spoiled. A friend, Ralph B. Steele, of New Haven, advanced him a little capital and Goodyear, returning to that city, with a few employees manufactured several hundred pairs of shoes from embossed cambric. In order fully to test the shoes before they were placed on sale, they were stored. When warm weather returned they were found to be a mass of melted gum. The failure of the shoe-making enterprise was especially hard to bear because Goodyear had given his friends such positive assurance of success. His backer became discouraged, would not lend any further aid, and those who had been keeping him in supplies refused further credit. In order to meet his bills he sold the little furniture he possessed. Then he found a boarding place for his family in the country, posted linen made by his wife as security for the unpaid rent of his cottage and went to New York to continue his experiments. A friend, John W. Sexton, provided him with a room in Gold Street and another acquaintance, druggist Simon Carle, advanced him the simple chemicals he required for his work.

Deciding that the decomposition of the shoes made in New Haven was due to the effect of the turpentine, Goodyear sought a remedy and supposed he had found one when he tried boiling the articles compounded with magnesia in quicklime and water. This, he said, 'appeared to have the effect of tanning the gum and destroying its viscous property.' One day he unexpectedly ran into former tutor DeForest whom he had not seen for a long while. DeForest was shocked by his appearance; he looked tired, his clothes were rusty and he bore all the marks of poverty. Goodyear invited him to his room, up three flights of stairs and filled with kettles, gum-elastic, magnesia, other chemicals. 'William, here is something that will pay all my debts and make us comfortable,' said the seedy inventor. The surroundings gave little support to his words and DeForest remarked that the rubber business was below par. 'I am the man to bring it back again' was the confident response of the enthusiast. He was so enamored of the stuff that he made it a practice to test his various compounds by always wearing some article of apparel made from them. This gave rise to the story, which he proudly recounts, that a man, asking how he might be recognised, was told: 'If you meet a man who has on an India-rubber cap, stock, coat, vest and shoes, with an India-rubber money purse, without a cent of money in it, that is he.' The inventor was much impressed (as he was with every experiment) by the results he got from the boiling process and he made up some rubber drapery which won him medals from the New York American Institute and the New York Mechanics' Institute in the fall of 1835. In a few weeks he was disappointed to find that the new effect was superficial and confined to the surface of the goods. Deciding that he had not used enough lime, he tried mixing the quicklime directly into the gum. Finding this mix too caustic to work by hand, he obtained permission to use Pike's mill in Greenwich and daily trudged the three miles, carrying on his shoulder a gallon jug of lime prepared in his room. The material was too powerful to be of value and its use was abandoned.

Disappointment was soon forgotten in enthusiasm for still another discovery. Attempting to ornament some of his lime-boiled drapery with bronze powder, he failed to produce the desired effect and tried to remove the metal with nitric acid. The specimen was discolored and thrown away. Thinking about the matter a few days later, Goodyear unearthed the shriveled sample again and found that the surface had lost its adhesiveness. This method Goodyear labelled his 'acid gas' process and patented on June 17th, 1837. Although the acid benefited only the surface of the goods, the process continued in use until the discovery of vulcanisation. In the patent, Goodyear covered not only the method he had accidentally discovered of first applying bronze powder and then washing over with acid but also patented the use of acid solutions of metals, preferably copper or bismuth nitrates. The value of the whole process is now known to have been slight. The new method attracted considerable attention, however, and the inventor was the recipient of testimonials from various eminent chemists and of medals from institutions. Goodyear, as usual, was enthusiastic. Proceeding to Washington to obtain his patent, he took along maps printed on thin sheets of his 'cured' gum, rubber calling cards and elastic bandages. These he presented to President Jackson, Henry Clay and John Calhoun and duly received their thanks. His enthusiasm also nearly proved his finish. Experimenting secretly, before the patent was granted, he came close to suffocating himself by generating a large quantity of gas in a closed room. During his subsequent illness, Goodyear was treated by Dr. Joseph Bradshaw, an Englishman, who took a great interest in the experiments. Some time later Bradshaw was returning to England on a visit and offered to take specimens of the material to manufacturers in that country. An assortment of samples of gum-elastic drapery was made up and in due time presented to an English manufacturer who informed Bradshaw that liberal compensation would be paid Goodyear if the invention proved of value. We quote Goodyear on what happened: 'The English patent was taken out by Mr. Hancock, without any communication with the inventor.

This may be correct according to the English patent laws, but is not according to the ideas of justice entertained by the inventor. The acknowledgement was however given, which is implied in the legal preface attached to English patents when they are obtained for improvements introduced from another country.' Hancock, in his book, does not mention this incident. Due to the publicity accorded his acid gas process, the Yankee had no trouble in obtaining a partner. William Ballard of New York supplied the capital and in a short time the firm of Goodyear and Ballard began manufacturing, first in Bank street and later in an abandoned rubber works on Staten Island. A warehouse was rented on Broadway in expectation of the business which was to follow. Goodyear's high hopes were merely a prelude to new misfortune. The panic of 1836 bankrupted Ballard and the factory had to be closed.

So penniless was the inventor at this time that, wishing to take the ferry from Staten Island to New York, he pawned his umbrella with the ferry master, later famous as Commodore Vanderbilt. Goodyear's family now joined him in New York and the struggle for a bare existence became a bitter one. Having access to the factory, he made a few piano and table covers and ladies' aprons. These bought a little food, but the demand for them was small. So resorting to the refuge of other dark days he began pawning personal property. New York held no encouragement for him and he decided to go to Roxbury, which had previously been the center of the industry. Here he met with friendship and sympathy from Henry Willis, who had served an apprenticeship with him in Philadelphia, and from Haskins and Chaffee of the Roxbury founders. Most of the countryside about Boston had suffered severe losses in the collapse of the India rubber mania and was not inclined to look with favor upon the subject. There were, however, a few men who, in spite of heavy losses, were, quoting Goodyear, 'from pride of opinion on the subject, very desirous to see the business rise again.' These men, and especially Haskins and Chaffee, obtained aid for him and gave him the use of the machinery which was standing idle in the Roxbury plant. For a brief time, unaccustomed fortune smiled on the inventor. In the winter of 1837 he again commenced to make shoes and had better success than he had had at New Haven. A new method of making shoes was patented, and this, together with an acid gas license was disposed of to J.W. Clark of Boston and Charles Jackson, of Providence, who started a factory at the latter place. Goodyear was now able to bring his family to Roxbury and the change from absolute want to comparative luxury was a pleasant one. The greater share of his income still went into additional experiments. Two licenses were disposed of in the summer of 1838, one to Luke Baldwin and John Haskins who began the manufacture of piano and table covers at Lynn, and the other to Luther Clark, who began making carriage covers at Northampton. The sale of the licenses brought Goodyear a few thousand dollars which enabled him to bring his aging parents and two younger brothers to Roxbury.

The really fortunate occurrence of 1838, however, was his introduction in September to Nathaniel Hayward. This was a happening as fortunate for the world as for Charles Goodyear since it was through Hayward that he became acquainted with a new material to be mixed into his rubber batches - sulphur. And sulphur, in the hands of Charles Goodyear, was to prove the key to the riddle of rubber. Hayward was born in Easton, MA, on January 19th 1808, and as a young man went to Boston to operate a livery stable. In the 1830S the youth became interested, as did most of the public, in the India rubber business, and he was among the first to become prominent in the rubber industry in the United States. After private attempts to manufacture a waterproof shoe blacking of rubber and to turn out rubber fabric, he became manager of a company formed at the Furnace, at South Easton, MA, to make sandals of waterproof fabric for the outside of shoes. When this enterprise failed he returned to his private experimenting. He next became general superintendent of the Eagle India Rubber company at Easton and moved with it to Woburn in 1835. This company also failed but Hayward kept on with his experimenting in the abandoned plant and in 1838 discovered — according to his own account as the result of a dream — that when sulphur was spread over the surface of rubber goods which were then treated by exposure to the sun the surface became tougher and more durable. Right there Hayward had the essentials of vulcanisation in his grasp — rubber, sulphur and heat. He did not suspect, however, that substituting a high degree of heat for the mystic sun rays might be the answer to the caoutchouc problem and neither did Goodyear, who about this time took over the Eagle factory and hired this experimenter after his own heart. At Goodyear's suggestion Hayward patented 'solarisation' on February 24th 1839 and assigned the patent to his employer for $200. The Hayward patent governed the use of sulphur in turpentine solutions of rubber, and also its incorporation directly into the gum either by mixing it on a mill or by spreading over the surface. 'The effect of the sulphur,' said Hayward, 'in whatever way it may be added to the gum, is to cause it to dry more perfectly and to improve the whole substance thereof.'

During the time he worked for Goodyear, Hayward and his employer turned out life preservers and other articles by both the acid gas and the solarisation methods. Both processes achieved a sort of superficial surface cure of the goods, of considerably less value than the inventors thought. The interior of the articles was not changed at all. Goodyear later said that, at the time, he thought the acid gas treatment had changed the gum throughout its whole mass and that there was no further improvement to be made upon rubber. The inventor was soon to learn that he had not been as successful as he had hoped. He received an order from the government for a 150 mail bags and took pains to noise it about, as this seemed like a final endorsement. The bags were completed in the summer of 1839. They kept their shape and promised to keep it permanently, but Goodyear thought it well to hang them up for a prolonged test and went away for a few weeks' holiday. Returning to Woburn, he was dismayed to find the bags decomposing and dropping from their handles. He had used chromes, white lead and vermilion in the compound in addition to sulphur to lend a leather-like appearance, and it was these substances, he believed, which caused the failure. Several thousand life preservers and cushions which had been made of the same compound and distributed throughout the country were similarly affected and were returned. The good Charles made no attempt to explain his views in regard to the coloring materials, for he knew that he had already been sentenced by public opinion. What he had ballyhooed as a great discovery, the public pronounced a total failure. Once again he saw his family stripped of the comforts they had enjoyed for so brief a time. Everything he possessed 'was brought to the hammer for the discharge of private bills.' In addition to being reduced to extreme poverty, he knew he could no longer expect any sympathy from his friends if he persisted in his course. He had ruined himself by the four years he had spent in fruitless attempts to improve a business which had previously ruined the entire community. Friends pointed out that he could do no more, that if he could not solve the problem after occupying himself with it exclusively for four years, it was insoluble.

The recollection of previous losses was refreshed in the public mind by the mail bag episode and the inventor knew it was no use to look for funds to be used in the manufacture of rubber goods. The men to whom he had sold licenses were so discouraged that he could not apply to them for assistance. Any other man might have turned back to respectable ways, but not Charles Goodyear. Spurred by his inner conviction, he persisted in what everyone else agreed was an idle and foolish enthusiasm. With the aid of his family he continued to make a few small articles and the sale of these together with frequent trips to the pawnbrokers were his chief dependence. Friends continued to importune him to give up his rubber folly and had 'the improvement sought for been one which depended upon any amount of capital beyond a few sixpence at a time, necessity would have compelled him to yield; but so long as these could be obtained or hoped for, experiment could be continued, and the discovery made, as it was, in the most humble sphere.' Again he set to work with 'unabated ardor and diligence' to see what he could do with rubber, and, as usual, 'he had hardly time enough to realise the extent of his embarrassment before he became intently engaged with another experiment, and his mind buoyant with new hopes and expectations.'

4.2 The Answerer

OVER 300 years after the men of Cortez first looked upon caoutchouc in Montezuma's city, a century after Charles Marie de la Condamine directed Europe's attention to *Hevea* milk, civilisation finally caught up with and passed the dullards of the Amazon. Before vulcanisation dawned upon Charles Goodyear in 1839, the world cannot be said to have done any more than prettify caoutchouc articles beyond anything ever achieved in South America. Basically, however, it did not improve upon them; rather, fell short of their quality save in such instances as the macintosh. Goodyear was the first to eliminate rubber's fundamental fault of hardening in cold weather and becoming soft and sticky in warm; but that was far from all he achieved by his process of heating caoutchouc with sulphur. In addition to correcting gum-elastic's great flaws, he added so many valuable qualities that, in effect, a new substance was created. Changing the hardened tree juice from a plastic of little utility to a material that is tough, non-adhesive and elastic over a wide range of temperatures, he brought caoutchouc's pioneering period to a close and opened the century of real utilisation only now drawing to an end. Only because of vulcanisation do we have the automobile tire, which otherwise would not hold pressure and would flatten out under the lightest load. Only because of vulcanisation do we have rubber articles of any sort which do not decompose and lose elasticity after the briefest periods of service. Exactly what vulcanisation is, the greatest scientists of the world cannot yet tell you, however. That the process consists of heating rubber with sulphur even the man in the street knows, but concerning exactly what happens we have nothing but theories. For long the chief word war was between chemists and physicists, one party claiming that vulcanisation was a chemical combination of rubber and sulphur and the other that it was a purely physical process of rubber absorbing brimstone. Of late years, the chemists have prevailed to the extent that some degree of chemical combination is generally conceded, but many scientists believe that the effects of vulcanisation are due to the combined action of chemical and physical forces.

These were things that never worried Goodyear either before or after coming upon the first clue to vulcanisation. Luck it was that brought the clue to his attention, but there was no luck about his instantaneous recognition of what had happened. About his accident — the kind which happens only to the researcher who exposes himself to the lightning time upon time upon time — many ridiculous legends have gathered and grown and been soberly perpetuated. In books, in rubber company pamphlets, in magazine articles, the reader comes upon the tales of sulphur being accidentally dropped into a cup of melted caoutchouc, of rubber balls being accidentally bounced on a red hot stove in the Woburn general store, of sulphur being spilled on a rubber apron. It is well, therefore, to give the one true story of the 1839 revelation in Goodyear's own words, accessible since 1853 but almost always disregarded. Says his third person chronicle, greatest of all rubber source books:

> 'While on one of the visits alluded to, at the factory at Woburn, and at the dwelling where he stopped whenever he visited the manufactory at Woburn, the inventor made some experiments to ascertain the effect of heat upon the same compound that had decomposed in the mail-bags and other articles. He was surprised to find that the specimen, being carelessly brought into contact with a hot stove, charred like leather. He endeavored to call the attention of his brother, as well

as some other individuals who were present and who were acquainted with the manufacture of gum-elastic to this effect, as remarkable and unlike any before known, since gum-elastic always melted when exposed to a high degree of heat. The occurrence did not at the time appear to them to be worthy of notice; it was considered as one of the frequent appeals that he was in the habit of making, in behalf of some new experiment.

'He, however, directly inferred that if the process of charring could be stopped at the right point, it might divest the gum of its native adhesiveness throughout, which would make it better than the native gum. Upon further trial with heat, he was further convinced of the correctness of this inference by finding that India rubber could not be melted in boiling sulphur at any heat ever so great, but always charred.

'He made another trial of heating a similar fabric, before an open fire. The same effect, that of charring the gum, followed; but there were further and very satisfactory indications of ultimate success, in producing the desired result, as upon the edge of the charred portions of the fabric there appeared a line, or border, that was not charred, but perfectly cured.

'He now removed with his family to Lynn, in order that he might have access to the steam power of Messrs. Baldwin & Haskins, for the purpose of trying experiments in vulcanising by steam.

'A few weeks after, he removed from Lynn to Woburn, where he now pursued his inquiries and experiments for some months quite alone, until the desired result was obtained. On ascertaining to a certainty that he had found the object of his search, and much more, and that the new substance was proof against cold, and the solvents of native gum, he felt himself amply repaid for the past, and quite indifferent as to the trials of the future.

'... It being now known that the results of the vulcanising process are produced by means, and in a manner, which would not have been anticipated from any reasoning on the subject, and that they have not yet been satisfactorily accounted for, it has been sometimes asked how the inventor came to make the discovery ... he was many years seeking to accomplish this object, and ... he allowed nothing to escape his notice that related to the subject. Like the falling of an apple, it was suggestive of an important fact to one whose mind was previously prepared to draw an inference from any occurrence which might favor the object of his research. While the inventor admits that these discoveries were not the result of scientific chemical investigations, he is not willing to admit that they were the result of what is commonly termed accident; he claims them to be the result of the closest application and observation ...'

> 'It may ... be considered as one of those cases where the leading of the Creator providentially aids his creatures by what are termed accidents, to attain those things which are not attainable by the powers of reasoning he has conferred on them.'

The great discovery was far from being the end of the inventor's trials. Considering the state of public confidence in caoutchouc, he knew that it was useless to try to interest anyone around Boston in his new discovery and that it would be equally useless to visit any other part of the country without specimens large enough to be used. Samples he had so far been able to make were of small size and though he felt confident that the process was applicable to larger articles he realised that it was unreasonable to expect anyone to attach much importance to a statement to this effect, since he had been making such statements about successive improvements for years.

The inventor confesses that he only subsisted at this period through charity. His own library had long since gone to the pawnbrokers and now he collected and sold the school books of his children, which brought him 'the trifling sum of five dollars; small as the amount was, it enabled him to proceed.' During the spring he made 'some specimens tolerably perfect' by heating them before a brushwood fire. These he used to persuade neighbors to help him build a six-foot square brick oven and also to help him make some large samples of fabric. These spoiled before they could be heated and Goodyear at last resolved to go to New York with what few samples he had. He had been assured by a friend in Boston of the loan of 50 dollars, enough to take him to New York and to keep his family in his absence. The friend failed to keep his promise. Goodyear remained in Boston a week hoping to obtain the sum from some other source and on Saturday night was presented with a hotel bill that he could not pay. After walking until late at night, he strayed into East Cambridge and stopped at the house of a friend. Next morning he walked the 10 miles to his home, to be met at the door with the news that his youngest boy was dying. The family was in dire straits, having been denied supplies promised by a dealer before Goodyear had left.

The man with the secret worth millions then appealed to another friend in Boston and received seven dollars, a severe reprimand for not turning his attention to some other occupation, and a barrel of flour from a total stranger who had been in his friend's office when the letter was received and was touched by its contents. With 50 dollars sent to him by his brother-in-law William DeForest, he was able to go to New York, where for the first time he met William Rider. Surprisingly enough, Rider, a merchant, was impressed with the value of the invention, and, along with his brother, Emory, agreed to furnish the capital necessary to carry on the experiments. More important even than the monetary aid, according to Goodyear, was the continual encouragement the brothers offered him during the next several years when obstacle after obstacle interfered with successful manufacture. Chief of the difficulties Goodyear encountered was in heating his goods uniformly so that they did not blister. Until the Riders' funds had made better means available to him he had resorted to all sorts of dodges to heat his compounds. He visited all the factories and shops in Woburn to ask the privilege of using an oven after working hours or permission to hang some rubber compound in the manhole of the boiler. The testimony of various foremen indicates that he bothered them constantly and smeared their works with sticky mixtures. All seemed to regard him as a lunatic, but at the same time they helped him willingly.

The singular ill fortune that dogged the researcher's steps made itself felt again when the Riders failed in business. Shortly before this happened, Goodyear had commenced operations at Springfield, MA, and had succeeded in overcoming most of his difficulties, making some yards of the elastic compound into sheets by drawing them through a heated iron trough. He had also recently invented a method of making shirred goods which attracted the attention of William DeForest, now a woolen goods manufacturer. The brother-in-law took the place of the Riders in supplying the capital for Goodyear's operations and over a period of years was to loan him $46,000 which had not been repaid at the time of the inventor's death. Once more, however, the man who should have had an immense fortune by the tail was to be thrown into an American jail. While working at his Springfield shop one morning, he was visited by a bailiff who demanded immediate payment of a debt. Unable to meet the demand, he was hauled to the Boston hoosegow jail. His resentment was keen since he felt that commercial success was in his grasp, and for the first time he took advantage of 'the odious bankrupt law.' After the tide had turned he paid back to creditors some $35,000 from which he had been discharged by certificate of bankruptcy.

It was on June 15th 1844, more than 10 years after he had started his experiments in rubber, five years after his discovery of vulcanisation, that he received his patent for the process. The specification for the patent had been prepared on December 6th 1841 and deposited in the Patent Office as a caveat or claim of invention with details not made public. Application for a patent was not made until 1843. Both the 1841 and 1843 claims were for compounding rubber, sulphur and white lead and subjecting them to a high degree of heat, but the inventor also mentioned that 'oxides of metals generally, or other pigments, may be substituted for the lead.' Indispensable constituents of the process were, of course, the rubber, the sulphur and the heat. The lead shortened the time of vulcanisation, but was not essential to the operation. Goodyear had several reasons for filing his claim and permitting application to ride. His financial state in 1841 did not permit him to manufacture articles to demonstrate the utility of the idea to anti-rubber American financiers and he felt that he should find out if there was any substitute for sulphur, then considered objectionable. Without so protecting himself, he feared that the patent, if issued, might be lost to him. Further, foreign countries required that a patent, to be valid, must be filed with them on the same date on which it was filed in the United States, and he did not have sufficient means to do this. Goodyear made his application in the United States, England and France, as stated, in 1843 but he was trimmed in both of the foreign countries. The French laws compelled him to manufacture and keep in use constantly the goods produced under his patent. He tried to do so as far as his means would permit, but officials took advantage of his circumstances to annul the patent on the allegation that he had permitted introduction of American-manufactured shoes. On John Bull's island, Thomas Hancock beat him out. Through the agency of a messenger from America, Hancock, the despoiler of Goodyear in the matter of the 'acid gas' process, was able to repeat his performance. After depositing his specification for a United States patent in 1841, Goodyear had entrusted samples of his vulcanised rubber to Stephen Moulton, a young Englishman visiting in this country. Returning to England in 1842, Moulton showed the samples to William Brockedon, an inventor, who introduced him at the Macintosh company. The English firm was unwilling to take up the process unless it was first patented in England to assure a monopoly. Moulton returned to America to confer with Goodyear, leaving some samples in Brockedon's hands. Goodyear explained his secret process to Moulton in December, 1842, but in spite of further negotiations with the Macintosh company in the following April the firm did not see its way clear to buy the process for the £50,000 Moulton asked. Meanwhile, the samples left with Brockedon had found their way into the hands of Hancock. This gentleman eventually succeeded in duplicating Goodyear's product by immersing rubber strips in melted sulphur. Brockedon suggested that the process be called vulcanisation, deriving the name from Vulcan of mythology. This, ironically, was the word that caught on in both England and America, poor Goodyear thus being deprived even of the satisfaction of naming his own epochal invention. 'Metallic gum-elastic' the American had chosen to call his product but 'vulcanised India rubber' won the vulgar endorsement and has survived save for the dropping of 'India' by the wayside. Hancock successfully defended his patent in the English courts in 1851 and 1855.

The basis of the decisions was that although the discovery first had been made by Goodyear in America the material did not disclose the means of making it and Hancock therefore was entitled to be considered an original inventor.

The battle over credit for the discovery of vulcanisation still periodically waxes and wanes. Americans have held out unanimously for Goodyear and in this they have been supported by a fair share of Britons, or if you prefer, a share of fair Britons. Other Englishmen have begged the question by dividing the credit equally between the two men and a number have been rabidly pro-Hancock. In this latter class falls John Hancock Nunn, grand-nephew of Thomas and inheritor of his rubber works, who might be expected to show some partiality but who can scarcely be pardoned for making such a statement as, 'Goodyear did not discover vulcanisation in 1839, or in any other year, and consequently his alleged discovery could not have revolutionised the entire industry.' Nunn voiced this inexcusable outburst in 1920 when his firm was celebrating its centennial. It is only necessary to refer to Hancock in order to establish the length of his grand-nephew's ears.

Under oath, in court, he admitted that the first vulcanised rubber he ever saw came from America. In his book, published several years after this admission was in the record, he gave himself the usual good break — and laboriously avoided mentioning Goodyear by name — but again conceded that the samples obtained by Brockedon from 'a person from America, who represented himself as the agent of the inventor ... certainly showed me for the first time that the desirable change in the condition of rubber of not stiffening by cold had been attained.' Right at this point he argues that the samples afforded 'no clue' to the mode by which the change had been brought about but elsewhere he tells of noticing that they had a dirty, yellowish color, a little dusty powder on the surface and a slight odor of sulphur. With this as a guide plus, he says, 'a strong impression that the rubber underwent the change either when in a state of solution, or when greatly softened by heat,' he went to work 'resolved, if possible, not to be outdone by any.' Months of experimenting, which allegedly produced the wanted change in portions of some of his scraps, convinced him that sulphur must be responsible. Acting on this belief, he took out a provisional patent in November, 1843, a few weeks before Goodyear filed. Under English law Hancock had six months in which to furnish his detailed specification and in that interval he mastered the problem, he said.

It is, perhaps, worthy of note that Goodyear's French patent had been published. By Hancock's own admission, he had been engaged in rubber work for 20 years without getting on vulcanisation's track and there is no slightest reason to believe that he ever would have done so if the American's finished product had not been slipped to him. It must be conceded, however, that he certainly was a sharpshooter worthy of comparison with the American manufacturers who took Goodyear into camp.

The invention of hard rubber (also known as vulcanite or ebonite) was, of course, an inevitable sequel to the production of vulcanised soft rubber. Completely dissimilar the inflexible and the immensely flexible products seem, but their manufacture differs only in the amount of sulphur used and heat applied. It is not odd, then, that Hancock produced the hard while experimenting in his effort to duplicate Goodyear's soft. American patriots unanimously credit Nelson Goodyear, younger brother of Charles, with fathering the stuff, but there seems little doubt that the Englishman was years ahead of him. The proof is the 1844 Hancock patent which states, 'By higher temperatures or by longer keeping in high temperatures, the caoutchouc gradually changes until it ultimately becomes black and has something the appearance of horn.' The first American patent was taken out by Nelson Goodyear in 1851. In this year, also, the elder Goodyear obtained a number of hard rubber patents and both

Hancock and his American rival exhibited rigid articles along with soft vulcanised gadgetry at the Crystal Palace Exposition. Noting in his book that 'The hardest of these compounds resembles marble; that which is less hard, ivory and buck-horn; that which is still softer, buffalo-horn and whalebone,' Charles Goodyear footnotes his marble reference by recording, 'For the modification which gives it its extreme hardness the writer is indebted in good part to his younger brother, Nelson Goodyear.' It is his only reference to Nelson.

As might be expected, there have been others besides Goodyear, Hancock and Hayward for whom claims to the discovery of vulcanisation itself have been made. The important ones we shall briefly consider. First of these in point of time is mentioned in that extraordinary 1791 article in *The Bee*. Here appears the startling statement, 'It is many years since Dr. Bergius, at Stockholm, made some experiments on this substance (rubber) in Papin's digester. By subjecting it in this way to an intense degree of heat it is said to have been converted into a hard, elastic horn-like substance.' Search through the published works of Petter Jonas Bergius, professor of medicine and pharmacy at the Carolinischen Institut, Stockholm, reveals no reference to such an experiment. Had Bergius heated rubber by itself he would have gotten only a sticky mass, and it is inconceivable that he would have made the statement attributed to him. It is possible, therefore, that some time before 1791 Bergius prepared hard rubber by heating caoutchouc and sulphur, but either did not recognise the part played by the sulphur or failed to mention the use of this ingredient to those whom he told of the experiment. Much less impressive are the bids for a substantial share of credit placed by German writers on behalf of Dr. Friedrich Wilhelm Ludersdorf, who in 1832 seems to have made the first published reference to the use of sulphur in rubber. Ludersdorf was troubled by stickiness of his rubber-turpentine solutions until he added sulphur and found this a preventive. In his 1832 book he also advised rubbing finely powdered sulphur on rubber articles to stop deterioration. All this is interesting but the Professor's failure to heat his mixtures disqualifies him from any serious consideration. In Holland, at about the same time, however, apothecary Jan van Geun of Haarlem is reputed to have marketed rubber tubing which did not stiffen in the cold nor soften with heat. Van Geun left a diary which contains many cost calculations for rubber goods. The diary on June 16th 1836, records costs for solution for air cushions, made of one part of rubber dissolved in three parts of a solution of sulphur dissolved in turpentine. The entry lists the cost of 88 yards of cotton treated with this solution as well as the cost of peat and wages of workmen for heating. The Dutchman had all the elements needed for vulcanisation — rubber, sulphur and heat.

It is not difficult to understand why widely separated investigators all happened to hit on the use of sulphur. Their first impulse would be to use a powder which would absorb or 'dry up' the tackiness of the rubber. The commonest and cheapest materials of this sort carried on the shelves of the pharmacists of the time were lime, magnesia, talc and sulphur. Of these, the last named possessed the added advantage of being soluble in turpentine, in which rubber itself also dissolved. About the addition of heat, of course, there was always considerable of the accidental. Bergius, if he did hit upon the production of hard rubber vulcanisation, seems to have had almost no conception of the value of the product, however. Van Geun, if he actually achieved vulcanised soft rubber, would seem to have appreciated its value in an extremely parochial way — the magnitude of the finding completely evading him.

Had Goodyear been of their sort, his vulcanisation discovery might well have died in Woburn as the secrets of the Swede and the Dutchman - whatever they were — perished in Stockholm and Haarlem.

As it was, Goodyear's grasp of the full value and magnitude of what he had done and his tireless campaign of whooping up the virtues of the product proved virtually as important to the world as the act of discovery. Vulcanisation's conquest of England directly resulted from the unanticipated enrolling of Hancock, of course, but here it must be remembered that it all began with the evangelical Goodyear's chasing samples over seas and that, had Hancock never entered the picture, the same effect would have stemmed from the American's English patent action. In America itself it was only through the fanatic's efforts that manufacturers dubious of all pertaining to discredited caoutchouc were first persuaded to tackle the business they soon were jostling, lying and cheating to enter. Even at the moment when the scoffers who had been ridiculing him for more than a decade were forced to admit that he was right and they were wrong, Goodyear did not pause to enjoy his triumph. And for all his remaining days he busied himself with such frenzied researching into new applications of his discovery, that the manufacturers required decades after his death to catch up with him. Friends of Goodyear's remembered him 'sick, meagre and yellow,' constantly running to them exulting in some new use he had thought of. Once it was a rubber cane, for which he predicted wide popularity. Another time he would be deviling workmen in the comb factories to get them to try their tools on rubber. Again he would be lost to sight for months, having shut himself up to make a rubber sailor perfect or rubber tiles for floors. At other times he would be proudly displaying cutlery with rubber handles or a watch with a rubber case. Of caoutchouc was the doorplate of his office and his portrait was painted upon and framed with the sovereign substance.

His *Gum-Elastic*, written in 1851, published in 1853 and now a museum rarity, consists of two books, a 246-page one sufficing for his autobiographical, historical and argumentative stuff, and a 378-page one being required for the expounding of the invention's uses. Showing himself the traditional absent-minded inventor, he permitted the volume to be peppered with scores of blanks by the usual literal-minded printers at spots where the author quite evidently had intended to fill in forgotten or otherwise troublesome dates, names and references to chapters and pages. Not to any applications of soft or hard vulcanised gumelastic does the forgetfulness extend, however. Goodyear lists over 1,000, and in the inventor's mind all of them fitted neatly into a gigantic and logical system of improvements and inventions promising the rubberising of the world. Chief value of his book, he thought, was that it now first presented the subject as 'a complete system.' His only reservations were that rubber was not recommended as an article of food and 'should not be worn next the skin, nor should one sleep enveloped by it.' For the rest, the range of his recommendations, nearly all supported by actual production of an article by himself or other inventors, can best be indicated by listing the sub-divisions under which he marshaled them.

These included: educational; carpeting, tents and awnings; house, ship and camp ware and utensils; mechanical; springs; hydraulic; military; naval and maritime; medical and surgical; philosophical, optical and mathematical instruments; musical; gymnastics and calisthenics; toys and trifles; sports and games; horse trappings; harness; fancy and ornamental uses; air-work; miscellaneous; appendages of wearing apparel; wearing apparel; traveling apparatus; articles for the preservation of life and property. These, it is needless to say, cover very nearly the whole range of caoutchouc's employment today. Most important of all were the several that Goodyear overlooked but they became so only with the rise of a motorised and electrified civilisation rubber's passionate advocate could not foresee. This proves no more than that he was not altogether the prophet he thought himself; it diminishes no whit his feat in covering the world as he knew it from almost every possible rubber angle. Nor does it alter the fact that nearly every year sees introduction of some 'new use' of rubber advocated in this book of 80 years ago; that in its pages lurk other still forgotten or never exploited applications which the manufacturers currently hunting diversification might do well to consider.

Immensely absorbing to anyone at all interested in rubber is Goodyear's whole catalogue volume; vastly amusing even to the non-rubber minded are the whimsically or uproariously funny uses which he evolved with so straight a face. Accounting for many hundreds are the familiar and semi-familiar items, some familiar before or at the time Goodyear wrote of them, some only recently having come into general or experimental use. Of machine belting, he writes, inkstands, hay rick covers, knife handles, diving suits, engine packing, hose, bandages, pontoons, boats, letter bags, floats, decoys, hot water bottles, breast pumps, 'dissecting gloves,' syringes, gas bags, dolls, footballs, stereotype plates, life preservers, overshoes, abdominal supporters, air cushions.

Then there are those, most of them practical in the inventor's own time, which the passing years, have outmoded or reduced to unimportance: bellows and bellows coverings; windmill sails; whale springs ('The use of the spring is to ease off the strain upon the cable, by which the whale is secured from breaking loose from the ship in rough weather, while taking the blubber on board the vessel'); elastic tape ('among other uses for it, is that of tying ladies' hair without danger of cutting it'); horse collars; daguerreotype boxes; pantaloon straps; imitation buffalo robes; 'hat and cap springs to prevent hat or cap from blowing off'; shoe springs ('Whoever knows by experience the difficulty of teaching a family of children to keep their shoes tied, to say nothing of the neatness and convenience of the articles for adults, will hail this as one of the great improvements of the age').

One of the most entertaining things in the book is the good man's ballyhoo for his line of fording and baptising dresses. Ordinary fording dresses, he records, were made 'in the form of pantaloons with boots attached, and a large tube surrounding the top, which is inflated with air. They are used for fording rivers, and by the aid of a cord to pull the dress back across the stream, a party of any number may cross a river with one dress.' The full fording dress draws a bit of history. 'This is a similar article, differing from it only in the upper part by the addition of sleeves, gloves, etc., so as to cover the whole person. Being drawn closely about the neck, a man may ford a stream without removing his garments. While wearing under the sheet of water at Niagara Falls one of the open, uncouth over dresses of oil cloth, that are afforded visitors at Niagara, at the price of half a crown, the writer imagined that the public might be better accommodated with a dress, such as the one here described.' Goodyear's sales talk for baptising dresses is perhaps his literary masterpiece. 'They have been found useful, and have been highly approved of by Baptist clergymen, who are obliged to stand a long time in water in cold weather. They are made similar to fishing pants, except that they are cut higher in the waist. The full fording dress, either with or without the life preserver, is also well adapted to this use.' Rubber pants for the water-walking clergy still enjoy some sale, incidentally, the most of them going to the smaller towns in the South.

The 'system' included, also, articles easily made of rubber but which never have amounted to much because the substance offers no outstanding advantages in these instances. Conceivably, however, such things as rubber furniture might at some future date come into favor.

Of the completely cockeyed inventions, outstanding examples are: air-inflated saddles; improved boxing gloves with 'an air chamber upon the back of the glove, instead of one that is stuffed'; boxing jackets ('when inflated with air the hardest blow has very little effect upon the person wearing it'); skating caps inflated with air ('Considering the numerous hurts received by boys falling upon the ice, this will not be considered an unnecessary precaution'). Had someone suggested rubberised cat's pajamas or bathtubs lined with caoutchouc in imitation of fur, Charles would soberly have listed them, too. The 'inflated batclub' probably was designed for use by clowns of stage and circus which would remove it from this classification. 'This article is . . . inflated by a self-acting valve tube at the end. They form a weapon, the hardest blows from which are quite harmless.' A honey of an idea was the household reservoir. 'These tanks, or sacks, are designed to be placed under the roof in the attics of dwelling-houses, with hose connecting with the water spout, and other hose extending downward to the different apartments of the building. The convenience of this arrangement for the supply of rain-water in the rooms, and also for extinguishing fire, must be apparent.'

Deaths by drowning constituted one thing which Goodyear believed he had been divinely raised up by God to combat. From the day when he fingered a life preserver in the Roxbury store on down the years that idea marched with him. While in England in 1858 and in extremely poor health, he suffered a severe physical setback as a result of tearing into new life preserver projects after reading that 20 people throughout the world drowned every hour. According to one of his Parson Weems biographers, he was unable to get together with Morpheus of an evening and told his wife, 'How can I sleep, when so many of my fellow creatures are passing into eternity every day, and I feel that I am the man that can prevent it?' This too-pious anecdote has no very convincing ring but it is certain that the man was obsessed by the subject of life preservers. All through the 1851 manuscript this motif threads. Other forces of nature were less on his mind but in putting on the record for chair 'shoes,' he bore down on gum-elastic's being so completely a non-conductor of electricity that 'a person having on India rubber shoes, sitting upon a chair of this kind, or lying upon a bed provided with similar shoes, need have little apprehension of danger from lightning.'

His firm belief, however, was that rubber ship sails constituted 'perhaps the most extensive and important of all the applications of gum elastic... The first gum-elastic ship's sail was a top-sail, which was made by the writer in 1844.' According to the inventor, Captain Popham of the Stephen Whitney commended the sail after a first trial and after six Atlantic crossings. Its virtues were that it remained pliable and clear of ice and was not liable to mildew. The United States Government gave orders for a number of them, he adds, but they were not executed, the manufacturers under the Goodyear licenses being too busy making money from smaller articles. Nor were rubber sails produced later on, objections to their heaviness being sufficient to sink the idea.

Goodyear was not, of course, the inventor of all the contraptions he listed, although he ultimately did have some 500 patents to his credit. The shot plug 'for stopping holes made by cannon balls in sides of ships' which he so greatly admired was, for instance, the work of Lieutenant Seely (Goodyear also spells it Seeley), formerly of the United States Navy. An interesting partial roll call of early American rubber patentees is provided by the others he lists along with their findings, improvements or first American registrations. To a Dr. Stearns, we find we are indebted for foot holders to 'prevent slipping on ice' and to Henry Barnard of Hartford, CT., for blank globes on which pupils could do maps to be afterwards washed off or fastened on the spheres by means of transparent varnish. Charles Goodyear, Jr., invented a method of manufacturing sectional globes from rubber; Nelson Goodyear, Hayward and Dr. E. Pratt of New York had valves and stops to their credit. Chaftee and John Lewis of New Haven were car spring inventors and Fowler M. Ray of New York conceived the one utilising alternate discs of rubber and metal.

Hancock in his 1857 book lists 150 or so articles, most of them mentioned by Goodyear but many manufactured by Hancock before the discovery of vulcanisation. Vulcanised tires (solid) for carriages was one the American did not have. One deluded inventor Hancock had to sidetrack from keeping an appointment with the Duke of York to demonstrate that rubber would resist the force of a bullet. The experimenter had fired at a piece of rubber and found no bullet; the explanation being that the hole had closed up after the bullet passed through. Of hard rubber, Hancock noted, '. . . if it were economical, or in any way advantageous to do so, it would make good houses, ships, wagons, and carts. . .' One of his most interesting passages deals with his developing of pneumatic articles in pre-vulcanisation days. He made a rubber bed, but found that no sleeper could keep his balance on it. So, too, with rubber pillows. The solution was found by dividing the articles into compartments. Pneumatic rubber pillows of such partition construction but morocco covered — were used by George IV in his last illness.

A fertile inventor was Hancock, and a good businessman also — a manufacturer first, in other words, and a discoverer second. Hayward, too, did well as a manufacturer under Goodyear license and was in no need of sympathy for having parted with his rubber-sulphur-sun patent for a measly $200. Basis of the Goodyear invention it was, but worthless in itself. Vulcanisation was, however, worth millions even with England and France lost. Yet Goodyear failed to collect these millions even though he had title not only to the master invention but to a great share of the uses and applications growing out of it. The explanation in part is that it was a matter of choice with him: he preferred tearing into new inventions by the wholesale to piling up a fortune through manufacturing the things achieved. In pre-vulcanisation days, Goodyear has recorded, he wanted to carry on as a manufacturer of rubber goods and 'establish a reputation for himself' in that field of activity. But with his great discovery, there opened before him the vista of what remained to be done to perfect familiar articles and extend applications. Now he was pulled two ways. There was his wish to 'pursue the calling of a manufacturer of gum-elastic, as a means by which he might hope to establish a better reputation for the manufacture than others will be likely to establish for it.' On the other hand there was the urge to devote himself to 'making application of this substance to the useful purposes and inventions, which would otherwise probably escape the notice and attention of others, as the original discovery might have done had it been sought after with less enthusiasm by the writer.'

The making and perfecting of inventions in the service of caoutchouc won and the American renounced all idea of himself engaging in regular manufacturing. With many a premonitory shudder over what would happen to the fair name of vulcanised rubber at the hands of other manufacturers, he farmed out to them the rights to make articles then in demand and thereafter devoted himself to devising, launching and financing new applications up to the point where businessmen could be induced to take over and make the money. Then he would throw himself into developing still more inventions successful and otherwise. By nature he was a poor bargainer. Also, he was in debt to begin with and had to have money for the investigations with which he was obsessed. His deals were those of a man in need then, and the manufacturers took advantage of this and of his innocence to drive hard bargains and set royalties far below their real value. And this continued to be his story throughout the whole of his life. Some of the manufacturers lived up to their royalty agreements, others only did so when the law was invoked, and from first to last there was a tremendous amount of business done by outlaw users of the process. As United States Patent Commissioner Holt once phrased it, '... no inventor probably has ever been so harassed, so trampled upon, so plundered by that sordid and licentious class of infringers known in the parlance of the world, with no exaggeration of phrase, as 'pirates.' The spoliation of their incessant guerilla warfare upon his defenceless rights have unquestionably amounted to millions.' Even so, a good many thousands of dollars flowed to Goodyear through his inadequate license fees — sufficient to have established him as a reasonably wealthy man for those days had he cared to hang on to it. Instead, however, he sunk every dollar into rubber exhibitions and inventions ($5,000 into those sails, for instance) as fast as they came in, borrowed many thousands more to add to them and lived and died a shabby, debt-loaded and jail-menaced poor man. Nearly always there was a considerable measure of popular support for the unending series of attacks upon the security of his patent, the sentiment rising from popular belief that he was the manufacturer of all rubber articles. This sprang from the use of the name Goodyear in so many corporate titles assumed by the caoutchouc makers (a practice which delighted him although he held no interest in any of the companies) and from the patent stamp on each legal article.

Most implacable of all the industrial troublemakers was one Horace H. Day. He entered the picture shortly after Goodyear had sold a license for the manufacture of shirred goods to David Suydam of New York. Day started in the same business without a license and, by allegedly overstocking the market with an inferior article, discouraged Suydam to such an extent that he refused to go on with his contract. The inventor brought a string of suits and after extensive preparations by both sides, Day proposed a settlement which was accepted. The infringer offered to confess judgment in one suit, paying Goodyear $5,000; to buy from Goodyear an exclusive license for the manufacture of shirred goods on a royalty basis; and to refrain from making any other article of vulcanised rubber. Failing thereafter to pay one cent of royalties, Day also disregarded his contract by launching into the manufacture of rubber shoes. Not satisfied with that, he shortly was snapping at Goodyear's heels with scandalous allegations.

These last were cooked up after Goodyear in 1849 applied for and obtained a re-issue of his 1844 patent in order to correct certain imperfections of language in the original document. Both patents were for the same invention and the method of correction was one duly provided for by act of Congress, but Day and some other manufacturers immediately seized the incident as an opportunity to gang up on the inventor in still-current businessman style and attempt to destroy the patent. In March, 1850, then, there was presented to Senate and House of Representatives a petition signed by some 200 rubber manufacturers, dealers and workers. Assailing the Goodyear patents as fraudulent, denouncing him for entering into 'a combination and association with a few individuals who have agreed to prosecute all who infringe their illegal patent' and attacking the Commissioner of Patents, they whooped for laws and amendments. Supporting their cries was a sheaf of learned opinions from counselors who had cudgeled their poor lawyers' wits to show that the 'artful' Goodyear's patents were illegal, void and invalid; that he had practiced 'gross fraud' upon the Commissioner of Patents; and that his re-issued document was 'a palpable attempt to defraud the public, by appropriating to himself, what in truth is the common property of all.'

This move got nowhere, but Day very soon thereafter obtained that decision by court of law so lustily yammered for in the petition. The several rubber shoe manufacturers operating under Goodyear license picked on him as the most obnoxious and important of the infringers and in 1851 filed suit in Goodyear's name for an accounting of articles made and sold and a perpetual injunction restraining him from further infringement and violation. To argue the case the licensees hired Daniel Webster, then as usual a candidate for Presidential nomination by the Whig party. Webster's fee was $25,000, highest ever paid an American attorney up to that time. Day, not to be outdone, retained Rufus Choate, who alternated with Daniel in the United States Senate, and a battle of the titans was on.

The 'Great India Rubber Suit' (so contemporary pamphlets called it) came on for hearing on March 23rd 1852, at Trenton, N.J, before Judges Robert C. Grier and Philemon Dickerson of the United States Circuit court, who promptly rejected Day's prayer for a jury. Associated with Webster was James T. Brady and with Choate, Francis B. Cutting. Four bound volumes of evidence resulted - evidence showing, incidentally, that Day had made large profits by his shirred goods monopoly and paid nothing for it. One of Day's allegations was that Goodyear had failed to protect him in his monopoly. In a previous court action a jury had allowed Day damages as a set-off against royalty demands although the evidence was that Goodyear had done all in his power to protect the license and that the injury to Horace was effected by others whose acts neither plaintiff nor defendant could control. Chiefly, however, the Circuit court defense took the form of an attack from every conceivable angle upon the validity of Goodyear's patents. It was alleged that Goodyear's 1844 patent had been abandoned to the public. It was alleged that Hayward was the discoverer of vulcanisation and that his patent also had been abandoned. It was alleged that Hancock was the inventor. It was alleged that John F. Nash and William Beers, two of Goodyear's workmen, were the inventors. It was alleged, and crushingly disproved, that Day himself had vulcanised rubber as early as June 15th, 1842, two years before the granting of Goodyear's patent. One Elisha Pratt testified that prior to Goodyear's patent, he had vulcanised hats by hanging them in a room heated by a stove — an impossibility, of course, and especially so in view of the fact that Pratt told of being in the room for intervals of a quarter of an hour or more while the heat was on. Another so-called vulcanisation inventor introduced in person was Richard Collins. This gentleman claimed, without previous knowledge of rubber, to have vulcanised the substance by exposing it to a temperature of 2000 to 2500 of in a wood stove in his 12 by 14 sleeping apartment, continuing the process for 24 hours on end. Judge Dickerson, a judicial wit, commented, '... it is quite manifest that if he had kept the stove heated to the highest degree estimated, it would have produced upon his own body, the very effect that was desired to be produced upon the rubber — rendering it insensible to the effects of heat and cold.' The blunter Judge Grier said that the Collins testimony contained 'much internal evidence of falsehood.'

One Day allegation there was, however, which, if well founded, would have rendered immaterial the question of who was the original inventor. This was his customary assault on the legality of the re-issued patent.

Choate's argument does not appear to have survived, but biographers of a century easier on authors assure us that the speech was very powerful stuff even though they had no idea of what he said. In 1862, for instance, biographer Samuel Gilman Brown reported, 'In March (1852) he made a powerful argument in an India-Rubber case in Trenton, N.J ... Mr. Choate was said to have surpassed himself in learning, strength, and brilliancy; but of the argument, as of the great majority of speeches at the bar, absolutely nothing remains — *ipsae periere ruinae* (the very ruins are perished and gone).'

Choate himself allowed that Webster was in supreme form at Trenton, which is one way of saying that he naturally had to be good to take Choate. This was the 71 year-old Webster's last appearance in any courtroom, and it could hardly have been the godlike Daniel of old who stood up on his hind legs and hollered for Goodyear.

Seven months later he was to go to his rest with his hand on his breast and a deathbed hearts and flowers oration quivering in the air.

Webster's $25,000 oration fortunately survives and we can judge it for ourselves. It opened with an exposition of the Government's right to grant patents and then got down to business with a brief history of rubber in which Daniel demonstrated his perfect mastery of scientific principles by transplanting *Ficus elasticus* to Brazil. This was followed by a résumé of rubber-factory failure in the '1830's and of Goodyear's entrance into the business. Now it was time for the humor. Relating the objections to rubber of the pre-vulcanised era, counsel proceeded, 'I well remember that I had some experience in this matter myself. A friend in New York sent me a very fine cloak of India Rubber, and a hat of the same material. I did not succeed very well with them. I took the cloak one day and set it out in the cold. It stood very well by itself. I surmounted it with the hat, and many persons passing by supposed they saw standing by the porch, the Farmer of Marshfield.'

Webster next told of the issuance of the Goodyear patent and cut in with the elaborate sarcasm in order to 'relieve' the judges of 'distress' caused by Choate's picture of Day's unhappy condition. Analysing Day's claims that the invention had been abandoned to the public and that anyhow Goodyear was not the inventor, Black Daniel then hit a high spot with:

'We have reached that point in this discussion where the great question of the case rises up before us. . . . That great question is, the truth or falsity of the claim made by Charles Goodyear to the invention of the process of vulcanising India rubber. Did he make such an invention? Is he who sits here before us the man known now, and to be known forever, while the history of art remains, as the individual who introduced to the knowledge of his country and to the knowledge of the whole civilised world, this extraordinary phenomenon? ... Is he the first man upon whose mind the idea ever flashed, or to whose intelligence the fact ever was disclosed, that by carrying heat to a certain height it would cease to render plastic the India Rubber and begin to harden and metallise it? Is there a man in the world who found out that fact before Charles Goodyear? Who is he? Where is he? On what continent does he live? Who has heard of him ? What man among all men on earth has seen him, known him or named him?

'Yet it is certain that this discovery has been made. . . . It is certain that 10 or 12 years ago it was not knowledge. It is certain that this curious result has grown into knowledge by somebody's discovery and invention. And who is that somebody? If Charles Goodyear did not make this discovery who did make it? ... Is the discovery so plain that it might have come about by accident? It is likely to work important changes in the arts everywhere. It introduces quite a new material into the arts, that material being nothing less than elastic metal. It is hard like metal, and elastic as pure gum elastic. Why, that is as great and momentous a phenomenon occurring to men in the progress of their knowledge, as it would be for a man to show that iron and gold could remain iron and gold, and yet become elastic like India rubber. Now this fact cannot be denied; it cannot be discredited; it cannot be kept out of sight; somebody has made this invention. That is certain. Who is he? I say that there is not in the world a human being that can stand up and say that it is his invention, except the man who is sitting at that table. I believe that the man who sits at that table — Charles Goodyear — is to go down to posterity, in the history of the arts in this country, in that great class of inventors, at the head of which stands Robert Fulton.

'. . . We want to know the name and the habitation, and the location of the man upon the face of this globe, who invented vulcanised rubber, if it be not he who sits before us.. . . Mr. Hancock has been referred to. But he expressly acknowledges Goodyear to be the first inventor.... There was a time, I admit, when there was a disagreement between Goodyear and Hayward and Mr. Hayward was foolish enough to set up some pretences of his own, but was soon ashamed of it, and his chief merit is that he had the manliness to disclaim it. . . .'

'I know but one man in the world who denies the originality of this invention... It is Horace H. Day.'

Decision was handed down on September 28th 1852 and was a complete victory for Goodyear. The accounting was ordered, a perpetual injunction was issued against Day as prayed for and both judges took occasion to hail Charles as vulcanisation's inventor and uphold his reissued patent.

Judge Dickerson commented on the interesting point that Hayward's superiors at the Eagle plant had, prior to their business failure, prohibited the further use of sulphur as injurious to their goods. Hayward and Hancock, he noted, both accorded the merit of the invention to Goodyear, and, after working out on the stooges pushed forward as inventors, he expressed himself as entirely satisfied that Goodyear was the original inventor of the vulcanisation process and not only entitled to the relief asked 'but to all the merits and benefits of that discovery.'

Small enough were all the benefits that vulcanisation's discoverer ever derived from it. Little more than a moral victory was the triumph over Day so far as he was concerned. With that in the bag he promptly sailed for England to look into his reputation there and in Europe. The year before he had topped Hancock and knocked out the British eye by sinking $30,000, chiefly borrowed, into that Crystal Palace exhibit. 'Goodyear's Vulcanite Court' it was called and it was a suite of rooms in which everything — walls, roof, cornices, carpet and furniture — was caoutchouc. In England he met Hancock and in England he lost his wife and found a new one. Fanny Wardell of London was the second Mrs. Goodyear and three children were born of this union.

Paris was planning an International Exposition for the following year that was intended to surpass that of London. In November 1854, Goodyear and his family moved to France in order that he might plan his part in it. Patents had been taken out in several European countries and a son and daughter of Goodyear's went on to Vienna to instruct workmen of one of the Continent's largest factories in methods of manufacturing shoes after their father's patent. In preparing his Paris exhibit, the proud Charles outdid his previous efforts, spending $50,000 in assembling every product of vulcanised rubber known. His purse was emptied to the last dollar and his indebtedness mounted. Goodyear's agent, Charles Morey, had awakened a great interest in the rubber industry in France in 1853 and three large companies had been formed to manufacture under Goodyear licenses. A suit involving his right to his French patent was first judged in his favor and Goodyear received considerable sums in notes of these companies in payment for patent rights. The decision was later reversed. Goodyear had endorsed the notes to help pay for his exhibit and was unable to meet them when they came due. Having sampled a fair share of the jails of the United States he now had a chance to compare them with the French bastilles, for he was immediately tossed into Clichy, the debtor's prison of Paris.

Napoleon III, bridegroom of the Eugenie hat, visited the exhibit and was greatly impressed, so much so that he awarded the Grand Medal of Honor and the Cross of the Legion of Honor to Goodyear, who was even then behind the bars of his majesty's prison. The unconvincing but typical anecdote in connection with the Emperor's sightseeing is that, spotting a mound of round footballs, he remarked, 'always thought that rubber could be used in warfare as a means of defense, but it never occurred to me that it was suitable for projectiles; but there you see a pile of rubber cannon balls.' No one corrected the speaker, it is recorded.

Goodyear got out of Clichy on December 21st 1855 and immediately returned to England. In February he was again arrested on French demands and had a taste of an English gaol. He soon proved that his arrest had been obtained by fraud and was released. The strain on his naturally weak constitution resulted in an illness even more severe than usual and it was not until April that he could be removed to Bath, where he remained until he returned to America in May, 1858. He was in very poor health during the entire period and seemed in constant danger of dying. His business received even less attention than usual. In America his patent was due to expire, some of his licensees were still avoiding their obligations and a trusted attorney embezzled his funds. Goodyear pawned his wife's jewelry to pay their passage home.

Hearing on the application for a seven-year renewal of his 14 year patent was before the Hon. Joseph Holt, Commissioner of Patents of the United States, in June. Money, political pressure and a certain amount of press clamor was brought to bear against the inventor. Foremost among the contestants, of course, was Horace Day and prominently aligned with him was one Charles F. Stansbury, reputedly representing Macintosh and Co. and Thomas Hancock. There were attempts to depreciate the ingenuity of the invention by representing the discovery to be solely an accident, there were claims that the inventor had cleaned up millions and there were intensive hair-splitting efforts to show that the patent had expired in January and before the application for renewal. The technicalities were easily disposed of. As to remuneration, Holt found that the receipts of Goodyear's assignees and licensees, admittedly amounting to 'many millions,' should not be charged to the patent. Goodyear's own ragged accounts provided a field day for the lawyers of the opposition. These accounts, once amended, showed him personally profiting to the amount of $33,358.63, receipts being set at $162,894.09 and expenditures at $129,535.46. The Examiner then excluded some of the deductions, which tilted the amount to $I14,128.09. Holt restored two of these and commented:

'Restoring then, these items, and adopting the other corrections of the Examiner, there will still remain to the credit of the invention a clear profit of $54,733.63. The applicant, in his amended statement, acting under the promptings of the same high sense of honor which led him to satisfy an indebtedness of $35,000 from which he had been discharged by a certificate of bankruptcy, shrinks from debiting the patent with any expenditures, the particulars of which he cannot recall with some degree of certainty, but, while doing so, unhesitatingly expresses the belief that they were quite as large as the sums set forth in gross in the first account. It is probable, indeed, in view of the whole testimony, it is my firm conviction, that, if it were possible to extract from the tangled mazes of the multifarious and now half-forgotten transactions connected with the invention, all the moneys expended therein, it would be found that, instead of there being a balance to its credit, the balance would be on the other side. I am justified in arriving at this conclusion from the fact that, although the applicant has had no other occupation or business, yet, instead of having now in hand this sum of $54,733.63, he is admitted to be penniless and overwhelmed with debt — and this, too, notwithstanding his life is shown to have been temperate, frugal and, in all respects, self-denying.'

Had Goodyear been peacefully permitted to enjoy his rights, Holt estimated he would long since have realised an 'ample remuneration.' Noting that the American trade in India rubber fabrics now amounted to some four to five millions of dollars annually, with seven million profitably invested and 10,000 workers employed, he scored 'unscrupulous' infringers, socked the 'predatory' Day (not mentioned by name) for his 'remorseless' pursuit of Charles, and ended up by granting the extension in a burst of oratorical fireworks surpassing Webster.

The Patent Commissioner's picture of a Goodyear with the 'winter of age creeping upon his shattered constitution' proved to be rather accurate. Buying a home in

Washington the year after the decision, the inventor let it be known that he was taking life easy. So he was, perhaps, by his standards, but the man was incapable of resting. One room of the house was fitted up as a workshop and here he had a large tank built for testing life preservers and models of life-saving craft. Under his patent extension he took in about $30,000 in 1858-1859 and the same amount in 1859-1860. Learning of the illness of a daughter in New Haven, he immediately set off to see her. He was, however, too weak to make the journey by rail, and so took a steamer to New York, becoming severely sea-sick enroute. He was met at the dock by his son-in-law, who informed him of his daughter's death. Goodyear, nearly exhausted, took quarters at the Fifth Avenue Hotel and there died the following day, July 1st, 1860, some $200,000 in the hole.

Due wholly to Goodyear was the then current American crude rubber price of over 60 cents a pound, a price that would hit $1.20 for fine hard para within the next half decade. When he launched his experiments it had been selling at five cents a pound (a fortunate circumstance then, since it was only because of the cheapness of the material that he had been able to obtain sufficient for his purposes). In the America and Europe of 1860 were 150 rubber factories and sales were running to many millions yearly. Again this was wholly due to the man jailed by three countries, for England's couple of factories were very nearly the only profitable ones in the world when he began his affair with caoutchouc.

Accumulating his indebtedness while making fortunes for so many others, this man with no nose for money had kept his liabilities in suspense by partial payments and renewals and never actually knew the amount he owed. The audit after death showed just $191,100.73 of undisputed debts and important amounts in dispute but not proved - all incurred in the pursuit of improvements on rubber. Dependent upon the patent for their maintenance were five of his children (the youngest now but four), his widow, an adopted child and a grandchild. With the estate still a mess of liabilities in 1864, Charles Goodyear, Jr., as executor, petitioned Congress for another seven-year extension for the benefit of family and creditors. To Washington rushed the old foes of the American to oppose the request and in newspapers and magazines the presence in the capital of the widow, the son, a friend and their professional lobbying agent was denounced as the campaigning of 'a powerful lobby, male and female' which, having wrested $20,000,000 from a noble and exploited public, was out to nail $20,000,000 more. The petition was rejected and the Goodyear patent expired June 15th 1865. Today virtually the whole colossal, world-wide rubber industry still rests upon the basis of the Charles Goodyear discovery of hot vulcanisation which will be a century old in 1939. Several special methods since devised are used to limited extent, but it is safe to assume that not even these relatively insignificant processes would have been invented save for Goodyear's finding. The most important of them, 'cold curing,' invented by the English chemist Alexander Parkes in 1846, stems directly from the Goodyear-Hancock episode. It was seeing the results of Hancock's experiments that set Parkes to trying the effect on caoutchouc of all the sulphur compounds he knew. Hancock claimed also to have directly aided him in his work. At any rate the chemist found that thin strips of rubber immersed in a solution of sulphur chloride in carbon disulphide were vulcanised in a few minutes. The process is used today for the manufacture of surgeons' gloves, toy balloons and other dipped goods. The vapor cure-exposure of rubber goods to sulphur chloride in the form of vapor instead of in solution — was introduced by Abbott in 1878 and is still used for very thin rubber sheet such as dental dams. In 1915, the Russian chemist Ivan I. Ostromyslenski claimed to have achieved vulcanisation of rubber by means of trinitrobenzene, cousin of TNT. In 1917, C. R. Boggs was granted a United States patent for vulcanisation by use of selenium, an element similar to sulphur but a hundred times as expensive. It has some slight use among manufacturers for very special purposes. In the same year S. J. Peachey of London patented a method of cold vulcanisation which calls for exposing the rubber compound to sulphur dioxide, which it readily absorbs, and then to hydrogen sulphide, which it also soaks up. The two gases react to produce a form of sulphur so active it vulcanises without heat. The Peachey method is only applicable to very thin goods and has never become of commercial importance. As to unvulcanised rubber, it survives as little more than crepe-soled footwear, adhesive tape and rubber cements for adhesive purposes.

4.3 Reclaimer

CHARLES GOODYEAR solved one problem for the rubber industry and dumped another in its lap; for the question of what to do with old rubber goods came into existence the day the first vulcanised rubber article was made. Before that time there was no problem. Hancock, confronted in 1820 by a growing accumulation of small scraps and cuttings and snips of rubber, devised a masticator by means of which he worked these up into useful form again. It is possible to do this with raw rubber, but not with the cured, which is a tough substance which is without plasticity. Confront a chemist with anything resembling waste and you are searing his soul with white hot irons. And so not many years had followed 1839 into limbo until the experimenters were hot after ways of reclaiming vulcanised rubber. More than half a century was to pass, however, before young Arthur Hudson Marks, only a few years out of college, devised the method which has not been improved upon to this day. Due to his ingenuity, the discarded tire, belt, hose, overshoe and all the various other articles which have lost only a small part of their rubber content through wear now undergo a reincarnation which restores them to a useful second life.

The reclaiming process is not a reversal of vulcanisation; cured rubber does not become crude rubber again; sulphur which has combined chemically with the rubber is not removed. These things do happen: the cotton fabric contained in nearly all rubber articles is removed, the free or uncombined sulphur is eliminated and the tough, non-plastic vulcanised material is transformed into a soft, workable substance that can be handled and mixed in the same fashion as crude and is capable of being vulcanised again.

Originally this 'reclaim' was produced solely to utilise manufacturing waste and scrap goods. Driven to more and more experimentation during periods of excessive prices for crude, the rubber industry learned so many things about the scrap that it acquired unexpected economic importance — and something more. Its economic importance may be judged by the fact that in one year of high crude rubber prices more reclaim than crude went into rubber articles, that at no time in the last score of years has reclaim amounted to less than a fifth of all rubber used in this country. And no longer is it necessarily just a cheap substitute for raw rubber; it has, to a certain extent, taken its place with the crude, with cotton, with powders, oils and tars, as a raw material in its own right, one with distinctive and valuable properties. All manufacturers employ it.

First of the would-be reclaimers of vulcanised rubber was the indefatigable Hancock who in 1847 joined with Reuben Phillips in patenting an invention which called for reducing the scrap to shreds or sheets and boiling them in turpentine. The process never amounted to anything. Six years later Goodyear patented a use of waste vulcanised in which he simply mixed the finely ground scrap with raw rubber. All that he achieved was an inexpensive filler and where fiber was present it was of even less value. In 1855 a forerunner of the Marks process cropped up when one, Sigismund Beer, was granted a patent for boiling ground rubber in lye until the free sulphur is removed. He might have followed through to success — but didn't. More successful was Hiram L. Hall of the Beverley Rubber Works, in Massachusetts, who subjected ground scrap to the action of live steam and then mixed it with tar. By this process the rubber took on a certain measure of plasticity. Hall tried boiling scrap in water containing sulphuric acid, and so did the brothers C.H. and D.E. Hayward in 1863. Apparently, they either did not use a sufficiently strong acid or else did not boil their mixture long enough, for otherwise they would have anticipated Lieutenant-Colonel Chapman Mitchell, who was to evolve the best of all reclaiming processes prior to the Marks discovery.

Mention of the Haywards brings us to 1870, at which time there was still practically no old rubber in use. Such scrap as came to them, the individual rubber companies used as they saw fit, but there was no real commerce in the stuff. One firm was sending out old shoes to women who tore them apart by hand and were paid by the pound for all the rubber they could strip from the cloth; at Ansonia, CT, Austin Day was grinding car springs and shoes; at Boston, E. H. Clapp was performing a similar service at the Perkins grist mill. That was about all. The trouble was that there was no process which removed cotton fiber from rubber. Clapp partially solved that one in 1871 by adding to his process a treatment with an air blast. This method blew the fiber away from the rubber and the subsequent application of live steam produced the commercial product which came to be known as 'shoddy.' Its defect was that the separation was incomplete and so a part of the cotton always remained with the rubber, a part of the rubber always was blown away with the cotton particles. The 'shoddy,' therefore, was little more than filler.

It had to serve, however, until Colonel Mitchell, brother of the author S. Weir Mitchell, hit on his method for freeing rubber from fiber more completely. Mitchell, a Philadelphian, began his rubber experiments about 1876. His first success was in reclaiming some uncured trimmings from rubber shoes by placing them in an iron pot and boiling them in a caustic solution for three or four hours. Why he did not pursue this investigation further is not known; instead he turned to the use of sulphuric acid and succeeded where Hall and the Haywards had failed. In 1881 he patented the acid process of reclaiming rubber which still finds slight use today. To exploit it, he organised the Philadelphia Rubber Works Co. This concern, which was acquired many years later by Goodrich, was the first organised for the exclusive purpose of reclaiming rubber. Mitchell's process consisted of two steps. The finely ground rubber and cotton were first boiled in a dilute acid solution for several hours in lead-lined wooden tanks. This destroyed the cotton, leaving the rubber still firm and tough. After being washed to remove the acid the rubber was heated for a day to a day and a half at a temperature of 300 °F degrees to restore it's plasticity. Free sulphur in the rubber was not affected, however, and consequently this process was only satisfactory for uncured factory waste such as trimmings from belts, hose and tires or for old boots and shoes which contained only low percentages of sulphur and were not very highly vulcanised.

There was no way of re-shaping or re-using the vast bulk of rubber articles until Arthur Marks appeared on the scene. Marks heated ground rubber waste in a dilute caustic soda solution for 20 hours at a temperature of around 360 °F. This treatment simultaneously destroyed the fabric, plasticised the rubber and dissolved the uncombined sulphur and some of the lead compounds. It gave an amount of rubber equal to half the weight of old tires put through the process. Most important of all, it worked equally well on all kinds of scrap. The reclaimed rubber made by this method and that made by Mitchell's process were vastly different. Acid reclaim is seldom without small traces of acid which cause the rubber to age poorly; its tensile strength and other properties are poor. The Marks alkali reclaim is strong, it blends well with crude rubber, and the traces of caustic which it contains preserve the rubber and exert a slight accelerating action during the subsequent cure.

Marks was not at first granted a patent for his method of reclaiming. In May, 1899, samples of the new product were shown to patent examiners, who only then agreed that something different from any previous reclaim had been made. Marks was granted a patent for his alkali process on October 17th of the same year.

The Marks method is still in use today, unchanged except for minor details. Eighty per cent of the reclaimed rubber produced is made by this method. Another 10% is made by the 'heater process,' used only on articles such as inner tubes which contain no fabric and which do not have to be treated with alkali or acid to destroy cotton. Three per cent of the total is reclaimed by Mitchell's acid process and the balance by miscellaneous special processes.

Old automobile tires furnish the bulk of the reclaimer's raw material. One part of the tire, however, refuses to be restored to useful citizenship. This is the bead, the combination of rubber, fabric and wire which holds the tire to the rim. The beads can only be cut from the tires and thrown away. In the rubber center of Akron, they are trucked to the outskirts of the city where they are burned in great piles, to the accompaniment of huge volumes of black smoke, the fierce indignation of all the downwind housewives and the dismay of town councilmen who are trapped between fear of the big industries and of their irate constituents.

In the reclaim plant the tires, inner tubes, shoes and old hose lengths are sorted into two classes, those containing fiber and those free from it. Tires are debeaded and then carried by conveyors to chopping and grinding equipment. Leaving the cracking rolls — two corrugated rolls running at uneven speeds — the finely ground mixture is then passed over screens and a magnetic separator removes all nails, tacks and similar foreign objects. Into the Marks-method tanks it next goes, to be heated and stirred in the presence of the caustic soda solution and finally 'blown off' into a washing vat. Here the cotton, made water-soluble, is washed out and the rubber mass is ready to be dried and worked into sheets.

Many times a millionaire was Marks as the result of evolving this gadgetry. And his whole meteoric career in industry is of interest. Born in Lynn, Massachusetts, in 1874, he graduated from Harvard just 20 years later. He then became assistant chemist at the Boston Woven Hose and Rubber Co. and in 1897, head chemist of the Revere Rubber Co. A year later, at the age of 24, he was made superintendent of the Diamond Rubber Co., and took his reclaiming process to Akron.

With crude rubber selling for $1.08 a pound at the time, it is not surprising that this first really successful reclaiming method was an immediate and sensational success. Within a short time the Alkali Rubber Co., the Northwestern Rubber Co., in England, the Pan American Rubber Co., and the International Process Co., were formed to exploit the Marks patent. Marks was president of all these companies while retaining his connection with the Diamond. Licenses were granted to the great Continental Caoutchouc company in Germany and to companies in Italy, Canada and France. Marks collected on every pound of reclaim that was manufactured and his royalties were enormous.

With Marks as the directing genius, the Diamond also prospered greatly. Everything the man did was a spectacular success. Marks was one of the first to use guayule successfully. Marks developed a woven wire fabric base to make a better solid tire. A chemist himself, Marks was the first man in the rubber industry to realise the value of chemical research and he started George Oenslager on the path which resulted in the discovery of the organic accelerator, a discovery by which the Diamond again profited enormously.

Although the value of Marks' other ideas can be measured concretely in terms of millions of dollars, it may well be that his greatest service to the rubber industry was in forcing it to accept the chemist as a result of his own dazzling success. The attitude of the leaders of the trade a few years before Marks flashed on the scene is typified by the statements of three company heads: E.S. Converse, who said he had employed chemists but found the cost to his company greater than any value received from their work; Joseph Banigan who flatly stated that he had no use for chemists; and Henry C. Morse, who said he would give more for the guess of his old superintendent than for all the certainties produced by the best chemists on earth. If Marks did not make these men eat their words, he at least prevented them from ever repeating them publicly. When this youth flaunted one brilliant achievement after another in their faces, the older manufacturers awoke to the fact that the chemist could be something besides an impractical theorist and a mere tester of materials.

Marks' Diamond record secured for him the post of vice president and general manager of the combination resulting from the 1912 Goodrich-Diamond merger. This post he held until 1917. He dropped out of rubber to serve the government in ship-building and was commissioned a lieutenant commander. After the war he bought an estate at Croton Lake, NY, assumed the presidency of the Skinner Organ Co., and later joined the board of directors of the Curtiss Aeroplane & Motor Co., To semi-active participation in rubber affairs he returned in 1930 as a Goodrich director and a year later became an executive committee member in a shakeup which reduced that powerful board from six management representatives to a five-man group including three non-management men. That position on the present (six-man) committee he still retains.

The popular notion is, of course, that reclaim is a cheap and inferior material used by unscrupulous manufacturers who wish to adulterate their goods at the expense of an innocent public. Cheap it certainly is, but not necessarily inferior. For some uses it actually is 'just as good' as the crude, for some it is superior. In still other instances, its inferiority is of a sort that does not penalise the purchaser. And it could be wholly eliminated from the industry only at the public's expense. Reclaim, amounting currently to 24% of the rubber used by manufacturers, has represented as much as 56.7% and consistently has done its part in holding prices of both raw rubber and finished rubber goods to a substantially lower figure than we otherwise would have known. Ban it entirely and the consumer would receive a better article in some cases, an equally good one in others and an inferior one in still others. But he would pay a higher price in every instance.

Crude rubber today is selling at 16 cents a pound; a modern reclaiming plant producing at capacity can turn out its product for slightly less than three cents a pound. Add to this the factory economies resulting from the fact that reclaim is more easily mixed than crude and is handled more rapidly in the tubing machines by which many rubber articles are shaped, and you can see how important a saving it represents. In 1929, to cite a year when reclaim was in heavy use, the American rubber industry used 462,101 tons of crude rubber at a cost of $209,458,746; nearly half as much reclaimed rubber (206,091 tons to be exact) at a cost of but $26,864,425.

The repossessed stuff ages well, gives good service in resisting abrasion, and there is no reason why it should not be judiciously used in all rubber articles save those demanding maximum tensile strengths. Its lack of this last quality does not mean that it is inferior to crude, but only that it is different. Pure crude rubber alone, for instance, would be as bad a choice for an article where resistance to abrasion is required as reclaim would be for one requiring tensile strength. And so peculiarly is the treated scrap adapted to certain uses that it would be retained in these compounds even should its price mount beyond that of crude.

In the making of certain lines of mechanical goods, crude rubber could not replace it; these articles would he impossible of manufacture without the use of reclaim. And so it is that 30% of all reclaim steadily continues to be used in these lines. Again, take the tire beadsteel wire with fabric and rubber protection to keep it from rusting out.

Here, reclaim's low water absorption quality makes it invaluable. In the making of rubber soles for shoes, the mixing of reclaim with the proper proportion of crude gives an article that wears substantially better than would soles made of crude alone. In some of the articles where reclaim admittedly is not as good as crude, the latter is apt to be too good for the service demanded. Typical of these are rubber auto parts, which need last no longer than the life of the car; to make them otherwise would be sheer waste. Rubber overshoes constitute another example. These wear out in the sole, in the course of time, and the other parts of the shoe must be discarded simultaneously. It would be pointless to make these other parts of a more valuable material than is necessary.

So much for the vulgar belief that reclaim necessarily can be put to scoundrelly uses alone. Having disposed of that libel, however, it must be admitted that there is a certain element of justice in the average citizen's suspicions. During the periodical balloonings of crude rubber prices, the manufacturer does change his compounds by including a far greater percentage of reclaim than is employed when, able to disregard the price factor, he uses the formulas he has found best. Statistics convict him of this and in a manner which leaves no room for wriggling.

Let's go back to 1917. In that year war-enforced economies and 72 cent crude rubber resulted in a record American consumption ratio of 56.7% for reclaimed to crude. With 48 cent rubber in 1919, that ratio dropped to 36.3%; with 17 cent rubber in 1922, it was down to 19.2%. In the year 1925 Great Britain's Stevenson restriction act really made itself felt and the average rubber price for the year was 72 cents, followed by 48 cents for 1926, by 37 cents for 1927, and by 22 cents for 1928. The belated 1925 rush to utilise reclaim saw it mount to a 35.3% ratio in that year and continue climbing to 45% in 1926, to 50.8% in 1927, to 51 in 1928. Thereafter, with crude prices continuing their downward course, use of scrap rubber fell off until the ratio was 40.8% in the 12 cent year of 1930 and 35.1% in the six-cent year of 1931. Thereafter it was 23.3% in the 3½-cent year of 1932 and 21.2% in the six-cent year of 1933. Under recent 12 to 16 cent prices, the ratio again is mounting.

This sensational drop can only be accounted for by the manufacturers having used vast quantities of it where they knew they should have used the crude. We can see but one defense: the contention that only the multiplied use of reclaim prevented a sharp hoisting of prices and that the public prefers some debasing of goods to such price lifting. Perhaps the public does, but the point in this case is that it never was informed, never had the opportunity to make a choice. Expect an undue amount of scrap in your rubber purchases, then, if the current five-nation rubber restriction scheme sends crude prices beyond the 20 cent mark. Do not fear that tires will be included, however. Up through the years, the standard casing brands have offered a consistently increasing mileage that leaves no room for doubt concerning the correctness of the proportions of crude and reclaimed going into them.

Necessarily, the annual production of thousands of tons of reclaim means that the scrap is an industry in itself. Where the crude stems back to the native tapping a tree, the reclaim goes back: to that familiar anti-Hitlerite, the bearded junk dealer making his rounds in a rickety wagon and trading for rags and bottles and shoe and tire discards. From the lone skirmisher the casings and tubes and busted hot water bottles and worn out rubber shoes pass on to the large dealer, sometimes a concern handling scrap rubber alone, again an all-around enterprise salvaging anything and everything and boasting membership in the Scrap Iron Institute of America. This corporation sells to the reclaiming plant and the elastic junk is renovated and put back into circulation.

These activities are limited, of course, to areas within economic hauling distance of the reclaim plants and these last are located near the rubber factories or at the shoe companies maintaining rubber manufacturing departments of their own. Usually, these days, the reclaiming works is a subsidiary or mere division of the large rubber corporations.

What happens to the old tires not reaching the reclaimer's grinding machines? For years thousands of them have been annually exported to foreign lands where they are cut into soles for primitive types of shoes worn by the natives. Macedonia, for instance, imports some 50,000 used casings annually and the shoes made from them have almost completely replaced the leather *tcharik* because they give greater economy and protection from the weather. In Spain, *arbacas* with soles made from old tires are popular, and in Portugal the shepherds use old inner tubes as overshoes for their wooden sabots. China is another good customer for old tires. In Akron they are cut into small blocks, welded together and placed on the market as door mats.

Recently there has arisen, too, a new worn casing enterprise delighting the junk dealer and considerably postponing the automobile tire's inevitable journey to the grinder. It is the retreading of casings, a logical outgrowth of tire repairing which, boomed by the depression and the upward journey of crude prices under the current restriction plan, has grown to a definite industry with a bearing on rubber consumption (the retread uses only half as much rubber as a new casing), tire production, tire merchandising.

Balloon tires gave it its first lease on life. Since their introduction, the carcasses of first line tires commonly outlast the best treads and may be very little affected even though the entire non-skid surface of the tread has been worn away. Why not a new tread on the same old tire, then? And especially since the tire manufacturers themselves had demonstrated its feasibility?

Originally, repairs to casings had been sectional ones devoted to patching the defective tire so that it would give the amount of service for which it was designed. Manufacturers themselves did repair work at their factories, and seven or eight years ago they got into the business of servicing and fixing tires under mileage contracts they had negotiated with bus-line customers. That led them to discover that such tires could be retreaded at great savings to themselves.

It was not long, however, until bright citizens decided to get into the game on their own account. In California, it spread so rapidly that by 1929 the annual retreading figure was 250,000 tires; four years later the fancy name 'rebuilt' had been substituted for retread and the figure was fully a million. From the coast it spread to the Northwest, the Mid-West, and the East, but it has not yet really caught on in those areas. Nationally, the estimated 1935 figure was two million tires. The average tire life's rise from two and a half to three years is partly due to the appearance of retreads; federal and state governments are experimenting with them on their trucks and official cars; private bus and truck fleet operators as well as thrifty individual motorists are buying them; the industry has acquired its own trade magazine; one Chicago retreader has expanded into operation of an export business. Some retreaders are guaranteeing their handiwork to furnish as much as 75-85% of the mileage given by new tires, and a Gotham department store has found that a percentage of casings are worth even a second retreading. Naturally, of course, the durability of the 'rebuilt' wholly depends upon the capability and integrity of the retreader, and a good part of the new industry's ultimate progress will hinge upon the weeding out of concerns guilty of incompetent retreading.

Cost of a moderate-sized retreading plant is estimated at $5,000 to $15,000, depending upon the range of sizes serviced. The retread stock used is a rubber compound known in the trade as a 'camel-back' and originally was supplied by the small rubber-manufacturing companies on the West Coast which found the business a life-saver in the early days of the depression. By 1934 the largest tire companies were forced to hop in. A glance at statistics shows why. 'Camel-backs' are classified under 'tire sundries and repair materials.' In 1932 some 3,288 tons of rubber went into tire sundries; the next year it was 4,464; the next 7,876. For the first three months of 1935 the figure was 2,218 and the sales value of products included under 'sundries' was $3,188,000.

Dealers found their profit on a retread was higher than on a new low-price tire. And manufacturers, naturally, found the sales of their low-price lines were being noticeably reduced. One such corporation opened a half dozen retreading plants as fast as it could get them going and arranged for 25 more. Others fell in line and today they are, almost without exception, operating as retreaders in connection with their company-owned stores and service stations.

And all that, it would seem, is just the beginning. The great bulk of the American populace does not even yet know of the values to be obtained in retreads. Let advertising do its work and the stampede will be on. There is, of course, a definite limit to expansion — the number of suitable worn casings availing for retreading. Sorting, selection and sale of such tires to the retreaders has been the chief source of scrap dealer profit in recent years, and so great has been the demand that already there is talk of a scarcity. Especially is this true in the case of the smaller tire sizes used on the low-priced cars. Partially, such shortage can be averted by educating the individual driver to have his tires retreaded the moment they have worn smooth but before the fabric shows through the rubber. Once this last has happened, retreading is out of the question. And always there will be injuries to tire carcasses similarly foiling the retreader.

So sharply, however, has the retread cut into sales of second and third-line tires that certain low price lines are being discontinued and there are reports that several of the 'rebuilders' and smaller tire factories are toying with the idea of building new carcasses to supply retread demands. Logically followed through, this trend would finally result in retreads and first-line new tires being the only kinds on the market — and new tire sales would amount to little more than the original equipment deliveries to auto manufacturers. Whether retreading will ever mount to such heights is a matter for the future to decide. It remains, at the least, an important new industry and one which shortly may find itself in the healthy position of inspiring a demand for suitable worn tires greater than the supply.

4.4 Accelerator

SECOND ONLY to the discovery of vulcanisation in the annals of rubber research is the discovery of the organic accelerator. The first made the rubber industry possible; the second, coming 67 years later, saved untold millions of dollars, permitted the tremendous expansion called for by world demand and worked almost incredible improvements in the quality of rubber products and the methods by which they are manufactured.

Left to itself, sulphur is extremely slow about working the magic of vulcanisation on raw rubber. Accelerators speed up the mysterious change, whatever it may be. One of them — white lead — was present in the first mixture ever vulcanised; some simple oxide of lead, calcium or magnesium was used in practically every batch of rubber processed thereafter. All of these inorganic accelerators were, however, relatively slow in acting, and they did nothing towards improving the quality of the compounds in which they were used.

The organic accelerators — rapid-acting compounds of carbon with various other elements — revolutionised the industry. They cut vulcanising time to a mere fraction of the former period required for curing. They bridged the gap in tensile strength between different varieties of rubber and freed the manufacturers from the necessity of relying on the restricted and expensive stocks of fine, hard para for the majority of their products. They went beyond bridging that gap, even, for they gave to compounds a strength beyond that formerly achieved by such mixtures as employed fine para as the only rubber ingredient. They increased rubber's resistance to heat, abrasion, oxidation, ageing. They gave to rubber a wider range of properties which multiplied its uses by permitting the manufacturer to invade fields he had never previously dreamed of entering. They made possible the recent important development of articles directly from latex. And they were given to the industry by a man — George Oenslager — whose name means so little to the nation that, like Marks, he has failed of rating with the clergy, newspaper editors and hayfield college professors immortalised in Who's Who In America.

We have observed that before Marks wrought his own miracles and sponsored the greater one achieved by Oenslager, the chemist had no place in the rubber factory. The job of rudely approximating his functions belonged to the superintendent. We are told by W. C. Geer, author, scientist and former rubber company executive, that this versatile gentleman served also as inventor, power engineer, labor chief, production driver, cost estimator, price fixer, first-aid doctor and sometimes as a salesman and promoter. To the superintendent the testing of the factory's rubber compounds was a matter of applying fingers to twist and pull, teeth to bite. By trial and error he had learned the varying properties of goods manufactured from different grades of rubber, and from such observation and experience each factory had evolved its own closely guarded formulas.

In the superintendent's mind, all grades of rubber came to fall into two broad, general classifications. Under innumerable designations, the crude, varying as it came from different trees or from the same trees in different districts, was dumped at the factory door. Only a few grades, however, could be vulcanised into a stock having the high tensile strength demanded of the best rubber goods. All the others, regardless of the skill with which they were compounded, fell into an immensely inferior class. Between the worst of them and the best of the para the difference in tensile strength amounted to 1,800 pounds a square inch.

When you consider that out of a world rubber production of 50,000-60,000 tons annually during 1900-1905, less than half the total was of the sort that gave high-tensile compounds and could be used for inner tubes, for holding tire plies together, for surgical supplies and any number of other articles, you can appreciate how the expanding industry would have been held in by lack of necessary supplies if some means to convert low-tensile rubber to high-tensile had not been discovered. And when you remember that prices of the fine para from the Amazon ranged from a low of 78 cents to a high of $1.37 a pound during the 1899-1906 period which saw all rubber average but 49 to 68 cents, even with the para figured in, you can appreciate what that discovery meant in immediate terms of dollars and cents.

It was the then current 92 cents to $1.12 price of para that set Arthur Marks nosing after the secret in 1899. It was the Brazilian squeeze sending the price to $1.50 and holding it above $1.25 for nearly a year that caused him to loose Oenslager on the trail in 1905. Not chiefly was he seeking to speed up the time required for vulcanisation; his primary aim was to find something that would make the low tensile rubbers interchangeable with fine para, that would enable him to make all grades of wild rubber relatively uniform and industrially important. Plantation rubber was not important in the picture at that time. Once found, however, the organic accelerators were ready for use in its case, and there is no question that the Oenslager discovery had much to do with the ready and wholesale acceptance of cultivated crude which accompanied the record Brazilian squeeze in 1909-1910 and speedily and permanently flattened the Amazon's rubber supremacy. To begin with, American manufacturers had been extremely wary of the plantation stuff, investing in it grudgingly and at prices far below those paid for the Brazilian product with which they had become familiar. Experimenting with it, they found it of uneven quality and much of it low-tensile. Improvement in plantation methods long ago resulted in a rubber of uniformly high quality, but Oenslager, years before that happened, had found his way to control quality in the factory. Without his aid in the critical days, it is possible that the tremendous burst of planting which carried *Estrada* acreage from 750,000 for 1908 and all previous years to 1,050,000 by 1909, to 2,250,000 by 1912, to 3,000,000 by 1915, never would have occurred.

The story of accelerators after Oenslager is the story of the rubber industry; the story of accelerators before Oenslager can be told in a few paragraphs. It begins with Charles Goodyear, as we have noted. Had white lead not been present in his rubber-sulphur mixture the day he touched it to a hot stove and noted that it charred instead of melting, he might never have observed the long sought 'change.' Goodyear was a tremendously observing experimenter and it is not surprising that he commented on the effect of this inorganic oxide on vulcanisation. Before discovering the 'change,' Goodyear had used lime and magnesia for 'drying' the surface of rubber articles. Similarly, Andrew Ure in 1840 described the addition of lime as a means of preventing surface tackiness. It was not long until these substances were also found to act as accelerators of vulcanisation, as did litharge, another oxide of lead.

All that they did was hasten vulcanisation, remember; rubbers vulcanised with these substances emerged from the process bearing substantially the same relation to each other as would have been noted had no accelerators been used. They cut the time of curing cheap rubber as they cut the time of curing fine para, but they did nothing whatever to bridge the gap in physical strength between them. And that remained the case until Oenslager discovered the nitrogenous organic accelerator. Other men were ahead of him in adding these organic nitrogen compounds, just as Hayward and others preceded Goodyear in mixing sulphur into rubber. All of them, however, were blind to what they were doing and none realised what he had accomplished. They detract no more from Oenslager's achievements than the Goodyear predecessors do from his.

Interesting it is to note, however, that the action of ammonia (an inorganic nitrogen compound) in speeding vulcanisation was recognised by the Englishman Thomas Rowley as early as 1881. In this year, Rowley was granted a patent for the use of the gas in the vulcanising chamber. Using various salts, and even urine, as sources of ammonia, he noted the accelerating effect of curing in its atmosphere. The patent is of historical interest in as much as a great number of present day accelerators are organic derivatives of ammonia. Other early experimenters were not so observant. Although they used substances later proven to be accelerators for other purposes in rubber batches, they did not observe the effect broadly. N.S. White, for instance, as early as 1884, used aniline to brighten the color of pigments used in rubberised fabric, but did not notice the effect on vulcanisation. In 1891 C.A. Fawsitt found that the bromides and iodides of mercury, antimony, tin, zinc, lead and bismuth were accelerators. He also used aniline - trying to obtain a transparent product — but he did not report any acceleration.

A step forward was announced by C.O. Weber in 1904. At the time he was working as consultant to a company employing A. A. Glidden in development work. Hoping to increase the rate of cure of the poor rubbers, Glidden tried the effect of adding different soaps. One of them had a marked accelerating action and Weber, analyzing it, found free oleic acid. He tried the acid by itself and discovered that it supplied a deficiency in low grade rubber. Oleic, stearic and other fatty acids, as they are called, are widely used in rubber compounds today and in amounts far in excess of those occurring naturally in high grade rubber. They are valuable because they soften the rubber during mixing, cause various dry powders which are added to disperse more readily and react with some of these powders to make the true organic accelerator more active.

Weber, as we have said, was the first to report on the acids. Actually, however, he was years behind the ubiquitous Marks in using them. At the Boston Woven Hose and Rubber Co., in 1896-1897, the youthful Arthur had used both oleic and stearic in the regular commercial manufacture of rubber stamps and bicycle tires. A year after his removal to Akron and the Diamond plant, he resumed his quest for the missing link between low-tensile and high-tensile. One of his first steps there had been the installation of a process for the daily extraction of 10 tons of rubber from Borneo jelutong at a cost of 50 cents a pound. The jelutong was churned with acetone and naphtha, the resin dissolving and the rubber settling out as a cement. After the resinous solution was poured off, the solvent in the cement was driven off with steam and a reasonably pure rubber was obtained. Pure enough this rubber was, but it was by no means interchangeable with high-priced para. There was something missing, the lack of which gave vulcanised rubber products of poor quality. The extracted jelutong could only be used in cheap goods or blended in small amounts with good rubber in the more costly ones. Marks knew that fine para, collected when fresh and artificially hardened, was much superior to coarse para, obtained from the same tree by spontaneous coagulation. He reasoned that there was something in the latex that was destroyed by natural coagulation or else left behind in the serum. He extracted some fine para with acetone and obtained a rubber that was poor in quality when cured. Cheap rubber, to which this extract was added, was found to be greatly improved. He became convinced that his cheap jelutong could be made much better, possibly even as good as para, by the addition of some other substances. Discovery of the nature of this missing something meant a daily saving of about $21,000, as he would be able to use 50 cent rubber while rivals paid $1.50.

In 1905 Marks hired George Oenslager to find out what this material was and told him he could take six months or six years to do it. Sandy-haired, spectacled, six feet two inches tall, and thin as a test tube, this son of a Harrisburg, Pennsylvania, clock-maker, had graduated from Harvard with Marks in 1894. While the latter had immediately entered the rubber business, Oenslager had remained at Harvard for two years of research work in the graduate school, obtaining his master's degree in 1896 and then entering the pulp and paper industry as a chemist on the staff of S.D. Warren and Co., at Cumberland Mills, Maine, MA.

In the new job, a policy of the strictest secrecy prevailed. Oenslager was and is a bachelor and no other chemists were employed to help him.

His assistant in the laboratory was a man who had operated a cheese factory. Among other duties, this aide invented plausible stories about the work that was being done in order to mislead stray visitors. Oenslager has retained his secretive attitude to this day; it is seldom that his colleagues in the Goodrich laboratories know what he is doing.

The first question that Oenslager considered was whether the cheap rubber had the same chemical composition as fine para. Analysis showed that both contained carbon and hydrogen in practically the same proportions. Then he decided to carry out a systematic study of the effect of adding various inorganic compounds to a mix of a poor rubber, sulphur and zinc oxide. It did not seem logical to him that the oxides of three elements — lead, calcium and magnesium — were the only compounds capable of speeding up vulcanisation. He intended to try at least one compound of all the common elements. With this in mind he experimented with materials such as the sulphides of barium, calcium, silver, mercury and antimony, the sulphates of zinc and barium, tin oxide, red phosphorus, aluminum powder, tin and zinc dust, lead powder and many more. All proved to be of no value. On March 22nd 1906, he tried mercuric iodide and at last thought he had found what he was seeking. Two and one half per cent of this material raised the tensile strength of a cheap rubber from 1,200 to 2,600 pounds per square inch. Oenslager continued to experiment with mercuric iodide until the end of May and at the same time was testing the chloride and bromide of mercury, metallic antimony, bismuth and arsenic, and slipping over into the organic field to test bromoform and iodoform. The lanky Diamond chemist did not overlook the fact that mercuric iodide, which appeared to be so satisfactory, might act as a carrier for oxygen as well as for sulphur and thus hasten the deterioration of the products in which it was used. It appeared so promising, however, that, disguised under the name 'Chinese vermilion,' it was used in tire treads on April 6th and in solid tires on April 30th. From then until June 22nd it was used in many compounds. The vulcanising time of compounds containing the new ingredient was cut in some instances from two or three hours down to five minutes. As had been feared, the mercury salt did hasten the ageing of the goods and they failed badly in service. The factory went back to the use of good rubber and long cures.

The thin man of the laboratories felt encouraged in spite of this, and turned to a systematic study of organic compounds. At this time he changed his mode of attack. Previously he had thought that the metallic part of the inorganic accelerators was responsible for their action. Now, considering that those in use (lime, litharge and magnesia) were all alkaline, he decided that perhaps their activity was due to their basic nature. He concluded that an organic base, soluble in rubber, might be the answer to the problem. Aniline is one of the simplest, cheapest and commonest materials of this type and, naturally enough, it was the first he investigated. Oenslager's laboratory record book shows that on

June 1st 1906, he mixed 15 parts of aniline into 250 parts of rubber, 20 parts of sulphur and 15.5 parts of zinc oxide. Samples were cured for periods of one, one and a half, two or three hours. In his book the results for the one hour cure are marked with the single word, 'Excellent.' He had found the answer to his problem. From June 2nd to June 17th, aniline was tried in many mixtures along with numerous other materials of a similar nature. It is interesting to note that one of these was tetraethyl lead, the familiar constituent of 'ethyl' gasoline, which was found to be of some value.

The results with aniline were so good that the Diamond factory began to use it, masked by the name 'Residium,' in inner tubes and in the carcass part of tires. On June 22nd it was used in solid tires and from that time it began to appear in factory formulas with steadily increasing frequency. Aniline is a poisonous liquid and consequently was not entirely satisfactory. Wondering if he could combine it with anything that would give a less toxic product, Oenslager remembered a solid derivative of aniline called thiocarbanilide which he had prepared in college. This he tried and found even more satisfactory. Three per cent, added to a cheap rubber, raised its tensile strength to 2,500 pounds, at the same time cutting the cure from two hours to one. Six per cent raised the tensile strength to 3,000 pounds and reduced the time of cure to ten minutes. On September 19th, tires were built with thiocarbanilide (masquerading as 'Lagos') in the tread and aniline in the rubber with which the fabric was coated.

The toxic material was not replaced in the latter use because it was found to give a desirable 'tack' which was helpful in building the tire. Because of the unfavorable experience with mercuric iodide, careful ageing tests were made on these tires and it was discovered that the new ingredients not only did not cause the goods to deteriorate but actually exerted a mild preservative action.

The use of these substances as regular factory materials brought up many new problems. The workmen, of course, had no knowledge of accelerators. Thiocarbanilide was apt to 'scorch,' that is, cause the rubber to vulcanise on the rolls of the mixing mill if the machine was too hot. Few believed that satisfactory goods could be produced with such short cures. It was necessary for Oenslager to be in close touch with the workmen to tell them how to get the best results with these new chemicals. These men had never come into contact with a trained scientist before and they were impressed by his meticulous habits as well as exasperated by his constant demand for careful adherence to details. They were also baffled by his desire to stand so closely to the batches as they were being mixed that the fumes brought tears to his eyes.

In the fall of 1906 Oenslager started a small factory on the Diamond grounds for the production of thiocarbanilide. This is made by reacting aniline with a yellow, mephitic liquid-carbon disulphide — which is extremely inflammable and blazes enthusiastically on receiving the slightest encouragement. Consequently the Diamond manufactory was segregated from the rest of the plant. It was a rude shack of corrugated iron sheets, held together with steel hooks which could be pulled out in case of fire, permitting the roof to fall and smother the flames.

During this time, Oenslager was also conducting experiments on scores of other compounds. He found none better than thiocarbanilide, however. By the end of 1907 it was being used in all of Diamond's products and Oenslager's active interest in the discovery of new accelerators ended. In 1912, aniline was largely replaced at the Diamond plant by a material with the jaw-shattering name of para amino dimethylaniline. This was discovered by Dr. David Spence, whom Marks brought to this country from the University of Liverpool in 1909.

Marks and his attorneys for months studied ways of patenting Oenslager's invaluable discoveries. Reluctantly they decided that no description could be evolved which would entirely cover the field. Secrecy appeared to be of greater value than a patent which would only reveal to rivals the great possibilities of the discovery and stimulate them to find similar materials, not covered by the patent, which would have the same properties.

It was impossible, however, to keep the secret from the other Akron factories for any great length of time. In those days, as later, it was common practice for the factory hand desiring a change of scene to check out of one factory and hire in at another the same day. Through this regulation flow of workmen from the Diamond to other plants it became known that new chemicals were being put into rubber. The nature of these materials finally leaked out and their use became general. The popularity of aniline continued almost until the World War. Hard-bitten, tobacco-chewing mill men who mixed it into batches turned blue at the lips — were forced to down their quart of milk every day to counteract its effect.

News of the great discovery did not reach Europe. Wolfgang and Walter Ostwald, German scientists, came close to it in 1908. They applied for a patent covering the use of organic bases to increase the durability of rubber compounds and prevent them from becoming hard and fragile. Their own laboratory not being fitted with rubber machinery, they instructed a well known German laboratory to investigate the idea. Upon receiving a negative report, they dropped the patent. In 1912 Gottlob and Hoffman, of the Bayer Company, were trying to prevent the rapid ageing of synthetic rubber. They found that piperidine would retard the oxidation of their product, and, studying its effect in natural rubber, they discovered that it had the properties for which the Americans had started their successful search nearly a decade before. The discovery was patented and immediately hailed all over Europe as a great scientific achievement. The Bayer company took out patents on other substances and in 1914 was granted an amazing patent which did not name any particular substance, but was based on a physical constant which covered thousands of materials, both known and unknown. Because of the war and subsequent unsettled conditions, the validity of this patent was not tested until 1926, when the Grasselli Chemical Co., assignee of the Bayer patent, sued the National Aniline Co., for infringement. An unfavorable verdict would have cost the defendant company, and others not involved in this particular suit, scores of millions of dollars. The suit was lost. Judge A.N. Hand, in 1927, held that the claim was too broad, covering an enormous class of substances and based on experimentation with only a few. He also ruled that substances with identical characteristics had prior use in America. Testimony of Dr. Spence at the trial proved that he had used piperidine and many other accelerators as early as 1910. No mention of Oenslager's work was made at this time.

In the meantime, in 1914, a British chemist, S.J. Peachey, had discovered an accelerator in the formidably named para nitroso dimethylaniline which did not come under the terms of the German patent. Jealous as the British were of anything German, it is little wonder that in 1916 we find Peachey proclaiming in print: 'It is quite true that the Germans were led to the discovery . . . but it is equally true that by far the most satisfactory organic accelerator was discovered in a British laboratory.' He elaborated on the wonders of his material and cried down the German discovery of piperidine because it did not cease its action when the goods were taken from the mold. In a few years he was to learn to his chagrin that his vaunted discovery had been tested by George Oenslager in 1906 and found powerful, but discarded because it stained yellow everything it touched — including the workmen, to whom it gave dermatitis.

The rival drum-beating of the Germans and English at first amused the Americans, who had been using all of these substances for years.

At last it became irritating and finally Spence could stand it no longer.

The composing of letters to the editor is one of the first symptoms to appear in such cases, and in 1917 Spence was writing to the leading British chemical journal. Uproarious controversy ensued. Finally, credit for the discovery of the organic accelerator was placed where credit was due when in 1933 George Oenslager was awarded the Perkin Medal, one of the highest honors that can be bestowed on a chemist. He is the only rubber man ever so honored.

Figures aid in arriving at an appreciation of the magnitude of Oenslager's achievement. In the case of tires, the strength added by the use of organic accelerators means an annual saving to the user of a sum which, a decade ago, was estimated at $50,000,000. In their manufacture, Oenslager's discovery has cut the curing time from three hours to 30 minutes. Thinner articles, such as inner tubes, are cured in one-fifteenth the time formerly required. 'Ultra' accelerators are available which will vulcanise articles in two or three minutes. This speeding up of the curing process has made possible a vastly increased turnover in factory equipment. One rubber technician not long ago estimated that organic accelerators had saved the industry an investment of $200,000,000 in buildings, presses, molds, and power plants which would have been necessary to maintain the present volume of production without them.

One of the three chemists boasted by the whole of the rubber industry was Oenslager when he started on the research that ended so triumphantly. Today there are over 400 of them. It was Marks, as we have seen, who brought about the passing of the factory superintendent's rule-of-thumb methods and paved the way for this regiment of researchers toying with some 1,500 different materials that enter into rubber compounding. But it was Oenslager who equipped these technicians with the several hundred organic accelerators which are the most potent of all the tools used in producing innumerable different rubbers — each specially adapted to the work for which it is wanted — where in the old days rubber was pretty much rubber and there were but few basic divisions under the heads of soft and hard.

Compositions to contain air, compositions to resist heat, compositions to resist abrasions; these basic ones have expanded until we have others designed to fill the bill where the demand is resistance to steam, oil, acids, the action of light and oxidation, special ones where the use of color is desired, hundreds and hundreds of them — one, almost, for every product made of rubber. Rubber from two grades has become half a hundred different types; simple rubber, sulphur, pigment and inorganic accelerator compounds have given way to intricate mixtures with every single substance justifying its presence by serving a particular purpose. The old formulas called for 5% to 15% of inorganic accelerators by weight; the organic ones required less than 1%, yet one Akron factory, and that not the largest, uses 76,500 pounds of them a month. Insignificant, of course, is this figure beside the corresponding 10,050,000 pounds of pigments, but it must be remembered that in every one of the thousands of formulas it is the fractional portion represented by the organic accelerator that performs the most vital functions of all the chemicals, sulphur alone excepted.

And in the history of pigments, too, the name of Oenslager is notable. Innumerable they are, as the succeeding chapter will disclose, but in developing the king of them all, Oenslager was one of the pioneers. Carbon black, obtained by the incomplete combustion of natural gas, is that substance and it has increased, immensely, the life expectancy of tires. In the old days zinc oxide was the chief material that went into the tire compounds along with rubber, sulphur and the accelerators. To color these gray or white casings black, the manufacturers took to adding small quantities of carbon black. At about the same time in 1912, however, several observers including Oenslager discovered that this coloring stuff was not only a reinforcing medium, but the best one that ever had come the tire maker's way. Gradually amounts used were increased, and today 20% by volume of carbon black goes into the long-wearing treads to aid in resisting the abrasion of road upon tire.

Most recent of the Oenslager accomplishments stem from his studies in the chemistry of vulcanisation, and they affect products ranging from tires to tank linings for chemical service.

So much for the developments. As to the man who turned them loose in the rubber field, he remains today a working chemist. To Arthur Marks, rubber was an easy and generous mistress from whom he lightly parted with millions in his pocket. But in George Oenslager there is something of the deep-seated passion for rubber that marked the single-minded life of Charles Goodyear. Because of the impossibility of patenting the organic accelerator find, there were no royalties in it for either Marks or Oenslager. There have been rewards, however, and the 63 year old dean of Akron chemists could long ago have retired from work in the laboratories had he seen fit. Not as a millionaire, of course, but as one more than well able to take his ease after 40 years with the test tubes. He chooses to remain among the rubber smells, daily reporting at the same hour to carry on his carefully guarded transactions with gum and powders and retorts, so absorbed in learning new things about rubber that he has not even time to learn the names of fellow chemists, whom he universally addresses as 'Mr. Man' when he finds it necessary to recognise their existence. Nor is it any dream of repeating his accelerator triumph that keeps his nose in the test tube. Rubber research has now reached the stage where such revolutionary discoveries in new materials are unlikely, he holds, and the job is to find new applications of those now at hand. It is that job that pulls him to floor six of the Goodrich factories every morning and seldom will he delegate even the routine portion of his work to other chemists.

Greatest of the rubber chemists and a good deal of a character is George Oenslager on his home grounds. Away from rubber, however, this gruff, secretive, knowing fellow is transformed into a trustee and past president of the YMCA and a vestryman of the Episcopal house of worship respectfully referred to by the religious editors as 'the church of Harvey S. Firestone.'

4.5 Pigments, Mills and the Liquid Invasion

HIGH ABOVE the sullen rumble of giant mills sounds a shrill shrieking and crackling. An Odyssey that had its beginning near the edge of a sweltering *Estrada* jungle has come to its end in the dusty mill room of a mid-Western gum shop and the traveler — crude rubber — is being made into a useful citizen. The immigrant from the East is being mauled and bruised about on huge steel rolls and is making a grimy welkin ring with ear-splitting clamor. This is no pain filled screaming of a wheel-broken wretch, however, for rubber is a tough fellow; he is merely shouting for food.

The outcry is heard in the pine forests of the South, the oil fields of the Southwest, the copra sheds of the Pacific isles, the asbestos mines of Quebec, the clay pits of Georgia and the Carolinas, the stock yards of Chicago. Picks swing in the zinc mines of New Jersey, Kansas and Oklahoma; the lead mines of the Ozark hills; the talc mines of the Green mountains of New England; the barytes beds of Missouri. Gleaming in sooty benediction, hundreds of thousands of tiny flames burn in sheet iron sheds in the gas fields of Texas, producing carbon black. A naked African clambers up a palm to fling down the oily nuts; in Texas, an enema of hot water is injected into the bowels of the earth and molten sulphur gushes forth; in India, millions of lac insects go about their business of encasing twigs in a shiny sheathing; all this being done at the command of rubber. Blessed with an extremely catholic appetite, rubber relishes such things as asphalt from Trinidad's sticky lake, tar from the coke ovens of Ohio River steel towns; hardwood pitch from the Northern forests; pitch and pine tar and resin from the Southern woods. A blubber-loving Eskimo might well look with jealous eye upon rubber's daily menu which includes such delectable items as grease from sheep's wool; paraffin, vaseline and lubricating oil from the oil fields; waxes from every source-mineral, such as ozokerite from Utah and Colorado-vegetable, such as carnauba, laboriously obtained by rubbing the leaves of a Brazilian palm-animal, such as spermaceti from the sperm oil in cachalot's head cavity — and insect, such as beeswax; greasy lauric and oleic acids; oil from castor bean and cottonseed and cocoanut; stearic acid made from beef tallow; and brown grease made from, pardon us, garbage. These are but a few of the things consumed by the rubber industry. We might also mention fossil flour which is the ground skeletons of microscopic prehistoric animals; mica, ground quartz, whiting, sawdust, charcoal, potato starch, wheat flour, and zinc stearate, the fluffy powder that is dusted on baby when he commits a social error. The complete list runs into several hundred items. Add to this list of natural products the scores of materials which issue from the chemists' queer-looking devices — accelerators, age resisters, organic colors and perfumes.

No other industry in all the world uses so many or such vastly different raw materials. 'Everything in Rubber,' proud boast of an Akron factory which turns out some 30,000 items, is a motto that is truthful in more ways than one. To the rubberman, all of these materials are 'pigments,' no matter what they are or what their use. The total amount of pigments used by the American industry every year very closely approaches the total amount of rubber used, in each case the figure being nearly a billion pounds. Perhaps by this time you have decided that the things you have so long regarded as 'pure rubber' may be something else. You are right. The contraceptive device, the toy balloon, and the rubber band fit the description of 'pure rubber' more closely than anything else, containing only a little zinc oxide, sulphur, and accelerator in addition to the caoutchouc.

What about some other common things? Your rubber heel might well have been turned out of a pottery — it's half clay. The sole on the same shoe is more than half tooth powder — magnesium carbonate — with a good shot of shellac in addition. At least a fourth of the tire on your auto is just plain, old fashioned soot, the kind that smudges your white collar and begrimes Junior's nose. The red rubber tube inside that tire owes its color to that great everyday necessity of feminine life - rouge. There's enough chalk — the rubber man calls it whiting—in the rolls of the wringer on your washing machine and in the hose out in the garden, to supply Faulkner words for every wall in your end of town. You need not go into a saloon to walk on sawdust if you have rubber tiling on your floor. The baking powder that raised the breakfast biscuits did the same thing for your bath sponge. The lime with which suburban lawns are coated every spring is a not uncommon ingredient of rubber goods.

In addition to the rubber there are two other materials common to every rubber article. The fillers, the accelerators, the coloring agents, the oils may vary in every piece but there are two great common denominators — sulphur and zinc oxide — which are present in all; the first to do the curing and the second to make the accelerator more active and to act as a reinforcement.

The third great rubber pigment, carbon black, is not used in every compound, but its usage is the heaviest of all, accounting for 25% of the total pigment consumption — 250,000,000 pounds a year in the United States. Black is the best reinforcing agent because its particles are the finest, being sub-microscopic in size, and the strength imparted by any pigment to rubber depends upon the fineness of its particles. Black was long used in paints and printing inks before it found its way into rubber as a reinforcing material. The British like to take credit for the discovery, but records in this country, forwarded from England in 1912, show that British tires contained only enough black to give them a distinctive color.

Of all the many pigments, few are added to the compounds for the sole purpose of cheapening them. If tires, for instance, were to be made of pure rubber with only enough sulphur for vulcanisation and nothing else, they would run no more than a few hundred miles. And numbers of the things which the compounder mixes into his rubber batch are more expensive than the rubber itself. Zinc oxide is an example of this sort. More costly than an equal volume of rubber, it has to be used in countless articles because it adds so much to the quality of the product. Carbon black, on the other hand, is doubly valuable, for it is cheaper than the rubber it replaces and at the same time it increases tremendously the strength of the compound in which it is used. Great, however, is the variety of purposes served by pigments other than those upping the physical properties of the finished article 100% or more. Some are cheap materials such as clay, whiting and mineral rubber, which merely add bulk to the rubber stock. The tars and oils and greases are used not because they add desirable qualities to the finished product but because they make the tough, stretchy rubber more tractable on the mill, make it easier for the dry, powdery ingredients to be mixed in. Softeners such as the stearic acid from the Chicago stockyards and the lauric acid from the cocoanut serve a double purpose since they also link hands with the zinc oxide to 'kick' the accelerators. In addition to these softeners, the rubber man may add stiffeners such as glue or shellac to the same batch to give a stiffer cured article. Accelerators we have already mentioned, but their opposites, the 'retarders,' are worthy of a word. These (benzoic acid is one of them) keep the accelerators from going to work before they should — from going to work on the mixing mills, that is. After the rubber is safely mixed and placed in the molds to be cured, the higher temperatures destroy the effectiveness of the retarder and the accelerator does its job. Age resisters, complicated organic materials added to help the rubber combat the effects of air and light, developed into a distinct class of compounding ingredient from the observation that certain types of accelerators gave better ageing stocks. In the class of vulcanising agents, sulphur stands alone both in effectiveness and cheapness. One or two compounds of sulphur are used for special work and selenium finds occasional use.

On the testimony of two great pioneers, Goodyear and Hancock, the use of other materials in rubber is as old as the industry itself. In his Gum-Elastic the former makes mention of colored earths and oxides, white lead, litharge, magnesia, gums, oils and lampblack. Of the last named item, he states that 'it is often used to cause the gum to endure the effects of sun and weather.' In Hancock's caoutchouc chronicle, mention is made of the use in rubber of whiting, Fuller's earth, ochre, Stockholm tar, plumbago, natural asphalt, coal tar and glue.

Rubber men of a century ago apparently were no different from those of today, being willing to mix anything and everything into the gum to see what would happen. Before 1839, of course, an attempt to bring about the dreamed of 'change' was involved. After Goodyear's discovery of vulcanisation, the chief aim was to increase the bulk of rubber, which was then an extremely expensive material. Discovery of the valuable properties of the compounding ingredients which they used was welcome, although incidental to their other aim.

The manner in which these other ingredients are mixed into rubber is, as a rule, a profound mystery to most of the populace. Visitors to Akron's 'gum mines' invariably express surprise when they learn that the caoutchouc is not melted in a' pot, the various fillers stirred in and the resulting mixture poured out into molds much after the fashion in which steel castings are made. The process is not so simple.

Rubber arrives at its American destination in the form of 250-pound oblong blocks, encased in neat jackets of woven straw matting or burlap, or in a plywood case sternly labeled 'Stow Away from Boilers!' The blocks are cut into smaller pieces, not to remove sticks and stones encased by wily natives, as was the case when the wild product was still widely used, but to facilitate handling. Unlike its wild brother, plantation rubber is fairly clean, and the wash room in the factory is not as important as it once was. For most purposes, even plantation rubber is washed, however, to remove the dirt which cannot be entirely avoided and the splinters from the wooden cases.

The cleaning is accomplished on a washing mill, two huge corrugated steel rollers turning at different speeds. The rubber forms a sheet on the slower roll and the excess piles up in a twisting, squirming mass between the two. Water is sprayed on and as the embedded dirt is torn loose, it floats away, for wet caoutchouc is not sticky. The rubber is cut away from the mill in long, lacy sheets which are hung in dryers. Dried, it is not yet ready for the mixing operation. It is still too tough to accept the pigments readily and so it is thrown on a smooth-rolled mill and torn and kneaded until it is softer and more plastic. The 'masticated' rubber then goes back to storage until needed in the compound room. Here a dust-caked compounder weighs it, and also the pigments required for a particular batch, and places them in a round metal can, which is sent to the mill room.

At this destination, the former tree milk is yanked from the compound can by a gaunt, bulging-cheeked mountaineer and flung into the slow moving, seven-foot long rolls of a rumbling mill. The rubber shrieks in protest, but gradually it quietens down and becomes a smooth sheet on the slower roll. Into it the millman pours the heated oils and melted waxes that he has obtained from his counterpart who presides over a long row of steam-jacketed kettles in one corner of the room. Now the caoutchouc is ready for its diet of dry powders. Sulphur is added first, that it may get the maximum amount of mixing. Then the other materials are shoveled on top of the revolving rolls. Part of the pigments fall through into the pan below but, armed with brush and scoop, the millman relentlessly hurls them back on top of the rubber. Finally, aided by an occasional squirt of tobacco juice from the ex-ridge runner when he knows he is unwatched, the powders are all ground in. With a sharp, short-bladed knife, the millman slashes the mixture back and forth from side to side to insure that all parts are uniformly mixed. Then dexterously he wields his knife, cutting slab after slab of stock from the moving roll. Thus is rubber made ready for conversion into anyone of 30,000 articles.

Fully equipped with fast-acting brakes and clutches, guards and other safety devices, the giant mills are not the dangerous monsters of former days. Before the law made its safety demands of not always too willing employers, it was no uncommon thing for a careless mountain boy to have a hand trapped by the treacherous rubber and be pulled, screaming, into the rolls up to his elbow or his shoulder, and, in a few cases, beyond.

The heavy roller mills are not the only mixing machines which are used today. For mixing some batches, especially those containing gas black which is light and flies badly, recourse is had to the 'Banbury,' an internal mixer which bears the name of its inventor. In this device the rubber and pigments are placed in a chamber and buffeted about by rotors turning in opposite directions.

Having escaped the Spanish Inquisition methods of the millroom, the batch may be formed into useful articles in various ways. Blanks of definite weight may be stamped out, placed in molds, and heated under pressure until vulcanisation has taken place. In such fashion are most of the rubber parts of automobiles, wringer and typewriter rolls, erasers, rubber soles, and hundreds of other items of not too intricate shape made. Thus are made rubber heels, invented in 1891 by E.C. Critchlow, who then proceeded to spend years in overcoming popular objections to their 'sneakiness.'

Articles of a more complicated cross section are shaped by forcing them out of a 'tubing' machine. This might be likened to a giant sausage grinder — a revolving screw squeezing rubber out through a die of any desired design. In this way are all kinds of tubing and hose, strips for auto windows and insulated electric wire produced. Too long to be cured in molds, the articles, after being shaped by the tuber, are vulcanised in hot air or steam. Rubber bands and jar rings are first made as long tubes and, after vulcanisation, are placed on lathes and cut by sharp knives.

The rubber batch may be dissolved in benzol or gasoline and the resulting cement spread upon cloth, later cured by exposure to vapors of sulphur or sulphur chloride and then fashioned into raincoats or sheeting. Balloon cloth and the upper parts of boots and overshoes are similarly spread, then shaped, varnished and vulcanised in air. From rubber cement also come toy balloons, contraceptive devices and rubber gloves made by dipping forms of the appropriate shape into the dissolved gum and drying and curing in sulphur chloride.

The making of a tire is the assembling and 'curing' together of these three things: the wire bead, stiff ring which holds the tire to the rim; the carcass, a varying number of plies of rubber covered with cotton cord which give the tire strength; the tread, outer rubber compound which protects the carcass from the abrasion of the road.

In one part of the factory are clattering looms which pull steel wire from huge reels and weave it into a strong tape. The tape is fed into a machine which plays a never-ending game of quoits with hoops of its own manufacture. This machine coats the wire tape with rubber compound, bites off a piece of the proper length, coils and fastens it into a ring which it picks up in a mechanical hand and flings dexterously over a nearby rack. The wire hoop is covered with a protecting strip of fabric and is placed on a conveyor to join the march of other parts which are converging on the tire builder.

The creel room is the only quiet spot in the factory. In it demurely sit thousands of cones of stout cotton cord, each silently delivering its eight mile long content at the behest of a nearby giant - the calender. Through a central gathering comb, a wide flat band of many cords is pulled over a heated roll and in between the bottom rolls of the calender. This consists of three heavy vertical rolls operating at different speeds. Rubber compound is fed between the two top rolls, is carried around on the middle cylinder and is wiped off by the cord passing through at the bottom. The sheet of rubbered cord is rolled up in cloth treated to prevent its sticking, and taken to the bias cutter, where, as it is unwound, a cutting knife lops off strips at an angle. These strips, of the proper width to form a ply in the carcass of the tire, are joined together, then cut to proper length and carried off to the tire builder. Not all manufacturers use the calender for coating cord with rubber. One shouts the merits of 'gum-dipping,' in which the cords are passed through a bath of rubber cement; another ballyhoos 'latex-dipped,' in which process the natural juice of Hevea replaces artificial cement. Tests indicate that there is no difference in fabric prepared by any of the three methods.

A hulk like a heathen idol squats in a corner of a third dimly-lit room. Up to it rolls a string of little cars, filled by attendants at a distant point. The hulk reaches out with iron hands, lifts a car to its yawning metal mouth, pounds it to make sure no morsel has escaped, puts it down and gulps other loads until its belly is full. Then it snaps shut its mouth and begins to masticate the offering. This mixing machine is the Great

God 'Banbury' and its food is chiefly rubber and carbon black. In a few minutes a door at the bottom of the machine opens and it excretes nearly half a ton of the black mixture onto a moving belt which carries it to nearby sheeting mills. Here sulphur is added and the mass has become the compound from which tire treads are made. The compound is stuffed into a tubing machine and from the other end it issues in a continuous strip, thick in the center, tapering to a thin edge. The strip is chopped automatically into lengths which are equal to the circumference of a tire. These are weighed by the machine, and, if satisfactory, sent on to the tire builder.

The tire builder is the only artisan of the rubber industry. True, he works at a tire building machine, but this only assists him at his labor; the tire is still hand made. The machine is little more than a collapsible revolving drum equipped with one or two semi-automatic tools which help press the various parts of the casing together. At the back of the machine and slightly above it is a set of trays which a helper fills with the parts of the tire in the order in which they will be needed. The builder places the two beads in position on his machine, gives the revolving drum a coat of cement, pulls the first ply from the tray, slaps it on the drum and splices the ends together; treats the second ply in the same way and rollers press the two plies together. The beads are placed and the edges of the first two plies folded over them. The remaining plies — the passenger car tire usually has four or six are added in the same way with the cords in alternating plies at right angles to each other. On top of these go a cushion strip of high rubber content, a 'breaker' strip of coarser cord and finally the combined tread and sidewall. The builder collapses the drum and pulls a wide flat belt from his machine; this is the completely assembled tire though it looks like almost anything else.

Long rows of tire builders stand at their machines, working at incredible speed. Every four minutes a single builder completes one of these belts of rubber and cotton and wire, flings it on a conveyor, and has another started in almost the same motion. Shaping the flat band into the semblance of a casing is a simple operation. The band is placed in a vacuum box which, when closed, forms a tire-shaped compartment. A water bag, which looks like a heavy inner tube and which plays a part in the subsequent curing operation, is fitted inside the band and when the air is suddenly withdrawn from the box, the band envelops the bag and is sucked into the tire compartment. Removed from the box, it has the look of a tire at last; clinging to a conveyor hook, it moves toward the final step-vulcanisation.

The great majority of present day tires are made in this fashion.

Unusually large sizes for trucks, airplanes and farm implements are not built on these flat drums, however, but on cores in the shape of a finished tire — as all casings once were. The making of a core built tire calls for a vastly greater amount of skill than is involved in drum building and is, of course, much slower.

Down to the 'pit' go the tires to be cured. Aptly named, the 'pit' is not the inferno of a few years ago, although in the older factories it is still a very satisfactory imitation of purgatory. It is a place of heat and seeming confusion, where the ear is assaulted by the sibilant protests of steam and the hammer and clang of metal, where the throat is gripped by acrid fumes and parched by dust filled air and the eyes are befogged by steam and soapstone. In the haze it is not hard to mistake the automatons who labor in this dismal place for parts of the clanking machinery.

Work in the pit has been lightened in recent years, but it still demands broad backs and tough muscles. Out of these lower regions came Bob Martin, heavyweight boxing champion of the AEF, and K.O. Christner, who as a rookie pug of 34 — the doddering age for most prize fighters — spoiled Knute Hansen, pride of the Madison Square Garden millionaires, and pounded out a win over Jack Sharkey, although the record books tell a different story in the case of this last bout. Christner especially was idolised by the gum workers, all of whom, incidentally, were baffled by references of *Estrada* sports writers to the 'Akron rubber puddler.'

Tires are cured in steel molds which are engraved with the desired tread design and the maker's name. As the tires come into the pit they are fitted into the bottom half of a mold passing on a conveyor. The top half, moving on another conveyor, is automatically lowered and fitted into place and the complete unit passes under a hydraulic press which hammers the two halves together. The tire in its mold then moves along to the vulcaniser, where it is dragged from the conveyor. A connection is fastened to the water bag inside of the tire and the mold is lowered into the vulcaniser — a vertical steam cooker capable of holding 25 to 40 molds at one time. When the vulcaniser is full, it is closed and steam is turned on. The water bag, filled with boiling water under pressure, forces the heat-softened rubber into every groove of the mold. After vulcanisation has taken place, the steam is blown off, the heater opened and the molds dragged back onto the conveyor. Breaking the molds apart, formerly a back-wrenching task performed by men with crowbars, is now easily accomplished with a pneumatic tool. The mold moves along to a point where the upper conveyor chain descends and picks up the top half. Revolving wheels close in on the tire in the bottom half and pummel it until it gladly leaps out and onto an inclined conveyor which carries it away. The bottom of the mold moves serenely on, is sprayed with soap, takes another 'green' tire and again meets its upper half from which it separated only a short time before.

The cured tire goes to a machine which reaches forth steel fingers to yank out the water bag. This, too, was formerly one of the harder jobs calling for crowbar and brawn. The tire is washed, trimmed of the mold overflow, painted, inspected, wrapped and is ready for the consumer.

In the most modern factories, the pot heater is giving place to rows of 'watchcase heaters,' in which each tire is cured separately. Resembling watches with hinged cases, these devices are electrically timed and controlled and a single operator is able to tend as many as a hundred vulcanisers. Because of the tremendous saving in labor they eventually will wholly displace the old style pit and thus contribute their share to technological unemployment. The expenditure on equipment required to make the change universal is one thing holding it off. The current bellicose mood of the gum worker is the other.

A great deal of expensive power is required by the mixing operations, for the machinery of the rubber industry is massive. A great deal of expensive heat is required by the curing operation, for vulcanising temperatures run as high as 34 °F. Power and dust and steam. Will these continue to be the symbols of the rubber industry? Will the thunder-filled mill room continue to be the heart of the factory as it has been for a hundred years past? Will the manufacturing process continue to be one of grinding dry powder into dry rubber and heating the resulting mixture at relatively high temperature? The 'practical' rubber man says yes. The owner, with huge investments in machinery, hopes that it will be so. The scientists of the type who are willing to say 'I don't know' when, as a matter of fact, they do not know, readily admit their ignorance.

Instead of a room filled with massive Stentors of steel, where every stray beam of sunlight is sharply etched in the dusty air, picture a clean, quiet room where the equipment consists of light churns and tanks of liquids. Such may be the rubber factory of the future.

The process which may yet cause the manufacturers to scrap part of their heavy mills and calenders is so old that it is almost new — the direct utilisation of latex. Up-to-date factories are turning to a consideration of the centuries-old methods of Amazon morons who were enjoying the comfort of protective rubber footwear while city dwellers in medieval Europe were tramping quagmire-like streets on wooden pattens upheld by iron hoops or long pegs.

For uncounted years, as we have noted in this book's first chapter, the South American jungle dwellers have made rubber shoes and bottles by dipping rude day forms into the milk of the *Hevea* tree. Basis of the native Indian rubber industry, latex did not become of importance in the early days of European and American rubber manufacture because of the difficulty of transporting it without coagulation taking place. The industry might have developed along entirely different lines had the knowledge of how to preserve the tree milk been widespread a century ago. Within the past 15 years, however, there has been a great rebirth of interest in latex. A layman, looking over the pile of patents issued in this period, might well be excused for assuming that mixing mills, calenders and vulcanisers have already been scrapped and that a present day rubber factory is hardly distinguishable from a dairy. But for pure chance, incidentally, that might long have been the case, for the industry well could have been built around the use of the milk instead of the coagulum from the very start.

In earlier referring to Edinburgh James Anderson's 1791 suggestion that caoutchouc be brought into Europe in its liquid state, we also noted that a French chemist was a year ahead of him. This experimenter was Fourcroy, maker of 'fat' varnish, briefly mentioned in the story of the rubber experiments before Goodyear. Nor was his recommendation that latex be imported for direct manufacturing purposes a mere expression of wishful fancies. Fourcroy had discovered that, unlike the acids which coagulate *Hevea* juice, the alkalies preserve it for considerable periods of time. It was because the Fourcroy observation went unnoticed that the rubber industry started off on the wrong path and only today is making tentative excursions down the logical one.

Howison, discoverer of Asia's native rubber, worked direct from latex in the 1790s, but, as in the case of Fresneau 50 years before, he was right at the source of supply and there was no thought in those days of building manufacturing establishments in the tropical caoutchouc lands. Writing in 1798, Howison tells, however, of immersing wax molds in latex to make boots and globes, of spreading the milk over fabrics with a ruler, of a dipping machine which might have been a model for those used today.

Those not so fortunately situated with regard to latex supply dismissed his remarks just as they had overlooked Fourcroy's proposal to transport the supply, and so it was more than a century before the findings of these investigators were put into commercial practice. Hancock, Goodyear and others flirted with the idea of manufacturing direct from milk and gave it up owing to the apparent impossibility of regularly obtaining supplies in an unfermented condition. The first of these actually imported a small amount of latex from 'Guatamala' in 1824. It was shipped in vessels 'formed of two or three joints of the larger kind of cane, four or five inches in diameter' and Hancock took out a patent on the manufacture of a leather substitute which he made by saturating felt, cotton, hair, flax or hemp with the milk, drying the material and pressing it into molds. The following year he patented the idea of impregnating ropes with latex to make them more resistant to air and moisture. Five years passed, in which even the invention of his 'pickle' did not drive the fluid out of Hancock's mind, and in 1830 he submitted a very modern-sounding patent for a latex-fiber composition used for making sheets, gaiters, galoshes, boots, shoes, hats, caps and shoe soles. He also made thread by dipping a grooved cylinder into tree milk, scraping off the excess on the surface, drying and repeating the process until the groove was full of dried latex. The Englishman apparently expected a boom in the latex business, for he arranged with an agent at Tampico, Mexico, for regular shipments. The liquid was shipped in 'good sound barrels, well stopped,' but in practically every case it was found to be coagulated on arrival and Hancock was forced reluctantly to give up the scheme.

In America, Goodyear, not knowing of Hancock's difficulties, was also struck with the simplicity of the latex idea. His experiments however, were unsuccessful, and so it was that he veered to vulcanisation and to founding the rubber industry as we know it today. And the latex idea remained dormant until the 1850s. In those years an American in Brazil patented an invention for latex exporting (it didn't work) and the Society for Promoting Christian Knowledge pamphleteer hit on the idea of transporting Mahomet to the mountain. Commenting on the 'desirableness of manufacturing waterproof cloth from the fresh milk,' this writer suggested establishment of a factory for this purpose on the banks of the Amazon and added: 'A material thus prepared would, of course, not fall within the prohibitory limits of any existing patent, for it is as old almost as the discovery of America itself.'

It was in 1853, too, that forgotten Fourcroy's idea was hit on by English William Johnson who obtained a patent for the preservation of latex with ammonia. Strangely enough, no better preservative has been discovered to this day.

In 1865, just 40 years after Thomas Hancock's latex patent, his brother Charles, with Stephen Silver, was granted a patent for the manufacture of molded goods from tree milk by methods which have not yet been much improved upon. In spite of the success of the process and the fact that means were known for the preservation of latex, the experiments did not get beyond a preliminary stage, however. Reactionary Thomas, who had long since given up latex as a raw material, wrote: 'Although rubber in this state would be very useful and many things could be done with it, the loss of weight by evaporation, being nearly two-thirds of the whole, the expense of vessels and the freight of so much worthless matter (water) will probably prevent its ever being used extensively.' The argument appears a valid one. Later discoveries made it otherwise.

Three years after Charles Hancock achieved his success and didn't know what to do about it, the first American latex patent was issued to one Bishop. It was for producing waterproof collars, cuffs and banknotes by coating the paper with the para fluid. From this time until after the World War, the use of latex was practically forgotten. So recently as 1921, for instance, caoutchouc chemists at the London Rubber Exhibition looked upon it as something quite out of the ordinary. A lone voice crying the virtues of latex to a coagulated world was that of Kelway Bamber, of the Ceylon government service. One of the first to learn how to mix compounding ingredients into latex without causing coagulation, he carried on extensive experiments in the period 1905-14 which coincided with the development of the *Estrada* plantations. Bamber showed the natives how to make belting, flooring and mats by treating cocoanut husk fiber with latex and sulphur and then vulcanising. Concerning the Bamber work the editor of the *India Rubber World*

(United States) pedantically sniffed, 'to take the latex of any rubber tree and make it up into marketable goods is not to be thought of.' Something like that is what the neighbors said about Charles Goodyear, his experiments.

There is in the records, however, one important defense of latex during this period. In a pamphlet, *The Revolution of the Rubber Industry*, printed in Paris in 1908 and signed by 'Le Syndicat du Latex, Paris, June 1905,' appears a statement which at that time was almost heretical enough to justify burning at the stake. The statement reads:

'It cannot be denied — and this is the opinion of the real scientists who have studied rubber latex - that the problem of rubber was on wrong lines from the very beginning.... The industry was the slave of errors from its birth, developing processes of so complex a nature and so much against logic and nature.'

The Frenchmen had their justification in the early part of 1923 when manufacture direct from imported *Hevea* juice first began to come into a small part of its own. It was the discovery of various methods of concentrating latex so that it was no longer necessary to transport huge quantities of water along with small quantities of rubber that brought this about.

By one method, the latex is run through a machine like a cream separator and is divided into a 'cream' containing 60% rubber and a 'skim milk' containing only 5%. The concentration can also be effected by adding creaming agents, such as Iceland moss, which causes the latex to separate into two parts, one containing almost all of the rubber.

Thick maple syrup is made from thin maple sap by evaporation, and by a somewhat similar process a heavy latex paste can be made from thin *Hevea* juice. This was the first large scale method to be used and is still one of the most popular. Latex and protective agents are placed in a revolving drum which is heated from the outside. A current of air removes the evaporated water and a paste containing 85% of solids is formed. E.A. Hauser, inventor of the process, calls the material 'Revertex,' for the paste is reversible and by careful mixing with water can be turned back into latex for manufacturing purposes.

Most of the compounding ingredients mentioned at the beginning of this chapter can be incorporated in latex mixes if first made into dispersions or emulsions with water. For making spread or dipped goods, latex is used in the same fashion as the gasoline and benzol cements previously described. For these uses it offers several advantages; cheapness of apparatus, no loss of costly solvent, no fire or health hazard. In the case of toy balloons, contraceptive devices and gloves, articles made by dipping forms in the milk are much stronger than those similarly made with cements. Impregnation of tire cord with latex was the first major use of the material in this country and is still one of the chief selling points for the tires made by one member of the Big Four.

Round rubber thread, familiar to feminine readers as 'lastex,' and far superior to the old style square cut thread, is made by squirting a fine stream of latex into a coagulating bath. Automobile tubes, garden hose and tire treads have been made by squirting the milk from nozzles of suitable shape, but these products are not yet commercially practicable.

Upholstering material made from latex offers many advantages over sponge rubber. The process, as patented by Britain's Dunlop company, consists in beating the liquid to a froth in giant egg beaters, or, as the English insist on saying, in 'egg whisks.' Just as the lightness of a cake depends on the amount of time it is beaten, so does the lightness of this English product depend on the length of time the latex is whipped. The frothy liquid is then cured in properly shaped molds. Formed of millions of interconnected air cells, the new type of cushion does not cause the local heating which makes sponge rubber so uncomfortable, and it is finding increasing use in autos and in hospital mattresses.

Latex is now being incorporated in paper to give a stronger product and in cardboard for waterproof food containers. Its use as a binder for vertical tufts has greatly simplified rug weaving and given a non-slipping product as well. It has proved of great value as a can-sealing material, since it does not impart odors to the contents, as rubber cement will. Other uses for this versatile milk include artificial leather, velvet, suede and chamois; use in photographic paper to give a special finish; in making insulating board from waste products such as sawdust; in sticking plaster; in paints and lacquers; as an adhesive in the shoe industry; as a preservative coating for fruits, cheese, eggs and other foods; to sea-proof hawsers, fish nets and lines; as a preserver for building stone, cement, plaster and concrete, and in combination with asphalt in paving compounds.

Worthy of particular notice is the discovery made by Dr. Philip Schidrowitz, in London in 1923, that latex could be vulcanised without coagulation taking place. After more than a decade, the possibilities of this discovery are not yet fully recognised by rubber men. Pigments are incorporated in the milk together with sulphur compounds and suitable accelerators. The whole is heated for some time at a temperature considerably under the boiling point of water. There is no visible change in the latex, but it is only necessary to evaporate the water from it to obtain fully vulcanised rubber. The advantages are obvious. For spreading, the 'Vultex,' as it is called, is simply spread on the fabric and dried and the article is finished. Since there is no subsequent vulcanisation at high temperature, fine fabrics such as silk and many dyes and colors which would be destroyed at vulcanising heats can be used. Forms can be dipped in 'Vultex,' and, when dried, the rubber article is complete.

Since no treatment beyond drying is required, it is difficult to imagine the manufacture of rubber goods reduced to any simpler terms.

Most striking of all latex uses to most laymen is in the electroplating of rubber. Accustomed to thinking of plating only in connection with metals, they find something incongruous in the thought of using rubber in similar fashion. Rubber, it has long been taught, is a well nigh perfect insulator, a substance that does not conduct electric current, and to expect it to plate out like nickel or chromium is a little beyond reason. The answer lies in the fact that although rubber is not influenced by the electric current, latex is a suspension and in a suspension, the suspended particles almost always carry an electric charge. Consequently when electrodes are placed in the para milk and current applied, the tiny rubber particles flock to the oppositely charged electrode. The French scientist, Victor Henri, in 1907, was first to show that the particles in latex carried a negative charge and moved to the anode, or positive pole. S.E. Sheppard, American student who was studying at the Sorbonne at the time, indexed the fact in his mind and trotted it out 16 years later when working for the *Estrada* Kodak Co., on possible uses for various aqueous suspensions. Together with L.W. Eberlin he prepared an artificial latex, applied current and found that the rubber was deposited on the positive pole. Many pigments could be mixed into the latex and they would also plate out with the rubber, it was learned.

Independently and about the same time in Budapest, Paul Klein and Andrew Szegvari made the same discovery and applied for patents. Other patents were granted to the B.F. Goodrich Co., in the United States and to Dunlop in England. The four companies, Eastman, Hungarian Rubber Goods Co., Goodrich and Dunlop pooled their interests and formed the American Anode Co., now the leader in rubber-plating.

Being faster and avoiding the formation of layers, the anode process is displacing older dipping methods for the manufacture of many articles such as gloves, balloons, bathing caps, drug sundries, and medical specialties. It is especially valuable for coating metal parts of complicated shape which cannot be satisfactorily covered by dipping or molding; parts such as springs, fan blades, screws, plating racks, and milk bottle carriers. The process is not complicated. The sulphur, zinc oxide, accelerator, age resister, dyes and other desired ingredients are ground with water, ammonia and dispersing agents. The mixture is added to latex and placed in tanks for the electro-deposition. The metal part or form which is to be covered with rubber is used as the anode and a mild electric current turned on. Rubber will plate out on a zinc surface 500 times faster than zinc itself, so it can be seen that the process is not an especially lengthy one. When a sufficient coating is formed, the form and the deposited rubber are washed, dried and vulcanised. Vulcanised latex works as satisfactorily as the normal product, and when it is used, the last step in the process can be dispensed with.

The die-hard rubber man of the old school who still believes latex to be a laboratory plaything will do well to consider export figures which show 368,644 gallons of latex leaving Malaya and Sumatra in 1922 and 3,880,746 gallons a decade later. Since concentrated latex did not figure in the picture until about 1927, these figures are more startling when translated into tons of dry rubber. The 1922 gallonage represents only 420 tons of rubber and the 1932 figure 8,510, or an increase of more than 20-fold. All through the worst of the depression, when imports of crude rubber were steadily dropping from year to year, the imports of latex had mounted just as steadily. And even more sensationally did they climb in 1933-1935. For 1934 British Malaya alone exported 14,172 tons and for 1935 the world estimate was some 26,000 tons or about 3% of the total of all rubber. Already, then, latex has driven around wild rubber, which in 1934 represented less than 2% of the total. In just a few years it will amount to 25% of exports, a number of authorities hold.

For the commercialisation of latex, chief honors are due to the late Ernest Hopkinson, New York patent lawyer and chief of development work for United States Rubber Co. Goaded by Hopkinson, 'US' is the leading rubber-milk user in this country and owns tank ships and tank cars for transporting the fluid from its Sumatra plantations to its American factories. These tankers bring over 5,000,000 gallons of the fluid to this country every month and nearly 25% of the total goes to the company's tire plant at Detroit.

Hopkinson, born in Hyde County, England, began practice in Gotham in the 1890'S and made his first contact with the then rubber Trust through a subsidiary in 1897. Eventually he became patent counsel and vice president in charge of development for the giant combine as well as president, vice president or director of some 17 of its subsidiaries, including United States Rubber Plantations, which he served as an executive committee member. Latex attracted Hopkinson's attention while he was seeking to develop methods of preparing crude rubber other than by acid coagulation. The fruit of this research was US's 'sprayed latex' rubber. Continuing his studies in the milk after developing the spraying process, he sponsored latex-dipped tire cord and latex thread and took the first steps towards transportation of tree juice in bulk. To him the corporation with the George M. Cohan name owes its position as leading user of the liquid. Hopkinson is also one of those receiving much of the credit for the development of the flat band method of tire manufacture in universal use today. He died on May 3rd 1933.

Developed by the Rubber Powder Co., of England in the last two years, rubber powder, newest of the new, is an offshoot of the Hopkinson sprayed latex idea. No mixing mills are needed for incorporating compounding ingredients with powdered rubber; simple stirring of the various materials together suffices. Vulcanisation is accomplished by packing the powdered mix into molds under pressure and heating. Rubber powder is of value in the manufacture of molded goods only and its very newness makes it impossible to predict its possible effect on rubber goods manufacture of the future.

4.6 Synthetic

Almost from the moment that the tree milk from the Amazon began to make fortunes, the hunt to duplicate it artificially was on. All over the globe, fortune-seeking chemists had at the hardened tree milk in a race to take it apart. Once absolutely establish its constitution, they reasoned, a duplicate product could be built up out of constituents derived from cheap and common products. This was sound arguing but it encountered an unexpected stumbling-block — the fact that rubber proved to be so unfathomable a substance that its exact constitution could not be absolutely established, has not been to this day.

It is hard to make anything when you don't know what you are trying to make. Yet with Monte Cristo visions luring them on, these descendants of Raymond Lully and Roger Bacon and Silvester, the magician—pope, distilled and collected and peered and sniffed and compounded in a hundred laboratories in France, Germany, England, and the United States. When rubber prices tumbled a few of these sorcerers would desert, when prices soared an army of new recruits would rush to join the ranks of those questing for a synthetic product.

Regularly, uproarious clamorings would go up as some honest but stampeded scientist or some predatory impostor would let the world know that he had wrested the great secret from potato or sugar cane or petroleum and that the globe was his. Sometimes they would subside and be heard from no more, sometimes they would have an expensive substitute of a sort that would fail miserably in competition with rubber, but never did science circles or lay majority doubt that the man who produced something as good as rubber would have the world by the pocketbook.

Figure it out for yourself. Say that a fellow evolved a jim-dandy of a synthetic rubber and laid back until prices again went whooping up to $1.50 or $3 a pound. Then he'd just let it loose on the market for about $1 a throw, salt down the millions and retire as one of the richest men in creation. So went the universal reasoning. And in the laboratories great rubber chemists labored diligently, tenaciously, single-mindedly, month after month, year after year, decade after decade, in futile attempts to duplicate the baffling combination of hydrogen and carbon which rubber most certainly is.

Keep this last picture in mind. It is necessary to full appreciation of the grandly ironical scene staged some years ago in a little known corner of the world's greatest football university. Once upon a time, it seems, there was a young man who fell under the spell of acetylene. It was something like a modern dress revival of Charles Goodyear in that act wherein he was so overwhelmed with curiosity, surprise and admiration on first encountering a black, sticky, stinking slice of rubber that never thereafter could he reflect upon this ravishing substance without 'adoring the wisdom of the Creator.' Hydrogen and carbon it was, again, that played the siren, but this time as a colorless, gaseous compound with an evil odor derived from impurities in its immediate ancestry on the calcium carbide side. On paper, this symbol: C_2H_2. In the laboratory, a gas, a colorless liquid or a crystalline solid depending upon pressure and temperature manipulations. In history, the bicycle and auto lights of the gay 1890s. In common usage, the flame cutting through metal in the manner of hired hand *versus* blueberry pie.

Like many another mistress is this mistress of many ascetic scientists: comparatively safe when gaseous or dissolved; so dangerously explosive when liquid and subjected to heat or a stray spark that there are stern laws to regulate its appearance in that condition. Acetylene's new follower, the hero of this little story, grew familiar with all these forms; grew to know, perhaps, more of acetylene's moods and more about those moods than any other living man. Never was he led to stray after any other compound; to learn all that could be learned about this one, he found, was a job for more than a lifetime. Nor could he be bothered with following up the results of experiments; that he left to other men while he moved on to provoke new major reactions. As a college boy, for instance, he synthesised acetaldehyde from acetylene by a method identical in principle with today's commercial process. As a graduate student, he evolved an excessively poisonous product from acetylene and arsenic trichloride and went his way, leaving it to be pounced upon in the blood-mad days 14 years later and put to producing the most deadly war gas ever made.

There came a day, however, when the acetylene devotee, now a teacher, happened to pass his favorite gas into a solution of the chlorides of copper and the alkali metals. And with that random act, man, after those decades of vain searching, was on the track of synthetic rubber!

Nor does that exhaust the ironies of the situation. With thousands being rolled up by the artificial rubber, with millions a possibility, this man, it developed, cared less for all the money in the world than he did for one whiff of acetylene gas! From his discovery he received: one synthetic rubber base for a fountain pen set (which he promptly lost), one medal (which he never wore), and one set of synthetic rubber heels which wore out one pair of shoes and were transferred to another. That old familiar figure, the swindled inventor, once again? Hardly. He knew what his discovery was worth; his patents were assigned as he wanted them; his royalties were paid exactly as he desired and, at the time of his death a few months ago, he certainly was one of the happiest and most contented individuals on earth. So ended the hunt for the modern philosophers' stone. Ended, that is, to all practical purposes, but isn't over yet if you refuse to waive hair-splitting.

In the strictest sense of the word, the acetylene wizard's product is not synthetic rubber, for it is not something identical with the natural material in composition and physical properties. In one way, it is further from being synthetic rubber than are all the artificial compounds which preceded it for these, at least, are chemically like rubber, although they differ greatly in their physical structure. The acetylene product is not chemically like rubber for it contains chlorine, an element wholly foreign to the rubber trees and their milky juices. In other ways, however, it resembles rubber more closely than do the compounds chemically like rubber. It may, for instance, be emulsified in an ammoniacal soap solution to produce a milky liquid resembling natural latex in appearance and application. And in structure it does, when stretched 500%, closely resemble the structure of rubber under the same stress, the X-ray has disclosed. In these senses the acetylene product proves to be more like rubber than its discoverer and developers had expected, for such similarity was not deliberately sought.

What was sought was something economically practical that would do rubber's work as well or better. That was obtained, and certainly, therefore, the 'synthetic rubber' label is justified, whatever the hairsplitters may say. As to its being the 'first synthetic rubber,' that too expresses the higher truth. The others, even the ones produced by the hundreds of tons in Germany during the war, never duplicated rubber's performances in reasonably satisfactory fashion. They do have historical interest and for that reason we turn to them now.

The first man to propose artificial rubber was, it seems, the Chevalier de Claussen. Traveling in South America he became acquainted with that odd tree, the Hancornia, which yields both eatable fruit and rubber. Studying this oddity, the Chevalier sagely noted: 'It bears a fruit ... full of a milky juice, which is liquid India rubber. To be eatable this fruit must be kept two or three weeks . . . in which time all the India rubber disappears or is converted into sugar.... The change of India rubber into sugar, led me to suppose that gutta percha, India rubber and similar compounds contained starch. . . . A great number of compounds of the gutta percha and India rubber class may be formed by mixing starch, gluten or flour with tannin and resinous or oily substances.' Published in the *Journal of the Franklin Institute* in 1856, this preposterous finding qualifies the Chevalier as the original Raymond Lully of modern times.

Meanwhile, the really eminent scientists of England and Germany were still so busily engaged in trying to take rubber apart (a pastime launched in 1833) that they had not yet got around to thinking about putting it together again. Destructive distillation was the popular method - heating the substance in a closed container and collecting whatever might be driven off. In rubber's case, it didn't seem to be anything of value, however. Not, that is, until Greville Williams in 1860 made the first really thorough-going and systematic attempt to isolate and examine the products present in the stuff from the jungle. Repeatedly, Williams purified the foul-smelling distillate first obtained until he had a reasonably pure low-boiling fraction which he named isoprene. Historically and actually this was of great importance, for isoprene or some of its close relatives are the basis of almost every rubber substitute ever made. Nor did Williams stop at tearing down. From his isoprene he built up - polymerised - something. It was 'a pure white, spongy, elastic mass, having, when successfully prepared, but slight tendency to adhere to the fingers. When pure, it is opaque; but if allowed to become exposed to the air, especially when warm, it becomes transparent, first on the edges and subsequently throughout the whole mass. When burnt it exhales a peculiar odor hitherto considered to be characteristic of caoutchouc itself.' Was he the first man to make an artificial rubber? Half a century later, English and German chemists were to brawl about the point, but Williams, himself, never said that he had done so. That he was aware of some connection between isoprene and rubber is evident from his statement: 'I am anxious to call attention to the fact that the atomic constitution of caoutchouc appears to bear some simple relation to the hydrocarbons resulting from its decomposition by heat. The composition of caoutchouc coincides with that of isoprene ... to a degree which is remarkable...'

Gustave Bouchardat, a Frenchman, was the next to point out the close connection between isoprene and rubber. His experiments climaxed in 1879 when he permitted concentrated hydrochloric acid to act on isoprene for 20 hours in a sealed tube and by distillation obtained from the remaining product a solid residue which he said had 'the elasticity and other properties of rubber itself.' Over this method, too, the English and the Germans later warred, the English announcing confirmations of it and the agnostic Teutons bellicosely maintaining that it was impossible to produce rubber by the method.

Regardless of the merits of these experiments, however, by 1880 it was felt that but one step separated researchers from a satisfactory and commercially practicable synthetic rubber. That was the preparation of isoprene from elementary materials, the only source until then having been crude rubber itself. William Tilden, no tennis player, made it from turpentine in 1882, and perfected his process in 1884. Apparently, he made quite a bit of the material and then put it away and forgot about it. While things were happening in those forgotten bottles, Dr. O. Wallach in 1887 was telling readers of the German scientific journals that 'When isoprene is placed in sealed tubes and is exposed to the action of light for a long time, a product is formed, which, when treated with alcohol forms a tough mass, resembling rubber: this material becomes more or less hard on exposure to air.' It does not seem that he further investigated the product. Five years later Tilden happened to come across those bottles of isoprene out of turpentine and was surprised to find them entirely changed in appearance. A limpid, colorless liquid had become a dense syrup in which were floating large masses of a yellowish solid.

'India rubber!' Tilden whooped. That he really had some kind of a synthetic rubber this time has never been questioned save by a few die-hard Germans. But optimist Tilden became a thoroughgoing pessimist after years of vainly trying to speed up the process. By 1907 he was announcing that synthetic rubber was commercially impossible at any price and advising hearers to stick their money in plantations. That apostasy dissuaded no other alchemist. Even before its announcement they had abandoned turpentine, however. It seems that one of them happened to think that turpentine was expensive, relatively scarce and just as subject to market manipulation as rubber. And that if rubber had to be obtained by tree tapping, it might best come from the tree containing the actual substance.

Helpful, indeed, was the British consul at Philadelphia at this juncture. In an excited report he whooped it up for a secret process of making hard rubber from 'a cheap raw material' which was certainly cheap enough and plentiful enough at that time. The finished product could be turned out for nine cents a pound, the enthusiastic diplomat seriously averred. Had he been able to peer into the future, however, he would have found it becoming a lot scarcer than either rubber or turpentine. The 'cheap raw material' was horse dung. Somnolent Philadelphia of the turn of the century had its spoofers, it would seem.

Not always were the hoaxes so innocent. As alchemical slickers worked through the Bible Belts of Europe in the Middle Ages foisting gold bricks on the suckers, so phony rubber inventors roamed modern America and the public, like a faithful old *Hevea* tree, was given its usual milking. One of the impostors even took into camp that canniest of all Scots - Sir Harry Lauder in person!

Sir Harry was introduced to the rubber discoverer in New York and informed that Ford, the Dodges, Willys and others were on the verge of paying this new day Cagliostro a colossal sum for his invention. The comedian fought off temptation, came back for a hesitant nibble, retreated, debated, resigned himself to being conquered by caution. One evening, however, the synthetic genius invaded his dressing room and produced a chunk of the supposedly artificial material — a brick of 'black gold.' In that moment Harry was lost. 'It looked like rubber,' the woeful knight later mourned, 'tasted like rubber (I broke a false tooth on it) and, most wonderful of all, bounced like rubber.' To which he would sadly add, 'Yes, I jumped right in. Never mind how much; I hate to think about it.'

Sometimes, in the early part of the century, large-sized synthetic scares nearly produced panics in plantation securities. And, regularly, the newspaper 'science' writers had the lower order fit to be tied as a result of smashing exposes of the manner in which staple foodstuffs were to be turned into rich men's auto tires. In England in 1906 it was a tale of making rubber from grain that turned on the muttering. Periodically, thereafter, the yarn about transforming the potato crop into rubber would set patrons of the poor to yowling indignantly and energetically.

All the time, however, an earnest international race was on between scientists spurred by Dr. Karl Harries' notable work on the structure of rubber in 1904. Germany scored first when the Bayer chemists, Dr. Fritz Hofmann and his colleague, Dr. Carl Coutelle, in March 1909, produced isoprene from para-cresol, a coal tar fraction. Coutelle tried all sorts of likely and unlikely chemical and physical agents to hasten the thickening into rubber and in August 1909, obtained rubber by simply heating the isoprene in closed tubes. The discovery received wide acclaim at the time and has been cited by German writers ever since as the first synthetic rubber.

In the following year occurred an example of the almost simultaneous discovery of a scientific fact by more than one observer, the discovery being another method of causing the 'change' in isoprene. In England, Dr. Francis E. Matthews found that sodium metal caused isoprene to polymerise readily and applied for a patent on October 25th. Three days later, Harries, in Germany, communicated the same discovery to the Bayer Company. In the resulting patent suits, the Englishman's claim was upheld in both countries.

Hofmann started the manufacture of isoprene rubber on a small scale in Germany in 1910. The isoprene was difficult to make, however, and the Germans turned to the use of its cousin, methyl isoprene, which is quite easily produced from acetone with the aid of aluminum. A pair of synthetic tires was presented to the Kaiser and manufacture of the methyl rubber began to assume some slight importance in Germany, but it was halted by the crash in crude rubber prices which occurred at this time, the drop in Germany being from thirty marks to four marks a kilo. As narrated elsewhere in this book, the independent discovery of organic accelerators in Germany was one result of this work on methyl rubber.

The Englishman Matthews had been started on the synthetic quest in 1908 by E.H. Strange of the firm of Strange and Graham for which he was head chemist. Matthews first proposed a route to isoprene in which acetone was the raw material, but shortly afterwards devised another and better method, using fusel oil and obtaining amyl alcohol from which isoprene could be derived. William Henry Perkin, son of the great dye chemist, was working on the same subject. He was approached by Strange, who proposed an alliance. The group was soon joined by Sir William Ramsay and half a score of England's other leading chemists and a mass attack started on the problem of obtaining artificial rubber from various alcohols. The aid of Professor A. Fernbach, of the Institut Pasteur in Paris, was enlisted to find a source of amyl alcohol other than the scarce and costly fusel oil. The professor decided that fusel oil was scarce simply because the aim of all previous fermentation and distillation had been to obtain as little of the substance as possible, for reasons which should be obvious to everyone who became acquainted with the liquor of prohibition days. After 18 months work he found out how to make huge quantities of fusel oil by properly fermenting potato starch. From the fusel oil was wrung the desired amyl alcohol and an even greater amount of butyl alcohol from which another isoprene relative, butadiene, was obtained. Rubber was made from both the isoprene and the butadiene by means of Matthews' sodium process.

So enthusiastic were the Englishmen about their prospects that an ambitious undertaking, the Synthetic Products Co., was organised in 1912 to acquire rights for 'new and economical processes for the manufacture of acetone, fusel oil and synthetic rubber.' A long list of patents was taken over from the individuals in the group. Only men of the highest standing in the business and scientific world were associated with the company. The output of butyl alcohol and acetone kept the company going for some time. Serious attempts to produce rubber were never made because the price of crude had dropped to three shillings by 1913 and its manufacture on a large scale, even had it been possible, was not attractive. (The company was liquidated in 1926 with a record of never having paid a dividend.)

The World War, however, gave the German chemists more than ample opportunity to show what they could do. Did the German high command figure the thing would be over so quickly that a rubber shortage never would be felt? Did it assume that there would be no great trouble in getting this essential commodity from neutral countries? Was it taken in by some over-confident professor and convinced that the synthetic product would fill the bill? Or did it just forget about rubber altogether? This last conjecture seems to be the best supported one. Whatever the cause, however, the German failure to assure itself of an adequate rubber supply before the start of the war was a blunder of almost unbelievable proportions when otherwise meticulous preparations are recalled.

The Kaiser's Brain Trusters could have studied every past campaign in history without having rubber brought to mind; that, perhaps, explains their mistake. From the American Civil War on, rubber had, of course, acquired more and more military uses, but the World War was the first armed conflict in which it figured as one of the absolute essentials. This was the war of the gas mask, the war of the airplane, the war of the dirigible, the war of communication systems relying on rubber and, above all, the war of motor transport — transporting of munitions, transporting of supplies, transporting of armies to the front.

And now take a look at Germany. She awoke within the first few weeks of the war and with nowhere to turn. Her agents popped up in the New York markets, and England promptly slammed an embargo on all rubber from its *estrada* plantations.

After desperate pleas from American factory heads, the embargo was lifted when all manufacturers and importers guaranteed that no rubber would be exported from this country. Rubber brokers who violated the promise would get no more rubber. The American industry appointed a committee of control to see that pledges were kept. Smuggling was difficult because it meant violation of federal laws against defrauding the government through swearing to false cargo manifests, and few were willing to take the chance. In Brooklyn one friend of the German cause did. Acting on behalf of the Kaiser's secret agents he bought left and right and at a premium. This eagerness to go beyond the market price was what put the control committee on the trail. They found an empty warehouse. From it, fifty tons of rubber had gone out of the country packed in barrels and disguised by means of resin which had been melted and poured over it. The government reached out to seize the supposed shipment of resin in a neutral foreign port and half a dozen German agents went to jail. A short time later, government agents became suspicious of exporters who suddenly began sending shipments of cotton waste to foreign ports, although they had not previously dealt in this material. The bales were X-rayed and found to contain rubber. A number of shipments labeled 'stationery' turned out to be blocks of caoutchouc 'ink eraser.' More ingenious was the sending of rubber sausages in jars of pickles.

Enormous shipments of rubber to Baltimore early in 1916 aroused the suspicions of the control committee, but nothing could be done except to prevent the material being smuggled out of the country. The Germans had no such intentions, however. They got the rubber and left a startled world gasping at the method they employed. Poking a gray nose out of the waters off Norfolk on July 16, the Deutschland became the first submarine to cross the Atlantic. An admiring America watched as the German craft crawled up the Chesapeake to Baltimore, discharged a cargo of dyes and drugs worth more than a million dollars and took aboard nickel and 500 tons of rubber. The Deutchland came as an unarmed merchant ship and the wrathful British could not prevent the departure of the rubber, although the broker who was involved was barred from future dealings. The English could only hope to intercept the craft on its return journey. This they were unable to do and the courageous Germans pushed their way safely through the depths, arriving in their home port after a 22 day trip. So desperate was the German need that the daring trip was repeated in November. This time the Deutschland appeared at New London, Connecticut, stowed away another 500 tons of rubber and slipped into the depths to reappear in home waters nearly a month later.

These occasional windfalls did little to relieve the rubber famine in Germany, however. Every scrap of old rubber was reclaimed. Crews of foreign ships stopping at American ports purchased raincoats, overshoes, water bottles and all sorts of small rubber articles that could be smuggled out of the country under their clothing. Carried abroad, these brought fabulous prices when sold in Germany. As soon as it was realised that the conflict was going to last much longer than anticipated, everyone in Germany was required to register all automobile and bicycle tires. These were inspected and, if it appeared that the army could use them, were confiscated. This supply did not begin to meet army requirements, however, and recourse was had to all sorts of makeshifts. Wooden tires were used, canvas — and leather — covered wood were tried, and a score of spring tires, consisting of varied combinations of steel springs inside steel bands, were devised. Pleasure cars, of course, nearly disappeared from German streets. Attempts were made to use worthless tires by refacing them with steel-rivet studded leather and filling them with a jelly-like material in place of inner tubes, but these could only be used for a short time before the jelly melted under the heat that was generated and squirted out of the tires. No wonder that real rubber was priceless and synthetic brought $30 a pound!

In the earliest days of the war, no efforts were made to manufacture the synthetic rubber, since pre-war plant trials had convinced rubber men of its inferiority. It was not long, however, until Germany became desperate and then the shout went up for the chemists to do something. Besides the need for tires, there was an urgent need for hard rubber cases for the batteries that drove the V-boats and were vital to other forms of transportation and communication. All kinds of substitutes for hard rubber were tried, but none were satisfactory. A small amount of synthetic rubber had remained in the apparatus of the partly dismantled methyl isoprene plant of 1910, and the Hagener Accumulatorenfabrik decided to use it. The material proved to be of sufficient worth to warrant manufacture on a large scale.

This brought the Germans up against another problem. The raw materials — acetone and aluminum — were lacking. The former was needed for the manufacture of explosives and the latter for zeppelins and aircraft motors. Acetone can be obtained from wood, but these stocks were soon depleted and so was the small pre-war stock of imported acetate of lime. Acetone can also be made from acetic acid obtained from fermenting grain or potatoes, but with foodstuffs so badly needed, these sources were very nearly out of the question. Only those potatoes so putrid that starving people could not eat them were available to the chemists. They found, too, that the process was not successful on any sort of a commercial scale because the potatoes had to be fermented by a much too sensitive bacillus. Resourcefully, then, the Germans worked out a process using lime and coal, both of which were plentiful. The method illustrates the roundabout way in which chemists work. The lime and coal were converted in an electric furnace into calcium carbide. Every motorist of an earlier day knows that it is only necessary to add water to this substance to get acetylene. The gas and more water, heated with a mercury salt, produce acetaldehyde, the stuff that makes hard cider drinkers mean. This was oxidised to acetic acid, familiar to all of us in vinegar. Heated with lime, the acid yielded calcium acetate and then it was only necessary to dry distil this substance to get the all-important acetone. Treated with aluminum and caustic soda, acetone gave an aluminum salt of pinacol and this, distilled under pressure, gave methyl isoprene.

An enormous number of experiments were conducted on ways of changing the methyl isoprene into rubber. Any sort of polymerisation required several months, and, as the demand was urgent, the chemists could not wait for the results of one set of experiments before they started others. Two methods were chosen, producing two kinds of rubber. By allowing the liquid to stand in tin drums for six to ten weeks at 30 °C, H-rubber (H for hart, hard) was made. This was found to serve for hard rubber goods, for the H-rubber had even more electrical resistance than hard rubber made from crude. By the spring of 1918, 30 tons a month were going into battery boxes and more was being used for battery separators and parts for magnetos and wireless sets.

By storing the methyl isoprene in iron drums for three to six months at 70 °C, the so-called W-rubber (W for weich, soft) was produced. This was used in soft rubber goods, but was far from being satisfactory. It was hard to vulcanise and the resulting product had elasticity only when warmed. Solid tires were made of it, but they were hard and inelastic in ordinary outside temperatures and in winter chunks would drop off them when a car was started after being left out in the cold. Where heated garages were not available, machines equipped with them had to be jacked up when left standing. At least, however, they were better than those made of wood and steel and by early 1918, ten tons of the W-rubber were being used for solid tires each month.

A third form, B-rubber, was made by allowing methyl isoprene to stand in contact with sodium wire in a carbon dioxide atmosphere. This was found to serve for insulating wires.

Near the end of the war the Germans were making rubber at the rate of 150 tons a month. In all, nearly 2,500 tons were manufactured and when the war ended, two plants were under construction which would have boosted production to just short of 8,000 tons a year.

The miserable inadequacy of the stuff was well demonstrated by the speed with which it was dropped once the armistice was signed. The German experience became, in fact, a horrible example to which experts pointed in dismissing plantation substitutes from serious consideration. The ancient artificial rubber bogie was laid at last and planters no longer trembled at the mention of the word 'synthetic.'

It was half-heartedly and with no real expectation of success, then, that a number of American and German industrial concerns again interested themselves in the quest when the British restriction squeeze of the 1920s boomed prices and chased Harvey Firestone into Liberia, Henry Ford into Brazil and Thomas A. Edison and the United States Department of Agriculture into high-geared experimenting with the hybridising of common plants. With falling prices in 1926 all but two corporations again abandoned the idea. They were the I.G. Farbenindustrie, German dye trust, and the E.I. du Pont de Nemours & Co., of the United States. Independent chemists stayed on the job, however, as did the Russian government.

The Soviet assault on the problem took a characteristic direction: a prize competition for the best synthetic rubber launched by the Supreme Council of the National Economy. For butadiene rubber from grain alcohol, the prize and the Order of Lenin were bestowed on C.V. Lebediev and an awe-inspiring ballyhoo went up. A few years later Russia discovered that it had its own rubber-bearing plants as previously recounted. To their finder went the Order of Lenin, to his discovery was diverted the ballyhoo and Comrade Lebediev became just another Forgotten Man. In June 1931, it was modestly announced that the manufacture of an artificial rubber from oil had reached the total of half a ton, but the cultivation of real rubber continued to be the big thing.

In Germany, meanwhile, I.G. Farbenindustrie activities resulted in a flood of patents. Methyl isoprene, chosen during the war only because the large scale manufacture of isoprene at that time involved great difficulties, was counted out from the start and it was with isoprene that the chemists labored. Eventually they were claiming that with their new processes rubber of a sort could be made in a few days as compared with the months formerly required. There was, however, no noticeable improvement in quality. They were still on the old track, still trying to duplicate rubber's composition.

That was the error into which the first seekers after synthetic rubber had fallen, the burdensome misconception which every subsequent researcher had accepted as a matter of course. These were learned men, so completely so that in that lay their failure. Confronted by rubber and informed that, despite decades of the most intensive analysis, its exact composition was unknown, a simple man might well have reflected that the things demanded of rubber are exceedingly well known. Would it not be easier, then, to work out something which would serve all those known purposes instead of fumbling in the dark to duplicate an unknown composition?

And while it eluded them, while the synthetic rubber dream was being finally relinquished by hard-headed manufacturers, while the whole business was being rapidly consigned to the realm of the philosophers' stone and perpetual motion, out of the du Pont laboratories came the long-sought discovery. It was 40% chlorine, which is not an ingredient of real rubber, at all. As good as rubber? For some uses better. So much better that single rubber factories have been consuming as much as a ton a month or more although it has cost them from nine to five times as much as the natural substance.

What had set the du Pont rubber experts on the right road? One man, Reverend Julius Arthur Nieuwland of the University of Notre Dame, the acetylene enthusiast of this chapter's opening paragraphs. Had Father Nieuwland at the beginning of his career dedicated himself to finding an artificial rubber, it is quite likely that he would have fallen into the same trap as all the others. Because rubber meant nothing in his life, because he had the problem half solved before ever he knew that he was engaged in solving it, he carried no handicap of false ideas and success was assured the moment du Pont discovered him.

Born in Hansbeke, Belgium, was Julius Arthur Nieuwland. In 1880, when he had reached the age of two, his family came to South Bend, Indiana, close to the gates of Notre Dame. As the boy grew older he naturally enrolled at Notre Dame. He did not play football nor did he learn to know acetylene from *Hevea* milk. He studied Latin and Greek, botanised indefatigably, drew covers for the school programs and twanged a mean guitar. Also he prepared to turn his collar about and don the black of the Catholic clergy.

Nor did he have chemistry in mind when he began his graduate work at Catholic University of America in Washington, DC. He had intended to major in botany under Edward Green, but, when Green left the school to take up special work at the Smithsonian Institution, he switched to chemistry. His professor, Dr. Griffin, was interested in two things — the condensation products which can be made from acetone and the study of acetylene. With young Nieuwland's introduction to this last, botany immediately and permanently receded to the proportions of a pleasant hobby which had the advantage of taking him out into the fields after long sieges of close confinement in the laboratory.

As a Notre Dame senior he had taken down the English essay prize with a piece on John Keats which was issued as a university publication, a considerable tribute considering the financial status of the South Bend school in those pre-Four Horsemen days of 1899. As a priest, ordained in 1903, and a candidate for his doctorate at Catholic University, he wrote 'Some Reactions of Acetylene.' It was in this 1904 thesis that he called attention to the poisonous product resulting from acetylene reacting with arsenic trichloride in the presence of aluminum chloride. Until 1918 it shared the oblivion that overtakes nearly all such dissertations.

In that year, it was chanced upon by W. Lee Lewis, now making the world safe for democracy as research director of the Institute of American Meat Packers, but then a captain in the United States Army stationed at Catholic University and engaged in a hunt for the most poisonous gases possible. The reaction which Nieuwland had scorned to investigate further, Lewis pursued until he had perfected the process of making the poisonous constituent. Lewisite — the stuff beside which mustard gas becomes a sissy's scent!

Meanwhile, Father Nieuwland was at Notre Dame busily teaching botany and making it finance the acetylene experiments which constituted his one overmastering interest outside the church. Botanical slides he turned out by the hundred and, sold throughout the country, they built up a chemical library from beginnings hardly discernible on the day the young priest had come home from Washington.

It was in 1906, two years after that homecoming, that the experimenter passed acetylene into a solution of copper and alkali chlorides. Father Nieuwland now was in the rubber business and didn't know it. All he did know about that experiment was that a peculiar and extremely noticeable odor had been evoked and yet no solid or liquid had been isolated. It had to be a gas, then.

The years passed with the botany professor returning again and again to this experiment. In 1918 he was made professor of organic chemistry at Notre Dame. The new title evidently did not impress the mysterious gas for it continued to defy all efforts to isolate it. Fourteen years in all this went on — the same period of time as had elapsed between Nieuwland's discovery of Lewisite and Lewis' discovery of Nieuwland's discovery. And then one day in 1920 the unwitting experimenter in rubber hit upon a means of increasing the concentration of the catalyst mixture and the rate and extent of acetylene absorption. Substituting ammonium chloride for the alkali metal chlorides effected the first; altering the acidity of the mixture did the rest.

With success apparently at hand, the good Father arranged appropriate apparatus to trap the gas. To his astonishment, he obtained a yellowish oil in addition to the gas. A year later this unexpected visitor was identified as divinyl acetylene — a compound having a degree of unsaturation even greater than isoprene or butadiene. Father Nieuwland now had synthetic rubber in his grasp — still without knowing it. He did know that he had an extremely dangerous substance on hand. Left to itself, it had a habit of thickening into a jelly and then into a hard resin which could not be handled without exploding. The Notre Dame acetylene team continued to experiment with the oil, nevertheless. On a day in 1923 the Father and his boys treated it with sulphur dichloride and produced an elastic substance resembling rubber. It was too plastic for practical use, but now Nieuwland knew, nevertheless, just what he had. Characteristically, he failed to grow excited.

Two years later, however, he did condescend to make casual mention of it while delivering a lecture before an assembly of organic chemists at Rochester, New York. At least one member of that audience came to startled attention — Dr. Elmer K. Bolton of the du Pont laboratories, which had been actively searching for synthetic rubber. They had experimented with acetylene, only to meet with the same disappointments that all other likely sources had brought. Patents and assignments were arranged; E.I. du Pont de Nemours & Co., the corporation with the $600,000,000 enterprises, formally took over commercial development of the Nieuwland work; and a squadron of 28 chemists headed by Wallace H. Carothers and advised by the South Bend priest dug in to grind out 'black gold.'

First specimens of divinyl acetylene rubber products were a sad disappointment all around. The material varied every time it was made, no two specimens retained their elasticity for comparable lengths of time and no means could be found to correct the fault. Foiled by the yellow oil, the chemists turned to the odorous gas which always accompanied it into the world. It proved to be a sort of little brother of divinyl acetylene traveling under the ponderous name of monovinyl acetylene. Treated with hydrogen chloride at Nieuwland's suggestion, the gas turned into a thin, clear liquid promptly christened 'chloroprene.' The name chloroprene is reminiscent of the name isoprene and for good reason. Look at a chemist's picture of the two materials and you will have to give it more than a hasty glance to notice any difference. Chloroprene, you will finally note, has a chlorine atom dangling from one of its carbons; isoprene has none. Though the difference appears slight on paper, actually it is enormous, for chloroprene will polymerise into a rubber-like substance 700 times as fast as will isoprene.

The resulting product is very similar to fully vulcanised rubber. If the polymerisation is stopped at an intermediate stage, a material similar to soft, uncured rubber is obtained. This is 'DuPrene' (trade name), the dark brown substance with a peculiar and penetrating odor which is being sold to rubber manufacturers today. Various ingredients are mixed into the chloroprene on mixing mills in the same way that rubber is compounded. The mixes are cured simply by heating for chloroprene has the advantage over rubber in not requiring the addition of sulphur to vulcanise it.

Modestly, indeed, did duprene burst upon the world. At a 1931 Akron, 0., dinner of the rubber division, American Chemical Society, three du Ponteers, Carothers, F. B. Downing and Ira Williams, read technical papers, answered questions, exhibited samples and announced that a plant was being built at Deepwater, NJ, for its manufacture on a commercial scale. Reporters took it more than calmly. And when they communicated with sub-editors their own nonchalance was exceeded. To the *Cleveland Plain Dealer* it was six paragraphs for the Akron edition next morning. To one Akron newspaper it was an inside story - 'Claim Chemists Develop Method To Make Rubber' heading five paragraphs on the local session, five A. P. paragraphs from South Bend on Father Nieuwland. The Scripps-Howard opposition started the story on one — with the flat assertion that 'Synthetic rubber is a reality' — but gave it less than a column and dismissed the good priest with a single line.

There was good reason for the generally blasé attitude. To news-hawks of the rubber belt the synthetic rubber yarn was a chestnut that had long before lost its flavor. Far too many times they had written up the pipe-dreams of the modern alchemists and, du Pont name or no du Pont name, the chloroprene announcement impressed them as the same old hokum. Add to this the fact that the real old *Hevea* milk was selling at five cents a pound and you can understand why the rubber industry was not much more interested than the Fourth Estate.

Nevertheless, the du Ponts quietly went ahead with their Deepwater plant, farmed out a mess of Duprene to Dayton Rubber Manufacturing Co., Dayton, OH. At Dayton tires were built from the synthetic, tested by skidding, by sharp acceleration on hill climbs, by hot weather and by sharp turns at high speeds. In June, 1934, announcement of their completely satisfactory performance went out. Next year Father Nieuwland was called to New York to receive the William H. Nichols medal of the New York section of the American Chemical Society. Now the ballyhoo really was turned on for the first time and a good part of the nation discovered that something actually had happened.

Only about the methods of commercial manufacture is there still any secrecy. Ostensibly it is a relatively simple process. The du Pont company buys calcium carbide from which it makes acetylene. The gas is polymerised into monovinyl acetylene and treated with hydrochloric acid. The resulting product changes by itself into synthetic rubber. Actually, the manufacture of artificial rubber is not so simple. There are a score of tricks in each of these steps which make the operation successful and these are not being publicised. Although they retain a monopoly of manufacture in this country, the du Ponts are not trying to keep the product for the exclusive use of du Pont-controlled rubber factories. Duprene has been on sale to all corners for several years and the rubber barons actually bought substantial quantities of it despite a price first of $1.05 and then of $1 a pound. At this rate production naturally exceeded demand, and, effective April 1st 1936, the price was lowered to 75 cents. At the same time the du Ponts indicated that they are aiming at a 50 cent price for the not distant future.

There are numerous reasons for the not insignificant purchases at $1.05, $1 and 75 cents while genuine para has been ranging from 12 to 16 cents. Compared to real rubber, Nieuwland's creation is much more resistant to oil, gasoline, kerosene, acids and alkalis; it is more resistant to ageing caused by air and light; it is more resistant to penetration by water and it is less permeable to gases such as hydrogen and helium. It can be used to better advantage for most purposes that natural rubber serves and for many uses where natural rubber is not satisfactory. Emulsified into synthetic latex, it can be used for every purpose to which *Hevea* milk is put and for some purposes it is superior because of its greater ability to penetrate materials such as leather and wood.

Many a citizen still believes, however, that synthetic rubber is a laboratory nurseling that plays no part in his daily life. He does not know that many of the 'rubber' parts of his automobile never came from any tree and that the tubing with which he is served his daily quota of gasoline came out of the coal mine and the limestone quarry.

The chloroprene product is finding many other uses in industry. Gaskets, washers, packing, transmission and conveyor belts which operate under oily or greasy conditions are better made of Duprene than rubber. Printing rollers, blankets for newspaper and lithographic offset printing, and printing plates for use with ink and paint are being made of this laboratory creation. Pump pistons and cups in oil slush pumps; grease gun and paint spray hose; hospital sheeting which must contend with severe disinfecting chemicals; balloon fabric and diaphragms for gas meters; and gloves and aprons for use in chemical plants are other applications in which this synthetic rubber is finding an increasing usage.

Auto tires of Duprene offer no outstanding advantage, however, and its cost therefore bars it from current consideration there. It has an important bearing upon the tire industry, nonetheless. At today's selling figure it stands as a permanent barrier against real rubber ever again passing the 75 cent mark. It may mean that rubber never again will hit 35 cents for any prolonged period.

Right now the accepted commercial source of acetylene is coal and limestone. It is a comparatively expensive system and the product is in demand for so many uses that the synthetic rubber industry can be spared only a modest share. Methods for obtaining a cheaper supply from petroleum are being steadily developed, however, and a more reasonable price is in sight. Du Pont itself is doing a great deal of the acetylene from petroleum work and obtaining as a by-product of the process the most valuable of all the rubber re-enforcing pigments — carbon black. No wholesale competition between Duprene and rubber did Father Nieuwland see for the immediate future should the natural product stick around 16 or 20 or 25 cents a pound. But it was his belief, as expressed to the writers shortly before his death, that the acetylene from petroleum will come to mean synthetic competition with raw rubber in every line should the price of the crude mount to 35 cents or so.

Nor did he, by any means, see Duprene as the last word in synthetic rubber. In the investigation of acetylene itself, he felt the chemists were far from having covered the ground in an adequate manner. And he predicted that we will have not only cheaper and better but innumerable substitute rubbers, each, perhaps, with some special field of use which it can occupy more capably than natural rubber.

If Nieuwland's discovery means the eventual passing of the rubber plantations; the ending of the wasteful, unscientific and relatively primitive business of relying upon the plant life of a limited portion of the world for a product which has become a modern essential; the manufacture wherever needed and in any quantities of an artificial rubber better, cheaper and more uniform than nature's product — then he achieved a colossal feat, indeed.

As to the Duprene rivals, Nieuwland looked for a flood of them within the next 10 years. Some will undoubtedly be fellow alumni of the du Pont laboratories. Reports are that a new synthetic rubber already has been made by du Pont chemists from monovinyl acetylene without treating it with hydrogen chloride. The compound is being separated from the acetylene by a process employing temperatures of 100° C below zero, the practicability of using these temperatures having been proved after research extending over several years. Being a non-chlorine derivative, the reported product would be closer to the natural rubber hydrocarbon in chemical constitution and molecular structure than is Duprene. In England, Imperial Chemical Industries, Ltd., is manufacturing Duprene under a du Pont license, but Russia and Germany are claiming new ones respectively christened 'Dovprene' and 'Buna.'

The Soviet now definitely asserts that the manufacture of artificial rubber, rather than the production of rubber obtainable from domestic plants, is to be the important feature of its 'emancipation from imported rubber.' Relegating considerations of cost to a secondary position, it has four factories manufacturing synthetic rubbers on a commercial scale. To assertions that cost of the primary materials entering into the synthesis far exceeds cost of the natural rubber obtained by the cheap labor of the Middle East's coolies, it retorts that this refers only to capitalistic economy and that under socialistic conditions these problems are solved in a different manner. Boosting tire production by a fourth while reducing its rubber imports from 28,607 tons in the first eight months of 1934 to 21,700 in the first eight months of 1935 is the accomplishment it proudly points to. 'Foreign currency which we economise in this connection amounts to about ten million gold roubles for the year 1935 alone,' said the Director of the Chief Administration of the Rubber Industry, People's Commissariat for Heavy Industry of the USSR, in the September 2nd, 1935 issue of the *Moscow Izvestiya*.

Four types of the artificial product are involved: two kinds of butadiene, 'Sovprene' and the so called 'Resinit.' The butadiene variety first in the field accounts for most of the production but construction of a new Sovprene factory near the Turkish border is under way.

One of the butadienes is an ethyl alcohol rubber stemming back to potatoes and grain and the other, still in an experimental stage, has a petroleum base. The dominating alcohol product represents, of course, the Russian forwarding of the German wartime experiments. Originally the Soviet butadiene product cost 25 to 35 times as much as natural rubber, but with the introduction of potato alcohol this figure has sharply dropped. The potato alcohol costs only a fourth as much as the grain.

The 'Resinit,' it has developed, is actually 'thiokol,' an American discovery. Without the tensile strength of rubber, it is a rubber-like compounding ingredient useful when oil resistance is desired.

American chemists have been unable to obtain samples of 'Sovprene.' Nieuwland believed it to be Duprene shamelessly appropriated and re-christened by a nation paying no attention to patent rights and he hazarded a guess that it probably was below standard because of Russian lack of success in operating the process. He was, of course, completely and admittedly prejudiced against the Soviet. He supported his rubber suspicions, however, with an account of a visit from two Soviet representatives who invaded South Bend in 1935. They sought to pump him on Duprene, he alleged, and they besought him to visit Russia and lecture on chloroprene. The Nieuwland word is not to be doubted. Whether he honestly mistook the motives of the unwelcome guests, whether they actually were officially connected with the Russian government: these are matters to be fought out between New Masses and the Hearst press.

The German 'Buna' product, like 'Sovprene,' has not been available for inspection in the United States. In estimating the accomplishment, it should be remembered that Der Führer, like the Russians, has the habit of annually announcing to the outer world the perfection of synthetic. Back in October 1934, for instance, Adolf's Minister of Economy, Dr. Hjalmar Schacht, made a speech announcing 'the problem of synthetic rubber production in Germany has been recently solved' and promising production on a commercial scale. Cost, he said, was 'considerably below that attained in the United States or other countries producing synthetic rubber and was only 60-80% above the price of natural rubber.' In September, 1935, Hitler announced at Nuremburg that Naziland had 'definitely solved' the problem and that commercial production was about to begin. This attracted very little attention, although The *Times of London* did pause to point out that there was no reference to the price at which the synthetic article could be sold. In February 1936, however, the Chancellor used the same speech in opening the German automobile show at Berlin and, news play vagaries being what they are, made the front pages the world over.

According to the chief of the Nazi men, the synthetic tires, tried out by army and postal service for over a year, have shown themselves 10-30% superior to para ones in point of utility and durability. This is applesauce in any language. Of real interest, however, is the fact that Germany's first post-war factory for producing synthetic rubber did go into operation this year. Located at Piesteritz near Wittenberg, it is an I. G. Farbenindustrie plant, of course, and the trust's officials have announced the placing of tire orders by the army, the postal service and the state railways which operate extensive bus services. As to costs, Dr. Shacht's 1934 reference is the only one. In this connection it must be remembered, however, that because of the shortage of foreign currency in Germany the Reich can afford to pay a higher price in marks. And that all of the Nazi work with synthetic rubber has been a part of the comprehensive program of reducing German dependence upon foreign raw materials in preparation for the next war. Reduction in the imports of rubber, textile fibers, non-ferrous metals, mineral oils, timber, hides, iron ore and wood pulp originally was specifically called for and it was made clear that where substitutes had to be developed their relatively high costs should not be taken into consideration. Synthetic gasoline as well as synthetic rubber has since been announced.

Meager reports from Germany on the method of manufacturing

'Buna' state that the process involves production of butadiene from acetylene. From this, by polymerisation methods which have not been revealed, comes the synthetic rubber.

The matter of war-time rubber supplies is, of course, a subject as important to the United States as to Germany, Russia or any other nation.

And it can be flatly stated, that, whatever the ultimate success or lack of success of chloroprene in industry, its creation has forever insured this nation against being cut off from rubber in event of war. Guayule rubber could not, of course, be multiplied swiftly enough to answer the problem. Chloroprene — so long as we have salt for hydrochloric acid; coal, limestone and petroleum available for acetylene — could be and would be. Nieuwland's Lewisite, untried in combat, may be open to doubt. But waive the expense question and Nieuwland's Duprene in quantities sufficient for the national defense is a sure thing.

So much for the man's work. As to the man, visitors could be pardoned for failing to realise that the football stadium was only the second biggest thing at Notre Dame. Modest, self-effacing, he had to be tracked down in science building or chemical laboratory. He never addressed the student body, he hid out in inconspicuous corners at the university functions and he was known to few undergraduates outside of the science and engineering departments. He frequently ate in the laboratory; more often slept there, stretched out upon a table with a rolled-up coat for a pillow.

His non-participation in Duprene royalties is explained by his vow of poverty taken as a Catholic priest in the Congregation of the Holy Cross. The University of Notre Dame was founded by this order nearly a century ago and is still its headquarters. And to the Congregation will accrue all of the inventor's share in the profits of synthetic rubber.

A little man, Father Nieuwland's appearance was rather that of a knotty kobold (German sprite) despite the half-moon spectacles over and through which his keen eyes glinted. Proudly he exhibited to fellow chemists the Duprene heels which wore out a pair of shoes in tramping over Europe — Russia excepted — in 1934. For that trip he had the Duprene fountain pen set made. It was to have been a present for the Pope. Father Nieuwland forgot it, however.

It is with reason that we note down such minutiae. They need to be preserved, for when this man created 'chloroprene' — something never before known in heaven or on earth — it well may be that he became the greatest figure in rubber history, one of the greatest figures in the industrial era of invention. Goodyear's discovery of vulcanisation, Oenslager's discovery of the virtues of the organic accelerators, we have ranked as the greatest rubber research achievements on the strength of accomplishments to date. A hundred years from now, it may be that historians will be ranking Nieuwland above Goodyear.

Charles Goodyear discovered how to make certain kinds of rubber commercially useful. George Oenslager discovered how to make nearly all rubbers commercially useful and how to improve the quality of nearly all rubber articles. Julius Nieuwland may have discovered something that will replace rubber altogether — and to the benefit of all mankind.

To the major press associations last June, however, he was just a hazily identifiable news figure of some sort who had the misfortune to drop dead at Washington while so epochal an event as a G.O.P. national convention was being unreeled at Cleveland. So they filed a paragraph apiece on him, one identifying him merely as 'head of the chemistry department at Notre Dame university,' the other as 'world-famed scientist and inventor who developed a process for making synthetic rubber.' Just so the newspapers of Akron, 0., rubber capital of the world, carried the dispatches and without elaboration. Unintentionally these were the appropriate farewells for the most modest and retiring of scientists, one of whose few woes was the year or so of banqueting that accompanied and followed his medal-winning. And even more appropriate than the obituaries was the manner of his taking. His heart stopped while he was visiting the Catholic University chemical laboratory where first he had encountered acetylene.

5 Big Business

5.1 Small Fry and Octopus

THE HISTORY of rubber-manufacturing in the United States is, to a great extent, the history of rubber manufacturing in the world. From the early 1850s down to the World War, this country imported 50% of all crude rubber produced. Then there was another jump and for 12 years, 1915-1926, the American consumption was 70%. Thereafter increasing production of manufactured rubber goods in other countries (by domestic and American companies) was such that with United States production still gaining, its percentage of total imports of raw caoutchouc yet began steadily to fall away. Not until 1932, however, did the percentage drop below 50.

The story of pre-vulcanization manufacture in England and America we have told. These, to all practical purposes, constituted a different industry and one that, in the United States, was fast breathing its last when the momentous find was made. The real chronicle of American rubber goods manufacturing begins, then, with the 1844-1845 industrial applications of the 1839 invention under licenses granted by a Goodyear - so busy with new discoveries that he had no time to cash in properly on the old. His great grief of this time, in fact, was that the would-be manufacturers were interested only in making "suspenders, shoes and elastics, with some descriptions of clothing." These were the articles that had been made of the unvulcanized gum in the period of America's first abortive rubber boom, the public was acquainted with their utility and to the businessmen there seemed no margin in going broke in the Goodyear style by messing around with a thousand and one rubber gadgets when there was real money to be made by grinding out gum shoes. "This was prudent as related to their pecuniary interests," Charles sadly observed, "but was a serious hindrance to the development of the whole subject."

Among the first of his licensees were: Union India Rubber Co. of Naugatuck, CT, and Harlem, NY, which had the wearing apparel franchise; the rubber shoe concerns of Leverett Candee of New Haven, CT, and William DeForest of Naugatuck, the accommodating brother-in-law who had advanced that $46,000; and B.F. Lee, a New York merchant who grabbed the United States rights to rubber-doll manufacture. By 1851 Goodyear had some 20 establishments licensed and was fairly well satisfied with their product, although he still yearned for them to start knocking out rubber sails, cannon-ball hole stoppers, floating trunks and air-cooled boxing gloves. It was in this year, too, that the Goodyears licensed the first of their hard-rubber clients, India-Rubber Comb Co., and American Hard-Rubber Co.

From the start, however, boots and shoes dominated the industry and, odd as that now may seem, their domination lasted well into the present century. It was the Civil War that gave the caoutchouc industry its first tremendous impetus in all lines and by 1865, year of the Goodyear patent's expiration, we find United States imports of the crude soaring to 3,000 tons, by 1871 to 5,200 tons with a declared value of $4,228,926. And in that year rubber boots and shoes were manufactured in this country to the amount of $8,000,000. Runner-up to footwear was the mechanical rubber goods trade, which seems to have had its roots in the supplying of a number of vulcanized caoutchouc pontoons to the Government for 1847-1848 use in the war with Mexico (unvulcanized ones had been made for the Government by the Roxbury company in 1836). It was, however, the post-war acceleration of industrial development that finally established this branch of the industry. In 1865 one factory alone was selling a million dollars' worth of engine packing annually. Next to mechanical goods came waterproof garments. In the South-west and in Mexico, rubber ponchos had been in demand from the beginning, but it was the Government contracts for Civil War rain-shedders that really started the rubber garment business on the way to success. Fourth of the principal divisions in this diversified system of manufactures now beginning to approach the setup Goodyear had dreamed of was the druggists' sundries business. Finally yielding to the impassioned entreaties of Charles, the Union Rubber Co., had pioneers in this field by turning out syringes, hot water bottles, bandages, air-pillows and air-cushions at Harlem in the early fifties. And a decade later the atomizer became a hot number. In the realm of hard rubber, fifth of the important branches of the industry, the two licensees of the Goodyears enjoyed a monopoly until 1871. A far more important caoutchouc event of this year, however, was the first manufacture of rubber articles at Akron, Ohio.

Geography made Para the old raw rubber capital of the world and Singapore the new. Akron, rubber manufacturing capital of the world, was made by a booster pamphlet. Issued by a volunteer board of trade towards the end of the 1860s, it announced that this town of 10,000 was on the make for industries, was ideally situated for the carrying on of such affairs and would be lavishly hospitable to any corporations prepared to do business. A copy of that pamphlet reached Melrose, NY, and Dr. Benjamin Franklin Goodrich, a losing bucker of the caoutchouc manufacturing game. The doctor pulled up stakes in a drastic move to escape from competition, and established the first rubber factory West of the Alleghenies.

Born in Ripley, NY, two years after Goodyear's vulcanization discovery, Goodrich had known the North-eastern Ohio country as a student at the medical department of Western Reserve College, from which he graduated in 1861. After service in the Civil War he established himself as a practicing physician at Jamestown, New York. There he met J.P. Morris, attorney-at-law. A year later both professional men gave up Jamestown as a bad job and descended upon New York City to make their fortunes in the real estate business. In one of their deals they traded a chunk of ground for a mess of stock in the Hudson River Rubber Co., founded at Hastings-on-the-Hudson some years before as a licensee of Goodyear's. Becoming interested in rubber, the partners bought out the stockholders with $5,000 of Morris' money and Goodrich was installed as president of a reorganized corporation. He lasted two years and, undiscouraged, persuaded Morris to buy him another factory for $10,000. This was the Melrose plant. The year 1869 found the partners losing money fast and convinced that they would have to go a long way off to lose the terrible New England competition. At this psychological moment amorous Akron made its advances.

With the Ohio Canal from Lake Erie to the Ohio River under way in 1825, General Simon Perkins of Warren had dreamed a city to stand on the highest point along this man-made waterway. Surveyors drew him a plan, and his friend, Charles Olcott of Medina, a student of the classics, supplied the graceful Greek name referring to the town site's altitude. A Warren crony, grumpy old Judge Pease, immediately got off a good one. "Acheron, river in Hell, hey?" he wheezed. Slab shanties inhabited by Irish laborers on the canal rose on the maligned location and shortly wily Americans were drifting in from other parts of this territory that had been the Connecticut Western Reserve. The first mild boom came in 1840 with arrival of boats from New Castle by the new East-and-West Pennsylvania & Ohio canal and it was as a crossroads of this mule-drawn traffic that Akron, population 1,664, began clamoring for a place in the sun. A local bard, Milo Fuller, was its prophet and its mouthpiece, and his "Hail lovely city!" epic saluting the community as "The great and mighty Lowell of the West," yet survives as a gem of purest ray serene. The "Lowell" reference was inspired by the five woollen mills which Horace Greeley approvingly noted when he dropped in 1843. The bride of the canals was to be variously known as the sewer-pipe city, the match city, the oatmeal city and the mower-and-reaper city before becoming The Rubber City. At the time of the negotiations with Goodrich the town's chief industries included the German oatmeal and Empire barley mills of Ferdinand Schumacherj, the Buckeye reaper and mower plant of George W. Crouse and Lewis Miller, who was to found Chautauqua and become the father-in-law of Thomas A. Edison; and the Empire reaper and mower works of J.F. Seiberling, owner of Akron's first bathtub. The sewer pipe and clay product concerns were scattered through Akron and Middlebury, the ancient rival it was to annex in 1872. Also in Middlebury was the match works of Ohio Columbus Barber. Schumacher, incidentally, became the first president of American Cereal Co., vast combine now known as Quaker Oats Co.; Barber the promoter and long-time president of the great Diamond Match trust.

Head of Akron's hopeful board of trade was Colonel George T. Perkins, banker, grandson of the city's founder, son of Ossawatomie John Brown's onetime partner in Akron sheep raising. Goodrich investigated Akron, Perkins was sent East to investigate Goodrich, and the result was that the Colonel persuaded 18 other civic boosters to join him in investing $1,000 apiece in the enterprise. With Goodrich came his brother-in-law, Harvey W. Tew of Jamestown, $5,000 supplied by the still optimistic Morris, and the Melrose plant's machinery. This was in 1870. A small building was erected near the canal basin in South Akron and by 1871 the 25 workmen employed by the partnership of Goodrich, Tew & Co., were grinding out fire hose and wringer rolls. Belting, packing, and billiard cushions were added to the line and the concern limped shakily along until 1874, when it became apparent that, without fresh funds, it was headed in the same direction as the doctor's previous companies. George W. Crouse, reaper baron, banker, and Ohio political power, now came to the rescue. Tew sold out his quarter interest for $7,500 and, under the name of Goodrich & Co., the reorganized hose plant staggered on until 1879, when Crouse had to be appealed to again. Inadequate working capital, inadequate raw-material supply and inadequate working force-these were the troubles. Crouse and Perkins came through once more and the other original backers were given their choice of stock or payment as the B.F. Goodrich Co.,was incorporated with an authorized capital of $100,000. The gentlemen took the cash and let the shares go and thereby cut themselves out of the millionaire class. Goodrich was elected president of the new organization, which flourished from the start. Alanson Work, factory superintendent since 1879, was vice president and Colonel Perkin's secretary and treasurer. With the death of Work in 1881, Crouse became vice president and Goodrich added the superintendency to his duties. Work had built well in his short time in the factory and the doctor proved more than capable of carrying on.

A democratic soul was the founder of the Akron caoutchouc industry and it was well that he was so for the Western Reserve proletariat of that time was composed of sturdy citizens who considered themselves as good as the next one and probably a damn sight better. In the 1880s the doctor made a trip to Europe and the men, women, boys and girls of the staff — all one hundred of them — knocked off work and went down to the Akron depot to greet their president on his return. Goodrich invited them up to the house, chased a couple of maids into the cellar with pitchers and the Bourbon was passed around. Such were employer-employee relations when the Akron rubber business was young.

In 1881, as sales touched $300,000, first addition to the tiny original plant was built. In 1887 Akron editor Frank Pixley, who had not yet thought of the Prince of Pilsen, was able to note a half million in sales — all in mechanical rubber goods. Goodrich had gone West for his health by this time. He died in 1888 and Colonel Perkins succeeded him in the presidency. As employees came to total 700 in 1892 and the colonel announced sales of more than a million dollars for the preceding year, Akronites began to decide this caoutchouc business might be worth looking into. The two important rubber events of the year, however, were the founding of the Rubber Trust and the feat of Nancy Hanks in clipping more than four seconds from the supposedly unbeatable Maud S. record. Cause of the trotting queen's triumph was the re-invention of the pneumatic tire by John Boyd Dunlop; result was the eventual downfall of the Rubber Trust as trust. Because of the pneumatic it was to become just another corporation.

What Dunlop re-discovered was the 1845 idea of 23 year old Robert William Thomson (sometimes "Thompson") "in the county of Middlesex, Civil Engineer." Thomson not only had the idea of wrapping springy air within a rubber tube, but carried it out. He patented the pneumatic tire in England, France, and the United States, and built pneumatics and sold them to brougham owners — all this between 1845 and 1847. There is record, too, of a set running 1,200 miles without the "slightest symptoms of deterioration," but the public of the 1840s refused to accept them as anything more than a novelty. So Thomson turned to making solid rubber tires and very shortly his invention was completely forgotten.

An entirely independent invention was Dunlop's in 1888. Since the bicycle was in need of it, it became a commercial success and this Scotch horse doctor is usually accepted as the father of the pneumatic tyre. Located in Belfast he was when in 1887 he began brooding over the idea of adding speed or ease of propulsion to vehicles by means of a triple tube of rubber, canvas and rubber distended with compressed air. At this time, solid rubber tires for bicycles were an inch thick and were heavy and slow. For racing purposes, narrow 3/8 inch tires which looked like bootlaces were used. Dunlop made a disc of wood 16 inches in diameter; with thin sheet rubber he constructed an air tube. He was used to working with rubber as he made his own veterinary gloves and other rubber articles which he used in his profession.

There was no tubing to be had of the kind I wanted, so I got a sheet of rubber and some solution and made a tube for myself. . . . Presently I had an air tight tube around my wheel with a bit of the tubing of a baby's feeding bottle sticking out for a valve.... If I had pumped air into it in that condition, it would have blown out like a balloon until it burst. I appealed to my wife. Fortunately she had exactly what I wanted. She got me a strip of an old dress of fine grey linen that had been fashionable in its day. A strip of this linen I pressed over the rubber tube and tacked it neatly and tightly to the wooden wheel. Then I pumped the air in with my boy's football pump, folded and tied the end of the bit of feeding bottle tubing.

The first test was carried out at night in the yard of the veterinary premises as Dunlop was anxious to keep the idea as secret as possible. At the close of business, the large gates were locked. There were only about six people present, including the Dunlop family. The inventor took the air-tired wheel and the front wheel from his son's tricycle into the yard and asked his assistant, John Caldwell, which he thought would be faster. Caldwell thought that the solid tire, being smaller, would likewise be the speedier. His master then threw the solid tired wheel along the yard but it did not travel the full length. The air tire, thrown with the same force, traveled the full 15 yards and bounced back from the gate. Unconvinced, Caldwell tried the experiment himself with the same results. Dunlop knew now that he was on the right track. So did his 10 year old son, Johnny, who immediately started to campaign for a set of the tires so that he could beat the boys with whom he raced in People's Park. Before the end of December, the horse doctor procured two strips of American elm which he fashioned into three-foot wheels. The air tubes were made of the finest and thinnest sheet rubber he could buy. Each air tube was drawn into a canvas tube, a small air supply tube inserted, and the ends of the air tube cemented together. The valve was of the simplest type, being a strip of rubber secured across the inner end of the air tube, thus forming a non-return valve. The canvas cover was made in two pieces in the form of a tube with flaps. The flaps were cemented to the rim, the entire jacket covered with a sheet of rubber, and two additional thicknesses of rubber were cemented to the tread. The tires and rims were completed and fastened with copper wire to the rear wheels of Johnny's tricycle late in the evening of February 28 1888. Unable to wait until morning, Johnny pedaled out into the streets of Belfast at about ten o'clock. His speed trials were interrupted by an eclipse of the moon which sent him scurrying home, but when the shadow passed, he was back out for another sprint. The report of this first test fleet was that the tricycle was fast on all surfaces and examination the following morning revealed no signs of wear or damage.

Dunlop's next step was to order a new tricycle without wheels from W. Edlin and Co., of Belfast. He then proceeded to make some extra large wheels in order to gear up the machine and take advantage of the lessened resistance. The completed cycle proved so satisfactory that Dunlop applied for a provisional patent on his pneumatic tires on July 23 1888. Meeting Edlin one day, Dunlop found the merchant wondering why he had not ordered wheels for his tricycle. Learning that the veterinary had made them himself, although he had never made any wheels before, Edlin expressed the opinion that they must be very poor wheels and laughed when Dunlop stated that the machine was faster than anything the expert could make. When it came to a trial, however, Edlin, who had been a racing man, found himself unable to outpace young Johnny Dunlop and immediately became a pneumatic enthusiast himself. Dunlop's boy soon became a well known figure on the streets of Belfast, pedaling the grotesque machine which got the name of "the mud cart."

In August, Dunlop ordered two air tubes and rubber covers from the Edinburgh rubber firm of Thornton and Co., and had the order refused with the statement that it was the most impractical idea ever heard of. The Belfast agent of the firm soon set his principals right and the air tubes were duly delivered, thus resolving Dunlop's doubts as to whether reliable tubes could be made easily. Meanwhile, a bicycle without wheels had been ordered; Edlin made steel rims, and, when the tubes arrived, Dunlop fitted them, using the best yacht sailcloth he could buy for the covers. The cycle was ridden for more than 3,000 miles. The front tire, which was never punctured nor removed from the rim, is on exhibit today in the Royal Scottish Museum in Edinburgh.

The rear wheel of the machine was lost while on loan for legal purposes in Paris.

Pneumatics speedily mopped up in the racing field and to exploit the patent rights Harvey Du Cros, Dublin promoter, organized the company which has grown into the great Dunlop corporation of today, a rival in every respect of the American Big Four. Early forced out of this corporation, Dunlop himself never gathered more than an insignificant reward for his invention. Perhaps he was fortunate to obtain that much, for his tire was never really subject matter for a valid patent as was shown when a minute search of patent office records, inspired by Du Cros' monopoly, turned up the Thomson patent.

Very slow was the United States in taking up the air tire until Nancy Hanks drew her pneumatic-tired, bicycle-wheeled sulky in that record trot and thoroughly established the virtues of the new-fangled idea. Steel tires disappeared from the Grand Circuit in immediate reaction to Nancy's demonstration; and solid tires also disappeared from American bicycles. The huge new United States Rubber Co., was not much interested, however, for caoutchouc for footwear obviously was of far greater importance. Du Pont controlled today is U.S. (as this company shall be called from now on), chemically pure and apparently ashamed of its swaggering piratical past. So anxious to disassociate itself from old glories has it been, in fact, that its history is the least known of any present-day Big Four members. Not in the classy pamphlets, the magazine articles and the stiff-covered subsidized books favored by its upstart rivals will this history be found, but it does exist — in yellowing newspaper files, in plushy business success anthologies of the days when financial buccaneers were proud of the designation, in gabby annual reports to stockholders of the same frank and swanking era. We mean, of course, the good old days when a national hero's rise was from watering stock on the farm to watering stock on Wall Street; the days of small companies walking the monopoly plank in crowding procession; the days when "trust" was a word on everyone's tongue, when Rubber Trust, Steel Trust, Oil Trust, Beef Trust and The Common People were the daily characters in the F. Opper cartoons; the days of Benjamin Harrison, Grover Cleveland and Mark Hanna's McKinley, the days when the voice of the original Roosevelt was loud in the land and without benefit of microphone.

VS, then, was the Rubber Trust. About its current desire to forget that phase there is a quaint air of prudishness, for the scarlet past is one it shares with all of the huge

American corporations in all save the most recent lines of gainful endeavor. That it aimed at and came close to complete monopoly let no man doubt, for by 1900 it controlled in one name 80% of the rubber boot and shoe business of the country, in another name 85% of the rubber mechanical goods.

Sire of this octopus, as we have mentioned before, was Charles Ranlett Flint, the international operator with a name reminiscent of a comic strip villain. Bossing the Trust's finances for a decade, he also operated as merchant of death (warships and planes) on his own account during all that time and later. He acted as an organizer in the formation of American Chide Co., and Sen-Sen Chiclet Co., (rubber industry cousins), American-Hawaiian Steamship Co., American Woolen, International Business Machines Corp., National Starch, VS Bobbin & Shuttle and a dozen other combinations in these and street railway, coal mine, vending machine and what not lines. At 73 he elucidated his trust-founding technique in an autobiography titled *Memories of an Active Life* (first understatement of the Flint career); at 77 he took a second wife half his age; at 84 he departed this world (February 13th 1934). One of his last honors, you may or may not recall, was chairmanship of the American Committee for Encouragement of Democratic Government in Russia!

VS Rubber was Flint's first job of trust making, and the sterling patriotism exhibited in his selection of that name was demonstrated over and over again as consolidations featuring the tags "United States" and "American" continued to flow from his merger mill. It was by the clouting together of nine Rhode Island, New Jersey, Massachusetts and Connecticut companies manufacturing a third of the country's rubber boots and shoes that US Rubber was formed in 1892. Included were the first Goodyear licensees, and so it is that the history of US stems directly back to the inventor whose main measure of immortality has been gained through the borrowing of his name by Goodyear Tire & Rubber Co., a corporation without ancestors which sprang full-fledged into being 38 years after Goodyear's death and 33 years after the expiration of his patents.

As a smaller running mate for US, the "father of trusts" put together another 1892 combination designed ultimately to dominate the whole field of soft rubber goods apart from footwear. Geographically this Mechanical Rubber Co., a merger of five companies owning six plants, was an even more extensive chain, for while the boot and shoe industry had confined its expansion to the original Atlantic seaboard states, belting, hose, raincoat and drug sundry concerns now studded the country as far West as Chicago. The tire business Flint virtually overlooked, and so this subdivision that was to become far more important than all other rubber enterprises combined spent its youthful years outside his fold.

Under both of the Flint setups the master corporation owned or controlled the stock of the manufacturing subsidiaries, but these latter retained their individual identity both as to management and sales. The announced idea, then, was abating of former "extreme" competition; eliminating overlapping products; eliminating (in the case of US) duplication of the vast number of expensive lasts required in the production of a thousand varieties of boots and shoes; pooling patents and secret processes; and obtaining raw materials on more advantageous terms. Out of these savings, the US prospectus affirmed, the combination expected to declare "handsome" dividends on its stock "without encroaching upon an ample margin for the improvement of the quality of its goods" and "without increasing their price."

Chartered under the laws of New Jersey and capitalized at $50,000,000 divided equally between 8% non-cumulative preferred and common stock, US began business as one of the largest corporations then in existence. Incorporated on March 29th 1892, it completed organization and elected officers on October 16th, with Flint becoming treasurer and executive committee director of finance. In 1893 it bought the capital stock of Woonsocket Rubber and Goodyear's India-Rubber Glove, automatically jumped from control of a third to a half of the country's caoutchouc footwear and set the vigilant (and Democratic) New York World to yowling for United States Attorney General Olney to "smash the Rubber Trust."

Some truth there was in the World's assortment of charges and a good deal of journalistic imagining. Set the prices for rubber articles the Trust did to considerable extent, but it wasn't altogether a matter of thriving and growing rich. Earnings in its best years were two to three million dollars; it paid its preferred dividends regularly and a few common dividends as well—but that was all. Far from controlling the world's crude rubber, as the newspaper boys had it, it did not even control its own and in setting exorbitant prices at least one eye had to be kept on both independent competition already in the game and on more threatening to come in, for it required no fortune to launch a new boot or belting plant on modest lines. This in itself was sufficient evidence that rubber goods did not constitute a product ideally suited to the needs of would-be monopolists. And in the case of US there were unpredictable year to year weather vagaries as well as small but stubborn opponents to contend with. And while there were important advantages inherent in consolidating the purchases of raw materials for all manufactories so far as practicable, failure to extend this central control to management went a long way towards nullifying them. In an effort to remedy the clumsiness inherent in the organization — or lack of organization rather — US modified the original idea in 1894 by making the master corporation selling agent for all the plants. Still there was a heavy drain resulting from the making of a large number of brands in isolated factories of the Trust, brands so little in demand that organization and machinery of these plants did not begin to be fully utilized.

In 1895 the 150,000 workers of all American rubber companies (total capitalization $85,000,000) produced finished goods valued at $70,000,000 to $75,000,000 as the industry consumed 18,646 tons of crude and substantial quantities of reclaim. The Trust found, however, that not one of its most productive plants had been utilized to its fullest capacity or even operated continuously at moderate production throughout the year. That meant reducing the number of brands, closing the least advantageously situated of the factories and concentrating production in the others to the tune of write-offs that were substantial even though not as great as they should have been. And once US even departed so far from its original footwear ideals as to convert its abandoned New Brunswick shoe plant into a bicycle tire factory — and took a licking in a patent suit. Meanwhile one mild winter followed another, former stockholders in various of the subsidiaries had to be paid, there were price reductions and rebates to jobbers, and all in all the Trust existence proved to be no thornless bed of roses. Even so US continued its march towards that longed-for monopoly — forcing some rivals out of business, absorbing others, boosting its percentage of the nation's total rubber footwear output to 75 by acquiring the stock of the large and important Boston Rubber Shoe in 1898, elevating that percentage to 80 by the turn of the century. Translating production capacity into sales now became a real problem, however, and the surviving independents, Hood Rubber at Boston, factories at Mishawaka and Grand Rapids in the felt boot belt, a scattering of other concerns holding forth at Hudson, MA, Bristol and Providence, RI, Philadelphia and St. Louis made it very tough for the Trust.

Of these independents a contemporary Dow, Jones & Co., (*Wall Street Journal*) pamphlet recorded, "Securities of these companies are not actively traded in and public financial statements are not made." So far as the general investing public was concerned, it noted, US and its newly-expanded sister in the mechanical field represented the rubber industry. For both of them it had honeyed words and especially for

US it praised its economy in manufacturingts beneficial and more uniform system of selling which had put down "injudicious forms of competition," its raising of management standards and its general contribution towards making "perfection" of the boot and shoe branch of the rubber industry "greater than in any other." And it pointed out: "Upon the basis of a $100,000 subscription to the underwriting in 1892, there has been paid over 9% in dividends, and an original subscription for that amount shows an additional profit of over $20,000."

On these things the expert financial writers apparently had no information: A large floating indebtedness in existence since the inception of the US organization now risen to $12,000,000 through the master and subsidiary company habit of borrowing on their notes for temporary requirements; a property account still inflated by the inclusion of items of more than doubtful value; sales falling in such a tail spin that the report for the year ending March 31st 1901, would show gross sales (measured by list prices) of but $32,224,216, net sales of but $20,853,633, a total income of but $265,621, a net income of $62,605 as against $164,531 losses and depreciation, and a surplus reduced to $25,013 with nothing reserved for dividends!

It was in October of 1900 that the applauding pamphlet was issued, incidentally. By January 1901, US was slashing prices 23% in a desperate move to sink competitors and stem falling sales, in February it passed its preferred dividend, in April treasurer Charles Ranlett Flint passed out of the picture along with the Trust's president number three.

At the time of the mammoth's organization away back in 1892 some few iconoclastic observers had rudely claimed that none of the incorporators, directors or officers were really rubber men (for the record, it may as well be noted that some of them actually were) and even more rudely had predicted that the whole thing was intended as a basis for stock-jobbing operations in Wall Street where profits sometimes were made out of the ownership of factories not profitable in themselves. Whether these outsiders or the informed gentlemen of Dow, Jones were closest to the truth, the reader may decide.

It was in 1899 that Flint set going another $50,000,000 rubber corporation (equally divided between common stock and seven percent cumulative preferred), the one foreshadowed by that Mechanical Rubber Co., of 1892. Incorporated in New Jersey on January 26th 1899, and organized March 17th 1899, this Rubber Goods Manufacturing Co. aimed at monopoly in tires as well as the traditional belting, packing and hose lines. Charles Ranlett was chairman of the executive committee and another member of the house of Flint, Wallace B, was company treasurer. General counsel (and an executive committee member) was William M. Ivins of the Amazon steamship monopoly scare.

The new Flint trust started out with these properties: 91% of the capital stock of Mechanical Rubber; 75% of the capital stock of Morgan & Wright (tires) at Chicago; entire capital stock of India Rubber (tires) at Akron, 0.; entire capital stock of Peerless Rubber Manufacturing Co., of New York. In its first year, it acquired all but 0.25% the remaining stock of Mechanical Rubber and the entire capital stock of Sawyer Belting at Cambridge, MA, Hartford Rubber Works (tires) of Hartford, CT, Indianapolis Rubber (tires) of Indianapolis and Peoria Rubber & Manufacturing of Peoria, IL, and completed negotiations for taking over the entire capital stock of American Dunlop Tire of Belleville, NJ. Solid tires, chiefly for carriages, now were a $4,000,000 a year business in the United States. Bicycle tire business was running to about $7,500,000 a year and was virtually the monopoly of the various companies controlled by Rubber Goods Manufacturing and holding the important patents. Pneumatic vehicle tires amounted to $1 ,000,000 a year and cushion tires $100,000. A little over $12,500,000 was the annual business in tires of all kinds, $3,500,000 under the total the rubber boot and shoe industry had reached 20 years before and still regarded as a subdivision of the mechanical goods business rather than a specialized branch of the rubber industry in itself.

It was because of Rubber Goods Manufacturing Company's taking over this tire business that had grown up outside the limits of the original Flint trusts that US withdrew from its one-factory manufacture of bicycle casings in 1900. Neverthe1ess it is somewhat startling to find the then world's greatest rubber company (by far) entering the automotive century with the formal announcement, "This company has retired from the manufacture of tires." For five years that retirement endured, being ended by the merger route as US took over Rubber Goods Manufacturing in 1905. Nor even then were automobile, bicycle and carriage tires the leader; they still were just an item in the caoutchouc catalogue which featured footwear and included belting, packing and hose, flooring, clothing and drug sundries.

A history somewhat similar to the US one is that of Rubber Goods Manufacturing. Controlling 85% of non-footwear soft rubber output as it really got under way in 1900 (preferred stock $7,62 1,300, common $15,134,6(0), it tried to operate with subsidiary companies maintaining their manufacturing and sales identities, made its preferred dividends and something more (first year earnings, a million and a half), but saw its monopoly fade away without ever knocking off such profits as a real Trust should. With the tire business still climbing, still bulging up under independent auspices every time the mechanical goods Trust turned around, this sector of rubber was proving far more difficult to control than the footwear end. For the US that absorbed Rubber Goods Manufacturing in 1905 really was a Trust again, one doing twice the business of its sister monopoly, one with a net profit of 3¼ million for 1904, one bearing little resemblance to that reeling hollow shell of 1900-1901. With the disastrous 1900 showing Flint's era had ended, and Colt's era begun. In its fourth president, Samuel P. Colt, the company had found the man who could make the Trust machinery work.

Nephew and namesake of the revolver inventor who pioneered in the international munitions racket later embraced by Flint, this Colt was former attorney-general of Rhode Island and newly named president of the newly organized Industrial Trust Co., of Providence at the time National India Rubber fell to him. That was one of the important companies going into the formation of V. S., and Colt became legal adviser, director and executive committee member of the octopus. In 1896 he was made secretary and five years later succeeded Frederick M. Shepherd as the latter closed a three-year run as president. Ruler of V. S. as dominating president in all the golden years down through 1918, chairman of the board from then until 1921, director in 40 corporations directly or indirectly connected with rubber, Colt invested some of his caoutchouc wealth in Rodin sculptures and became Ethel Barrymore's father-in-law when one of the heirs to that wealth made big news for the Winchells of the day with a 1909 blending.

Colt as president began by consolidating selling agencies and the company's own branch stores wherever possible; ditching the "contract system" under which a good part of the V. S. product had been sold without much effort by its sales department but with a notable loss of advantage to competitors; funding the whole floating indebtedness; charging off so many items of doubtful value that the property account was left with a deficit of over a million dollars; holding prices at or near cost until he had built sales back to the old figure and sent a flock of competitors to the scrap heap. Only then did he start thinking of "how to manufacture and market the large product of goods at a fair margin of profit."

In the company's old reports are found his frank declaration that prices too long had been "maintained at a figure which stimulated competition, and the formation of new companies and investment of new capital"; that in consequence of this, cuts to cost had been necessary to regain trade. For the year ending in March 1902, he was able to show gross sales back up to $45,917,536.84; in 1903 he was able to report $51,888,756.92 and to note triumphantly, "As an illustration of the low prices for manufactured goods the past two years it may be mentioned that during that period five companies manufacturing rubber boots and shoes have failed, or have retired from business." In 1904 he could note production of over 48,000,000 pairs of boots and shoes and gross sales of $64,000,000, largest volume of business in the Trust's history. Touching again on "the abnormally high prices for our manufactured products which had prevailed for some years" and permitted "many new rubber companies" to come into existence, he reviewed the history of the price slashes then still in effect and cheerily concluded, "under these conditions, as always, the evil supplied its own remedy, and only the stronger companies found themselves able to survive the strain of such competition."

This, of course, is how a Trust should be run. And with 1903-1904 gross sales double those of 1900-1901, the Trust master now was able to turn his thoughts to methods of snaring that "fair margin of profit." Apparently this was a simple problem, for by 1905 he was able to resume preferred dividends, first at 6% but retroactively to 8 as at end of the 1904-1905 year he showed net profits of $3,761,922.63, largest in the company's history. In that indecent rush to resume full preferred dividends is the first hint of how Colt, who had swung the corporate ship away from the reefs, was to leave it among worse ones after a long voyage very pleasant to preferred shareholders. For along with simultaneously kicking the preferred back to the old figure, the management extended for another three years $8,000,000 of funding notes while optimistically predicting that this debt would never have to be funded again. In regular payment of its fat preferred dividend and alarming growth of that funded debt is to be found the history of Colt and successor. Out of borrowings came all of the company's tremendous expansion. Out of the profits came dividends and interest payments. It was a great life while it lasted.

We have mentioned the 1905 absorption of Rubber Goods Manufacturing Co (RGM). It was done by increasing capitalization of US to $75,000,000 and exchanging first preferred stock for RGM first preferred share for share; 6% second preferred for RGM common on an approximate basis of one share second preferred for one and three-fifths shares of common. Over 80% of RGM stock was acquired from the very start and dissolution could have been effected immediately, but was held in abeyance for several years. Outstanding on March 31st, 1906, were these US shares: first preferred, $35,067,000; second preferred $9,586,300; common $25,000,000. As to relative importance of the two component parts of US it may be noted that in Colt's first year of tandem-driving, the footwear properties had net sales of $39,715,730, RGM properties $19,737,121. Purchasing, selling and management operations all had been united and the day of a commune of independently managed plants was gone forever so far as this Trust was concerned. As to the Trust idea, it survived and has survived in the footwear field, for US still controls more than half of that business; it never got anywhere in the auto tire field, for in the magnificent mushrooming of the casing business the upstart Akron companies outstripped the unwieldier mammoth. For the time being, however, the mammoth kept right on expanding. In 1907 it acquired control of Canadian Consolidated Rubber Co., Ltd., as much a leader in Canada as US was below the border. In 1908, still paying regular 8% preferred dividends, it refunded $12,500,000 in funding notes and debentures by issuing $20,000,000 collateral trust bonds. In this year, too, it began the enlarging of all RGM tire plants and bought control of Revere Rubber, rising tire manufacturer. In 1909-1910 profits passed $5,000,000 for the first time as tires consumed half of the crude rubber used by US In 1910-1911 RGM sales first passed those of combined subsidiaries owned by US prior to the RGM deal (net sales for all US, $76,077,019.65). In October 1911, common stock dividends were resumed at 1% a quarter, the first since the occasional ones of the nineties. In 1912-1913 profits first passed $7,000,000; a 20 storey office building on Broadway was completed; authorized capitalization was increased to $120,000,000; a 20% dividend in common stock was declared; and the common dividend was increased to 1.5% in the last quarter.

Out at Akron during all this time even more of rubber history was being written. First town boys to compete with the Goodrich company were Jacob Pfeiffer and his two partners in a drug store. They dropped $50,000 in their first bout with rubber in 1892, borrowed money to start again and began the manufacture of gloves in a 20 x 50 factory which eventually developed into the Miller Rubber Co., sixth largest in the industry. Cash resources on the day the wheels began to turn amounted to but $250, however, and that went into buying the first crude. In 1894, the year after John

Fullerton Palmer's patenting of the cord tire in England and America, Ohio Columbus Barber's Sherbondy Rubber Co., came into being — chiefly because the match trust millionaire had a vacant factory building on his hands, and thereby hangs a tale.

Seat of Summit county is Akron. And county auditor of Summit in the late eighties and early nineties was Charles W. F. Dick, who was to become Mark Hanna's right-hand man and fall heir to his seat in the United States Senate. County officials were operating on a fee basis in those years; an Ohio government running a deficit was trying to smoke out unreported intangible property for taxation, and Dick was especially active. By listing himself in the city directory as "Ohio C. Barber, capitalist," the president of the new trust was, perhaps, asking for what happened. At any rate the hardest wallops were landed on Barber and the Barber enterprises. A raging controversy developed — and a match king vow to pull out of the county seat and, anticipating Herbert Hoover, watch grass grow in the streets of Akron. So was born the city of Barberton on a 640-acre tract south-west of Akron, previously optioned by Barber in connection with an unsuccessful salt well venture. The move did not take Barber out of Summit county, but trustworthy old-timers, including close associates of Barber's, depose that this is how it happened.

The match king incorporated as a land company; took in Crouse, who had had experience in promoting the Pennsylvania glass town of Charleroi; and set up the world's largest sewer pipe plant, a railroad, bank, hotel, strawboard factory, ammunition factory, boiler factory and half a dozen others. Of all but two of these he was president. Then, to complete his nose-thumbing at the city of Dick, he moved the Akron plant of Diamond Match to the new made-to-order community, and projected the laying out of a nearby personal estate which eventually spread to several thousand acres and became the most startling farm ever seen outside the film musicomedies.

With Barber as founding father, Sherbondy Rubber was certain to be a success. The abandoned Akron building of the Diamond Match Co., in which the trust master installed it as thoughts of business triumphed over thoughts of revenge was immediately adjacent to the Goodrich plant and the Barber-Crouse tie-up kept relations friendly from the start. The 1894 of Sherbondy's incorporation at a modest $50,000 to manufacture drug sundries and bicycle tires was also the year in which Goodrich commenced the manufacture of solid rubber buggy tires for the Rubber Tire Wheel Co., which did the attaching in its own plant at Springfield, OH. In 1895 the rubber-manufacturing brothers Michelin, Edouard and Andre, found that they were wasting their collective breaths in trying to convince the makers of whatever motor cars France had at the time of the importance of pneumatic tires. As a result, they built their own motorcar, equipped it with the first pneumatics ever constructed for such service, and entered it in the Paris-Bordeaux race. It finished the trip at a cost of 22 inner tubes. And the next year at Akron the world's first commercial transaction involving auto pneumatics took place as the million dollar Goodrich company had an order for a set forced on it by Alexander Winton of Cleveland. The pioneer auto man was compelled to pay for molds as well as tires and to pay in advance, Goodrich arguing that it would have the casings on its hands forever if he failed to call for them.

Appropriately it was in this same 1896 that Akron's Seiberling family first got into rubber. The Empire Reaper and Mower Co., after spectacular success had gone down in crashing failure as had J.F. Seiberling's earlier Excelsior Mower and Reaper. Within a decade

Crouse's Buckeye would smash up in the same way and the man of innumerable investments would have good reason to be thankful for the best one he ever made — the few thousands stuck into Goodrich which yielded him some 9000%. Associated with the senior Seiberling in the Empire had been his sons, Frank A. and Charles W., and the father and this last son now launched India Rubber Co., (not to be confused with the currently liquidating India Tire & Rubber Co.,) in one of the Empire buildings. This tire company was taken over by Flint's new Rubber Goods Manufacturing Co., in 1898-1899, but the Trust operated in Akron for short years only. The plant burned down. It is interesting to note, however, that the octopus flipped a tentacle into this Ohio community and withdrew the singed feeler without realizing that it was pulling out of the rubber manufacturing capital of the world.

And it is also interesting to note that the Seiberling family's connection with rubber did not begin with the founding of Goodyear as most persons, Akronites included, believe.

The year of India's founding also saw the Sherbondy with 250 employees being renamed Diamond Rubber Co., another concession on Barber's part, and the issuance to Arthur W. Grant and assignment to Edwin S. Kelly's Rubber Tire Wheel Co., of the patent for what came to be renowned as the "Kelly" tire. Fastening the tires to the rims in satisfactory fashion was the problem with the carriage trade and the Grant patent was immensely superior to anything then on the market. Goodrich, which had been making all of the wheel company's tires, now became only one of many as factories throughout the country were licensed. Harvey S. Firestone of Chicago, a horse-trading Ohio farm boy who had started his business career as a peddler of Wild Rose lotion and Arabian Oil horse liniment, had a small plant for turning out wheels crudely tired with rubber. He had been buying his tires from the Morgan & Wright Co., of that city. Rubber Tire Wheel took him over and as an official of this company he once placed an order in Akron for half a million dollars' worth of Diamond and India tires, which was big stuff indeed for those days. Harvey dropped out of the picture as Consolidated Tire bought up Rubber Tire Wheel Co., to obtain the Grant patent.

It was into this monopoly and the Tillinghast pneumatic bicycle tire monopoly that Frank A. Seiberling banged shortly after bringing Goodyear Tire & Rubber Co., into being in 1898. Goodyear got going in an abandoned strawboard plant and an abandoned woolen mill on an eight acre patch across the street from the cemetery and near Krumeich's saloon in the Middlebury that finally had grown accustomed to thinking of itself as East Akron. Seiberling, meeting the owner in a Chicago hotel, had fallen to discussing these vacant buildings and had beaten down the asked price of $50,000 to $13,500. Of this he was to pay $3,500 down. The former secretary and treasurer of the great Buckeye works grabbed a train back home, rooted a friend out of bed at 6 AM. and before the morning was over had made the down payment to the Chicagoan's Akron attorney and received his deed. So was born the world's greatest rubber company.

Frank Seiberling's first act was to line up brother Charlie; his second was to start hunting the funds for the incorporation papers, the machinery, the caoutchouc, for everything he needed. The first man he hit was David E. Hill, East Akron sewer pipe baron. The promoter talked this prospect out of $10,000, but weeks of additional canvassing and button-holing brought his total to but $43,500. A quarter century later, $10,000 in original Goodyear stock would be worth just a trifle under

$1,000,000 after having averaged cash dividends of $13,400 a year over the entire period. Which explains why the wistful game of figuring up how much they'd have made by entrusting "F.A." with a few thousands in 1898 still is a favorite indoor sport among the older generation of Akronites. Saloonist Krumeich was not among those given to such vain imaginings. Into his till was to pile a take from the gum worker shifts that left no room to sigh for more.

Capitalized at $200,000 and with $90,000 paid in was Goodyear when it got going at the end of the year. Hill was president, Frank Seiberling general manager and Charlie Seiberling secretary. The next year R.C. Penfield, of the army of Seiberling brother-in-laws, bought out the interest of the Hills and assumed the presidency. New capital of $100,000 was put into the business at this time and a stock dividend paid out of the $35,000 profits made the paid-up capital $200,000. Seiberling's corporation was getting away to easily the most luxurious and profitable start of any caoutchouc company in Akron history.

It was not to last for long. Goodyear, threatened with suit by Consolidated Tire, came to terms, was promised $50,000 worth of carriage tire business a month; tied up its cash in materials and in whooping the working staff to 176; received but half the expected orders from the monopoly; had to bring suit to collect for tires delivered; and saw its license taken away. With Consolidated refusing to mount Seiberling's tires, the Akron concern now had to work out a method of its own before it could sell to carriage makers or dealers. This was easy, and to get around the Grant patent the Goodyearites evolved a "departure" from it which had side flanges to keep sand from working in between tire and rim. Consolidated promptly sued and won a desist order. Seiberling appealed, succeeded in having original bond of $1,000,000 reduced to $2 50,000, talked Penfield into posting it. Now Goodyear could proceed during the life of the litigation — but with all its carriage tire profits held in escrow.

That left it up to bicycle tires to carry the plant. They did in a fashion — as Goodyear sweated to make payrolls, traded tools for rubber, took a carload of bicycles in payment from a client in the same jam and peddled them around the town. Once in a while some Akronite would kick in with a loan when rubber arrived COD and the company had to lay cash on the line. Offered their choice of Goodyear stock or a 60 day note they usually accepted the note and thereby took on a load of regret that would last them for the remainder of their lives. In the bicycle tire field, Goodyear started off with a license under the Tillinghast patent (acquired by Rubber Goods Manufacturing Co., in 1899) and operated on that basis into 1900. Then it was accused of selling under the stipulated price and the license was withdrawn. In this emergency one of the East Akron geniuses concocted a variation by inserting a strip of tissue paper between the two plies of the tire in order to keep them separate when vulcanized. Patent was applied for. Sued for infringement, Goodyear kept right on building and peddling and stepped production up to 3,000 and then to 4,000 and tires a day. The matter never came to a showdown, for the legal action was withdrawn. The tissue paper had served its purpose.

Its sister, the hoped-for patent on the alleged carriage tire improvement, had been thrown out, but this was not long to be cause for regret. The United States Court of Appeals in May 1902, found the monopoly's Grant patent void and Goodyear tied down the factory whistle to celebrate the release of its funds from escrow. Goodrich and the other licensed companies that had confined their carriage tire activities to manufacturing for Consolidated now leaped into the open market competition and new companies mixed in until 1907 saw 25 in the field with Goodyear first as it had been since 1905. As its monopoly blew up Consolidated itself had gone into manufacturing by taking over the Buckeye Rubber Co., plant erected in Akron at the turn of the century. A sales company was set up under the name of Kelly-Springfield Tire Co., and that eventually became the title of the entire setup. Last year, incidentally, fate pulled one out of the hood as giant Goodyear pocketed the fair-sized but busted Kelly-Springfield in an unremarked sequel to that episode in which the giant carriage tire monopoly of the early years of the century tried to slap Seiberling's midget company into oblivion.

Chronologically we have gone ahead of our story. Returning to 1900 we find Firestone arriving and the census takers according Akron a population of 42,720. Little of this gain was due to the rubber factories nor were they the leaders in this city of diversified industries. From now on, however, the tale was to be all caoutchouc and the city would reap the whirlwind of reward or punishment (take your choice) it had sowed by sending forth that windy pamphlet of 1869. The pamphlet brought Goodrich, the Goodrich profits attracted Pfeiffer and Barber to the rubber business and the success of all three caught the eyes of the Seiberlings as they were casting about for a new one to tackle in the place of farm machinery. They also caught the eye of the head of an Akron twist drill company who brought in Firestone. So was the story of the beginnings written.

Well located for distribution of finished products was Akron, but not for delivery of crude rubber. Favorably situated with regard to the center of the carriage building trade it was, but far less favorably than many another community. And it was not at all close to the bicycle makers. Water power was unexceptional and would be inadequate. And the region had nothing to offer any rapidly growing industry in the way of labor supply. When the Goodrich success caused local emulators to spring up in the nineties and early 1900s, however, the city did have one tremendous advantage over the old rubber centers of New England, New Jersey and Rhode Island. This was the very newness of its caoutchouc enterprises. Absorbed in their boots and shoes and, to less extent, in their belting and packing were the *estrada* concerns, while Akron, alert for something not already sewed up by the established trusts, was meeting rubber-tired transportation more than half way. On bicycle and carriage tires it rolled into the new century and when the automobile arrived it was first aboard for the great industrial joy ride that left boots and shoes plodding so far behind on the dollar highway. Built on footwear was US Rubber and built on mechanical goods was its sister. But Goodrich was swift in becoming tires as well as mechanical goods; Diamond was tires above mechanical goods — Goodyear and Firestone were tires first, foremost and almost exclusively.

Firestone, feeling rich as the result of receiving $45,000 for his share of the Rubber Tire Wheel business, was unable to get along with the Akron drill manufacturer who had hired him to head a tire department. So he took an option on the department and just about then encountered lames A. Swinehart, Akron school teacher, contractor and inventor. Harvey in later years would be unable to remember the correct spelling of Swinehart's name, even, but in 1900 he was quick to recognize that the man had hold of the best fastening device for solid tires yet to be evolved. It was a side-wire application and the first to make possible the safe fastening of the larger sizes. Swinehart already had interested James Christy, Jr., manufacturer of harness leather, and several other Akronites. The group, in August 1900, organized the $50,000 Firestone Tire & Rubber Co. under West Virginia charter. Harvey put up $10,000 in cash and his option on the twist drill company's tire business; the Akron boys shoved in a like amount of cash and the Swinehart patent. Each group received $25,000 in stock and the Chicagoan was made treasurer and general manager at a salary of $3,000 a year plus a $600 bonus conditional upon the company's earning 20% or more on its issued capital stock in its first year. Christy was president and Swinehart vice-president. To round out a board of directors Firestone not only drafted his attorney but the first salesman engaged by the new corporation. For two years it amounted to little more than a buggy tire jobbing business, the tires being made by the city's factories. First year sales were $110,000 and second year $150,000. A stock dividend of $50,000 was paid as capitalization was increased first to $150,000 and to $200,000 (1902). All over Akron Harvey peddled stock, for his idea was to set up in manufacturing. Will Christy, banker, traction magnate and brother of James, was the principal catch. He yielded $10,000 and assumed the presidency that the firm's credit might profit by his name. A year later (1903) Harvey took on the presidential title, banker Will became vice-president and inventor

Swinehart departed from the company as it took over an old foundry in South Akron, which already housed Goodrich, Diamond and Miller. Swinehart developed a type of clincher tire and set up a factory within spitting distance of Akron's downtown. Never attaining to important size during or after the inventor's reign, Swinehart Tire failed for half a million dollars in 1929 and was gone before ever the big bull market crashed. Firestone, filling his foundry with second hand machinery, began as a solid tire manufacturer with a force of a dozen men. Characteristically, however, he had Harvey, Jr., actat 5, on deck to turn on the steam and get the factory off to a showman like start.

While Harvey, Sr., was working up to this tableau, his rivals were scrawling rubber history large. Goodrich in 1901 first passed the $5,000,000 sales mark. Diamond swept into 24 hour a day production. Goodyear upped capitalization to $1,000,000 and declared a 100% stock dividend. More importantly, all three companies were in small production of those rubber articles Alexander Winton had demanded of Goodrich in the later nineties — pneumatic tires for automobiles.

All of these were of that clincher type productive of so much motorist blasphemy, as the soft rubber bead had to be stretched over the rim in mounting or demounting. Basic patent for this was controlled by Rubber Goods Manufacturing and its US Rubber successor, which had licensed five rivals. US itself, Goodrich, Diamond, Fisk at Chicopee Falls, MA, and Michelin, which had reached out from France to buy International Automobile & Vehicle Tire Co., at Chelsea, MA, received the important production allotments from the Association which set the percentages. Goodyear, soon to become incomparably the world's largest auto tire producer, was supposed to content itself with r 3/4~ per cent of the American volume. And tiny Firestone was completely snubbed.

Flatly refused a license when he applied to the Association, Harvey in the early years of the automobile era found himself, then, in much the spot Seiberling had occupied as he sought to crash the buggy tire and bicycle tire manufacturing game. He could invite a law suit by going ahead in defiance of the monopoly or he could try to develop a different tire. What he picked up in 1904 was a stray inventor with a rim having side rings or flanges bolted together and holding the tire in place — "essentially the same as the straight side tire of today," Harvey claims. Goodyear, pooh-poohing this contention, insists that it was no such thing, that everyone experimented unsatisfactorily with flanges, that nothing entitled to the name of straight side auto tire appeared until that day in 1905 when Seiberling's volunteer consulting inventor, Nip Scott, wandered in from his barn-workshop at Cadiz, OH, with the idea of braided wire built into the bead so that it could be locked on and unlocked with comparative ease. As old as the clincher is the "wired on" type of detachable tire, but there is no question that Nip had an improvement. And it was this tire, equipped with a ring to make it interchangeable with the clincher on the standard rim developed for the clincher, that Goodyear advertised and pushed while continuing to make its small allotment of the type of casing then most in demand because most familiar. Good for many years thereafter was the braided wire idea, but it has given way to the woven tape wire.

Firestone, trying to market his flange job in 1904-1905, had the thankless task of attempting to convince individual motorists that he offered a tire so superior that it warranted their changing their standard rims to his peculiar ones. In the last of these years he heard that Henry Ford was planning to produce 2000 cars to sell at $500 and decided to go after the original equipment order. Tires were good for but a few hundred miles in those days, and Harvey's chief thought was sewing up the replacement business of 2000 autoists who would have to continue using his tires or change rims all around. Ford then was being sued by the Selden patent monopolists — automotive industry's equivalent of the tire industry's clincher association — and so it befell that the first meeting of Harvey and Henry was the beginning of a beautiful friendship. Nor was the Detroit outsider's liking for the Akron outsider lessened by the fact that Harvey offered his casings at $55 a set as against the monopoly's take-it-or-leave-it $70. Back to Akron hustled Firestone with the order for the 2000 sets. Ford began to take delivery in early 1906 and, once the cars were being peddled, awoke to the painful realization that US and the other monopoly members had dealers and branches throughout the country but that small Harvey did not. Ford buyers were stuck, that is, with a peculiar tire which they could replace only by writing the factory. To Harvey went an ultimatum from Hank — the Firestone company would have to supply clincher tires or lose the Ford business. The Akronite took a trip to New York and pleaded again with the Association head. Rebuffed once more, he went home to go into the infringement act and retain that Ford connection which has been the backbone of his production to this day. Never sued was small fry Firestone for the Pennsylvania Rubber Company's courtroom victory over the Association in 1907 terminated the monopoly. As not seldom happens this came as the clincher patent was losing its primary value, anyhow. Despite the straight side's 1907-1909 conquest of the American industry, however, the clincher did continue as standard equipment on Ford and Chevrolet until the early 1920s so that the passing of the patent was vastly important to Firestone at least. It was, incidentally, another pool of leading tire companies that forced Firestone into the steel business. Controlling the rim manufacturing of the nation, they cut Harvey off from his source of supply and he had to start making them himself.

The steel plant was erected at Akron for contrary to the US Rubber policy of scattered plants about the land the Akron companies unanimously believed in doing their expanding in one spot. And so it was that several years before the passing of the clincher monopoly, the tire city already was calling itself the world's center of the rubber industry. Already the Big Three of Akron rubber companies had shaped up out of the dozen now boasted by the city of less than 60,000 population. Goodrich and Diamond (due to be consolidated) were in first and second position, respectively, with Goodyear — where Seiberling had assumed the presidency — third, and Firestone fourth. Only as Goodrich-Diamond slipped back of Goodyear was that rating to change.

Not yet was the automobile business really under way although the first premonitory rattle of the mass production era was heard as the original Model T Ford slid down the ways in this same year of 1907. Akron was less interested in this launching than in the retirement of Colonel Perkins as Goodrich president and the election as his successor of Bertram G. Work, son of Alanson Work, that early superintendent of Dr. Goodrich's. Young Work, checking out of Yale after his freshman year, had joined Goodrich in 1890, stepped into the superintendency in 1892 and the vice presidency in 1902. His 20 year reign was marked by an early shift of control to New York City, a move which speedily would have reduced Akron to the status of an absentee-directed mill town had it been emulated. First and (to date) last man to hold presidency of an Akron rubber factory for any length of time and not identify himself with the caoutchouc capital, Work cleared out of the city as swiftly as possible, came back as seldom as possible thereafter. Two years after he took office Goodrich occupied its new million dollar office building at Broadway and 57th Street, New York, and that became executive headquarters. Work acquired residences in New York City and *Estrada* Bay; became a member of New York Yacht, Calumet, Piping Rock and a dozen more; died at St. Moritz in 1927. When he visited the tire city, the program was to dash to the gum mine offices for the day, spend the night at the Akron home and get off to Gotham the next evening. He was a nephew of Henry Clay Work, who wrote "Father, dear father, come home with me now." In 1910 Diamond acquired American rights to the Silvertown cord tire developed in England; Firestone reorganized as a $4,000,000 corporation; Goodyear opened a plant at Bowmanville, Ontario, in the first outside expansion by an Akron company; Akron attained to a population of 69,000. A year later Goodyear did its first job of dirigible building for Melvin Vaniman who had accompanied Walter Wellman, former Akron newspaper editor, on the first spectacular attempt to cross the Atlantic by air in 1910. Wellman and all his crew had been picked up at sea after a record flight. Vaniman and crew perished when his Akron blew up in the air over Atlantic City.

Augury of Goodyear's future Zeppelin record that was. And augury of future tire dominance was its amazing forward leaps in the casing field. Its straight side tire sold 35,282 in 1908 and 1,084,1 34 in 1912. Dollar sales leaped from $2,190,000 in 1908 to $25,232,090 in 1912. Earnings jumped from $651,687 in 1909 to $3,001,295 in 1912. Up to $10,000,000 was capitalization now; over 35 acres sprawled the plants; number of employees varied from 3,500 to 6,000. Seiberling started a mechanical goods department; bought a cotton textile mill in Connecticut. Of Goodyear preferred he personally held none, of common a third of the $5,000,000 issued. Dividends he had been dishing out until speculators at the beginning of 1913 carried common stock with an intrinsic value of around $125 a share to $465. In the five years 1908-1912, Goodyear paid $280,566 cash on the preferred, $443,292 on the common, $3,689,100 in stock dividends on common. Boasted its president in early 1913, "Our company has not one drop of water. We have $10,000,000 of capital out. If we had $100,000,000 of capital out we could not have done more business. I can cite a competitor who has $60,000,000 in common stock who has not done any more business or earned any more money than we have earned with our little capital."

This last was a series of extremely hot socks at the new Goodrich. Neck and neck with Goodrich (capitalization $16,000,000) and Diamond (capitalization $10,000,000), in the race for tire business was Goodyear in 1912 and all three right up with or ahead of the US which by 1911 had acquired enough casing plants to justify a boast of being "the largest manufacturer and distributor of rubber tires in the world." Coming fastest of all at the moment was Diamond with its Silvertown cord. Goodyear had yet to develop its own cord, but it was a sweet race for casing leadership all the same. And then came the merger and a combine shooting ahead of all others in tire totals.

Extra friendly from the start, Goodrich and Diamond had continued to play in closer from year to year. In 1904, for instance, the two teamed up in a reclaiming business under the Marks patents. And, it was generally understood, they had a "gentleman's agreement" on price matters. The Goodrich lists supposedly approved by Diamond were, incidentally, the ones generally followed by other manufacturers in hoisting or lowering prices on tires. Not to the extent that US bossed boot and shoe prices did Goodrich establish casing figures but it took a lot of pride in seeming to set them even though the varying discounts given by the assorted companies made selling prices considerably different from list prices.

If there were any lingering doubts about Goodrich's ceasing to be a home-town company they were set to rest by the manner in which the combine was incorporated, financed and headed up. The two Ohio companies became one New York corporation, Goldman, Sachs handled the refinancing, and Work took over the dual presidency. Goodrich of Ohio surrendered $16,000,000 capital stock; Diamond $10,000,000 capital stock and $4,000,000 undivided profits. Thereupon Goodrich of New York was incorporated for $90,000,000. The $60,000,000 additional stock, minus $6,000,000 in common stock handed the bankers for underwriting the new organization, was distributed among the stockholders. It represented the difference between the par value and market value of the stock of the two old companies at the time of consolidation. And together with the combine's sales of $37,533,861, earnings of $3,719,334 and suspension of common dividends in its first year, it perfectly explains all of Seiberling's cracks. A city of 9,000-13,000 caoutchouc maulers had been created as the two adjoining groups of factory buildings came under one banner, a city with a $6,000,000 payroll, one with 80 acres of floor space and 20 acres to grow on.

In Firestone's case, as in Goodyear's, increased capitalization represented reinvesting of earnings and investing of stock sales proceeds in plant expansion. Harvey had produced 28,000 tires in 1906 and 168,000 in 1910. The 1910 figure represented sales of $5,000,000; profits of $1,394,835, more than double the year before. For 1911 the sales were $7,462,581. Thirty acres of tire and steel plants the young company had by 1912 for it was producing nothing but casings and the rims it had been forced to make through being high-hatted by the rim monopoly. At top production, 2,900 men were employed on the day and night shifts as against the even dozen of a dozen years before. Dividends of half a million the company had paid while accumulating a surplus of nearly $2,000,000 in the 28 months since reincorporating. . Not long before it had been a case of United States Rubber and its sister doing 80% of the nation's total caoutchouc business. Already auto tires had so changed the picture that Akron's combined small fry had caught up with and passed the octopus. The 1912 quarter billion rubber business in all lines was divided as follows: $95,000,000 to Akron's ten noticeable and several minute companies capitalized at a total of $113,000,000; $91,000,000 to a US Rubber capitalized at $120,000,000 and with 27 factories scattered around the country; $64,000,000 to the 250 rubber plants independent of US and outside of Akron. Of the Akron business, Goodrich, Goodyear and Firestone accounted for about 75%. Small wonder, then, that the caoutchouc city thought its Big Three had carried it into the midst of a boom. Actually, it was only on the threshold.

5.2 Boom Town

The time stars fell on Alabama was as nothing compared to the time the Alabamans fell on Akron. With West Virginians, Tennesseeans, Kentuckians, Pennsylvanians,

Magyars, Scottishmen and Serbians they rained down on the overgrown town on the canal summit; choked its sidewalks and streets in a living tide of humanity that three times every 24 hours rolled in and out of East and South ends; washed over corporate boundaries and spread across the countryside for miles in every direction. In 1910, three out of every seven Akronites were invaders arrived since 1900. In 1920, two out of every three were members of the army of occupation that came in after 1910. There had been, of course, strange portents of the coming of this horde: gigantic smoke signals writhing into the skies; black snow pelting a sooty citizenry in summer as in winter; the mystic stink of caoutchouc in heat wafting over South and East Akron; acre after acre of block-long buildings suddenly materializing to add to the cloud by day and the thousand-windowed glowing of unearthly light by night; the roar and growl and mutter as of metal monsters caged in the colossal brick prisons swelling to the mighty voice of a new city.

These mammoth factory conglomerations - several of them cities in themselves - were the answer to the demand for auto transportation that changed the face of the nation after 1910 and more particularly from about 1914 on. Total passenger car and truck registration of 1,258,062 in 1913 and 9,231,941 in 1920, total car and truck production of 485,000 in 1913 and 1,905,560 in 1920: there lies written the story of the tire need. Rubber factory output of a value of $300,993,796 in 1914 and $1,138,216,019 in 1919; there lies written the story of what supplying the need did for the industry founded by Charles Goodyear. Akron going into that era as producer of little more than a third of the total value of rubber goods, coming out as producer of a half: there lies written the story of what concentration on casings did for a city that from first assumption of the title of rubber manufacturing capital of the world always was more accurately its tire capital.

Title of world's largest tire manufacturer was seized by Goodyear with a production of 4,II 8,300 casings in 1916 and never thereafter was relinquished. The year saw also the founding at Akron of General Tire, destined number five rubber company, by W. F. O'Neil, son of an Akron merchant prince. Sent West for his health, O'Neil had signed up as a Firestone distributor with headquarters at Kansas City and five states as his exclusive territory. As fast as he built up business, Harvey would take territory away, however. So the incensed Irishman came steaming home to found his own tire factory. The next year Goodyear, with sales of $111,451,000 drove around Goodrich (sales $87,155,000) and became second to US Rubber among caoutchouc manufacturers of the world. For 1919 it was Goodyear, $168,915,000, as its pneumatic auto tire production passed eight million; Goodrich, $141,343,000; Firestone, $9 I ,079,000.

Two, three and four these Akron companies stood in the nation and, to appreciate what casings meant to them, remember that Goodyear and Firestone were nearly all tires, Goodrich 55% so. As to the sort of profits found in them, note that Goodyear paid stock dividends of 100% in 1909, 1910 and 1912, of 20% in 1914, of 100% in 1916, and of 150% in early 1920 as its stock stood at over $400 a share. Note that its earnings on common, including stock dividends, had averaged better than 50% a year for the 12 year period up to 1920. For 1919 Goodyear netted $15,331,000, Goodrich $14,247,000, Firestone $7,995,000. And speaking of Akron's small companies, $2,000 invested in Mohawk as it came into being in 1913 would have accumulated a value on the books of $28,886 and brought in a total cash dividend of $II,981 in little more than a decade. Nor was it the most successful of the new school of small fry.

Immense fortunes were made on the stock market, of course; substantial ones by Akronites who didn't even buy or sell, merely raked in the regular, extra and stock dividends on shares fortunately purchased before the great boom hit. Some idea of what was done with smart money can be gained from the fact that this Ohio town was estimated to house 80 to 120 millionaires as 1920 dawned. And that several of the barons were reputed to be paying income tax on $2,000,000 even before 1916. Getting down to smaller but still authoritative coin, the town anthology bulges with tales of mournful widows left with stocks and little else who considered themselves well-nigh penniless before waking in a few years to find themselves among the Akron wealthy. Of schoolteacher roommates chipping together to buy a few shares from suave salesmen, remorsefully sticking the certificates away in an effort to forget, retiring on their profits not long after. Of those who, urged by clamorous families or inner promptings, got rid of their investments before the heavy dividends started and for the remainder of their lives went about ticking off on their fingers the number of thousands or hundreds of thousands they had thereby lost. Once the real boom was on, of course, every storekeeper, shop owner and petty professional man in the city was tossing all he could scrape together into the caoutchouc gamble. Many bought the handsome certificates of factories that never came into existence, of companies that stuck up a building just in time to fail and leave premises to be taken over by some newer promotion, of concerns that got going and hung on for years but paid no dividends or extremely insignificant ones. Thousands, of course, bought in too high and hung on too long in the case of the major companies but were far from losing everything. And just to prove that super-opportunity hadn't thundered on the door for the very last time in those years when Harvey Firestone was sweating to sell his earlier stock issues, General, launched so late as 1916, went on to hang up one of the finest long-time showings in the Akron record. Harvey Firestone, incidentally, is still quite a stock salesman. In 1917 he decided that every employee should be given an opportunity to buy one to ten shares of common at 75 cents a week per share deducted from the envelope. Later, purchase of shares was made a mandatory condition of employment, this stipulation prevailing until just recently.

Through the boom, as before, all of the Akron companies did their expanding at home. Sales branches and distributors throughout the world they had, raw rubber buying agencies at all the Singapores and Colombos — but nothing more. Exception proving the rule was Goodyear, which however, had also done the most spreading out within the tire capital. Goodyear built a tire plant and a textile mill at Los Angeles, as well as a new Canadian tire plant at Toronto. And it acquired coal mines in Southern Ohio, rubber-growing acreage on the island of Sumatra and cotton acreage in the Arizona desert. The last is of peculiar interest since Goodyear's 37,500 planted acres near Phoenix still constitute the only cotton plantation operated by a rubber company. Overwhelming proportion of all American cotton is short staple, while for auto tires it is the stronger and more flexible long staple that is desired.

Medium staple is the Mississippi delta product. Genuine long staple comes from Egypt's Nile and the coastlands and islands of Georgia, Florida and the Carolinas. Egypt long had made a nuisance of itself by operating restriction schemes along the line of that Stevenson rubber restriction act still in the womb of time. In 1914-1918 Britain at war was needing the Nile stuff for its own uses and the boll weevil was helping itself in Georgia. Following up United States Department of Agriculture work, then, Seiberling's company brought irrigation, Indians and Mexicans to the Phoenix region and did well until the 1920 of tumbling cotton prices. It still grows cotton there, but it also engages in the oddest assortment of businesses ever tackled by a tire manufacturing concern. Cattle are fattened on the alfalfa that has to be alternated with cotton and are sold by Goodyear Tire & Rubber Co., for beef. Oranges and grapefruit occupy some hundreds of acres and the classy structure stuck up for the use of visiting officials from Akron is operated as a winter tourist hotel complete with tennis, golf, swimming pool and saddle horses.

Minor, indeed, however, was Goodyear's Canadian, West Coast and Arizona building compared with the construction work being thrown up by the rubber companies in Akron. In 1911 the Akron construction volume exclusive of private residences was $1,474,000. For 1920 this industrial and business building total was $15,653,000. Top for residential construction valuation prior to 1916 was the $2,812,000 of 1912. For 1919 it was $13,871,940. Indirectly responsible for all of the commercial building were the rubber barons and directly for many an important plant in addition to their own. Foundries and machine shops came into being specifically to service them with molds and calendars and tire building machinery, and the endless circle widened as newer rubber companies found the presence of so many of these essential suppliers one of the potent reasons for locating in the Rubber City. In the home-building the rubber companies played an even more direct part, both Goodyear and Firestone getting into the allotment business and selling houses to their workers in a Goodyear Heights and a Firestone Park now long since taken into the city. Elaborate clubhouses and athletic fields for the toilers these paternalists built too, and some of them before the period of excess profits taxes. Goodrich kept on turning out rubber goods and, when visiting magazine writers wanted to know the date when it would get into these sociological gambles, it took refuge behind the plea that a plant located in the heart of Akron (it almost is) prevented its doing home development work.

Caoutchouc company subdivisions and all, however, the housing problem never came close to being solved in 1916-1920 Akron. Fastest growing city in the world during the 1910-1919 decade, it tripled its population without benefit of annexing anything save once-vacant land populated by its own crowded-out children. For 1910 the census figure was 69,067, for 1920 it was 208,435. And this was within the city proper.

Cuyahoga Falls, almost purely a residential suburb for Akronites, grew from 4,020 in 1910 to 10,200 in 1920. Kenmore, laid out between Akron and Barberton in 1901, mushroomed from 1,561 in 1910 to

12,683 in 1920 with workers in the South Akron factories contributing the larger amount of the increase. Barberton housed a number laboring in the gum mines of Dick's city and its population increased from 9,410 to 18,811. Kenmore submitted to annexation by the Rubber City a decade later but Barberton and Cuyahoga Falls cling stoutly to independence although no more than imaginary lines separate them from an Akron that has definitely stopped growing at 255,000 (probably less now) and needs those mergers if it is ever to give the booster boys something to talk up. Sad indeed has been the fate of these patriots, for in the brave days of war and post-war, when they should have been in their glory, things were happening too fast for the tongue to overhaul them.

By this roundabout way, then, we work back to the story of what did take place in the City of Opportunity (Chamber of Commerce) between 1910 and 1920; to the human and picturesque angles of a boomtown rush as colorful as any of those mine-town, oil-town and wartime munitions-town stampedes which have furnished Akron with its only inexact parallels in the America of our time. Boom Akron was not, incidentally, a war baby. The upshot had started before the war and continued after it. The chief effect of hostilities was to inaugurate gasmask fashioning and multiply balloon and solid truck-tire building while actually restricting production in some other lines.

Between 1909 and 1914 wage earners in the nation's rubber industry increased 89% in number. Between 1914 and 1919 they increased 139%. In Akron the increase was from 22,500 in 1913 to 75,000 in early 1920. Irish, Yankees and Germans — in that order Akron's settlers had arrived. There grew up a Dublin and a Goosetown separate in spirit from the main outpost of New Englanders, and the Irish, the "Dutch" and the 100 percenters still pretty much held their sections of the town as caoutchouc-processing got under way. First to be drawn from beyond were the Ohio farm boys, who came hopefully in around 1903 or earlier to see about that there 10-15 an hour wealth they had heard was being tossed around. The supply of these prospects was wholly insufficient and by 1910 the rubber barons were drawing on West Virginia for men to hire in at 16 cents an hour and maybe make 30 cents on piecework if they could step fast enough. By 1912 they were advertising for men in newspapers all over the country and were paying beginners 17½ cents an hour. As unskilled and semi-skilled labor the toilers were classified. Some of the jobs required tremendous brawn, but the mechanics of anyone of them could be picked up in a few weeks' time.

In 1913 a strike gave the boys a holiday and there was a slackening down from other causes, but by 1914 the hunt for muscle was on again. Over the Ohio border from West Virginia and Kentucky the recruits came pouring; newspaper advertising was redoubled; Frank Bucks were outfitted at the factories and chased South to comb the hills and brush for husky lads and bring them back alive. Manpower scarcity was sending the wages up and now the most effective of all advertising—word of mouth—was working for the gum kings. In the Slav lands of Europe the mispronounced name of Akron began to be bruited around. Through the black belt of America ran the rumor of the princely coin to be picked up by shines working as handlers of materials for Massa Seiberling and Massa Firestone. In the hills of Scotland the Akron wage news was told, the hills of Albania, the hills of Tennessee. In the years 1916-1917, better than 20,000 workers actually were funneled through the $20,000 miniature union station built to serve the 27,601 Akronites of 1890 and still doing its best for their quarter million heirs and assigns today. Bringing their womenfolk in some cases, this batch of newcomers also towed 3,000 children of school age. Thirty rubber factories the Akron district now had and by June 1920, Goodyear alone was employing 33,000.

Into this Ohio town strategically located where two abandoned canals crossed, Singapore, Colombo and Para sent a third of all the world's rubber; to work that rubber the world sent men in from its every corner. The metropolis of Simon Perkins became almost as cosmopolitan as the metropolis of Stamford Raffles. On any corner one might encounter a bewildered hunkie seeker after Case avenue's firewater wistfully inquiring the way to "Cassavenue Street." On every corner one saw the strange men who talked with their hands. Lack of speech and hearing was no handicap in certain of the rubber jobs and so the city came to house the largest colony of deaf mutes in the world.

Though quarterback and center had to line up tail-to-tail when signals were being called, the Goodyear Silents were as tough a football team as the state owned and they did not use too many such all-talking ringers as the pre-prize ring K.O. Christner. Pugs of their own the mutes had, each rejoicing in the "Dummy" nickname, and there was a downtown recreation center advertising that speech had been restored to one of the handicapped at the height of a session with a daughter of this house.

Plenty of Scots the city had to give the Magyars competition on the soccer field. Enough Turk rubber workers there were so that they had a coffee house of their own down among the Greek emporiums and the haunts of the Sicilians. Albanians, Montenegrins, Rumanians, Russians, Syrians, all of them worked rubber. Welfare workers in each factory had at them with Americanization programs and "first-papers" was as often as not the first phrase they picked up after "get the rag out" and "rawhide!"

Products of the American earth outnumbered the men from Europe about six to one. And topping the native stock division as the Hungarians did the exotics were the boys from the hills East of the Ohio river. Semi-seriously it long was claimed that Akron held more West Virginians than any city in West Virginia and certain it is that one state society was not enough to hold them all. Let an outsider cast any least aspersion, however, and they would present such a united front as the USSR might envy. Hard-up conductors of letters to the editor columns in the daily press were accustomed to drop an occasional sly hoot into their spaces and bring in sufficient red-hot missives to carry them for six months. Again, there was a judge of the court of common pleas who handed out a suspended sentence to a criminal case defendant on condition that he return from whence he came and never again set foot in Ohio. This was a routine condition, but it happened that the man was a West Virginian and it also happened that there was a reporter present whose depraved sense of humor always sought an outlet on a dull day. So he knocked out a lead about the judge sentencing a man to West Virginia for life and a helpful copy reader repeated the motif in the head. West Virginia in Akron started to boil the first day and in another 24 hours every newspaper in the state across the river was ablaze with red or black ink editorials flaying the innocent lawgiver as the black sheep of the Iscariot family. Candidate for elevation to the appellate court he was at this time and rejoicing in the reputation of champion vote-getter of all North Eastern Ohio Democracy. Fortunately, he hadn't surrendered the lower bench while aspiring to the higher.

So kind friends pulled him out from under the avalanche of ballots and toted him back to his own court. And even to hold that one he has had to loose profuse explanations, apologies and regrets every time election year rolls around. Able to get away with wheezes about "Akron, the capital of West Virginia" were the vaudeville comedians. And to tell the one about mountain children being controlled by the disciplinary threat of not getting to Akron when they grew up unless they mended their ways. These were privileged jesters, however, and politicians enjoyed no such immunities.

Favorite quip about the boys from all the Southern hill states was that they had had to be caught and held down while shoes were affixed prior to the trek North. Between Southerner and Southerner this one was standard. Actually there were any number of them who never had seen a street car, for instance, before heading for fabled Akron. Conductor and motor men jobs, incidentally, were the transition stage between backwoods and gum shops for hundreds. Green lads donned the blue and brass, thrilled to having the photo in uniform taken for the folks back home and in a few months made for the tire shops, grown great in confidence and eager for the big money. Good tire-builders they made and they dominated the high-priced labor divisions as dirtiest of the factory jobs went to the Cudjos and Sambos and the next least desirable to the patient Hungarians.

By 1916 the probationary wage while learning a rubber "trade" was up to 32½ cents an hour, ordinary iron-muscle labor to forty cents. Farm and mountain boys only a month or two away from field or hollow knocked off $5, $6 and $7 a day as expert tread room workers, rim painters, tire strippers who hauled the new casings from their iron cores with bull strength plus quickly learned knack. At the height of things pay of $60 to $70 a week was common in the tire departments. Daily wages with no overtime went as high as $15 and $16 a shift, weekly wages as high as $100 and more. The records show such payments as $28.07 to one man for a one-day double shift, $102.35 to a curing room hand who worked three double tricks that week. Hot for big money and eager for overtime were the workers and there were numerous cases of coin-crazed toilers traveling under two names and simultaneously holding man-killing jobs on different shifts in different factories. Not confined to the actual workers with caoutchouc were the big payments. Mammoth tire plants require skilled workers in a great range of trades and so we find machinists and bricklayers topping $90 a week.

Nor was it necessary in those days for the rubber worker to take any man's ragging. Nearly every factory was hiring every day and the worst that ever happened to a fellow was occasionally having to go as far as Kent or Mogadore in the next county. So from plant to plant they moved as they grew tired of the same old surroundings, or sought to land in the same department with some particular buddy or took a dislike to the color of the foreman's hair. And there were the wandering biceps boys seeing the country for the first time as they took a swing around the industrial circuit, working a few months in an Akron gum mill, then on to a Toledo auto parts factory and so over to the Detroit motor madhouses or East to Youngstown or Pittsburgh or back along their route to Cleveland. Goodyear found that keeping thirty thousand at work over a one-year period meant hiring in 60,000. And very costly that was, but the barons were rolling in such wealth that they couldn't be bothered. Nor did they begrudge the $60 and $70 pays handed the toilers out of fabulous profits, for the government would have taken it if the tire builders hadn't. So open-handed did the corporations grow that they even tossed out bonuses to white collar workers; $3,000 one year to a chemist is a case we happen to recollect. Nor did they spare the coin and spoil the government in the matter of entertainment for the gum workers. When track and field competition between two factories grew really marked, one of them brought in Joie Ray to hold some sort of mythical job and burn up the cinders at a princely salary. Minor league baseball the rubber shops killed by putting together industrial teams able to battle big league exhibitionists on better than even terms—as a couple of no-hit victories hung up by tire teams at the expense of the visitors from the majors amply attest.

Free handed spenders demanding the best of everything were the casing-makers, too; but one thing they couldn't buy was housing. At its worst was this situation during the first great 1916-1917 rush. For the single fellow a room of his own did not exist and in the rooming houses handiest to the factories, the beds worked three shifts the same as the tire departments. One gummer got out and another gummer got in and the landlady changed bedclothes when and as she felt like it and could find an opportunity. Among the proletarians one of the record cleanups of those years was made by a lad yanking down $6½ a day in each of two gum shops while at home the wife was doing $125 a week with boarders.

Out in the wastelands South of town Harvey Firestone in 1916 ran up barracks for 1,500 of his thrifty hunyaks. The setup was almost exactly like that of the coolie compounds of the Middle East. The tuans provided the quarters, paid the cooks, supplied the water and made periodical sanitary inspections. And the boys purchased their own goulash, in this case by pooling as a gigantic buying combine at a cost to each member of about $14 a month. One day Harvey got to worrying about the amount of money this left in their kicks. Out to the barracks was chased the manager of the foreign department of the newly organized Firestone bank and in one afternoon he rounded up $25,000 in accounts. Nothing like that could escape the Firestone eye. Once another of the partially-Americanized workers was knocked for a twister in the steel plant and his foreman took charge of the $1,800 in his money belt before shipping him off to the hospital. The man recovered and the day he returned he had to goosestep his way across the street with the foreman and shove the coconuts under a grille in the Harvey financial institution.

In nearly every case the workers making down payments on homes in the factory allotments figured on taking all the future payments out of the pockets of boarders. Whole houses just weren't for rent. The average family man could buy or he could build. And meanwhile he could install his brood in a chicken house, a garage, an attic or a basement for a mere $20 to $30 a month. Families rented city lots and tented out for the summer. Or shoved up shacks in the outskirts, a custom that survived the boom and has ringed Akron with hovel collections where Jeeter Lester would feel right at home. It was a case of the contractors just not being able to build fast enough and here again rubber wages played some part. Small contractors went broke while trying to get material through or keep employees from bolting to the gum shops. Large firms brought their own workmen in and often lost numbers of them.

Nor were rubber workers the only ones engaged in frantic quests for a place to lay the head. From coast to coast none too successful doctors, lawyers, small merchants heard of El Dorado in North-eastern Ohio and came galloping in for a share of the greenbacks and black silver (coins carried in gum worker pockets literally turn that color). The hotels, they found, were places where clerks were maintained only out of deference to tradition; there were no rooms to be had. Apartments might be rented if one got in a bid before the excavating was over. Real estate speculation ran wild, of course, and paper millionaires were made by dirt as by rubber shares while farms were being hacked into building lots at every point of the compass.

Floundering far behind a community that had run away from it in all directions was Akron's government. It was unable to supply sewers, water, pavements, even keep up the pavements it had. Transportation was as far back. Lined up for blocks would be the street cars at time for a shift to go off duty, but when 5,000 to 7,000 men came storming out of the factory gates to blacken streets and sidewalks in a sardine jostle, the conveyances were as inadequate as so many toys. Real buses were still of the future. Not yet had the tire factories developed and pushed the larger pneumatics that would make possible the trans-continental bus lines and the real city fleets while simultaneously cushioning truck-wheel slogging of pavements with saving air (pneumatic truck tires did not until 1926 pass the solid tires that now practically have disappeared). There were the jitney buses, however. Not in Akron did the idea of taking passengers for five cent rides in disintegrating Fords come into being, but it was the tire city that fell upon the inspiration with the most frenzied enthusiasm. Like a flight of June bugs the cars appeared overnight and soon the streets were literally crawling with them and a jitney inspector had been added to the army of bureaucrats at City hall. Runs were short, seven or eight paying fares could be packed into a flivver and there was money to be made in helping the inadequate street cars move Akron.

"Akron couldn't even get enough disorderly houses in those days," is the way the Hon. Gus Kasch, councilman, school board member, state legislator and town scourge once summed up the period. At least, however, there were sufficient of them so that the gummers did not find it necessary to form in sidewalk queues at their doors. This was true of no other form of enterprise catering to the diversion seeker. Lines formed for blocks at the three-shift (excuse it, please) burlesque house, the movie emporiums — the very cafeterias! Tire builders intent on getting "the works" and handing the tonsorial artists $2.50, took tickets and waited their turn — for hours on Saturday nights. It was the same in the pool rooms and the bowling alleys. Saloons near the factory gates packed them in until an unsophisticated observer might have thought mobs were out wrenching the buildings from their foundations and carrying them away. And when the saloons went out and raisin jack and corn came in, it was a case of six-deep at everything resembling a bar. Jewelers were accustomed to having the boys roll in and demand the biggest diamond on hand. Occasionally some of the

Southerners would have sparklers set in their teeth, too, but it was not the precious stone that was the caoutchouc worker's distinguishing badge.

It was the silk shirt. By the half-dozen they bought them, paying \$6, \$10, \$18, registering indignation when salesmen new to Akron tried to show them \$2 and \$3 stuff. "Is this the best you got?" was the voice of the city in the shopping district and a cigarette burn in a red or yellow masterpiece meant into the ashcan with it no matter what it had cost. Sometimes the prouder lads even wore the silks to work.

It was a city without women — so much so that Clark Gable used to consider himself in luck when he achieved an occasional blind date. This was in 1919-1920 when the Great Lover of the MGM lots was successively Miller Rubber filing clerk, Firestone rim plant timekeeper, haberdashery clerk and call boy at Akron's Music Hall. Also to be seen in Akron at this time were such odd spectacles as: Hart Crane, one of the inheritors of the Walt Whitman mantle, jerking sodas in a downtown drugstore; Nicholas Bessaraboff, translator (with Claude Bragdon) of Ouspensky's *Tertium Organum*, punching a gum shop time clock; Reverend Lloyd C. Douglas, standing up in his fashionable pulpit to denounce Akron as "a hick town" ere faring forth to captivate the hicks of the nation with super Harold-Bell-Wright novels currently menacing the sales records of *Ben Hur*; Wendell L. Willkie, minor member of the Firestone Tire legal department and future president of J.P. Morgan's Commonwealth & Southern Power Trust, impersonating Abraham Lincoln at a civic pageant in one of the town's parks; Helen Jepson, today's svelte, platinum eyeful of Met and movies, as a roly-poly schoolgirl, loosing the voice from the high school stage but lately trod by The Mikado of three-hundred-pound Jake Falstaff, shortly to be nationally renowned as one of the best of the newspaper columnists. Getting out the city's two leading gazettes were W. Kee Maxwell, star contributor to Judge in the last great days of American humor, and Charles Landon Knight, grand old personal journalist who coined "Saint Woodrow" and "The Wonder Boy" and editorially opined of Ruth Elder that "Ladies who go down to the sea in airships must expect to get their pants wet." It was that kind of a rowdy town.

With factories religiously closing from midnight Saturday to midnight Sunday, the Sabbath always presented the spectacle of 60,000 to 75,000 rubber workers loose in the town with nothing much to do. And so canoes jammed the dozens of lakes pretty well ruined by the Rubber City, roadhouses and amusement parks multiplied, dance halls sprang up beyond the corporate limits of a city too pure for Sunday dancing, Firestone-Goodyear baseball games drew whopping crowds, Akron football promoters bucked Canton's Jim Thorpe with Fritz Pollard and Paul Robeson years before New York or Chicago would deign to watch professionals on the gridiron. Week nights it was a city of public dance halls, roller rinks, the cabarets that sprang up only with Prohibition. This last oddity was accounted for by the fact that in Akron as in most

Western Reserve cities melody had been exiled from the drinking places by virtue of local ordinances which made the penalty for permitting music in a saloon almost as severe as the one for permitting minors there to hold forth. Only because of the Eighteenth Amendment, then, was it possible for those on pleasure bent to lift elbow, cup ear and shake hip within one and the same room. With shifts coming off duty at 2 pm, 10 pm and 6 am, there had to be something on tap at any time and gambling supplied a big part of the answer. Nearly every East or South Akron poolroom had its games and in a number of the downtown hotels there was real action. Faro dealers the town had of the caliber of gigantic "Cy" (for Cyclone) Hill who won his nickname from his speed in mopping up argumentative Texas bar rooms and who rolled into Akron by way of Colorado's mine camps. From Sistersville, West Virginia, oil boom town of the early part of the century, came "One Arm," one of the best stud dealers of his time, "Big Dick" Griffin and "Little Frenchy" to hold open house at the Buchtel hotel in the heart of the downtown. Often raided in its day was the Buchtel and so too with the Windsor, founded by bone-dry oatmeal baron Ferdinand Schumacher as a temperance house.

The scarcity of women resulted, of course, in girls fair or girls plain having their pick from among the silk shirt boys. Though the factory offices employed hundreds of them, the factories themselves used comparatively few and for the young woman drawn to the boom city it was tough going. Most of the boarding houses wanted men only — they had the money to spend and they weren't much trouble. So few were those accepting girls that when one attained admission she had to share cramped quarters with two or three others and still pay seven or eight dollars a week. Street car riding during the rush hours was almost impossible and this meant long walks to and from work. So acute did this housing problem become that Goodyear had the steel for a girls' dormitory building up when the bottom fell out of things financial in the last half of 1920. The skeleton structure still stands as a monument to the big boom. It was, of course, the extraordinary scarcity of laborers after the United States entered the war that really brought the women into the shops. Housemaids were impossible to keep as the factories hung up wages of $13 to $20 a week, and Amazons in jumpers became a familiar sight on the streets of Akron. Gumtown had an Amazon Rubber Co., at this time, incidentally.

How completely Akron had been a man's town was shown by the fact that, fifth in Ohio in population, it was second only to Cleveland in the number it sent to war. No women tire builders entered the picture to replace the men, but plenty of boys did. Youths of 16 and 17 made up to $100 a week and better by putting in plenteous overtime. So hard put to it to find policemen and firemen in the face of gum shop competition was the city government that it resorted to advertising for them with telephone pole placards. Not only motormen and conductors now but newspaper reporters, clerks, poolroom loafers fearful of work or fight roundups swarmed to the tire plants. Anyone who applied for a reporter's berth was hired on. When they were from out of town they seldom lasted for one assignment, yet none of them ever were discharged. Out on a story would go the newly-employed stranger, hear about the rubber mill wages and hire in as a factory hand without ever bothering to report back.

The toll of the influenza epidemic in the crowded South Akron rooming houses was one notable effect of the war on boom town. Another was the intensive building of aircraft for the Government that filled the air over Akron with dirigibles and balloons. Goodyear, in the balloon business before the war, stayed on with dirigibles and then with Zeppelins in the days of peace, but the other Akron companies bailed out with the Armistice.

Rubber workers coming back from the front, incidentally, were among those veterans who actually did find their jobs waiting for them. The gum shops still needed more men than they could find. And the veterans stepped right back into that silk shirt, three-shifts-to-a-bed life where they had left off. Through 1919 business boomed and the rubber barons recruited through Dixie with such record intensity that in early 1920 gum shop employment was up to 75,000 at the peak of prosperity wages. As with hill-billies, so with the other raw materials of tire manufacture. In early 1920 Goodyear every month was using $4,500,000 worth of 55 cent rubber, $4,500,000 worth of $1.50 cotton fabric and was contracted far ahead at these prices. A four months' supply of rubber had at all times to be in hand and on the way, for delivery from the East requires that length of time. And an eight months' supply had to be financed for so much time elapses between placing the order for crude and collecting from the dealer in tires. As to fabric, the supply was less than the demand. The giant companies were bidding against one another for it; the tiny companies were at wit's end trying to obtain enough for their needs; the New England manufacturers were taking the gamble of erecting new mills only as the tire companies would guarantee to take specified amounts under long time commitments involving years rather than months.

All the major rubber companies were involved in such contracts, of course, but none to the extent of Goodyear. It was supplying 50% of all the original tire equipment used by the auto manufacturers and from Detroit had come the word that it should be prepared for incomparably the greatest demand in history. Buy rubber! Buy fabric! These were slogans in the closing months of 1919. And 1920 opened with a terrific burst of business that more than seemed to justify the super-prosperity prophets. In 1908 Seiberling's company had done a business of $2,190,000; in 1919 it was $168,915,000 and for 1920 it seemed certain far to surpass the $200,000,000 mark and perhaps even reach $250,000,000 to $300,000,000. In the early months of the year the auto companies were shooting production skyward as they had promised;

tire dealers were turning over stock so fast that their orders were pouring in by wire instead of by mail; Goodyear was congratulating itself on a $20,000,000 profit in the purchase of raw materials as shown by the margin between current market and commitment prices. Almost it seemed that the plunging F.A. Seiberling had not been gambling in the least but merely riding along on a sure thing when he and his board of control authorized those unparalleled purchases to meet anticipated needs.

So with Firestone, the other tire specialist. In 1919 he had increased capitalization from $15,000,000 to $75,000,000 of which $40,000,000 was preferred, the issuance of $10,000,000 being contemplated. His biggest Akron building program was under way, he had just bought a boot and shoe factory at Hudson, Massachusetts, and by early 1920 he had borrowed $35,000,000 through note brokers for rubber and fabric purchases. To obtain a sufficiency of this last hotly sought material he had signed three-year contracts for the first time in his career. Secure in his belief that the boom would last forever, he then rented a mansion in England and took himself and family overseas to enjoy millions where they could properly be enjoyed.

US Rubber, despite a funded indebtedness of $71,600,000 (courtesy Kuhn, Loeb) felt itself to be in even plushier shape. With the end of 1918 Colt had moved over to chairmanship of the board and Charles B. Seger, office boy who had risen to presidency of the Union Pacific System and close friendship with Kuhn, Loeb, had quit railroading and stepped in as active head. For 1919 net sales were $225,589,465 and net profits after deductions $17,730,237. This was more prosperity than US could stand. Expansions at all five of its tire plants in the United States and Canada were started; authorized capitalization was increased to $300,000,000 (preferred from $75,000,000 to $100,000,000, common from $40,000,000 to $200,000,000, remaining balance of $10,000,000 second preferred retired); a $36,000,000 common stock: offering was made to common stockholders at par and instantly gobbled up; common was placed on an 8% dividend basis; $9,000,000 was sliced off the $52,310,163 total surplus and distributed as a stock dividend of 12.5% to the common stockholders. From 1892 until 1912 the common stockholders had waited for their first melon, from 1912 until 1919 for their second — and last. From October, 1911, until April 1915, they had received quarterly dividends and on October 1919, they began receiving them again — until January, 1921.

It was in the snapping of a finger that the whole structure of the nation's post-war boom business fell apart. The stream of tire orders dried up, cotton prices went into one long breathless drop, rubber prices came tumbling after. Heads dropped by thousands and by tens of thousands in the tire plants. By train and by auto the Southern army poured out of Akron in a retreat back to the hills and the safety of hog and hominy (name for the state of Tennessee) for the duration of this disaster. In the gum mines where 75,000 had been employed in June, 20,000 remained in December. Out West

Hill way 80 or more millionaires had become a dozen or less. On the sidewalks of the City of Opportunity stood beggars in $I 8 silk shirts putting the bee on passers-by for a thin dime. The boom was over.

On the way to Paris was Harvey Firestone when there came the cable that brought him home to be met at the dock in New York by company officials. Their report was that Firestone Tire now owed $43,000,000 and the banks wouldn't loosen up with another cent, that sales were vanishing, notes were falling due and couldn't be taken up, commitment materials were flooding in and couldn't be paid for. Two things saved Harvey at this time. One was that, being but half Goodyear's size and obligated for but half as much, he had no such huge amount to find. The principal life-saver, however, was of his own making. That was a decision not just to reduce tire prices but to whack off a fourth at one stroke and smear the nation with billboards, banners and full-page newspaper advertisements screaming the news of the 25% discount. It was a month before all the other tire makers had fallen in line. In that September and the succeeding month the graduate horse-trader moved $18,000,000 in tires. With factory, office and sales forces cut 70-80% and the banks lending a hand, that turned the trick. For 1921, net profit before inventory loss was less than a million dollars; after the write-down the deficit was $17,166,573. Harvey had taken up all the certificates of importation on rubber, however, had reduced his borrowed money account by $13,000,000. And so short-lived was this depression that he cleared five to seven million dollars in each of the years 1922-1924 and by the end of this last year owed nothing to the banks. So proud was he of his assumption of price-cutting leadership that in a 1922 letter to salesmen he boasted, "I realize that we have been the leaders in reducing prices to the consumer. This policy is based upon fundamentally good business, not only for ourselves, but for all dealers who want a permanent and profitable business." The leading slasher he remained for some years, but before the end of the decade the picture changed and through the depression we have had the spectacle of foxy Firestone lugubriously denouncing the vile practice of price cutting as carried on by others. Which goes to show, among other things, that the business of kicking prices to pieces is one thing when indulged in as an exclusive stunt, quite another when everyone starts taking part in the game.

Goodrich, so well connected in the East, navigated 1920 only by funding short-term loans through the issuances of securities having a high rate of return and unusually strong protective features. At least it got through, however. US Rubber, we have noted, continued its common dividends into 1921. It was a wholly unjustified boast. It meant that, although caught in the midst of a $28,616,616 program of expanding tire plants at Detroit, Hartford, Providence and Indianapolis and overbought on high-priced materials, the caoutchouc corporation with the really beautiful banking tie-ups chose superciliously to high-hat the panic. After all, 1920s early year business had been so vast that the year was the greatest in US history from sales, profit and every other standpoint; this downward flurry was passing and could be sneezed at. And so the mammoth complacently issued $20,000,000 in 7½ percent notes to help pay for the expansion program; dropped around to the banks casually and yet so regularly that its short-term loans skyrocketed to $61,205,000 (as against $14,000,000 it had in the banks); decided that $II,151,444 would do for a figure representing inventory write-offs; and set up an insignificant $6,000,000 reserve for future contingencies. Then it announced record sales of $256,150,130 and record net profits of $21,220,983 and, over $147,000,000 in debt, yet dished out $9,522,265 in 8% common and preferred dividends. Also it added Frank Vanderlip and John W. Davis, legal counsel, to the board.

In 1921 US had to take its beating and when Sam Colt died in August it was to leave the corporation in much the same straits as he had found it in 1901 though billions in sales and hundreds of millions in operating profits had passed through its coffers in the 20 years. In such circumstances, then, dawned the undivided rule of Charles Seger, one man who knew what to do with baronial power. He permitted no one else on the elevator when progressing to or from his twentieth floor office in the U.S building. Aside from Seger's ascent to his elevator throne, the US 1921 year was notable for a $40,681,798 reduction in financial obligations including forward commitments; an additional $24,821,289 write-down of raw materials and finished goods allocated to 1920 in which the operation actually should have been performed; a $91,443,509 drop in net sales and the payment of that last common stock dividend ever to come out of this corporation. Some $18,000,000 of the dollar sales drop represented price slashes belatedly made by US rubber in January 1921, slashes forced by the prompt actions of Firestone months before. In a depression January just 20 years before, US rubber itself had inaugurated the cutting. Not for years now had it been the price setter and in that single fact is implicit its failure ever to function as a tire trust.

So stout yet was the mammoth, however, that 1921s net profits of $492,81 I became $7,692,039 in the 1922 when there was no inventory rap to take; sales climbed slightly in dollar volume at still lower selling prices; short-term bank loans were reduced from $36,210,000 to $25,080,000. Funded indebtedness was up to $85,981,800, however, and requiring nearly $5,000,000 a year in interest charges and that other steady annual drain of $5,500,000 a year in stubbornly maintained eight percent preferred dividends was sapping the corporate strength exactly as it had in the 1890s.

For Goodyear as for Fisk of Chicopee Falls, MA (then almost regarded as number five of a Big Five), 1920 meant undergoing complete reorganization and surrendering management of the company to the financiers furnishing the necessary funds. That it was Clarence Dillon (born Clarence Lapowski) who controlled Fisk was claimed by counsel for Goodyear common stockholders in 1926-1927 litigation aimed at driving this sensational Wall Street operator out of the Akron company. It was also alleged that he had engineered joint purchases of properties by the two companies and it actually was proved that the corporations had bought fabric mills at New Bedford, MA, and Passaic, NJ, on a 50-50 partnership basis. What else Dillon did to Goodyear and how he was able to pocket it in the first place makes quite a story.

With the big crack-up, tires had piled to the roof at Goodyear; the $20,000,000 raw material price turned into a $20,000,000 inventory loss; those long-term fabric commitments suddenly loomed up as guide posts on the road to disasters. Stock that had been selling at more than $400 a share before the early 1920 dividend of ISO per cent, at as high as $125 thereafter, plunged to as low as $5. Seiberling slashed away at his million dollar a week payroll in wages and salaries; frantically and unavailingly tried to sell off raw materials under order; threw all his energies into attempts to root up the money to meet those diminished but crushing weekly and bi-weekly payrolls, payoff bills dismayingly become past due, satisfy his bankers with something. Sales of $4,500,000 in November after sales of $22,000,000 in April summarize the situation when considered in connection with the fact that the Goodyear of extravagant dividends had no reserve to live on, and had relied on income from current sales instead of maintaining adequate working capital.

Numerous were the companies in like case, but the thing that wrecked the whole structure of Seiberling's Goodyear was the Mid-West pride and independence of one who, always able to raise quick cash by local borrowings or stock sales, had relied on Akron, Cleveland and Chicago for his banking needs and allowed that he could get along without assistance from those he described as "the pirates of Wall Street."

Caught in the same sort of over-expansion as Seiberling, United States Rubber could increase its short-term bank loans to $61,205,000 in a completely off-hand manner. Seiberling had no friendly financiers to turn to and when he was driven to apply to the pirates scorned in the past it was naturally an icy reception that he encountered. After much heavy consultation, two Gotham banks finally offered refinancing to the amount of $25,000,000 to $30,000,000. The terms were harsh, but had the Ohioan accepted them the odds are that he would have saved himself. Instead he politely told the bankers to go to hell and himself went home to Akron to plunge again into the fight to meet payrolls by margins of hours, to dig up interest money and obtain the renewal of such bank loans as he had. With every possible connection made, with every last personal loan source tapped, with every resource exhausted, he finally applied again to New York and the upshot was that he virtually turned over control of his business to Goldman Sachs. Some $18,000,000 was provided under an arrangement for a revolving loan never to exceed $25,000,000 and the bankers took as security all liquid assets of the company and were to continue handling all its cash until permanent financing could be worked out. This was in November 1920. On the night before Christmas, the banking firm shoved a present into the Seiberling stocking. It was the announcement that it had failed to work out a satisfactory plan for permanent financing and would recommend appointment of a receiver. All hands chased attorneys to the Ohio state capital, Columbus, to watch one another for possible bankruptcy court moves and Seiberling made again for New York. Acting for him, Paul D. Cravath, New York attorney with high standing among the financiers, sought out Guaranty Trust and National City and was turned down. In January, however, Cravath was able to arrange a conference of creditors and bank representatives and, after a deal of talk about the gravity of a Goodyear crash which might pull other industrial giants with it, several of the bankers committed themselves to offering help if someone else would assume leadership and responsibility. So Clarence Dillon of Dillon, Read & Co., was called in and offered the opportunity to handle this unwanted one. He accepted, found the money without benefit of Morgan, and was off on the most spectacular ascent in the history of the Street.

Pretty much of an unknown was Clarence until the Goodyear opportunity whammed on his door. In 1919 he had had a hand in organizing Steel & Tube which later merged with six other concerns to form Youngstown Sheet & Tube and play a part in rubber as well as in steel history; but it wasn't a really big job. The Goodyear refinancing was. And with the profits and the reputation derived therefrom, Dillon was able to cut that swath through the 1920s which had the financial writers whooping for him as a Gould, Vanderbilt, Harriman and Morgan rolled into one. He flooded the country with Brazilian and Bolivian bonds which in a few years were to sell at prices little higher than his original commission; made loans to the Thyssen enterprises and bought some of the Stinnes properties in Germany; underwrote and sold $15,000,000 in bonds of a Japanese power company; handled loans for Holland and Poland; bought control of Union Oil of California from Royal Dutch Oil; grabbed Dodge Brothers right from under Morgan's nose for $146,000,000 spot cash in the form of the largest check ever written; underwrote in all $2,000,000,000 of new securities in the five years 1921-1925. And all this stemmed from what he did for and to Goodyear.

For finding the necessary financing for the rubber corporation, Dillon demanded nothing less than a control which not only involved the abdication of the Seiberling management but the disfranchisement of the 60,000 shareholders, a majority of whom were burghers of the Akron district. Without investing a cent in the rubber industry, without ever visiting Goodyear and Akron more than once, this Wall Street operator thus became a great overlord of the tire business and Goodyear, the home town's pride, became more truly a New York annex even than Goodrich. There has been a stock control concentrated in various banker hands almost continuously thereafter, but not since 1927 has the corporation been Dillon's. And Frank A. Seiberling is the chief cause of that.

Apparently finished for all time was 61 year old Seiberling on that day in May 1921, when he departed from Goodyear, his fortune of $15,000,000 gone, in debt to his old corporation in the amount of $3,744,729, loaded with other debts running the total halfway to eight figures. His debt to Goodyear resulted from the fact that for a long time Frank Seiberling and his corporation had been one. To Goodyear went free use of his personal patents; Goodyear notes he personally endorsed to the amount of many millions of dollars. And when he needed funds as an individual it was from Goodyear he borrowed without paying interest and without informing his stockholders. By standards of business ethics put out for public consumption, these last acts were indefensible. In a world of business as chaste as it pretends to be, they would be. Actually Seiberling did no more than an army of corporation executives was doing then, is doing today. As business is managed, however, those things come to light only when there is a control fight or a receivership. In the case of Goodyear there was a control fight and Frank Seiberling put himself on the spot, knowing that he would have to take it. He took it all right and those who dished it out were the bankers, then suddenly grown self-righteous after six years of keeping the transactions as secret as they had been before. What those bankers did to Goodyear on their own account, we shall see.

To the company the departing Seiberling had transferred great wads of real estate and railroad holdings in payment of his debt and as security for payment. To receive the transfer of all his other property and assets, principally 110,000 shares of Goodyear common, a holding company was organized and 2,500 shares of the capital stock of this Prudential Securities & Realty Co., was turned over to Goodyear to be held as still further security. Yet, thus straitened, handcuffed and bound down, Seiberling had a new tire manufacturing company going before the year was out and within five years had driven it past India, Mohawk, Mason, Lee Tire of Conshohocken, PA, and other such established corporations to take an approximate tenth place in the industry. Since then such larger corporations as Miller, Hood and Kelly-Springfield have vanished from the scene but Seiberling Rubber Co., still survives as an independent going concern.

For its survival, Edgar B. Davis is in good part responsible, the incident constituting a part of a lion - mouse drama proving the fabulists as right as rain. Meeting at a New York gathering of the rubber barons in the day of Edgar in caoutchouc land, they had become friends. Into dry holes in Texas went the Davis fortune, as we have recounted, and it was "FA," the multi-millionaire king of the tire industry, who responded to the plea of the trapped one. In the two years during which Seiberling's fortune of $15,000,000 was being reduced to less than nothing,' then, Edgar ran $57,000 borrowed from Seiberling to $12,000,000. And when he heard of his friend's situation, he grabbed a train for Akron. His figurative gnawing through of the ropes holding the old lion took the form of slapping down half a million in cash and underwriting Seiberling notes to the amount of more than $5,000,000. On the last day of 1927 a dramatic sequel was staged in a room of the Cleveland Trust Co. at Cleveland. Had there been a peeping janitor he would have been convinced that one of the heaviest poker games in history was in progress in those chaste premises. In the dealer's spot sat a short, square, moustached fellow counting out the money to be raked in by those gathered about the table in showdown formation. It was, however, no more than 67 year old F.A. Seiberling paying off his obligations—and in cash. When it came the turn of Edgar Davis to collect, there was a protest. Seiberling was trying to shove interest on him. Protested Edgar in surprise real or simulated, "That was no business proposition; it was just a friendly loan." Nor could he be prevailed on to accept the 6%.

Hardly so friendly were the loans Dillon arranged for Goodyear. What Goodyear obtained in the refinancing was about $5I,000,000 cash and credit. For this it paid not only 8% interest and the heavy expenses of reorganization, but a premium of over $I 1,000,000, a bonus of $6,000,000 in 8% prior preferred stock and a bonus of 275,000 shares of common stock worth even then at least $1,375,000 ($3,000,000, litigating stockholders later charged). To payoff bank loans and provide working capital there was, first of all, a $30,000,000 issue of 20 year, 8% first mortgage bonds sold by Goodyear at $90 for each $I 00 par and to be retired at $I 20. To raise additional capital there was a $27,500,000 issue of 10 year 8% debentures sold by Goodyear at $90 for each $100 par and to be retired at $110. With this went the 275,000 shares of bonus common representing about a third of Goodyear common. And then there was a $40,000,· 000 issue of 8% prior preference stock of the par value of $I 00 a share, retirable at $I 10, to be used in the settlement of merchandise creditors' claims on the basis of $80 per share. Further, the plan provided for the issuance of 10,000 shares of "management stock" and the creation of voting trusts for the new $40,000,000 prior preference stock, preferred stock and the no-par common stock replacing the old $100 par issue. The management shares of the par value of $1 each were to be issued to three men representing the three interested groups — Dillon, Read, bank creditors, merchandise creditors. And it was provided that this trio should have absolute control of Goodyear through power to name a majority of the board of directors. For years Goodyear had operated with the common the sole voting stock, voting power vesting in preferred shareholders only in the event of defaults. Under the new setup both prior preference and management issues were ahead of preferred in dividend matters, ahead of common in voting rights. And since the management stock would be surrendered only when the 20 year bonds and 10 year debentures were paid or otherwise retired it can be seen for how long Goodyear's stockholders were disfranchising themselves. That they voted to do this is not surprising when the alternative is considered. It was receivership, with the common stockholders being wiped out and the preferred stockholders subjected to assessments which many of them never could have met. Between the devil of Dillon and the deep blue sea of bankruptcy there could be only one choice.

Clarence himself was, of course, one of the three to hold the management stock as joint tenants. Representing the bank creditors was John Sherwin, board chairman of Union Trust Co., of Cleveland to which Goodyear had been indebted to the extent of a million and Frank Seiberling had been indebted to the extent of $900,000 secured by Goodyear stock. Nominee of the merchandise creditors was Owen D. Young, General Electric vice president who would become its board chairman in 1922. Young never was named in any of the stockholder suits, the effort being to dump everything into the laps of Dillon and Sherwin. Also named in some of the suits were directors they appointed.

How much, altogether, the bankers made out of Goodyear never will be known. Stockholder petitioners charged in court filings that profits "wrongfully made" amounted to at least $15,000,000 and they talked in terms of $50,000,000 in statements to the press. The accusations of fraud and breaches of trust were based on the operations of syndicates dealing in all classes of Goodyear securities-syndicates composed of Dillon, Read & Co., Oakmont Co., Inc. (owned by Dillon, Read), Union Securities Co. (Union Trust), Mid-Continent Securities Co. (largely Union Trust), and other companies in which Dillon, Sherwin and Goodyear directors allegedly were interested. A principal charge was that the Goodyear heads deliberately failed to have the company buy and retire debentures, prior preference stock and preferred stock at low market prices. Meanwhile the syndicates in which these officials allegedly were financially interested did the buying at the low figures and it was from these syndicates that Goodyear ultimately bought at prices immensely advanced. And during all this period, it was claimed, Dillon, Read had been borrowing large sums from Goodyear, the loans supposedly aggregating more than $7,000,000 at times.

To the historian of rubber it must forever be matter of regret that the truth of these legally lodged charges was never threshed out in a court of law. Fairness dictates recording that denials were made by all of the accused. It dictates also the recording that every action concerning these extremely serious accusations was settled out of court. And that prior to such settlements Dillon gave a very tolerable imitation of Red Grange's galloping ghost act in seeking to dodge service in any and every case. So much was this true that, with filing of the 1926 actions, all Dillon, Read offices in Ohio were hastily closed, firm names were scratched from the doors and managers were hustled off to New York.

So sound was this Goodyear that had been in such sharp distress for want of purely temporary credit in 1920-1921 that in the very 1921 of refinancing it cleared $7,307,876. To this was added $5,305,614 in 1922 plus $7,667,944 in 1923 plus $13,268,044 in 1924 plus $22,314,473 in 1925. To the preferred and common stockholders, however, this meant nothing as unpaid preferred dividends piled up to the amount of $25 a share. Over $26,000,000 had been paid out in interest charges on the new indebtedness; $15,000,000 had gone into retiring prior preference stock, $13,000,000 into taking up bonds and debentures. Small wonder then that there were stockholder suits in 1922, 1923, 1926 and 1927.

The 1923 out-of-court settlement of the earlier ones at a cost to Goodyear of some half a million dollars paid to attorneys for all parties provided for termination of one of the oddest contracts ever entered into by a corporation. This was a document undertaking in effect to transfer company management to a certain Leonard Kennedy & Co., Inc., for a five year period. It did not disturb Vice President George M. Stadelman, Goodyear's very able sales manager; Vice President P.W. Litchfield, very able factory superintendent; or any other Seiberling-selected lieutenants save for the treasurer who departed with "F. A." and "C. W ." It did result in Dillon's sending in a president, E.G. Wilmer; a treasurer, Herbert H. Springford; a secretary and a number of minor associates placed in finance, accounting and legal departments. Those who declined the Goodyear presidency immediately prior to the Kennedy contract included Gerard Swope, president of International General Electric and a bit later president of General Electric, and Herbert P. Howell, vice president of the National Bank of Commerce. Salary of $200,000 to $250,000 a year was talked about. Wilmer, a Milwaukee lawyer, came on for $50,000. On the face of it this was not an unreasonable figure even for one who frankly asserted "I don't know one damned thing about rubber," but the contract is worth looking into. For the extraordinary service of supplying president, treasurer and such other officials as it cared to name, the mysterious Kennedy outfit was to receive $250,000 a year plus 5% on Goodyear net earnings in excess of $10,000,000 per year. Salaries of president and treasurer would be paid by the Kennedy company, but Goodyear was to pay out-of-pocket disbursements of these men, including house rent and household expenses, and also to pay the salaries of all Kennedy nominees save the two key figures. Wilmer, we have seen, received $50,000 a year, treasurer Springford $15,000, a total outlay of $65,000 a year by the Kennedy concern. Yet in the 25 months the contract was in effect the Kennedy outfit raked in between $750,000 and $1,000,000 and had the agreement continued in effect for the full period would have averaged nearly a million a year! Owner of the largest part of the Kennedy company was another mysterious concern, the Nassau company. Owner of 80% of the Nassau stock was Mrs. Clarence Dillon.

Termination of the Kennedy contract did not terminate Wilmer's connection with Goodyear, however, and two months after the settlement of the suits he ceased being full time president at $50,000 a year to become part time board chairman at $125,000. Goodyear and not the Kennedy company now was paying the salary, of course. In 1925 Dillon acquired Dodge Brothers, and Wilmer and Springford went on to Detroit as president and treasurer although Wilmer continued to draw down his Goodyear salary until June 1926. All of which goes to show that management can come much higher than the $81,000 a year currently being paid Litchfield. It was not Litchfield, however, but Stadelman who became president as Wilmer ascended to the board chairmanship in early 1923. Elected a vice president at this same time was a now gone-from-Goodyear gentleman designated by Wilmer as his personal representative in the conduct of the company's affairs. Stadelman, a first rate executive, died in January 1926, and so it was that Litchfield mounted the throne at $100,000 a year in the year which saw the launching of the litigation which would end with the smashing of Dillon's control.

The 1922 suits Seiberling denied instigating. Testimony of the attorneys of the 1922 stockholders appearing as witnesses in 1926-1927 litigation was that Seiberling put up $10,000 in Liberty bonds as an original retainer in the case. Each and all of the 1926-1927 cases Seiberling also denied instigating, although he openly led the fight. Preliminary filings of 1926 were promptly followed by Litchfield's presenting to the stockholders a refinancing plan calculated to shake the intolerable 8% interest burden. Briefly it provided for the abolition of management stock and all voting trusts; for the issuance of 140,000 shares of new prior preferred stock having the same voting rights as the common; for the sale of $64,000,000 of new securities through Dillon, Read. It made no provision for payment or liquidation of the back preferred dividends, without which payment there could be no hope of common stock dividends.

The Litchfield plan was promptly rejected by the five-man group of common stock voting trustees of whom "P.A." and associates composed the majority. The claim was that the contemplated issuance of the new prior preferred with voting rights was intended permanently to fix the control of the company in the hands of Dillon, Read and associates. Common stockholders now organized a protective committee and in November a suit was instituted at Akron asking for removal of Dillon and Sherwin as management stock trustees and for an accounting and restoration in the matter of the $15,000,000 or more loss alleged to have been suffered by Goodyear through "breaches of duty" by the two. In court later on one of the plaintiffs testified that at a meeting in Cleveland just before the action was filed, Seiberling told the committee to go ahead but that it might mean his ruination.

Sherwin's answer to this suit first made public the Seiberling borrowings which had been matter of banker knowledge for six years. It was followed by a whole series of Goodyear suits against its former president asking interest on the borrowings, judgment for the difference between borrowings and amount realized by auctioning of securities, restitution of a million which Seiberling allegedly had caused the company to pay out in improvements at one of the Goodyear workers' allotments for which he was to have paid himself. Other filings pro and con brought to some 15 the number of actions in Akron, Columbus, Toledo and Cincinnati courts. And employed in deposition takings, arguments over motions, arguments for temporary restraining orders and other preliminary trifles was such an array of high-priced talent as seldom has been drawn into any legal battle in the country. Tops among the score of counselors from the celebrity angle, perhaps, were George W. Wickersham, former United States attorney general and future goat of a celebrated report on prohibition; Newton D. Baker of Cleveland, former secretary of war; Joseph P. Cotton of New York, but recently severed from his partnership affiliations with William G. McAdoo. No fitting sequel to the Webster-Choate clash was provided by the new version of a Great India-Rubber Case, however. For one thing, all of the later day Websters and Choates were on the same side, the side of corporation management and bankers, of Goodyear officialdom and Dillon, Read. For another, the assorted actions never advanced beyond the preliminary stage.

The principal suit having been set for trial at Akron in May 1927, Owen D. Young, forgotten man of the gigantic control fight, bobbed up as a mediator on the eve of battle and duplicated the success he had had as writer of the "Dawes Plan." Substantially this compromise was a victory for the common stockholders. It did away completely with the alleged joker of the Litchfield plan — the provision for a new prior preferred stock to be a voting issue. Dillon, Read was, however, to handle the $60,000,000 issue of 5% bonds to retire the outstanding 8% bonds and prior preference stock. Management stock was to be scrapped along with all voting trusts and there was to be an independent board of directors representing all classes of security holders. To be worked out was a method of satisfying accrued dividends on the preferred and automatically returning control of the company to the common stockholders. This last shortly was done when the stockholders ratified creation of a new preferred to be exchanged for the old on the basis of one and one-quarter shares for one. Back where it stood prior to the advent of the Wall Streeters was Goodyear but a great many million dollars poorer by reason of its experience with the bankers.

All the accounting and restoration actions against Dillon, Sherwin and the others were, of course, dropped in the settlement. The suits against Seiberling were not. In that connection his counsel issued this statement, "Throughout the entire negotiations for an adjustment of the Goodyear litigation, Mr. F. A. Seiberling at all times insisted that all claims between him and the company be left out of the negotiations and settlement and this has been done. It was his thought and desire that no personal interest of his should in any wise affect or influence the terms of the settlement." The suits last were heard of in 1935 when further continuation of these eight year old matters was asked in Summit county common pleas court by Goodyear attorneys who reported that amicable arrangements for settlement were being worked out as a result of a special master's hearing before Owen D. Young in which he ruled that Seiberling owed the company approximately $3,000,000.

Seiberling had been the largest single owner of Goodyear stock through the years of exile as well as the years of ruling. As the departure of Dillon sent Goodyear common into a market ascent, this situation did not long prevail, however. The stock went to Cyrus S. Eaton of Cleveland, an anti-Wall Streeter as bitter as "F.A." and one who was then in the midst of the rise to Mid-West financial and industrial domination that made him an even more spectacular figure than Dillon. Twenty-five percent of Goodyear common stock Eaton accumulated through acquisition of Ohio Goodyear Securities, Inc., holding company for the Goodyear interests of Seiberling and Edgar B. Davis, and buying in the open market. And in obtaining working control of Goodyear he obtained control of the world's largest rubber company. That it was, was because of the manner in which Seiberling had built, because of continuing his expansion program, because of the sales and production staff he had left behind in departing from a company that was in top shape in every respect save for its lack of ready cash or credit. Steadily increasing its share of total tire business as under Seiberling, Goodyear sold $230,000,000 in 1926 to drive around faltering V.S. Rubber, and hit a record $256,000,000 in 1929. In mid-1926 it had taken on a new kind of contract with a huge mass distributor, but even aside from this it was the number one rubber company.

The $256,000,000, incidentally, is co-holder of the all-time record for dollar sales of a caoutchouc corporation. To that same figure V.S. had attained in 1920, but it must be remembered that prices were immensely higher then. Not getting its proportional share of the business was V.S. in the later years of the decade, but for all of the larger and most of the smaller Akron companies the whole period was a record one in dollar sales as in production. It was the period of a still mushrooming auto production and consequent tire demand amazing beyond anything dreamed of even in the 1910'S. Goodrich sold $164,495,000 and Firestone $144,586,000 in 1929. And the whole American rubber manufacturing industry (some 500 companies) turned out products running beyond an average of a billion dollars a year for the entire decade.

From the sales production and physical expansion standpoint, then, it was a greater boom than the one of the preceding decade and especially for the Akron companies. It was, however, a boom that failed to make employment, a boom without profit to stockholder or tire dealer. In the tire capital for instance, employment recovered to, but little more than, half of the 1919 peak, but little more than twice the panic low of the end of 1920. And the Akron which had visioned a 1930 census rating of half a million on the basis of multiplying from 69,067 in 1910 to 208,435 in 1920 had to content itself with a sour 255,040, although its rubber workers now were in large part family men who had taken root instead of floaters and mere vacationers from the hills. The why of this is the subject of another chapter.

As to profit, earnings on that enormous billion dollar a year business of the entire industry were less than two percent. For Goodyear, V.S., Goodrich, Firestone, Fisk, Kelly-Springfield - six largest companies still in the ring in the early thirties — the average rate of profits on tangible assets had been 2.83% throughout the whole of the 1920s. For 191 2-20 it had been 12.5 percent. Nor was this the whole of the picture for we have not yet noted what the various refinancings accomplished. Between stockholders and profits had come bankers and bondholders whose interest requirements were I09.4 percent of profits instead of the 9.2% of the earlier period. Added to interest payments such dividends as were paid by the six during the 1920s amounted to I.95 times the amount of profits.

No market crash was needed, then, to take the profit out of the second half of the 1910-1930 rubber boom. Nor was it because of the November 1929, crash that the production boom ended, although the two events did come together. The reasons for these oddities we will now consider.

5.3 The Big Four

Patly into decades falls the nearly 30 years of rubber's history as big business. The 1910s constitute the period of the first wild boom with huge profits for promoters and huge dividends for shareholders. The 1920s comprise the era of financial reorganization and still greater production boom with profits and control passing to the bankers. The 1930's (to date) are the stretch of collapsing demand, vanishing companies and the readjustment of a billion-dollar industry on the fairly permanent basis of half size. Over lapping these last two calendar decades, however, is another organic 10 year period, the 1927-1936 of cut-throat price wars constituting the story we are now to tell. Whether that period will neatly close with the ending of the present year of comparative peace and mounting profits only time will tell. Conceivably it can be extended to 12 or 15 or 20 years. Hardly, however, can the next 10 be as rough as the 10 just closing, for so few are left to fail. The 5I 6 rubber companies of 1927 had become 351 by 1933. The 120,000 independent tire dealers of 1927 have become 60,000 in 1936. The 109 tire manufacturers of 1927 have become 28 (seven with principal or sole operations being carried on in the Akron district). Of the 28, just 19 are of some noticeable size, the others having daily production capacity of but 200 to 800 casings. There were, of course, any number of midget corporations passing out in the 1920-1926 period, but these were the always weak and unfit and it is only in the succeeding years that the large and once stout ones have been vanishing along with the bijous.

The campaign of extermination has been the work of the Big Four and its new allies: Goodyear and its partner, Sears, Roebuck & Co.; Firestone and its chain of factory-owned stores; V.S. (which was first very nearly a victim) and Goodrich with their Standard Oil and Montgomery, Ward associates. It is by company management — the barons that the battles have been fought, and in only one instance was management in the hands of those with a really substantial financial interest in the companies they head. In approaching the story of the 10 year war, then, it is of interest to examine into both the personalities of the actual battlers and the identity of the financial overlords who have been content to give the barons so free a hand in the matter of production, distribution and price cutting policies. In three out of four cases, it is worth noting, there was a change in management in or around the first year of the 10 that were to be so full of battling. And in two there was a change in identity of controlling interest.

Roughly speaking, management now is in the hands of a new order of rubber company chief executives — production men reared in the factories and with little if any experience with sales, distribution or executive problems before being transported to their front office spots. From their predecessor promoters and financiers they inherited extremely heavy debt burdens, US being the heaviest laden, Goodyear second and Goodrich third. To their successors they seem destined to leave an inheritance of mass distribution and commercial contracts guaranteed to produce no profits after the payment of interest.

The new deal dawned at Goodyear early in 1926 when P. W. Litchfield settled down in the president's chair after 30 years in the shops, 26 as factory manager for the Akron tire leader. Within a year the Dillon ownership that installed him was gone. Came then Cyrus Eaton's acquisition of stock control which passed to George T. Bishop, his successor as head of the collapsing Eaton enterprises, and then in part to Cleveland Trust of Cleveland and Chase National of New York, the institutions snapping up the bargains after the big crash. The report is that Chase National in 1935 talked turkey to Litchfield in the profitless operations matter, but there is no record of any of the bankers having interfered with his sales, distribution or production policies at any earlier date. Litchfield, president and chairman of the board, is the owner of 10,501 of the 1,493,870 shares of Goodyear common stock.

Management at Goodrich, a hot-bed of internal politics, twice changed in 1927-1928. As president, Harry M. Hough, former comptroller, lasted seven months. He was succeeded by factory manager James D. Tew (22 years in the shop) whose 10 years on top mean that his faction long ago consolidated its authority. Tew sticks to Akron and to getting out the rubber gadgetries while finances are looked after by Goldman Sachs and others in the East. He is the son of Harvey W. Tew, original Goodrich partner who sold out his quarter interest in the company for $7,000. In this case the son never will emulate the father, for his stock holdings are purely nominal, although his salary is $60,000 a year. Chairman of the board is David M. Goodrich, son of the founding doctor, and the company often makes proud reference to the continuity of its tradition by pointing to the fact that the sons of its two founders hold the highest offices today. Goodrich, a New Yorker, is not a rubber man and his office is chiefly ornamental. His, however, is the distinction of being the only officer or director owning any noticeable amount of stock. Nine of the 17 directors own no stock whatsoever and the sum total of officer and director interest amounts to but 1.1% of the preferred and but 0.5% of the common. Until 1930 the company's executive committee was composed of Tew, Goodrich and four of the minor executives at Akron. Today it is banker dominated with Tew as the only actual representative of management but with Goodrich also rated as such. There is no stock control, the widely scattered shareowners participating in company affairs by sending in their proxies while bankers and officers do the rest.

US Rubber control in 1928 passed to the du Ponts of munitions millions and General Motors, control. This was fortunate for US Rubber as otherwise there would be no such corporation today. It was not, however, a case of bankers to the rescue as at Goodyear in 1921, but of the Delaware dukes believing they were making a good buy. All through the 1920s US Rubber had been soaking up punishment administered by the price-slashing Akron barons who, as low cost producers, could both dish it out and take it. Up into 1926 this was the work of Harvey Firestone taking advantage of his freedom from the burdensome financing weighing on all rivals, taking advantage of an efficiency of operations challenged only by Goodyear. Not unreasonably rough did Harvey make the going for any corporation in any proper sort of shape, but US Rubber was not in such shape. And when the era of real price slaughtering arrived in 1927, this together with the terrific inventory losses of the period was enough to line up the ex-Trust for the knockout punch. Watery, inflated, over-financed was US Rubber and burdened with outmoded inefficient factories. Its share of total sales was decreasing and on this share its operating profit was but some 6% as compared to 14 for Goodyear and 16% for Firestone. Against the less than $600,000 annual interest charges being paid by Firestone, it was dishing out more than $6,000,000 as a result of having added layer after layer to its funded indebtedness. By helping itself to the profits of the plantation subsidiary, tapping the plantation surplus and postponing the writing down of inventory from year to year, it succeeded in regularly paying that 8% dividend, however. And to the casual eye it presented a not uninviting picture. So right at the end of 1927 and beginning of 1928 the du Pont Estate invested and into US Rubber common stock went $16,000,000 of the munitions millions of Pierre, Irenee, Lammot and those other boys old Charles Ranlett Flint hardly could have recognized as being in the same line of business. It was a logical investment, considering the du Pont interests in General Motors and chemicals and synthetic rubber experiments, and it looked like a good enough one at the time. Bought at $52 a share, those 300,000 shares (a fifth of the voting shares common stock and preferred then outstanding) speedily proved, however, to be the biggest collective headache ever acquired by the family. Right into continuous Akron warfare that year after year set new record lows for tire prices walked the du Ponts; right into the most terrific raw rubber nosedive of all time, the one that carried it from 20 to 22 cents in 1928-1929, to 12 cents in

1930, to6 cents in 1931, to 2.5 cents in 1932. In January 1928, the regular quarterly preferred dividend was declared. It was the last. And when Charles Seger on a day in December stepped into the elevator and dropped in lonely state for 20 floors, that was for the last time, too. Already a du Pont posse was at work on the $130,000,000 debt he left behind him. The first move was, in effect, an assessment slapped on the common stockholders and it was a necessary one indeed. Common, was changed from a par of $100 to no par and 728,412 shares were issued and sold at $35. Now the du Ponts had 600,000 shares (common) out of 2,000,000 (common and preferred), a 30% interest and control. As to the $26,000,000 it went into paying off short term bank loans instead of dividends. In January 1929, a du Pont management man arrived, F.B. Davis, Jr., (no relation to Edgar) as president and board chairman. While he exerted himself in centralizing operations and bringing operations to the current Akron standard of mass production, the financing problems were turned over to another du Pont appointee, William de Krafft, until then vice president in charge of finance for the Baldwin Locomotive works. The Davis salary is $125,000 a year. Since the company still is alive and the funded indebtedness has been lowered mightily, the consensus of rubber manufacturing opinion is that he earns it.

Firestone Tire, of course, is Harvey Firestone, although he has changed titles. Chairman of the Board instead of president he has been since 1932, but not one of his powers did he relinquish in taking on the new label which so often gracefully announces the shelving of a worn-out executive. The best guess is that he was about the business of establishing the succession well in advance of his translation to the heaven of the Episcopalians. His president at any rate is John W. Thomas, first a chemist and then for a long time, factory manager. Harvey paid himself $100,000 a year until 1930; since then he has annually transferred but $71,000 from one pocket to the other. His personal holdings of common stock amount to 260,000 shares or 11.6% of the total. The Harbel Corporation, a family trust, accounts for a like amount and the company itself holds 356,783 shares or 15.9%. Stacy Carkhuff, veteran secretary, has 46,360 shares; Russell Firestone, son and director, 29,512; Harvey, Jr., vice president and ambassador to Liberia, 24,920; Thomas, 18,510.

Figuring heavily in the profitless business of the 1920s and 1930s were the dizzy antics of raw material prices, but this was something the industry had been through before. Competitive price cutting was the chief trouble and more particularly after 1926 when it ran completely wild. And contributing importantly to the grief since 1929 has been the excessive production capacity resulting from the still continuing speeding up of workmen and the insane expansion of factory space that marked the later 1920s. Often attacked for over-expanding was Frank Seiberling immediately following his crash. Actually his every such move was justified and was shown to be so once the brief panic of late 1920 and early 1921 was over. For the building multiplication in which the companies indulged during the 1920s, there is no such excuse, however. Soaring demand there still was, of course; demand that ran sales from 22,000,000 tire units in 1921 to an approximate 70,000,000 in 1928 and again in 1929. So completely did the over-expanders overhaul and pass demand in this decade, however, that production capacity reached 106,000,000 in these years of 70,000,000 sales. In the panic year of 1930 sales were down to 54,299,000, and in 1931 to 48,870,000. For 1932 they were 39,817,000 and since then there has been a slow but steady climb bringing us to the 47,500,000 of 1935. For 1936 the estimate is 48,300,000 and there is sound reason to believe that they can't go much beyond 50,000,000 in any future not so remote as to be wholly unpredictable. Yet the Big Four alone has a production capacity of well over 50,000,000 today and the 24 surviving smaller companies raise the total to 68,000,000 — an excess capacity of almost 40% still remaining despite the rubbing out of plants with a 38,000,000 tire capacity.

That the major portion of the collapse in tire demand was inevitable and admitted of being foreseen is indisputable. With the basic situation the stock market crash of late 1929 and the resulting years of depression have nothing to do. They did, it is true, cut the number of tires sold to the auto manufacturers for original equipment from 17,689,000 in 1928 and 20,956,000 in 1929 to 6,000,000 in 1932. Since then, however, the automotive industry has recovered so completely that the rubber barons sold it 16,500,000 tires in 1935 and will sell it over 18,000,000 in 1936 — more than in 1928, more than in any year in history save record-shattering 1929. It is in statistics on replacement tires sold to the individual motorist that we find the tip-off on what has happened to the tire industry. These were the sales that not only fell off every year until the depth of the depression was reached but have been steadily falling off every year since. At the height in 1929 such sales were 49,498,000 units, at the depth in 1932 they were 32,725,000. In 1935, despite recovery in every line and a record gasoline consumption, showing that cars are being used more than ever before, they totaled but 29,500,000. For 1936 the estimate is 29,000,000 and the most optimistic estimates for the future do not go beyond 33,000,000.

The answer is a simple one - tires don't wear out so fast any more. Part of this is due to better roads and the advent of retreading, but in the main the answer lies in the fact that the factories have been building better articles. No one, of course, will argue that they should deliberately have held back improvements nor could this actually have been done save by a conspiracy in which every casing maker in the land participated. Study of the broad effect of their improvements should, however, have saved them from their over-expansion folly.

Between 1910 and 1916, and despite all the improvements wrought by the chemists, the casing makers had been selling the average motorist seven or eight high-pressure cord or fabric tires a year. With the cord being brought to a reasonable degree of perfection by 1916 and with the compounds steadily being improved, the tire engineers and chemists employed by the manufacturers succeeded in chopping this figure to three by 1919-1920. With more and more autos coming into use and the saturation point not yet in sight, continuance of such an average would have justified much of the 1926-1929 expansion. Prior to that time the manufacturers had introduced the low pressure balloon tire, however, and before the end of 1924 the sales to the motorist were down to about two tires a year. Today, with the steady development of the balloon into the still lower pressure tire now vended as the "air wheel," "low pressure" or whatnot the figure is between one and I.I 8 and the continuing growth of retreading should carry it still lower. And so it is that tire sales are off some 20,000,000 a year from the record and will remain there despite the general recovery that now offers 26,000,000 motor vehicles in use in this country as against the 26,500,000 of 1929 which mayor may not have represented close approach to the saturation point, practically speaking.

Nor has the motorist profited only by a tire life increasing from an average three 3,000 in 1914, to an average 8,000 miles in the early 1920s, to today's average of 20,000 plus the opportunity to add 10-15,000 more by doing business with the retreaders. Averaging the prices of all tire sizes) we find that in 1914 he paid $43.60 for his 3,000 mile casing or over $14 for each 1,000 tire miles. For the last several years he has been paying something over $10 for his 20,000 mile hoop or about 50 cents for each 1,000 tire miles. There was profiteering through the 1910s, of course. And with rubber amounting to 50% of a tire) it is patent that part of the reduction per casing represents the difference between 65 cent rubber of 1914 and the 16 cent rubber of today. Multiplied use of reclaim enters into the reckoning too, and lower cotton prices and multiplied efficiency in manufacturing operations. With all this taken into consideration, however, it has been the senseless cutthroat competition of caoutchouc barons at war that has brought and held prices to their present levels.

Between 1914 and 1920 tire prices fluctuated from the $30.50 low of 1915 to the $48.55 high of 1918. For 1920 the average was $45 and the panic and its price cutting brought this to $38.40 for 1921 (a 16 cent rubber year as is the present one). Thereafter and in spite of undue financing costs and steadily ascending crude rubber prices the new economy in operations which had succeeded boom-time carelessness kept tire prices trending markedly downward to the $24.50 of 1924. Then came dollar rubber and a jump to $29 in 1925 and 1926. With the end of the rubber squeeze and the start of the real tire price wars) the $29 charge tumbled to $21. 50 in 1927) to $18.75 in 1928, to $15 in the 12 cent rubber year of 1930. Thereafter it was $12.75 in 6 cent rubber year of 1931) $10 in the 2½-cent rubber year of 1932) $10 in the 6 cent rubber year of 1933. And $10 it has been in the 12-16 cent rubber years since.

So it was, then, that Hood and Miller stockholders had to take such dreadful beatings as Goodrich gobbled them up in 1929 and 1930; that Fisk smashed up in bankruptcy in 1931 and was picked up only in part by a new company organized in 1933; that Kelly-Springfield, fleeing Akron and locating at Cumberland, Maryland, crashed in 1935 and was sucked in by Goodyear; that Mason, Dwen, Northern, Swinehart, Murray, India and so many other once-familiar corporations wholly vanished in the late 1920s or 1930s, leaving nothing but abandoned buildings —and in many cases not even these to remember them by. And so it is that Goodyear has paid no common dividends since 1931 and earned none since 1929; that US Rubber has paid no common dividends since 1921, no preferred dividends since 1928; that Goodrich has paid no preferred dividends since 1931, no common dividends since 1930; that General has paid no common dividends since 1931; that stockholders in most of the surviving smaller companies have forgotten what a dividend check looks like. And that Firestone is regarded as a miracle man.

From 1927, the first year of super-warring, on into early 1934 there were just 17 price cuts and one small increase. Since then there have been more increases than slashes, but profitable business for the industry as a whole still is around the corner and the chances of further price boosting or a renewed frenzy of cutting are just about even. Even more impressive than the figures showing the plummeting drop of the average tire price, perhaps, are those for any specific one of the cheaper sizes. In July 1925, for instance, the Ford owner was paying $17.24 per casing to his retail tire dealer. In March 1933, it was $3.89.

The history of the super-warring begins with the scratch of a pen on that day in May 1926 when newly named president P.W. Litchfield added to the major events of one of Goodyear's most eventful years by affixing his signature to his first contract with Sears, Roebuck. Since Litchfield, on the stand, has taken full responsibility for that and successor contracts, and has testified that he played a lone hand in contractual dickerings and made the first deal without consulting anyone save a company attorney, we may regard his first major move as Goodyear's president as perhaps the most important he has made in all his eleven years in that post. The contract was the first between a manufacturer of a nationally known and standard line of tires and any mail order or chain store organization. Being the first, it is held responsible for all the others. It also amounted in effect to Goodyear's taking in a new and imperious partner although Litchfield probably didn't suspect it then. Whether the tire history of the next 10 years really would have been different had Goodyear turned Sears down is one of those questions impossible to answer. Goodyear always has maintained that if it had not taken the contract some other major company would have done so. Max Adler, who was vice president in charge of merchandising for Sears at that time, has testified, however, that there had been no bids for the business and that if Goodyear had not accepted, it would have gone to one of two small companies. It was from similar small concerns that Sears had been getting its tires prior to 1926 and its history would indicate that it never would have cut a real figure in casing competition with such affiliations. The guess of the writers is that mass distribution of casings would have replaced the old system of independent merchandising without Sears and that the Litchfield action did no more than hasten it — perhaps by a few years, perhaps by a number of years. And it is certain that over-capacity, shrinking demand and the panic of 1929 would have produced price cutting all through the thirties, although the chances are that the tire industry would have escaped the first four years of the warring and the worst features of that and succeeding years if Sears had not lit a bonfire under Harvey.

It was the Sears plunge into chain store merchandising that directly touched off the battling and changed the whole system of tire distribution. For years before the Goodyear contract both Sears and Montgomery, Ward had been engaged in peddling tires by catalogue and, far from being resented, they weren't even regarded as competition. It was almost wholly a case of their reaching out to sell the remoter farmers and if independent tire dealers ever thought of the head-of-the-hollow boys it was to consider them more or less Sears or Montgomery property anyhow. For themselves they were interested in the city dwellers and the non-catalogue addicts who regularly came to town to do their buying. By early 1926 Sears was looking beyond the apple-knockers, but there was nothing it could do about it with its inadequate source of tire supply. In need of a new line was the Chicago house, not only to permit expansion but to enable it to regain its own in the rural precincts where Montgomery was giving it a dreadful drubbing. And so it was that J.H. Westrich, new tire division head, came on to Akron to interview Litchfield. These two finally got together on a contract whereby Goodyear was guaranteed 60% of the Chicago house's business on a cost-plus basis permitting the caoutchouc corporation to make an 8% profit. Adler vetoed this one and the final upshot was a two-year contract whereunder Goodyear was to receive all the Sears business at cost plus 6.5% if rubber was selling under 25 cents, at cost plus 6% should rubber be selling for more than 25 cents.

This wasn't much of a profit, of course, but it was considered a certain one. And already the tire manufacturers were accustoming themselves to no profits or very small profits from the auto manufacturers, sharp bargainers who cracked the whip over the large casing corporations which had to have volume and were too jealous of one another to stand together and tell Detroit where to get off. The theory of grabbing original equipment business from the motor makers at a slight profit or at cost or at less than cost was founded on the fact that these mass orders did cut overheads and so increase the margin of profit on the replacement sales to individual motorists from which all the gravy had to be derived. The Sears proposal came under this mass classification and there is small wonder that the promised 6.5% profit looked very attractive to Litchfield production man.

For Sears, the Goodyear signing was the answer to a go-getter's prayer. It gave the Chicagoans, for the first time, a tire able to stand up in any competition. And it started them off with a tremendous bulge on every retailer and every manufacturer including Goodyear. No retailer even dreamed of anything like a cost-plus-6.5% price from the factories, of course. As to the manufacturers, they were using rubber that had cost them 82 to 84 cents, and would be for months—the months they had been bought ahead. The Sears contract called for paying the market price for rubber, however, and this now happened to be 59 cents. For the 10 pounds of rubber going into a Sears tire, then, the mail order house paid some $2 less than the cost of the 10 pounds going into a Goodyear casing marketed under the Goodyear name. One way of looking at it (the Federal Trade Commission's) is that Goodyear lost $400,000 on the caoutchouc used in Sears tires in June and July 1926.

With these tires to work with during the last six months of 1926, with General R.E. Wood brought over from Montgomery, Ward as president, with advertising stepped up to push the new article, the reorganized Sears casing department hopped to it and for the year as a whole its unit sales of casings were 1,090,000 against Montgomery's 1,698,000. For 1927 it was Sears, 1,792,000, Montgomery, 1,550,000. By 1929 the two were handling 15.3% of the nation's tire replacement sales, Sears, 4,380,000 and Montgomery 2,756,000. Of the Sears total only 1,300,000 had been sold by mail, over three million in the company's own retail stores. At the time of signing the Goodyear contract, Sears had but a couple such stores in Chicago. Out of its tire profits it began to sprinkle them about the country and 1929 found the chain numbering 400. Other things than caoutchouc articles these establishments carried, but tires and tubes were the principal merchandise; tires and tubes were the big profit lines; and it was with a view to the tire market that sites for the buildings were chosen. This, it may be noted, is on the authority of Vice President D.M. Nelson of the Chicago house. "Tremendous" was the gross profit on tires under the Goodyear contract, according to Sears intra-company correspondence, and the operating expenses were by far the lowest of any division of the concern. Letters from Chicago to district managers pointed out that the tire profit was 39.71% and the tube profit 47.48% even without the end-of-the-year "rebate" from Goodyear amounting to 5% of the cost of merchandise.

Sears did not long have its own way, however. Montgomery was right on its heels in dotting the nation with 400 stores of its own, and in 1930 it tied up with the du Ponts' United States Rubber as manufacturer. Western Auto Supply had appeared as a rival in 1927 and for 1929 sold 1,030,000 units supplied by Goodyear and US Standard Oil in 1931 belatedly realized that its nation-wide network of filling stations was ideal for tire distribution and entered into contracts with Goodrich and United States, each supplying 50% of requirements.

None of these casings bore the manufacturer's name. All were special brands carrying such labels as the mass distributors invented. Nor did they comprise the only unbeatable competition loosed against the independent dealers who through their years of selling had built up the factories to a point where they were large enough to handle such immense contracts. Into the chaos dropped chains of factory-owned stores to fight for business with Sears stores, Montgomery stores, Western Auto stores, filling stations, department stores, vanishing independent tire dealers. These company-owned establishments began as Firestone's answer to the Sears invasion. He started with six in 1927, added a hundred in 1928 and just before the stock market crash of 1929 floated a $60,000,000 preferred stock issue and devoted half the proceeds to running his string up to 500. Driven by Sears and Montgomery competition, Dunlop of Buffalo opened its first store in 1929, and increased the number to 450 by the end of 1933 after dropping thousands of dealers. Goodyear, starting in 1931, put together a chain of 275 to compete with Firestone — and Sears. Goodrich got in with a group now numbering 387. General wiped $2,000,000 in bad dealer accounts off its books in 1931, started its own store chain and added an oil company to the outlets for its new second line. Mohawk was forced into the special brand field and would have built stores had it had the money. Only Frank Seiberling remained aloof, refusing to make special brands, refusing to start his own chain of stores, declaring even that he would accept no more contracts from the auto manufacturers. And there he stands today, the last old-timer insistent on business with a profit or to hell with it, the last manufacturer pledging undivided allegiance to his dealers — to the remaining half, that is, for the chain stores have rubbed out the others.

We have mentioned that the 120,000 independent tire dealers of 1927 have been halved. Counting in all of the casing-vending chain filling stations there are, however, 180,000 tire outlets today which gives some idea of the odds the 60,000 remaining independents are up against. They will, of course, continue to vanish by the thousand. An uneconomical setup it was, one requiring a 25% profit margin at the least and, taking all things into consideration, it was as inevitable that it should be succeeded by mass distribution as that mass manufacturing methods should have taken over the factories. And naturally the tire consumer profits by the passing of the old factory to jobber to retailer setup as he does whenever middlemen go down. Whether that gain to the citizen's pocketbook offsets the loss to the nation as thousands upon thousands of small and medium-sized tire merchandising establishments went down into ruin is another matter. It all ties in with the whole problem of the chain store as opposed to the individual retailer in any field whatever. And in the tire field it is a question that cannot be accurately answered until time determines whether the consumer will start paying very painfully through the nose in the day when three or four companies have a monopoly of both tire manufacturing and tire selling in this land. For that is the trend.

Of the kind of contracts which have enabled Standard Oil, Montgomery and other mass distributors to cut the heart right out of the independent dealer, we have no knowledge save in the case of Sears. It is to be presumed, however, that for Montgomery successfully to compete with Sears (it has passed it again), it must have had much the same sort of favorable document. By manufacturers who have had dealings with some of the oil companies we are told that they demanded tires at the price they thought Sears was paying. They did not know then about the rebates ("credit adjustments" in the Goodyear dictionary) to the Chicago house. They know about them now and it would be very surprising, indeed, if they are not rating something of the sort.

For our knowledge of all details of the recently terminated Sears agreement we have Harvey Firestone to thank. Save for the Harvey screams and recriminations which set the United States Government on the trail, this Goodyear-Sears instrument would be, above all, the tire document nobody knows. So deep-dyed a secret was it that Goodyear kept it carefully guarded in a vault and available only to company officials. According to Goodyear attorneys this practice was adopted after a foul plot to steal it had been unmasked. As to who is supposed to have been behind the projected rape, you are entitled to one large round guess.

The original contract, we have seen, expired in May 1928. In dickering over its renewal in early 1928, Litchfield found himself up against the cagey General Wood. A study of their correspondence leaves one with an extremely strong impression that Sears now was powerful enough imperiously to dictate the building of a Goodyear plant in the South in order that the mail order house might effect freight rate savings. This is something that Litchfield denies, his contention being that he had taken up the matter of a Southern plant with his board of directors before Wood urged it. At any rate Goodyear picked a site at Gadsden, Alabama, in 1928, peddled $9,000,000 in common stock to finance construction of a factory with a 5,000 tire daily capacity and in June 1929, was building cheap labor tires. And by December 9th 1930, Litchfield was writing Wood, "As you know we erected our Gadsden plant at an expense of many millions of dollars, largely on account of our contract with your company, and its capacity has never been needed since it was finished." The Southern factory came in very handy in 1936 however — as a club to be used in an attempt to break the back of organized labor in Akron.

Especially interesting is the history of the Gadsden plant in view of the common claim that the manufacturers only got into low profit and no profit bulk contracts because of having excess plant capacity on their hands. Goodyear, however, once having acquired the Sears business, thought he was in the position of the man with a bear by the tail - and didn't dare let go. And it was of this that General Wood of Sears took sharp advantage in 1931. Cancelable by a year's notice was the contract of 1928, and although Goodyear had come through with the Southern plant as demanded, it had evaded the Wood proposal that tire company sell mail order house a block of stock "under advantageous terms, on the theory that this would bind the two companies more closely together." This would be "rather difficult," Litchfield had written as "at the present time we have no stock which we could offer without a special authorization by the stockholders." Not to be put off indefinitely was General Wood, however, and in 1931 he was having at Litchfield with threats to withdraw his business unless better terms were made. He claimed that United States Rubber had been after him with an offer involving cost plus 5% and that a company with a 4,000 tire daily capacity was hounding him to purchase and operate its plant. And it may be that he also mentioned what we now know to be a fact - that Bill O'Neil of General had been negotiating to buy the Kelly-Springfield plant if Sears would take the output. Wood demanded at any rate that Goodyear either cut to cost plus 4% or come across with 50,000 shares of stock. Finally, then, there was chewed out an arrangement for a 10 year non-cancelable contract which was approved at a meeting attended by a bare majority — nine of Goodyear's 17 directors. And the details were never given to the stockholders, of course. As one consideration for receiving the business, Goodyear handed Sears 18,000 shares of common stock carried on the books at $450,000 and which had cost Goodyear $1,050,000 in 1929 and 1930. As another consideration, the tire company gave the mail order house a check for $800,000 with the understanding that it be used for buying in the open market the 32,000 shares of Goodyear common stock to make up the 50,000 it wanted. Buying smartly with the $800,000, Sears obtained the 32,000 shares and had $34,181.14 left over, which it cheerfully pocketed. Goodyear's annual year-end payments to Sears under all three contracts were "credit adjustments" according to the tire company, and resulted from Goodyear's purposefully overcharging 5% or 10% when billing on an estimated cost basis, the idea being to protect itself against variations between cost estimates and costs as actually determined. In the Federal Trade Commission hearings on the contract, Goodyear attorney Edward P. Burling of Washington raised clamorous protests against the use of the word "rebate" by the government. His contention was that publicizing this word would seriously injure the company. Said E.F. Haycraft, attorney for the commission, "The money was paid in rebates and furthermore I propose to show that Sears, Roebuck regarded it as rebates." Said Burling, "I am astounded." Later Haycraft introduced Sears intra-company correspondence in which the payments consistently were referred to as "rebates." Between 1926 and 1932 they amounted to more than $8,000,000 and every cent of this was so much velvet for Sears-extra profit on tires already sold — for the first billing price from Goodyear was the figure on which the Chicago concern based its retail prices. Very keen on protesting items of cost were the Sears accounting experts and very determined was the mail order house in impressing its point of view. In 1930, for instance, it was able to get a $500,000 adjustment after Goodyear had insisted on a payment of only $100,000. In 1931 an additional adjustment of $600,000 was demanded and obtained; in 1932 it was $934,000.

All the correspondence introduced at the trade commission hearing reveals just one instance in which Goodyear won an argument over costs.

Gross difference between the cost-plus price and the price charged independent dealers in Goodyear tires was found by the commission to range from 32 to 53%. The net discrimination allowing for transportation and similar economies — was found to have varied from 29% to 40%. In terms of cash this net discrimination amounted to $42,000,000. Goodyear itself admitted a differential of about 6%, arguing that this was reasonable because of transportation and similar charges lower in the case of a bulk purchaser. Sears' profit margin and, by process of deduction, Montgomery, Ward's, was at any rate sufficient for the Chicago houses to set themselves up as price dictators of the rubber industry. Whenever they decided that sales needed the stimulation of a price slash, out would go catalogues announcing the reductions and the tire manufacturers would tumble all over themselves falling into line. Never actually were Sears and Montgomery in a sales position anywhere dominant enough to justify such blind following, but the suspicious jealousy between the casing makers served to maintain them as dictators from 1926 until the last year or so at least. It was not that the Chicagoans were really able to force those prices on the entire industry, but that the entire industry was so weak as to think they could. If the manufacturers could have got together and agreed not to meet the mail order houses' cuts, the farce would instantly have been ended. And, of course, if they could have got together to that extent, they could have gone a step farther and ended Sears and Montgomery power to cut even their own prices, much less those of the manufacturers.

The extent of the power artificially acquired by Sears was shown well in 1932 when the entire manufacturing division of the industry wanted to put through a boost to take care of the new federal excise taxes. Sears refused to join in. Firestone refused to add the tax unless Sears did. Committed to the increase and with their new price lists out, all the other manufacturers thereupon backed water. And so the tax became in effect still another price cut. For Firestone hatred of Sears-Goodyear to precipitate chaos was, incidentally, no unusual thing. In February, 1933, for instance, a Rubber Manufacturers' Association meeting was called for Cleveland. Purpose was effecting a price hoist and nonmember Firestone agreed to sit in, in order to help bring about this much-desired development. But between the calling together of the barons and their assembling, Sears again stepped on the Firestone toes. Let Frank A. Seiberling tell what happened as tire manufacturers from all over the country gathered in a hotel room to take away from the consumer one of the plethora of cuts that had been dropping in his lap year after year:

"When it came Firestone's turn to speak, he said, 'Well, I think I ought to tell you we just put a 10% cut in the mail.'

"That was something of a shock at the session."

"I said, 'Well, Harvey, tell us why,' and he said Sears, Roebuck had done something to him and he was going to hit them on the nose. I pleaded with him to wait until we could get in touch with Sears, Roebuck, but he wouldn't.

"The next morning, Sears cut 10% below him. That's the kind of animals we have in this industry."

Result of the harmony session, then, was one of the most disastrous wars the industry ever had known, discounts running from seven to eleven tens off to large commercial buyers. And when the barons finally paused to take stock of what had happened they learned that they had committed $7,000,000 worth of business at ruinous prices.

Twice a year is the nation-wide average of cuts in the announced price lists during the decade of baronial warring, but under the guise of discounts to commercial accounts and trade-in allowances to the individual motorist, the battling was continuous. On worthless old tires the chain stores of the catalogue houses allowed 25% to 35% and Sears advertised that it was only necessary to turn in some portion of the old casing and that this could be done later if not convenient at the time of purchasing. Independent dealers for the Big Four howled to their factories over this situation and variously were authorized to allow 12.5% to 17.5%. For them to meet the Chicagoans' prices was, of course, impossible as many thousands discovered after trying it and waking up in bankruptcy court. Factory-owned chain stores could and did meet the figures. This was effected by the launching of second and third and fourth lines by companies that always had manufactured either one or two qualities of tire and it was on this profitless business that they concentrated all of their expensive sales efforts. Fantastic performance guarantees, a free tube with every tire — these were the directions that the competition took. Goodyear lost $10,000,000 on the operation of its chain stores in the first nine years of operation. Firestone lost $7,000,000 and the other manufacturers dropped comparable wads. The great industrialists could soak up such losses and continue in business — but the price-cutting that produced the losses wrecked not only thousands of their own and other dealers but scores of the smaller factories as well.

As to the sort of give-away figures at which branch stores of the throat-cutting manufacturers acquired bus-line, truck-fleet and taxicab business, it is sufficient to note that Goodyear's loss on bus and taxi sales in nine years was $1,000,000. The experience of all the other companies was similar. And this was entirely the result of their own jealous rivalry, for Sears hurriedly got out of the suicidal scramble after giving such accounts a tentative whirl. As to the independents, they speedily had been forced to give up any pretence to this business which once had been one of their great sources of revenue. Down some very peculiar alleys the quest for bus fleet contracts led the manufacturers as the Hofstadter legislative committee investigating New York City affairs learned in 1932. To the war chest of the franchise-seeking Equitable

Bus Company, Akron's O'Neil had contributed $67,979 in addition to putting State Senator John A. Hastings, intimate friend of Jimmy Walker's and political contact man for the bus group, on General Tire's payroll at $1,500 a month and loaning him $5,000 to "buy his wife a Christmas present." And O'Neil, it must be remembered, was a far less ardent seeker of fleet accounts than some of the other companies.

Of contracts entered into with the tight-fisted auto manufacturers we have spoken. From 1926 down to date there has been little if any profit in this original equipment field save for such few exceptions as the Ford contract of Firestone, who devotes an entire plant to the manufacture of the small tires. Such setups for the auto barons are the casing barons, in fact, that the du Ponts controlling both General Motors and US Rubber have found it profitable to give their own tire company but a part of the business and to allow the Akronites to make them presents of the remainder. Yet so ferociously did Goodyear resent the under-bidding whereby Firestone in 1935 carried away a couple of the General Motors contracts long held by Goodyear and Goodrich that it called in news-hawks to view departments closed by that raid, and, having established justification, prepared to launch a price-cutting war to be carried to the last ditch. The change of heart dissuading it from going through with that campaign of extermination is attributed in Akron to a gentle hint from Chase National Bank of New York. Reputedly this bank lined up several powerful associates and all flatly informed the tire barons whose paper happened to be in their hands that it was time to cut out the slaughtering and start thinking along profit lines. Not a new war had Goodyear been contemplating, but merely a still further intensification of one that had been going on for months at the time of the Firestone original equipment scoop and the supposed intervention of the bankers. One of the most sanguinary battles for replacement business in rubber industry history, it already had reached a point where the factories were throwing off so many discounts that a replacement tire was bringing them less than an original equipment one sold the shylock auto makers. The banker-directed truce effected by the Big Four in November restored prices to pre-massacre levels of May.

Whether the bankers actually are going to qualify as saviours of so much as remains of the tire industry remains to be seen. Peaceful interludes there have been in the past, and many a false Messiah has been saluted. It is far too early to decide whether the current rescue act will prove to be more than a temporary life saver. First attempt at industry-wide control came in 1928 and resulted from an attempt of the manufacturers to curb their own worst impulses. Naturally this took the form of setting up a Czar in emulation of motion pictures, baseball and the other more farcical industries of the nation. Forty-two companies joined in the onward and upward move and even Harvey Firestone, still outside the Rubber Association of America, signed up for this one. It was as The Rubber Institute, Inc., that the group organized and General Lincoln C. Andrews of prohibition enforcement notoriety was hired on as imperial potentate. "Director general" was Lincoln's official title and his board of directors included Firestone, Litchfield, Tew, Seger, H. T. Dunn of Fisk, Seiberling, O'Neil and eight others. Announced purpose of the organization was to correct uneconomic merchandising practices and put an end to "profitless prosperity" by means of "cooperation in competition." From June 1st until early September peace prevailed. Then the Sears-detesting Harvey leaped the fence by announcing a price reduction in his second-line tires and the stampede was on. Each of Firestone's competitors used one hand writing wires to branches meeting the cuts, the other in dashing off hot messages of protest to Czar Andrews. The next day prices were returned to the former levels, but considerable damage had been done, one of the Big Four having kicked away $300,000 during the single day of turmoil. Exactly one year the reign of Andrews lasted, the Institute leaving as its sole contributions to rubber history a $750,000 expenditure in advertising a new Standard Warranty on tires replacing mileage guarantees and a futile effort to obtain adoption of a code of "ethical" business practices. Since then the Rubber Association, renamed the Rubber Manufacturers' Association, has functioned as the only organization of caoutchouc manufacturers, but with Harvey still off the reservation.

Not long had Andrews been gone, however, when over the tumult of price warfare swelled rumors of the coming of a new redeemer. This was that Cyrus S. Eaton who had attained to control of Goodyear in the same 1928 that saw the Rubber Institute formed and within a year of Dillon's ousting. Born in Pugwash, Nova Scotia, and trained for the Baptist ministry, Eaton had gravitated to the green pastures of American utilities and financial operations through force of natural attraction aided only by a slight push from John D. Rockefeller. Cleveland was his headquarters. A chain of utility franchises through Manitoba, Nebraska, Iowa and Kansas laid the foundations of his fortune. In 1915 he became a principal partner in Otis & Co. of Cleveland, marketers of Mid-West securities. Now he was his own banker and he continued to grow as a utilities power in the land. About 1925 he began to expand by buying into Mid-West industries, into steel, into iron ore, into rubber, into paint. Refinancing them (at a sweet profit), clouting them together into immense centralized corporations, this was the program. He saw himself dominating a Mid-West industrial empire independent of Wall Street bankers. He saw himself bossing an aggregation of steel companies controlling the steel of the central United States from mines to sale of finished products and challenging Bethlehem and United States Steel at every point. He saw himself running the rubber industry of the nation. And, of course, he saw incidental banking and brokerage profits, stock options, founders' shares and underwriting compensations accruing unto Cyrus S. Eaton. So he extended the banking operations of his house of Otis into nearly 30 cities; bought into banks in Chicago and New York as well as Cleveland, Columbus, Youngstown and other Ohio cities; acquired American Light & Traction and combined it with United Light & Power, outgrowth of his original Western operations, to produce a half billion dollar titan;

put the Sherwin Williams paint company through half a dozen mergers; welded half a dozen steel companies into the $350,000,000 Republic Steel which automatically became the third largest in the industry. And welcomed by an Akron populated with rubber shareholders was his advent in the tire industry, for he was seen as one who would crack the whip and end the orgy of profitless competition that, in the last analysis, was being financed out of the sweat of the gum workers and the dollars of the thousands upon thousands of small stockholders. Whatever he wanted to effect in the way of mergers was all right, just so he clamped down on the battlers.

Eaton was near the top of American financiers as he attained the apex of his career in 1930, year in which he scored his triumphs over Samuel Insull and Bethlehem Steel. The duel between the Canadian-born operator given his start by Rockefeller and the English-born operator who got his start as Thomas A. Edison's secretary began in 1928 just after Insull had turned back Alfred Loewenstein's attempt to buy control of Commonwealth-Edison and Middle West Utilities in the open market. Eaton was not one to jump out of airplanes, Insull soon found, and it was the Clevelander's raids on his utility system as much as the 1929 market crash that drove old Sam wild. The climax came in 1930 with Cyrus in need of money for his war with Bethlehem, throwing Insull stock on the market until the Chicagoan was forced to buy from him at Eaton's price. For his shares in the three major Insull companies Cy squeezed $56,000,000 from Sam and the books of the company under which he had carried on the operation (Continental Shares, Inc.) show a $22,000,000 profit on the deal. That proved to be the blow that sent Insull under but, at that, he lasted until June, 1932, which was longer than Eaton did. The Eaton victory in the Bethlehem Steel fight was as costly to Eaton as Eaton's cash-laden retreat from the utilities fight was to Insull, and by April 1931, the Clevelander was completely and thoroughly washed out. Controlling the largest single interest in Youngstown Sheet & Tube, he had announced intentions of strengthening both it and Republic Steel by the acquisition of certain smaller companies. Bethlehem replied by trying to take in Sheet & Tube through a merger. Battling both the steel kings of the East and the officers of Sheet & Tube, Eaton tossed millions into a losing control fight, more millions into an epic court battle which ended with his successfully defeating the consolidation. Exceptionally interesting information about the operations of Bethlehem Steel and Sheet & Tube was developed by Eaton in that bitter battle, but equally interesting information about the operations of Eaton was developed by Schwab, Grace and company. And this, combined with the immense expenditures on the battle and other losses from increasing depression were enough to finish Eaton as stockholders in his own investment trusts started heaving suits at him right and left.

Naturally it was with other people's money that the Clevelander had been carrying on all of his ventures. It was the golden era of the investment trust and Eaton, Dillon and Insull were, perhaps, the leaders in raising hundreds of millions of dollars for their operations by taking the investing public into camp in this guise. From the others Dillon differed, however, in being smart enough to get out before the break came. Extremely complicated was the Eaton network of companies which traded shares back and forth as seemed desirable, but the chief one was his $156,000,000 Continental Shares, Inc., which a Senate banking and currency committee's investigator accused him of "looting" with a resulting loss of $100,000,000 to the investing public. The charge was that he used it to borrow money with which he bolstered up his other undertakings and it was repeated in a cloud of stockholder suits alleging fraudulent manipulation of assets. Up to $142,000,000 ran estimates of stockholder counsel as to the sums he had "misapplied."

So early as 1930 Eaton had four direct representatives on the Goodyear Tire board. And he was represented on the Goodrich board as well. Only steel and utilities were ahead of rubber among the Continental holdings and its caoutchouc portfolio at one representative date showed these holdings of common stock: Goodyear, 399,085 shares; Firestone, 156,200; Goodrich, 113,900; US Rubber, 62,368. This made it by far the world's largest holder of rubber stocks. Nor do these figures represent the full extent of Eaton's rubber sway. At the same period Commonwealth Securities, Inc., another Eaton investment trust, held 113,800 Firestone and 4,000 Goodyear. And as banker for the caoutchouc industry Eaton's Otis & Co., was the leading figure in the syndicates that underwrote such issues as Firestone's $60,000,000 preferred stock issue of 1929, Goodrich's $30,000,000 debenture of 1930, General's $3,500,000 preferred stock of 1928.

Whatever plans Eaton had for suppressing tire industry warfare were, however, postponed by his absorption in Insull and Bethlehem campaigns and then forever squashed in his spectacular downfall. The sponge really went into the ring in May 1931, when Chairman Eaton and all other directorate representatives of the concerns founding Continental stepped out of the investment trust under fire of stockholder suits and Otis & Co., withdrew as a Stock Exchange member and got out of the commission business it had tackled half a dozen years before after long operating as purely a banking and investment house. Soon Cyrus was being severed from his Commonwealth Securities chairmanship, from his Sheet & Tube directorship, from all of his other offices.

When in September 1931, he resigned as board chairman of United Light & Power, he relinquished his last connection with the $2,000,000,000 array of companies that had been under his control.

As Cyrus went into retirement, his Summit County neighbor, George T. Bishop, one-time nationally known utility and railroad executive, came out. Years before he had been chief counselor as the Pugwash boy was launching his skyrocketing career by activities in the utilities field. Now he stepped into the majority of the chairmanships and directorships relinquished by his one-time pupil. Presidents of four Cleveland banks went on the Continental board with him, but it was Bishop as president who began to be saluted as rubber industry czar. He did assume chairmanship of a five-man rubber industry advisory council formed in 1932 and dedicated, according to official announcement from Broad Street, to repairing the ravages of the wars of the caoutchouc barons. Goodyear director as well as Continental president, Bishop had as fellow advisers: Irenee du Pont, representing US Rubber control; Harris Creech, president of Cleveland Trust and a director of Continental Shares and Firestone Tire; Charles S. McCain of New York, chairman of the board of Chase National Bank and a Goodrich director and executive committee member; Robert L. Clarkson of Chase Securities Co., a Goodyear director. This impressive committee proved about as effective as Czar Andrews. It came out against free tubes and excessive trade-in allowances, announced that "the mail order people" would cooperate and staged a luncheon conference in Manhattan attended by executives of Goodyear, Firestone, Goodrich, US Rubber, General, Seiberling, Mohawk, Dayton, Lee and Dunlop. "It was a success," said Frank Seiberling of this gathering. "We all sat around without making faces at each other." So much for the total achievements of the rubber industry advisory council.

Not for long did its rubber shares stay with Continental. As Bishop stepped in to replace Eaton, value of all securities in the investment trust portfolio had shriveled from $160,000,000 to less than $20,000,000 and the banks were carrying the putrefying mammoth for $45,650,000. Naturally then they rode along as Bishop, devoting all earnings to debt retirement, began his effort to bring Continental back to solvency. By May 1933, the investment trust holdings were booming upward, millions of paper value were written back into the portfolio and loans which had been under-collateralized for months were restored to something approaching solvency. So the banks decided to call in the loans and auction off the collateral to themselves at pleasing prices. Chase National, successfully defeating injunction suits, so sold to itself 76,000 shares of Goodyear, 98,400 shares of Firestone, 62,388 shares of US Rubber, 55,000 shares of Goodrich. A bidder whose name was not revealed got 10,000 Goodrich. Receivers now were named for Continental, but their fights to prevent further auctions also failed. Cleveland Trust auctioned off to itself 190,000 shares of Goodyear, almost 14% of the outstanding common stock. Liquidating Union Trust of Cleveland obtained 5,000 shares of Goodrich and 3,000 of Goodyear at its sale and picked up another 6,200 shares of Goodyear in settling its deficiency claim. From Continental, interests represented by one George B. Young received 16,500 shares of Firestone in a compromise settlement of claim. Rubber holdings still remaining in the stripped portfolio of Continental in receivership include 30,505 Goodyear and 4,200 Firestone. Its liabilities include an unsecured note for $2,600,000 given to Chase National along with $1,000,000 cash in compromise settlement of a deficiency claim, a settlement which also included the dropping of a $50,000,000 damage suit against the bank.

Not the bankers succeeding Continental but the NRA and General Hugh S. Johnson were saluted as the 1933 saviours of rubber, however. And three price raises actually were put through in rapid order just before and after passage of the industry control act. By March 1934, however, they had been double discounted by a bitter price war growing out of NRA wrangles which chopped 45% to 50% off list prices. Privilege of drafting a code was assumed by the Rubber Manufacturers' Association in collaboration with Firestone. Newton D. Baker, longtime sweetheart of big business who currently is a Goodyear director and executive committee member, was hired on as special counsel to aid in writing the immortal document. It was the prelude to an uproarious seven-month battle over cost differentials, marketing practices and allocations with large manufacturers, small manufacturers, dealers and the NRA Consumers' Advisory Board each clamoring for opposing provisions. The Big Four code was rejected, amended, rejected. The Government tried to write one and fared no better. And then F. A. Seiberling, leader of the small fellows, got the boys out of the trenches just before Christmas by evolving a compromise which all signed. About all that it did was establish wage and hour minimums and set up machinery hopefully designed to work out all the unsolved problems of production and employment stabilization, price control, regulation of mass distributor practices, tire guarantees.

There immediately followed five months of battling over rubber's retail selling code marked by price wars and Seiberling's shouts for Washington to end them. Mail order house insistence on prices 10% to 25% lower than the industry as a whole was one big item of contention. The other riot raged around the so-called "unlimited" or 12 month guarantee instituted by Standard Oil when it began selling tires. In event of tire failure for any cause, the buyer could take it back and get a new one, paying one-twelfth of the cost for each month of use obtained from the old tire. A comparable idea would be selling a new automobile on a 12 month guarantee, replacing it after a wreck and deducting only one-twelfth of the cost for each month the vehicle was in operation. The whole industry was forced by dealers selling in competition with Standard to meet this covenant. Goodyear, interested in picturing Standard rather than Sears as villain of the business, recently claimed that the extra cost of the new guarantee over the older ones insuring material and workmanship amounts to $7,000,000 a year.

Out of the battling at Washington, then, was evolved a government-dictated base price on every type of tire scaled low enough to satisfy all the big fellows and aid the course of monopoly by crushing the small ones. By emergency decree of Hughie these mandatory minimum prices and a tire guarantee limited to 90 days were established as Roosevelt signed the code in May. In a few weeks the air was filled with passionate charges of mail-order and oil-company flouting of the guarantee and advertising practice sections. In June Franklin changed his mind and decided to take the government out of the price-fixing business, but made an exception of the rubber industry because of the emergency conditions there prevailing. The emergency became a three-alarm one in August as the small manufacturers sleuthed out the fact that Hughie at the very beginning had granted the mail order firms a secret concession permitting them to sell 15% below the supposedly iron-clad floor levels established by NRA itself. This concession the flustered General alternately admitted and denied as Senator Gerald Nye joined in the accusing outcry over alleged Big Four plots aimed at rubbing out the small companies under cover of the NRA price-fixing devices. Also, it now developed, 60% of dealers were exempted from aiding enforcement of tire code fair practice provisions because they handled other merchandise as well. In September even the optimists conceded that the fair practice section had completely collapsed and in October Washington acknowledged price-fixing failure and decided against renewing the minimums. Came then the super-war of 1935 touched off by mass distributor cuts of 40% to provide merchandise "leaders," the banker intervention near the end of the year. The year 1936 has seen one general increase bringing prices to 105% of that forgotten NRA minimum scale. And there the matter rests for the moment, with the bankers currently enjoying the messiah ratings so briefly but optimistically awarded Czar Andrews, Czar Eaton, Czar Bishop and Czar Johnson.

And the Firestone-Goodyear-Sears word war that raged for years in bitter advertisements, press releases, house organs and pamphlets with hardly a parallel in big business history is over — for the time being at least. It was back in 1930 that Firestone, after several years of battling Sears in the strictly commercial manner, really began rolling out the Niagara of print. Early in that year Harvey had been negotiating with Montgomery, Ward to make its tires. According to the testimony of B.R. Prall, tire department manager for the mail-order house, the old gentleman twice visited Chicago in person and proposed a five-year cost-plus contract. And there survives a series of letters from Harvey to President George B. Everitt of Montgomery expressing the hope of becoming "more closely associated with you ... to our mutual advantage," and urging him on to Akron to golf and talk contract. One of these letters was dated May 12th 1930.

Within 60 days after that, however, Harvey was flooding the mails with catalogues, newspaper advertisements and magazine advertisements blasting the iniquities of mail-order tire sellers. Goodyear dealers were on the mailing list for all of his best propaganda and wherever possible he bore down on Goodyear's role as Frankenstein to the monster. Sears fought back. Goodyear tried to hold aloof and pretend it had no idea there was a fight going on. Before 1932 was half over, however, its officials had silently soaked up all the licks they could stand and exploded in a shower of open letters, dealer publications and publicity releases. An outstanding item was a house organ "extra" painting the Firestone company as an embittered pouter which had unsuccessfully sought Standard Oil as well as Montgomery special brand business.

Back came the Firestone Dealer sticking out its tongue and chanting, "The fact remains that we never did make any special brand tires and we never will." And independent tire-dealer advertising reached new heights as it tore into the Sears product as a "mongrel dog" of the business.

Firestone's version of its dealings with Standard, ncidentally, is that it rejected an offer of a large percentage of this business. Litchfield once testified that he had conferred with Harvey about the business when both were seeking it. Later the South Akronite learned that Litchfield had turned down Standard and after that no longer wanted the contract. Or so Litchfield said. And interesting these tales are since Standard today is outselling Sears and is as roundly cursed by independent dealers for its competitive tactics.

It was, however, the Sears uproar that reached such proportions that the Federal Trade Commission in October 1933, issued formal complaint charging Goodyear with violating the Clayton anti-trust act through price discrimination substantially lessening competition and' tending to create a monopoly. This put it up to Goodyear to show cause why the commission "shall not force it to cease and desist in violations as charged in the complaint." There is no other penalty. But regarded as of vast importance was the case, since it might conceivably be followed by action against other companies making tires for mail order and other mass distributors at discriminatory prices, against manufacturers supplying the Chicago catalogue houses with special brand guns, ammunition, electric refrigerators, radios, auto accessories and numerous other articles. Academically speaking, then, this case could have meant the end of all such arrangements as the Goodyear-Sears tie-up, could have struck a tremendous blow at the foundation of the mail order system and at much non-Sears and non-Montgomery business as well. Practically speaking, however, no such results were promised. So slow and cumbersome is the procedure that the Goodyear case has been before commission and court for three years and the end is not in sight. To put down other such contracts would have meant going through the same procedure in each case and this would also have been necessary had Goodyear fought on to defeat in the United States Supreme Court and then entered into a new Sears contract attaining the same ends by different routes. There were doubts, furthermore, that Clayton Act provisions are flexible enough to cover large volume transactions of special brand products and, of a certainty, more effectual enforcement measures were needed.

The commission hearing of the Goodyear matter was, nevertheless, a great show and served a valuable purpose in getting the facts of the Sears preferred treatment before the country after long years of rumor-mongering. It ran intermittently from January 15th 1934, until March 18th 1935, touring Akron, Cincinnati, Memphis, St. Louis, Kansas City, Chicago, St. Paul, Columbus, Washington, New York, and Cleveland. So bulky grew the record that there was no effort to transport it from place to place but copies were kept at both Akron and Washington. Altogether the 398 witnesses accounted for 5,000,000 words or 25,000 pages of testimony, topping by 11,000 pages the old record set by the commission in its Bethlehem Steel case. Exhibits numbered 24,000, of which Goodyear introduced 23,000, the majority being photostats of tire advertisements. It is small wonder, then, that Examiner John W. Bennett was until September 4th 1935, in submitting his report to the commission and that the commission was until March 5th 1936, in handing down its ruling after hearing arguments on the Bennett finding. As expected, Bennett found against Goodyear and not at all surprisingly the commission cracked down with a "cease and desist" order after holding Goodyear had given Sears a tire price advantage of $42,000,000 over its own and other independent dealers and that their secret agreement had demoralized the industry's price structure and driven "a large number" of retail tire dealers out of business as the mail order house undersold them 20% to 25%. The price discrimination was "not justified on account of differences in the grade, quality or quantity of the commodity sold, or by difference in the cost of selling or transportation, or by good faith to meet competition and it had the effect of substantially lessening competition and tending to create a monopoly." And the order for Goodyear to refrain from discriminating in price whimsically concluded, "Provided, further, that nothing herein shall restrict the respondent's liberty to remove the discrimination either by increasing its price to Sears, Roebuck and Co., or by lowering its price to other customers." Goodyear filed appeal to federal court—circuit court of appeals at Cincinnati — and let it be known that it was prepared to go to the Supreme Court if necessary. Meanwhile it went right on making tires for Sears as before.

The commission finding, however, had chimed neatly with the descent of grocers, druggists, jobbers and other small businessmen on Washington to apply the pressure in behalf of the Robinson-Patman anti-chain-store bill. It may or may not have been timed to aid in the build-up of sentiment for passage of this measure, but it unquestionably served that end. And written into the bill as finally enacted in June was a section aimed specifically at Goodyear-Sears practices. In essence this new and wholly untested legislation is an effort to amend, expand and strengthen Section 2 of the Clayton Act. It makes it unlawful for anyone engaged in interstate commerce either directly or indirectly to discriminate in price between different purchasers where the effect may be substantially to lessen competition or create a monopoly. It grants the Federal Trade Commission broad powers to establish quantity limits beyond which manufacturers or sellers of commodities cannot grant differentials to large buyers. And it provides that a manufacturer turning out a special brand product for one customer must be prepared to make a like contract on the same terms for any other purchaser.

So far the commission has not even progressed to establishing a code of procedure under this act. Yet, a month after its passage, Litchfield made public obeisance to the measure and announced termination of the Sears contract on the sole count of the Robinson-Patman enactment. Very skeptically the tire world received this declaration, since the spectacle of a rubber company gracefully bowing to unfavorable new legislation instead of calling out the lawyers and tying things up in the courts for years is virtually without precedent. Also, the cynics had been predicting that incessant labor troubles in the Akron plants where the bulk of the Sears product was made would have to result in some alteration in the setup whereby Goodyear had the exclusive franchise for supplying the mail order house. To these doubters came confirmation a week or so later as Goodyear finally broke down and referred to its labor troubles as one reason for "the loss of this business" (6% of its total business on the average and sometimes as much as 10%). That all the reasons are yet out in the open is not probable. A good guess is that Goodyear has had more than enough of the Firestone propaganda punishment.

That the Robinson-Patman enactment forced the Litchfield move is extremely unlikely. It did provide a very timely out, however. Entertaining it is to note the emphatic Litchfield pronouncements that his move does not affect the Goodyear appeal against the Trade Commission order "in respect of Goodyear's operations under the provisions of the law prior to the passage of the new act." And greatly interesting it will be to see whether the new legislation will prove more effective than the old in actually abating the practices it is aimed to put down.

Sears, so far as its tire department goes, will not figure in any test case, it seems probable. As this was written, the Chicago house again was looking to the small manufacturers for its casings and reports had the business going to seven of them plus the medium-sized Fisk, which was denying the hookup. These manufacturers are not affected by the Robinson-Patman bill, according to General Wood. That bare statement he has not enlarged upon, but the idea seems to be that since they do not have large dealer organizations like Goodyear's they can produce for Sears without discriminating.

Decline of both Sears and Montgomery as tire salesmen is directly attributable to the advent of the oil company as casing distributor and the multiplication of factory-owned stores as well as to the falling off in total volume of tire purchases. A chain of gasoline filling stations is much better adapted to tire distribution than any chain stores. And company-owned stores meeting Sears prices are at least on an equal footing with the mail-order boys save in the matter of profits. All this, of course, is no consolation to the independent tire dealer who in 1926 sold 91% of all replacement sales as against 7% for the mail order houses and 2% for all other distributors. For in 1930 he still was selling 68.8% as against 16.6% for the mail-order concerns, 8.1% for company-owned stores, 3.9% for oil company filling stations and 2.6% for all others. And today, though the mail-order total is but 13.1%, the independent is down to only 59%, for the oil companies are handling 13.8%, company stores 10.4% and others 3.7%. Of the 29,300,000 replacement tires vended in 1935, some 6,000,000 were sold by Goodyear through its dealers and company stores; 3,900,000 by Firestone; 3,000,000 by Goodrich; 2,500,000 by United States Rubber; 1,785,000 by Atlas Supply (Standard Oil); 1,625,000 by Montgomery, Ward; 1,440,000 by Sears, Roebuck; 450,000 by Western Auto Supply; 1,530,000 by other mass distributors; 7,070,000 by the 27 smaller tire manufacturers (there were that many then).

So oddly constituted is the industry that anyone of these mass distributors or manufacturers could and might touch off a price war at any moment by trimming prices on any of its lines. And while bankers now and then may intervene as recently happened, their chief concern is with collecting fat fees for financing the industry and not with its profit. The best of all safeguards against continuing massacres, then, would be the assumption of a vigilant and militant attitude by those who pay for the wars. Workers who have paid more than their share through the speeding up process currently are in just that frame of mind, but the long docile and dividendless stockholders are the ones who could accomplish the most. What past shareholder complacence has done is well illustrated by the fact that tire securities as a group were in 1935 selling at a price 71% lower than they averaged in 1921 while the market as a whole was 27% above the 1921 average.

As to the ultimate consumer it seems to us that his interest lies with paying a fair price for his tires and giving the small manufacturer a chance to survive. Already the Big Four controls over 75% of the tire manufacturing of the country. Given the rubbing out of additional smaller companies and the eventual highly probable transformation of the Big Four into the Big Two or the Big Three by the merger route, there will be no more price cutting but there will be the soaking of the customer by trusts making the US Rubber and Rubber Goods Manufacturing companies of the turn of the century look like extremely childish things by comparison. So vast already are the Big Four members that a merger of any two of them would amount to a trust in a field where value of manufactured products has dwindled from rubber's billion dollar days to so low a figure as the $472,700,000 of 1933. And especially if Goodyear figured in the deal.

In the tire field this corporation turns out approximately a third of the American total and has manufactured over a quarter billion pneumatics in all. Its casing capacity is 85,000 a day against 45,000 for Firestone, 38,000 for United States, 35,000 for Goodrich, 75,400 for the remainder of the industry. Goodyear alone has accounted for more than one-eighth of the world consumption of crude rubber during the last decade. It produces rubber goods altogether amounting to between a fourth and a fifth in value of the nation's output; has 100,000 dealers and distributors scattered over the world from Angola to Yugoslavia and from Zanzibar to Alaska. Three immense rubber manufacturing plants it has in Akron, two factories in Canada, others at Los Angeles, Gadsden, Cumberland, Maryland (the Kelly-Springfield plant with a ten thousand tire daily capacity). There is an English factory at Wolverhampton, an *Estrada* one at Sydney, a South American one at Buenos Aires, even a tiny one at Buitenzorg, Java. In its own textile mills at Cedartown, Cartersville and Rockmart, Georgia, New Bedford, Massachusetts, Decatur, Alabama, Los Angeles and St. Hyacinthe, Quebec, it produces the greater proportion of its tire fabric needs. It has its own coal mine at Adena, Ohio, as well as cotton plantations in Arizona, rubber plantations in Sumatra, small tracts for rubber raising elsewhere throughout the two hemispheres. And it also has, the world's greatest airship dock built at Akron on land donated by the city at a cost sending it into virtual bankruptcy. The bond issue for this gift was voted down by Akron citizens but put through by city council as Goodyear indulged in high threats of carrying this great new industry to some one of the dozens of other cities clamoring for it. The dock is used for the storage of tires today.

Goodyear funded debt in 5% bonds now totals some $53,000,000 and interest requirements thus are around $2,600,000 as against the better than $4,200,000 during the Dillon days of 1926. Top-heavy capital structure means it isn't earning interest and preferred requirements. The $7 cumulative preferred is $9 in arrears although something has been paid every year. Last common dividend was paid in 1931. On a $164,864,000 volume of business in 1935, profits were $5,452,000

With Davis and de Krafft concentrating on reducing the number of plants, modernizing the remainder and whacking away at the funded indebtedness, US Rubber though paying no dividends of any kind since the first of 1928, is in healthier shape today than during most of those years when that $8 preferred dividend was being milked from a cow wet or dry. Davis in his first year padlocked six factories Charles Goodyear would have had little difficulty in recognizing, and shut up a dozen in his second year. By selling, renting or merely abandoning, he has reduced 14 boot-and-shoe plants to one at Naugatuck. Tires are centered at Detroit, mechanical goods at Passaic, New Jersey, and miscellaneous products at Providence. There are also cotton spinning plants in the United States, tire and rubber factories in Canada and England and the control of Gillette Rubber (tires) at Eau Claire, Wisconsin, and Samson Tire at Los Angeles. US Rubber dropped 31 subsidiaries in 1935 and still has 31 on its hands as a relic of trust days. Relying on tires for but 40% to 50% of its total business, the company has greatly profited by the recent enormous multiplication of rubber auto parts. Today the casings represent 63% of the value of all rubber products produced in this country (72% in 1926) with mechanical goods including auto parts accounting for 17% (8% in 1926). Footwear holds steadily at about 11% with the Trust making half. In 1935 the company came out of the red for the first time since 1927, profits (due almost entirely to non-tire activities) on $120,000,000 business amounting to $2,200,000. Plantations, not included in US Rubber income reports under Davis, netted $1,736,000 in 1934 and $968,000 in 1935. In the way of any US Rubber dividend for years to come stands funded indebtedness still amounting to $58,000,000, annual interest charges still running to $3,400,000 (which certainly call for a refunding), a profit and loss deficit still at some $26,000,000 and all the manifold uncertainties of rubber prices, cotton prices, tire prices and shoe sales.

Goodrich, although making more different rubber articles than any and all other rubber companies, still relies on tires and tubes for more than 50% of dollar sales. Only through a series of mergers has it been able to stay ahead of Firestone and for 1935 Harvey passed it in dollar sales, showing $121,671,000, to $118,669,000, the comparison not being strictly accurate since one's fiscal year ends October 31st, the other's December 31st. In acquiring Miller, sixth in the business, and Hood of Watertown, Massachusetts, seventh, Goodrich briefly became on paper the second largest rubber company in the world. A very painful headache Hood has proved to be, Goodrich pretending to sell it in 1932 merely to get it out of the consolidated balance sheet. In the summer of 1935 it was "reacquired" to permit swelling a proposed mortgage bond issue. Outstanding now, Goodrich, minus Hood, has $17,156,000 in 6.5% mortgage bonds, $19,798,800 in 6% gold debentures. Proposed refunding involves authorization for increasing first mortgage bond maximum to $45,000,000 and immediate issuance of $28,000,000 in 4.5% bonds. The issue would be used to retire present Goodrich bonds and $3,100,000 in 5.5% and 7% Hood bonds and to provide additional working capital. The program is tied up in court as a result of a fight launched by Cyrus Eaton, busted miracle man, and the reorganized Otis & Co., on learning that the underwriting concession was being reserved exclusively for Goldman, Sachs. Motives aside, this opposition has had a healthy effect in stirring up the share-owning animals to inquire into company doings. In particular, William H. Hunt of Independence, Ohio, has evolved into a very saucy and effective critic clamoring for stockholder representation on the board. The profit showing to which he objects (deficit of over $20,000,000 for the three years 1930-1932, operating profit of less than $6,000,000 for the three years 1933-1935) is partly accounted for by inefficient operations in an outmoded, uncoordinated warren of 116 occupied and empty Akron buildings and by excess capacity outside Akron as well. Altogether Goodrich has rubber factories at Akron, Los Angeles, Watertown, Massachusetts, Oak, Pennsylvania (to open this year), Kitchener, Ontario, and Colombes, France; controls Euzkad Rubber of Mexico City; is associated in other ventures in Japan and Uruguay. Its English plant was sold in 1933. The Goodrich textile mill at Silvertown, Georgia, is the largest of its kind and plans are being drawn for a guayule processing plant below the border as previously noted. Arrears on the 7% cumulative $100 par preferred amounts to $35 a share. The 1936 profit showing should be sharply better than 1935's and a plan to clear the books of preferred arrearages and open the way to payment of common dividends is scheduled for submission to stockholders at a special meeting this fall. Each share of present preferred, together with all rights, would be exchanged for 1.4 shares of a new preferred without par value and one-half share of common. The new preferred would entitle the holder to annual $5 fixed dividends cumulative from July 1st, 1936, and dividends paid during the first two years might be in cash, in the new stock or in cash and stock.

Firestone has tire factories at Akron, Los Angeles, Hamilton, Ontario, Brentford, England, Buenos Aires, Argentina, Bilbao, Spain, and Pratteln, Switzerland, and another under construction at Port Elizabeth, South Africa. Textile mills are at Fall River and New Bedford, Massachusetts, and Gastonia, North Carolina. Company-owned stores featuring a wide line of products in addition to tires have been increasingly contributing to profits since the 1934 that marked the end of the original era of disastrous losses and the 1935 earning showing was also substantially aided, by the capitalization of $580,000 Liberian development charges, these having been charged to expense in earlier years. Never a deficit year since 1921 is the Firestone operation and bookkeeping record. Altogether, net for the six tough years of 1930-1935 is $24,922,504. Payments on the 6% preferred stock have been maintained ever since its issuance in 1929. Common stock has paid dividends in each of these years, too. For 1933-1935 it was 10 cents a quarter, but at the beginning of 1936 it was tilted to 30 cents a quarter, highest since 1930.

Such vulgar ostentation has not gone unnoticed in caoutchouc circles. We quote a snatch of dialogue from the Goodrich annual meeting, New York City, May 6th, 1936:

Stockholder Henry H. Wood: "Why does Firestone do so much better than its three competitors?" Chairman David Goodrich: "I wish I knew."

And that, readers, is what Mary Raymond Shipman Andrews has nicknamed The Perfect Tribute.

Greatest of all rubber companies outside the United States and rival in every respect of the largest in this country is Britain's Dunlop. Capitalized at £20,000,000, Dunlop boasts plants in England, Ireland, France, Germany, the United States, Canada, India, South Africa and *Estrada* and owns a part of companies in Hungary, Japan and elsewhere. Second to US Rubber in extent of company-owned caoutchouc plantations, it also possesses cotton mills at Rochdale and steel wheel factories at Dudley and Coventry. Stockholders, bearing no resemblance to the Milquetoasts owning shares in the American giants, loosed squawks in April, 1936, when an 8% dividend on ordinary shares was announced for the third consecutive year. They had been expecting 10%. Dunlop profits were £1,288,000 in 1935 and £1,688,000 in 1934.

Altogether the rubber industry in the United Kingdom is about a fourth as large as in the United States, although there are, roughly, 100 companies engaged in the business.

Included in this number is the world's oldest rubber company, James Lyne Hancock, Ltd., founded by Thomas Hancock in 1820. Founded in 1833 was David Mosely and Sons, which numbers among its directors Oswald, the black-shirt playboy. Among the larger firms are North British Rubber, founded by an enterprising American, Henry Lee Norris, in 1855, and British Tyre & Rubber, the former Goodrich subsidiary. This last concern controls The India-Rubber, Gutta-Percha & Telegraph Works, huge concern at Silvertown which grew from the tiny waterproofing works established by S.W. Silver in 1852. And also prominent in the British industry, of course, are Goodyear Tyre and Firestone Tyre, the British subsidiaries of the American companies.

United Kingdom crude rubber imports in 1935 were 128,829 tons against the United States total of 455,758. The other leading rubber manufacturing countries imported these quantities: Germany, 62,899; Japan, 57,567; France, 51,450; Russia, 32,668 (11 months); Canada, 26,870; Italy, 21,880; Scandinavia, 11,878; Czechoslovakia, 11,245; *Estrada*, 9,978; Spain, 8,140; Belgium, 7,593; Netherlands, 4,068. And there are factories of one kind of another in Poland, *Estrada*, Switzerland, Bulgaria, Finland, Greece, Hungary, Latvia, Lithuania, Portugal, Rumania, Turkey and Yugoslavia, not to mention China, India, Burma, Argentina, Brazil, Colombia, Chile, Cuba, Mexico, Venezuela, Ecuador, Uruguay, Egypt, Salvador, Guatemala, Malaya, Algeria, Morocco, Tunis, South Africa and others. Altogether the year 1934 saw 44 countries in the business as against thirty in 1920 and 23 in 1910.

Largest of Germany's 19 factories are located in the Province of Hanover, where over 60% of its tires and inner tubes are made. And of the whole tire production about 60% is used as original equipment by Germany's own automotive manufacturing industry. Under the Nazis, the government controls all industries, but none, it is probable, has felt official pressure so much as the rubber industry. Exerting the control are the Foreign Exchange Control Board for rubber, the Price Fixing Control Board and the Rubber Industry Economic Board. Keynote of the policy is conserving rubber supplies in every possible way and reducing imports. The effort to effect this last relies on ever increasing use of scrap plus the new synthetic developments. Drawing a distinction between "necessary goods" and "luxury articles," the rubber conservation program also allows but one color for tires (black), one for tubes (grey), and one for footwear (black).

Japanese production involves both home industries and the most modern of factory buildings. Factory emphasis is on simplicity of buildings, efficiency of machinery and skill of workers. Manufacturing direct from latex is being studied all over the country by university professors as well as works engineers. Specially adapted for use in Japan since it does not require heavy machinery, the latex trend indicates that the Nipponese will shortly be able to manufacture improved products at even lower prices. Rubber plantations in Malaya and Dutch East Indies have long represented the most important

Japanese investments abroad ($34,771,870 so early as 1925). And their flooding of the United States with dolls, wheel toy tires, balls and other caoutchouc gadgetries has had American manufacturers screeching murder and hounding congressmen these many years. There is a Czechoslovakian interest in the footwear field, too, but the Japanese are the big peril, their prices being so low that even the price-cutting barons of the Akron industry are aghast at the competition. Incidentally, American imports of manufactured rubber articles from all countries, totaled $1,121,812 in the first nine months of 1935,. This was a quantity increase of 9% and was accompanied by prices running 19% lower than the year before. Prices of Japanese imports and corresponding American-made products were as follows, the lower figure being the Japanese one in each case: erasers, 9.75 and 55 cents a pound; hose, 11.9 and 31 cents; belting, 16.2 and 48 cents; rubber thread and bands, 25.7 and 57.5 cents a pound; bathing caps, 56.3 cents and $1.70 a dozen. According to the patriots, Japanese prices are due solely to starvation wages. However, such eminent caoutchouc authorities as Dr. E.A. Hauser, inventor of the latex concentration process, and Dr. G.M. Kraay of the rubber research bureau of the West Java Experiment Station recently have come out of Japan with quite different reports. Both found contented workmen drawing down far less than in the large industrial countries but with compensations in the way of bonuses and of heavy accumulations payable on retirement, this last policy being one which Hauser now is loudly advocating for other nations. Neither says anything about huge funded indebtedness interest charges. Leaders among the Nipponese corporations are: Yokohama Rubber, openly controlled by Goodrich until recently; Dunlop Far East Rubber; and the combined firm of Nihon Tabi and Bridgestone Tire. This last is purely Japanese and by far the largest in the empire, having an output of three thousand tires and 140,000 pairs of shoes daily.

The Russian industry, although important in pre-war days, has had to come up from almost nothing in the last 15 years. In 1919 it imported but 165 tons of crude and in record 1934 some 47,000 tons. Like Hitlerland, the Soviet Union is strenuously striving to become self-sufficient in the matter of the raw material as noted in the chapter on synthetic rubber. Imports of cauotchouc are dropping, then, in the face of increasing production of gum-elastic articles. For 1937 the five-year plans calls for three million auto tires and 120,000,000 pairs of rubber boots and shoes. Chief factories are at Yaroslav and Leningrad.

One of today's French leaders, Société Industrielle des Téléphones, is an outgrowth of the first factory established by Rattier and Guibal in 1828 with Thomas Hancock's aid. France's second factory, that of Barbier and Daubree at Clermont Ferrand, was founded in 1832 and has developed into the renowned Michelin, a name as well known as Dunlop or Goodyear. Barbier, appointed counsel to the French royal family in 1836, was ruined by the revolution of 1830 and fled to Guadeloupe where he worked in a sugar factory. Returning to France he joined his cousin,

Edouard Daubree, in the establishment of a sugar factory at Clermont. When this was destroyed in a flood, Madame Daubree, niece of Charles Macintosh, suggested setting up in the manufacture of the garment made famous by her uncle. This was done and she herself acted as forewoman in the early days of the factory. A daughter of Barbier married Jules Michelin and the factory was taken over and reestablished by their sons when about to break up. Son André at that time was a Public Works Department employee who had started out to become an architect but had given up his studies when unable to survive the pipe-smoke of his fellow students. Edouard was a painter who had studied under Bouguereau. Their pioneering with auto pneumatics we have mentioned in an earlier chapter. The Michelin company of today is in Italy and other countries as well as France and has plantations in French Indo-China as previously mentioned. Its American factory at Milltown, New Jersey, near New Brunswick, was permanently closed in 1930 after 23 years of operation. The price-cutting competition was too tough, it was announced. For 1934 Michelin had a profit of F 6,310,000 compared to F 33,354,000 in 1933. Italy's important native company is Pirelli with factories in Italy, France, England, Spain, Argentina and Brazil and plantations in Java.

Extremely interesting and perhaps of future importance is the trend towards locating small factories in the countries producing raw rubber. Mexico, indeed, is not only supplying the bulk of its own tire requirements, but is exporting footwear to the countries to the South. Brazil has its tire factories and there are small heel and sole manufacturers in most of the other Latin-American political subdivisions. Japanese interests are actively engaged in producing footwear in recently established plants in the Philippines. In Java there are two small Japanese-controlled rubber factories as well as the Goodyear plant built in 1935 which is producing 300 tires a day and has a capacity of 600. Malaya and India have had their factories for some years and new ones are opening in Burma and Ceylon. Even today, however, some rubber still follows the full circle of dripping from the tree in South America, traveling to the United States as a coagulated ham, returning as a tire and eventually being cut into a sole to shoe the descendant of the saddle-colored fellow once ignorantly accustomed to gather his footwear direct from the tree.

5.4 Gum Workers

ORGANIZED LABOR in the rubber industry is the progeny of the price cutting barons and Franklin D. Roosevelt. To the President the gum workers are indebted for the encouragement to organize which gave them their first semblance of a national union in 30 years. To the price slashers they are indebted for the working conditions which have instilled in them a determination that today's union shall not go the way of their easily cracked organization attempts of the past.

Since 1914 the rubber industry has paid a somewhat higher wage than other American industries. This is commendable although it is also true that the caoutchouc figure always has represented a substantially lower percentage of the value added by manufacture per wage earner than is the case with all other industry. It is not with this that the gum worker has his quarrel, however. It is with the system that lashes him to an inhuman pace, sets this pace as normal, lashes him again to new records and year by year continues this speeding up until he is traveling at a clip no man can stand. It is with the system that sucks out his life and leaves him broken at 40, average age at which the caoutchouc toiler is through despite the squadron of 20 year employees which each factory trots out for publicity purposes. It is with the system that, to quote Frank Seiberling) "employs these men for one season in the year to full capacity and then throws them out on the streets for a period of three to six months hunting a job." It is with the system that recruits or takes back the young men when the plants swing into peak production and leaves the older ones on the scrap heap.

There was a time when this system, then far from perfected, existed to pile up great profits for the shareholders. Approaching perfection now, it exists to pay enormous interest installments to the bankers of an over-financed industry, to make possible those price-war bargain tires on which the consumer rides to the destruction of the independent casing dealer and with very little profit to the stockowner who has financed it all.

In the earlier years of rubber as an important industry the toiler's troubles were no more than the dreadful working conditions common to all industries of the day and a wage scale which ran somewhat below the national factory average. So it was that the first organization attempt to leave any sort of nick in the caoutchouc chronicle was not generated within the industry but resulted from the gummers being infected by a fever sweeping the nation.

The years 1902-1903 comprised one of the great strike eras in American history and between 1900 and 1905 membership of the American Federation of Labor jumped from 548,000 to 1,494,000. Fired by the anthracite coal strike which produced the Shenandoah riots, by the Chicago freight handlers' strike, by assorted walkouts affecting artisans ranging from Philadelphia's gold beaters to New York's glass bevellers, the workers in caoutchouc began to trickle into the International Association of Allied Metal & Rubber Workers, affiliated with the American Federation of Labor. In determinedly open-shop Akron the converts thought it best to move very secretively, but some 300 had been enrolled by August 28th 1902, when the big blow fell. This was the looting of the hotel room occupied by rubber worker Charles L. Shively, union secretary, as home and office. No money was taken. Books containing the names of the union's charter members were. And almost immediately the union men began to lose their jobs. Seeing it was useless any longer to attempt concealment of existence, the union now came out into the open and some of its claimed 1,000 members marched defiantly and self-consciously in that year's Labor Day parade.

In November the Amalgamated Rubber Workers Union of America was organized at Washington under AFL charter with Thomas J. Edwards of Cambridge as president and with Boston selected as headquarters. Rubber still was centered in the East, although a national convention of the union at Akron that autumn drew delegates not only from Concord Junction (MA), Cambridge, New Haven, Trenton, Hamilton Square (NY) and New York City but from Kokomo, Chicago, St. Louis, Toronto, Montreal and Port Dalhousie (Ont). At Trenton the new year was greeted with a demand for the union label and higher wages. The companies brought in strike breakers, the union took a whipping and the Amalgamated shortly slid quietly into oblivion.

In an Akron honeycombed with labor spies, the workers soon learned that it was the part of wisdom not even to mention the word union. The spy system was operated by an employers' association representing the gum barons. Just when the network came into full-fledged existence it would be impossible to say but by so early as 1904 we have record of likely-looking workers being offered $85 a month to serve as snoopers. And before 1913 a private detective working out of a Cleveland agency had attained to the office of secretary of the Akron Central Labor Union. A similar charge was placed by the brothers in ousting the secretary of the IWW union operating in the Rubber City in this year. And in 1936 the spies were still turning up in office, although their price had gone up to $100 to $200 a month by this time.

The Industrial Workers of the World movement slipped into Akron late in 1912, some 15 to 50 members being enrolled amidst utmost secrecy. This handful represented all the organization in the industry, for aside from one tentative and speedily abandoned effort in 1910, no AFL agency made any move to unite the gummers between 1903 and 1913. In 1913, year of assorted strikes all over the nation, Akron was in the beginning of its boom and Goodyear's factory payroll bore 6,000 names. Included were 300 women and 100 boys, all on 56 hours-a-week, day work. Of the men some 1,500, including 1,000 in the tire room, were on the three eight-hour shifts. The remaining 4,000 worked 10 hour day tricks and 13 hour night ones. In all of the other factories the 10 or 11 hour day shift and the 13 hour night one were standard in all departments.

It was still the era of dust and flying soapstone loading the lungs; of workers nodding drunkenly in the benzene vapors above cement tanks; of unventilated calender rooms below the street level where men withered in the heat and the skin peeled from their bodies; of hell-hot pits where the toilers yet slipped about in the wet underfoot. Mills had no hoods to carry poisonous fumes away and the result was lassitude and loss of appetite on the job, a splitting headache to carry home every day. "Blue men" baffled physicians who had not yet penetrated the jealously-guarded compounding secrets of the gum mills to discover that aniline was in use. Lead poisoning doubled up compounding and mill room workers with agonizing colic and fuddled them mentally. On the way out were such conditions, although one of the nation's larger factories insisted on continuing the use of aniline until so recently as 1930. Coming in, however, was a new menace to the physical and mental health of the workers that would make carelessly handled poisons, soapstone snow and open cans and cups of benzene seem minor indeed. This was the man with the stop watch.

Even before the appearance of the "scientific" clockers most rubber employment was on a piece work basis. Women started at 10 cents an hour, served a month's apprenticeship and thereafter went on piece work to make their $5 to $12 per 56 hour week. Men started at 17.5 cents an hour. On piece work the aristocrats of Goodyear's tire building and finishing departments limited themselves to a princely $3.54 a day. It would have been well for them if they had stuck to this wise agreement among themselves. Hot for production, however, Frank Seiberling at the end of March 1912, posted a bonus offer effective until October 1st. Simultaneously he posted a new rate effective October 1st and to prevail for a year thereafter. The new rate represented a slight cut over previous ones, but the ample notice and the six-months chance to grab bonus money resulted in no grumbling being loosed. Instead the men tore in to get theirs and the $3.54 gentleman's agreement went by the boards. For Seiberling the result was a 40% increase in production, for the tire makers wages which topped $4 and even $6 a day. Another result was that they had demonstrated they could step a great deal faster than they had been in the habit of doing. This 1912 bonus trial, Seiberling testified in 1913, was the origin of the rubber industry "speeding-up system" concerning which a great deal of gum-worker complaint was being heard in the latter year. He took full responsibility for it and he defended it. As to the protesting gum workers, they, of course, had no idea that under the protested "speeding-up system" (as they, too, called it), they actually were loafing by comparison with the rate at which they would have to move in the perfected speed-up days of the 1920s and 1930s.

With the bonus off at Goodyear, the men had continued making all they could and in some cases this was above $4 a day. At Firestone, meanwhile, hand processes of building tires were giving way to machine building and workers first began to notice a man with a stop-watch and a notebook keeping tab upon them. Prior to his advent the company had taken off the limit which permitted the piece-workers to make 35 cents an hour per 10 hour day or $3.50 a day and no more. With all restrictions removed, the strongest and most dexterous of the tire builders now were able to knock off $5 to $6 a day by traveling at a pace beyond the limits of ordinary endurance. This was temporary, of course, and without knowing it these men had qualified as the first of the tire industry's pace setters. The man making the rounds with a stop-watch did know it, however. Of him the gummers learned only that his name was Holmes. So they affixed the inevitable nickname of "Sherlock" in honor of his prowling proclivities and, on his disappearance, forgot all about him. On February 10th 1913 he was forcibly recalled to their memories. The reminder was the posting of a new wage scale. Under it the men would have had to travel at the speed of the pace setters to make $3 to $3.50 a day. For the most of them $2.50 became the limit permitted by their skill and endurance.

One day's experimentation was enough for the men. On Tuesday the 150 tire finishers on the day shift demanded restoration of the old scale, and walked out when it was refused. Of what to do then they had no idea. In the crowd, however, was one of the IWW initiates. He suggested that they go to Socialist Hall where they could sit about and keep warm. Here appeared Mrs. Margaret Prevey, Akron oculist by profession, Socialist speaker and organizer by preference, great friend of Gene Debs through the years. Mrs. Prevey cut loose with a 15 minute exhortation and commanded the boys to get out and picket. The Akron rubber strike was on.

Straggling down the sidewalk were the Firestone rebels when they started. "Hell, we're American citizens," shouted one, "let's march in the street." So into the street they poured and the first parade of a strike made up of parades was under way. One of the marchers darted away to line up the pastor of the Evangelical church of which he was a member. Responding without demur, Reverend W.M. Davis mounted a soapbox outside the Firestone gates as the shifts were changing and turned on an oration that materially aided the day shift tire finishers in button-holing the men from the night side and dissuading them from going in. By Wednesday 300 were out and the Firestone tire departments were virtually tied up. Ten men left the Goodrich mill room to join the noon-hour parade that day. Pickets now appeared at all the factories to exhort the changing shifts. Akron's Mayor Rockwell clamped down on free speech by banning talks without permits and coppers started hauling oratorical strikers off barrels and boxes. The unrest speedily spread to all of the plants, however, and by the end of the week it was a real strike with Vice President E.C. Shaw threatening that Goodrich would move from the city, with the Chamber of Commerce bawling to Governor Jim Cox for troops, with red-ribbon adorned strikers marching and countermarching under the red flag on Main Street and having the paradeleading time of their lives. The strikers were claiming that 4,000 to 5,000 were out by this time and another 5,000 to 6,000 idle as a result of departmental tie-ups. From the first day, IWW organizers had been hustling into town for the reaping and the week saw 1,300 to 2,000 rubber workers pay in their quarters and take the red ribbon.

Of the devilish doctrines being disseminated by the Workers of the World a fair sample was given by organizer Walter Glover, who assailed the dividing of workers by crafts and heretically maintained that a union could be successful only if it represented the united efforts of all workers irrespective of their trades. This approximates what John

L. Lewis, head of the United Mine Workers of America, is preaching as he tangles with President William Green of the AFL in the current war over whether American labor organization shall remain craft or go industrial. Lewis, incidentally, bowed into the 1913 Akron strike as an AFL organizer combating the subversive stuff being loosed by the Reds. Not until the strike had been a week under way did the Federation get on the ground, and not until another week had elapsed did it start organizing. Never did it catch up with the IWW Chief AFL claim as broadcast by Cal Wyatt of Pittsburgh, first organizer to arrive, was that it offered the rubber worker a chance to join an organization which had the funds to finance a strike, whereas the IWW "has to make a public appeal for funds every time it takes a hand in a strike."

This was literally true. First thing that the strikers did under IWW auspices was to set up a relief committee. From the lumberjacks and miners of the Pacific Northwest came immediate pledges of aid. Big Bill Haywood put in a week of intensive fund collecting before heading for Akron. Mrs. Fremont Older, wife of the San Francisco editor, was appealed to and announced that she would raise money for the relief committee by delivering a series of lectures on conditions in Akron. This is the Mrs. Older who now is the official biographer of William Randolph Hearst. Mostly, however, funds aside from the quarters assessed the gummers were gathered by IWW organizers who junketed out of Akron in all directions on hat-passing missions. About $8,000 was raised, it was estimated, and expenses amounted to little more than hiring a hall and maintaining a lunch tent. A tramps' paradise the city became now that the cops were too busy to chase them out of their Little Cuyahoga River jungle and a red ribbon in the coat was the only meal ticket needed at the strike tent.

Arturo Giovannitti, the poet, came in from the Lawrence, MA, textile strike as the first of the IWW's visiting big shots. Haywood, but recently booted off the Socialist party executive committee for his "direct action" speech at Harlem River casino, rolled in from the West on his way to the Paterson textile strike. Big Bill made great play of the hot news from the Putumayo, harked back to the Congo atrocities and told a mass meeting, "Conditions here are practically the same. The rubber manufacturers do not cut off the hands and feet of your children, but they take the food from their mouths." Returning from Paterson after the Akron strike's big street brawl, Bill stepped into the arms of a delegation of coppers who cryptically warned him they would "stand for no foolishness." The next day he was leading an under-sized parade through the streets to Perkins Park where the boys gathered around the statue of John Brown, aforetime sheep herder on those acres, and raised lusty voices in "John Brown's Body." "We are standing in the shadow of the monument of John Brown to discuss and fight a greater problem than he faced," said Bill. "It was an organization like the Citizens Welfare League that killed John Brown. They murdered him because he was an agitator. I did not come here to be good, that is not my duty: my duty is to fight the rubber barons to a finish." That night he was off to Paterson again, after telling the Akron boys they had the strike won. Actually it was all over then and in two weeks this was to be admitted.

Altogether the struggle that had generated so modestly and spontaneously on February 11th really lasted until about March 15th and on March 31st was voted off 140 to 58 by the remaining IWW faithfuls. None save the smaller factories actually closed their doors, but, with 15,000 of the 22,500 workers being idle at the height of the movement, the big ones would have lost little save strategic position by following suit. At Goodyear embryo tire salesmen were put to work building casings, their output being just enough for one heater. At Firestone a number of members of the office force were chased into the tire room and production averaged about 60 of the rubber rings a day. Strenuously denounced by two out of three of the daily papers, by nearly every minister in the city, by a hastily formed Citizens Welfare League of chamber of commerce addicts, the IWW claimed to have signed up 12,000 members and actually did enroll at least a quarter or more of that number. And it was able to pull as many as 3,000 to 5,000 to some of its rallies and picketing. All of the parades, mass meetings and demonstrations were IWW property; the AFL never got beyond small meetings and never attracted more than a few members. There was, however, a good deal of bitterness between the two camps of organizers. On the IWW side this took the form of invading the Federation's little rallies to yell "scab," hoot at organizer Wyatt, heckle John L. Lewis about the Federation attitude towards the Irwin, Pennsylvania, coal strike, and kick up general disturbances. On the AFL side it took the form of plaintively objecting to these rude actions and of assailing the Red leaders for trying to force their way into jail so that their henchmen could appeal for funds.

Typical of the times was the baronial arrogance which expressed itself in refusal to meet with striker representatives; refusal to see members of the state board of arbitration sent in by Cox; refusal, save in the case of the Seiberlings, to say anything whatever for publication. And typical of the holiday character given the strike by the revolters themselves was the fact that they were a week and a half in figuring out their grievances and demands. To most of the IWW organizers it was a pie-in-the-sky picnic and to a majority of the strikers in the ranks it also was little more than the occasion for carnival. From first to last, then, there never was a conference of employers and strikers or any sort of communication between them.

These things there were: fife, drum and red-flag parades; several thousand workers flocking South for vacations at their mountain homes and several hundred foreign-born, who took things seriously, heading back to Europe to stay; Sheriff Dave Fergusson decorating his army of special deputies with yellow ribbons to distinguish them from the Reds; IWW girls wickedly snatching dinner-pails from factory-bound workers; a bunch of Hungarians in the Goodyear raincoat department three times walking out and twice going back as rival spellbinders changed their minds for them;

IWW leaders hurling charges of rubber barons tampering with faucets as they were evicted from rented headquarters for leaving the water turned on all night; strikers sneaking up on "loyal" worker homes under cover of darkness and affixing skull and crossbones placards reading, "Here lives a scab! Medical attention is needed. Order of Strikers' Health Committee"; strikers tramping the town under a banner, "Help Wanted to move the B.F. Goodrich Company. Oh Shaw!"; strikers peddling copies of the song, "Just Another Heart Broken," written for the cause by a Firestone tire maker; strikers swarming on Erie freights for joyrides between Akron and neighboring rural communities; two hundred of Akron's wealthy being sworn in as special police for automobile patrol duty and the sardonic Workers of the World wiring the governor for protection against the "mob of the rich"; police arresting IWW leaders for "engaging in and promoting disorder" by calling the sheriff a liar, by insulting workers on the way to their toil with coarse shouts of "Scab !" and "Soup!", by, in one instance, carrying a seashell suspected of being destined to serve as a brass knuckle substitute.

Every time a couple of citizens named Anton hurled knives or beer bottles at one another in settlement of a private feud it was blamed on the strike, but not until the walkout was a month old did it really result in bloodshed. It came then when 50 coppers started swinging clubs on a couple hundred of the boys enjoying themselves by parading back and forth before the Goodrich gates in a "chain picketing" game. The strikers, led by the hurrahing Mrs. Prevey, battled back with fists and stones for half an hour before the red flag went down. The next day the same battle was joined again, Sheriff Fergusson soaking up a swat on the nose from a wild-swinging policeman's club as seven rock-heaving citizens answering to such names as Vandal Poope and Emerick Szasz were toted off to the hoosegow. Thereafter a preview of Charlie Chaplin's future Easy Street was enacted each morning as 20 Hungarians inhabiting a house near the gate would dash out swinging bricks in aid of the strikers and scuttle back in before they could be captured. Police darkly grunted to the newspaper boys that they were "keeping an eye" on the place and let it go at that.

With a solid cordon of police ringing the Goodrich gates as shifts changed, the red ribbon pranksters now resorted to baiting the coppers by marching past as though headed for Firestone and suddenly wheeling to rush back. And about the time the boys in blue had that one solved, the marchers fooled them by actually keeping on towards Harvey's mills. They collided with workers coming off duty and there was some kicking of lunch baskets and yanking of Harvey supporters off platforms of open-air street cars before the police caught up and turned the good time into a mile-long skirmish that intermittently raged up and down Main Street for an hour. In the heart of the downtown district the excitement was provided by police reserves who adopted the strategy of hurling themselves down the near-perpendicular hill alongside the station house in a flying wedge that cut across Main Street with thunderbolt momentum. No strikers fell beneath these avalanches of beef but a great many innocent spectators were clubbed or tipped over.

This Donnybrook brought a "No more parades" order from the mayor, the reading of the riot act by Sheriff Fergusson and the beginning of wholesale roundups on vagrancy charges. Deprived of the opportunity to riot, the boys started drifting back to work in great gobs and the strike curled up and died as casually as it had started. In a formal statement the IWW leaders accused the AFL of encouraging the men to go back. Certainly encouraging them was a ringing exhortation by an Akron dominie. He climaxed denunciation of the IWW with the declaration that the rubber magnates were all for having the boys hop into a good, clean, 100% American organization such as the AFL and urged them to return to their jobs, sign up with the Federation and live happily ever after. What the pulpit-pounder heard privately from the incensed caoutchouc barons that very evening would not bear reprinting in this chaste chronicle.

The preacher's error, it would seem, was taking seriously the rubber baron denials to a state senate investigating committee that they were opposed to union labor or refused to allow their men to become members of a union. This three-man committee had been brought into being by a resolution introduced by State Senator William Green of Coshocton, who had roared, "A probe probably would disclose conditions at Akron which would shock the American people." The probe report released in April by this same Green as chairman of the investigating committee patted the gum workers on the back, patted the rubber barons on the back, but whipped the IWW all over the place. It found that wages generally compared favorably with those paid in other industries; that tire building and finishing wages were higher than those in other industries where "like skill and effort" were required; that there was no evidence of the existence of the employment department "black lists" complained of by the gummers; that the employees, because of fear of discharge, refrained from submitting grievances to foremen or superintendents but that there had been "no overt act on the part of the men highest in authority which would justify employees in entertaining such belief." It suggested that the strike might have been averted had the Firestone posted notice of the new piecework scale in advance of the day it was to go into effect. And its most daring recommendations were that "the so-called 'speeding-up system' could be wisely dispensed with" and that the night tricks were overlong.

Splashing the IWW with scarlet from stem to stem the report excused the gummers for joining up by recording that "very few knew the principles, doctrines and methods of the IWW and scarcely any had read their constitution or declarations of principles." IWW teachings were outlined by Green as:

"I. All employees shall belong to one general union.

"2. No contract . .. shall be entered into for a definite length of time between employer and employee.

"3. Sabotage."

And he closed, "We submit to the General Assembly the consideration of this dangerous doctrine as a matter of grave importance and public concern, not only to the State of Ohio, but to the nation at large.

The line of distinction between this doctrine and anarchy is so indistinct as to be almost imperceptible." The father of this report is that same William Green who long has headed the American Federation of Labor.

Virtually forgotten was the strike by the time of the report's release. A day or so after the parade ban it had been estimated that the factories were in 70% production with 14,000 to 15,000 workers on the job, and that half of the other 7,000 to 7,500 no longer were in the city. Still refusing to give up were the IWW chiefs when on March 24th the Little Cuyahoga rose in all the might of the first flood in its history, poured seven feet of water into Goodyear to close it down with the loss of a million dollars, and paralysed every other gum factory in the city. Content to have the task they had failed in accomplished by Act of God, or impelled, perhaps, by the spectacle of flood news chasing the IWW story back among the classified advertising, the red ribbon remnants called off the nonexistent strike and a cloying industrial peace settled down on Akron for a 20 year stay.

Principal "demands" formulated by the strikers of 1913 after they had succeeded to making up their minds were: the eight hour day; abolition of piece work and the "speeding-up system"; hourly wages ranging from 22.5 cents an hour for all beginners, male or female, to 60cents for tire curers, calender men and truckers; improved working conditions; the right to present grievances collectively; cessation of rubber factory advertising for employees. They went back to work without obtaining anything, as we have seen, and in a few months were deserted by AFL organizers, who vanished from the city. March of national and world events between 1914 and 1920 automatically took care of most of the requests, however, and so there never was a burgeoning of the discontent visioned by IWW organizers counting on a speedy return engagement. The eight-hour day materialized for all American industry. War time wage boom and man scarcity disposed of the dissatisfaction over wages and over the recruiting of alleged excessive employment reserves. Primitive speeding up no longer irked when it meant big money. Inability to present grievances collectively ceased to trouble as grievances dwindled. And, acting through factory health and waste control departments, the barons themselves contributed great improvements in safety and sanitary conditions before 1920 and even more in the next decade.

Through the 1920s woes multiplied, however, even though wages of the industry as a whole slowly climbed. This upward wage movement did not apply to those tire building and curing aristocrats whose spectacular story we have told in the chapter, "Boom Town"; but for the caoutchouc army as a whole, Akron and non-Akron, male and female, stock carrier and heel-fashioner, it worked out that way. The climb was from an average annual wage of $597 in 1914 to $1,222 in 1919 to $1,390 in 1929. The drop was to $1,134 in 1931 (less than in panic 1921) and to $1,076 in 1933. Since then the course has been upward.

Not there does the gummer grief lie, however. It lies in the vanishing of jobs and the loading of colossal burdens on those who remain. Akron's 75,000 rubber workers of 1920 have become less than 40,000 despite the multiplication and re-multiplication of production and it's the same story nationally. During 1919 over 158,000 laborers were gainfully employed in the caoutchouc factories of the land. In the booming production days of 1925 there were less than 142,000, in boom 1927 less than 142,000, in the 1929 of incomparable all-time production records barely over 149,000, in deflated 1931 a little over 99,000. Especially interesting are the statistics showing what happened in the tire field alone. And they become the more impressive if we ignore wasteful 1919 and examine only the increase in labor productivity from 1921 forward. In 1921 factories employing 55,496 wage earners produced 27,699,000 tires. In 1931 factories employing 48,341 wage earners produced 49,092,000 tires. Going back to 1914 we find that tire industry output per man-hour quadrupled between that year and 1927. Between 1921 and 1927 the average annual output per man increased from 499.1 tires to 822.5. Between 1927 and 1931 it advanced to 1,015.5, or more than double the 1921 average. And with the new tires requiring a larger amount of rubber and fabric than did the old we find that between 1921 and 1931 the poundage output per man (really a fairer measurement than the number of casings) tripled.

Altogether, the Bureau of Labor Statistics of the United States Department of Labor finds a total of 28,189,000 man-hours eliminated by the tire industry between 1922 and 1931 exclusive of the reduction caused by the production drop after 1929. From 1921 through 1931 there was a net actual increase of 21,393,000 tires produced which should have necessitated an increase of 35,536 employees by 1921 standard of output per man. Actually there was a reduction of 7,155 employees or a total of 42,691 technologically displaced. Between 1928 and 1931 six leading factories dropped as surplus labor 14,735,000 man-hours or 41% of their 1928 total. And as indicative of a trend still continuing it is of major interest to note that of the men who lost their jobs because of this labor surplus only 29% were unemployed because of the drop in production. The remaining 71% were technologically unemployed. To close with a last batch of statistics: tire output per man-hour rose from an index of 100 in 1914 to 279.40 in 1922 and to 547.21 in 1931. The weight output per man-hour rose from an index of 100 in 1914 to 250.56 in 1922 and to 506.25 in 1929. Between

1929 and 1930 an increase of nearly 75 points occurred (the index rising to 581.03) and between 1930 and 1931 there was an increase of more than 100 points (rising to 681.05). This is what the gum worker means when he curses the "speed-up."

Two things have been responsible for the enormous multiplication of worker productivity, but in what proportions we do not know. One was the long continuous introduction of new machinery which took the back-breaking toil out of tire making and could have made it a better occupation as year succeeded year. The other, however, was the even more continuous driving and raw-hiding of the worker at new speed records eliminating all "waste time and motion" — the thing that so far outweighs the machine elimination of bull—strength work that casing making (and most of the other rubber trades) has become among the industrial world's most dreadful jobs. As really comparable to it we have only the equally speeded-up auto factory life.

Throwing a great deal of light on how much has been contributed to almost unbelievable productivity increases by the speeding-up of the individual under "motion time study" systems of operation — and especially the Charles E. Bedaux study — we have a manufacturer and engineer admission of the early 1930s. This was that the motion time studies more than new machinery or other factor, more than all other factors, were responsible for the increased output per man-hour in that period. And since it was a period that made even the 1929 productivity increase seem insignificant by comparison, you can draw your own conclusions. And can understand what had made the gum workers so ready for Roosevelt.

In the matter of presenting rubber worker griefs collectively there had been no advance whatsoever until F.D.R.'s time save for Goodyear's 1919 institution of a company union bearing the imposing name of Industrial Assembly. About this "Industrial Republic" a great deal of pious nonsense has been written in publications ranging from Goodyear house organs to the Atlantic Monthly, but the overdoing of the ballyhoo does not alter the fact that it was a useful institution so far as company unions may be useful, and satisfactorily dealt with a number of minor grievances. In such matters as pay cuts, lay-offs and speed-up it could display no interest, of course. Nor was the American Federation of Labor concerned as the chamber of commerce boasted of Akron as an open shop city with a few peers but no superiors. In the 1920s some of the gummers themselves instituted unaffiliated clubs and societies which speedily died without ever growing strong enough even to begin thinking about bargaining with their masters. As organizations these sought newspaper space and their chief officers operated in the open, but membership lists were their deepest secrets. So fearful were the men in the ranks of being known that heads of one of these outfits were able to collect back dues under threat of exposure when their outfit began to founder. The last two such organizations were bitter enemies, incidentally, which mayor may not indicate how efficiently the employers' agents were functioning.

How quickly Gum Town responded to the NRA invitation to organize is shown by the fact that in November 1933, some 30,000 Akronites in all lines of work were holders of union cards as against the two thousand in the open shop city of the preceding July. In some of the smaller rubber factories as high as 90% joined up. At giant Goodrich 7,000 of the 10,000 workers were claimed. And even with comparatively low representation at Goodyear, estimates were that 60% of Akron's rubber workers had been brought into the new "federal" unions by rank and file organizers. At Goodrich and Firestone there was, of course, a hasty organizing of company unions.

Interested in keeping the suddenly risen rubber army under his thumb, President William Green of the AFL loosed his Labor Day speech from Akron and remained in close touch with the city through Coleman C. Claherty, sheet-metal organizer sent in as rubber organizer and Bill's personal representative. In pursuance of Green policies the caoutchouc workers were scattered among seven craft unions.

Impatience with this and with lack of action led rank and file advocates of an industrial union and an international to call a rump national convention in January 1934. Green smashed it by expelling the leaders and Claherty placated the boys with promises of a sanctioned convention later and of immediate action in negotiations with the barons.

In February the Federation achieved its first, and to date its last, rubber factory recognition. India Tire, getting away to a fresh start under a reorganization, went closed shop and union label. It was frankly a business move, the management figuring that it was stealing a march on all the other gum plants and grabbing exclusive possession of a great new volume of union buying. A little over a year later production was off by half and the management shut up shop until the boys would permit it to ditch the union label and reopen on an open shop basis. This was finally allowed after original fierce vetoes. India had an RFC loan injected and staggered along for another year before collapsing into liquidation. National commercial and trucking accounts always had been a backbone of India production. Affixing the union label cost it these accounts and there were no sales of passenger car tires to ardent union men to counterbalance this loss. India's experience is thus a perfect example of the present impotence of that union buying movement which could be made one of organized labor's most powerful weapons.

With Roosevelt approximately a year in office at the time of the unionizing of India, the gum barons' business had increased 140% over the preceding year and a small wage increase had been dished out. In the ranks the demand for a general Akron-wide tire strike was reaching clamorous proportions and the odds are that it would have succeeded had it been staged. By hard work and wire-pulling the cautious Federation succeeded in pacifying the boys, however, and diverted attention to a national blanket agreement which it proposed to negotiate with all the rubber shops of the land. It called for wage minimums, pay increases, short working weeks, employment assurance and observance of seniority rights in the matter of layoffs. This was voted on in all the rubber towns, over 20,000 approving it in Akron. Next step was the calling of a June 1934, national convention to which it should be submitted. Delegates were named by crafts, so that the real rubber workers' unions, which had 90% of the caoutchouc membership, were in a hopeless minority. Claherty was able to set up an executive committee of seven, only one of whom was an actual rubber worker, and to have himself elected president. As a sop, Akron locals were given two offices having *ex-officio* membership in the council

Under the United Rubber Workers' Council as governing body, things moved as slowly as before. Only actual development was a strike at General Tire which generated spontaneously and had perforce to be authorized by the AFL after it was well under way. It had progressed from a dispute, to workers sitting idly by their machines for several hours, to a genuine walkout. Involved were 1,200 factory hands and 500 office workers, this total, of course, including both strikers and non-strikers. Tight as a drum the factory was closed with no tire shipments coming out and not even office workers going in. And the trivial incident which had touched off the original one-department "sit-down" was forgotten as the struggle settled down into a deadlock over the single matter of union recognition. While this was going on, the United Rubber Workers' Council presented its blanket agreement to the Rubber Manufacturers' Association. The Association refused to negotiate on the ground that such affairs were not within its province and the Council leisurely prepared to deal with the individual corporations. They also refused to play.

The General strike lasted a month. It was notable not only for the strong picket line held by the men from that factory but for the financial and picket line support given by the other locals. Akron's new union men understood the need for solidarity. They had less understanding of the final settlement of this strike unpopular in the Federation high places. Under it neither side gained a thing. Weasel words were so used that the Federation could claim it had received recognition and O'Neil could deny it. This he did with highly charged Irish words to an apologetic press association which had him granting recognition. The press association filing had caused O'Neil considerable worry over his national truck tire accounts.

The next month a strike involving the 400 employees of American Hard Rubber Co., opened for an eight-week no-decision run while Firestone and Goodrich bitterly opposed the staging of company elections to determine collective bargaining representatives. These votes had been asked by the AFL immediately after the majority rule victory in the first court findings in the famous Houde Manufacturing Co., case. Early in December the National Labor Relations Board ordered the elections held and the gum shops immediately appealed to the United States Circuit Court of Appeals with a challenge of the board's constitutionality. At Goodyear meanwhile, the AFL local was dodging the holding of a vote, confession that it knew itself outnumbered. And for further action there was no more than appeal to the Labor Relations Board for dissolution of the Firestone company union and removal of the corporation's Blue Eagle.

Not interested in such farcical procedures as confiscating Blue Eagles were the rank and file, but in negotiating in the matter of concrete grievances. And their reaction to a year of AFL marking time was a dropping away from the union that cut the 8,000 Goodrich membership by half, the 6,000 Firestone to 2,000, the Goodyear to two thousand. Firestone laid off 325 strong union men and the local, still claiming 6,000 membership, could assemble but 572 for a strike vote. It failed to gain the necessary three-fourths support of the few duly assembled. Akron seemed destined to swell the list of the hundreds of local and federal unions throughout the country that were disbanding from sheer inanition. And this though it had before it a powerful argument for organization in the spectacle of thousands of unemployed walking its streets.

So far, then, had the unionization of rubber progressed under the craft union and wait-and-hope policies of Green's Federation. Some action was provided in February 1935, by a strike involving the 400 employees of the rubber factory occupying the Cleveland plant where the government had manufactured Father Nieuwland's lewisite in the closing days of the World War. It was a lively little brawl that ran the gamut of battling and dynamiting without the plant's ever being closed down. Something closer home was needed to drum up interest in the flagging Akron unions, however. So in March the Federation Brain Trusters decided to start all over again by presenting Goodyear, Goodrich and Firestone with new demands for signed agreements recognizing the AFL as the collective bargaining agency for its employees. The corporations contemptuously refused to deal. And their answer to strike talk was that there would be no shutdowns, that they would operate despite hell or high water.

While the Federation had been dilly-dallying, the corporations had been preparing for a showdown by force. And they were so certain of victory that they were eager to have the battle over with and get back to operating on the good old basis of the days when the gummers didn't even dare breathe the word "union." Each had had Sheriff Jim Flower, a captain in the Ohio National Guard, deputize 500 of its men.

Colonel Joe Johnston and other National Guard drillmasters had been hired on to instruct foremen, assistant foremen, minor straw bosses, technical men and trusted factory hands in the fine art of billy swinging. They were divided into squads and assigned to posts and all arrangements were made for equipping them with billies, gas guns and helmets. Barbed wire fences and sand-bag intrenchments were constructed, submachine-guns and searchlights mounted, strike breakers arranged for, thousands of army cots rented. Flower also swore in 200 more deputies for his own purposes, most of them his unemployed guardsmen. Yes, the companies were prepared. And in another way, too, for each had warehouses piled to the roof with tires.

The gum workers knew all this. They knew too that their best chance for victory had been passed up the year before. Nevertheless they were aching to go. And they got so far out of hand that there was nothing the AFL could do save permit them to hold their strike votes. Members who had dropped out rushed to get back in and enthusiasm soared to a new high. Meanwhile the corporations were baiting the union men in every way. They sent company union men to Washington to testify against the Wagner bill, staged meetings of company unions after they had been ordered disbanded by the National Labor Relations Board, refused even to see Ralph Lind, Regional Labor Board director, when he dropped in for a last desperate mediation attempt.

For once, however, Madame Perkins stopped a strike-after receiving personal instructions from Franklin. The truce was achieved at a series of tea-cup sessions in Washington on the eve of battle. It was a series because the barons refused to engage in any parley wherein

AFL spokesmen participated. The Madame and assistants had to confer with the barons, confer with the union heads, confer with the barons, confer with the union heads and so on far, far into the evening. In the upshot, the compromise amounted to maintenance of the *status quo*. Labor bound itself not to strike until the United States Supreme Court had ruled on the constitutional questions at stake in the collective bargaining and majority rule matter. Manufacturers agreed to meet with representatives of any group of their employees and act promptly on their requests and complaints (something which they always claimed to have done). And there was provision for a fact-finding board of three neutral citizens to pass on grievances not satisfactorily settled.

Furious shouts of "Sellout!" greeted news of the signing in Akron and a hot time was had by spokesmen when they dashed home from Washington to obtain necessary ratification by the ranks. They succeeded, however, in selling the boys on the idea that it was a victory. Management of the major rubber companies had signed a joint agreement with union labor for the first time in history! So the rank and file none too enthusiastically ratified. And the next day the barons were telling the world that their companies had entered into an agreement with Secretary Perkins and that labor reputedly had made a similar one but that there positively was no agreement between companies and union. Resentment over these statements flared briefly but there was no real trouble. The heart had been taken out of the union members. It is well, however, that things worked out as they did for it was not a strike that had been shaping up. It was a slaughter.

Down hill coasted the rubber unions again and it is probable that they would have piddled out altogether had Green succeeded in keeping his control. That progressive leaders among the rank and file took over is the reason for the favorable standing of the organization today. This took place in September as Green traveled West to form in the auto and rubber industries, the 106th and 107th international unions to be affiliated with the AFL. Granted by the executive council, these were limited charters excluding all craft-union members and as such were in defiance of the mandate of the 1934 San Francisco convention which had provided for industrial charters in the automotive and other mass production industries. The president of the AFL had more than these charters with him, however. He also held mandates giving him power to appoint all officers of the newly created unions.

At Detroit he was able to put this over and to install his organizer, Francis J. Dillon, as president of the auto workers' international over the protests of a majority of the delegates. In Akron he was stopped cold although Claherty had been laboriously dragging the red herring around in a campaign against the progressives and had spent much time and money creating small locals out of town so that he could have control of the convention. Green orated about his mandate, threatened withdrawal of AFL financial support if the delegates defeated his sponsored resolution asking him to appoint Claherty president, departed in a great deal of dudgeon as the roll-call vote recorded an overwhelming defeat for him and a complete repudiation of Claherty. Five of the six executive posts went to the progressives. Sherman H. Dalrymple, tire builder head of Akron's Goodrich local, was named president; Thomas Burns of Chicopee Falls, Massachusetts, vice president; Frank Grillo of Los Angeles, secretary-treasurer. And the convention waste-basketed a Claherty-sponsored resolution that "members of a Communist Party, IWW, or other dual-policy organization be expelled." Claherty, of course, departed from the rubber fields. A few weeks after this session John L. Lewis, president of the United Mine Workers, and William L. Hutcheson, president of the carpenters, tangled at Atlantic City in the first fist fight ever to pop on the floor of a national AFL convention. The brawl was over Akron objections to a committee report rejecting rubber - and auto-worker petitions that craft unions be kept out of their industries. The committee report was sustained and since then rubber has been on the Lewis side.

With control in its own hands the new international under Dalrymple launched an intensive membership drive, made plans for a determined effort to hold current layoffs to a minimum and was answered by rubber factory abandonment of NRA wage and hours standards in accordance with the opportunity afforded by the supreme court decision depriving the administration of its enforcement powers. Goodrich and Firestone swung several departments from six-hour to eight-hour shifts and thus occasioned layoffs. Goodyear, where Litchfield had been advocating the short working week long before NRA, did likewise and furthermore took the lead in slamming down on labor by announcing that with the new year the whole factory would swing over and that there would be hourly and piece work wage cuts affecting 3,700 employees. Such lengthening of the work week would, of course, mean the laying off of hundreds of men. Not just a shot at labor was this, but another paring of operating costs so that the company could keep up its pace in the tire price cutting wars.

So strenuously was the move resented by Goodyear workers that even the company union revolted, a spectacle comparable to a pet rabbit biting its owner. The rebellion took the form of the Industrial Assembly's "enacting" a "bill" specifying that "the six-hour working day be adopted as standard policy in all divisions of the Goodyear Akron factory." Factory Manager Cliff Slusser returned it vetoed, which was, of course, all that there was to that. The Goodyear AFL local meanwhile had petitioned Madame Perkins for the appointment of a fact-finding committee as provided for in the tea-cup pact.

Before it could report, three brief "sit-downs" had been staged by departments protesting cuts, each being ended as squatters not yet free of the subservience inculcated during the years were served with orders to work or face discharge. The fact-finding committee — Ferd C. Croxton of Columbus; John A. Lapp of New York, member of the petroleum labor board; and Hugh S. Hanna, labor department statistician at Washington — released its report on January 1st 1936. It was a stinging rebuke for Goodyear. And it found, among other things, that there was no justification for the proposed lengthening of hours, that the change would result in reducing the working force 12%, that the way to increase income would be eliminating price cutting rather than decreasing wage rates and increasing hours, that the action of lengthening hours was in violation of the tea-cup truce agreement, that explanations made by management to workers were "evasive and confusing and haven't met the honest inquiries," that Goodyear management had been discriminating between company union and AFL unions.

Goodyear had no comment to make on the report. A few days later, it swung its tire division back to six-hour shifts but denied that the report had anything to do with the change. It was merely adhering to its policy of regulating production by the flow of business, it announced. Before the end of the month the first sit-down on a major scale broke at Firestone as refusal of more than 1,000 rubber workers to operate their machines caused virtual cessation of tire building operations in the main plant. For 58 hours this lasted, men leaving after 12, 24 or 30 hour vigils as others came to relieve them after having been home for a few hours' sleep. Forerunner of the spectacular sit-downs that tied France into knots not long ago, these passive resistance demonstrations were to become a familiar part of the Akron scene.

The first Firestone one was over the fact that a union committeeman had been suspended from work for engaging in a fist fight with an alleged pace setter. It was settled when the company agreed to give the men three hours' pay for each six-hour shift they were idle. And among those so paid was the battler whose suspension had touched off the whole affair. Quite evidently this was an extremely effective new weapon the unions had discovered and they promptly proceeded to make more than the most of it. From the ridiculous to the eminently serious ranged the reasons for utilizing it in the months that followed. Abuse of newly-found power by those long held in servility describes it. And it has been an extremely dangerous thing to organized labor since it amounts to the indiscriminate calling of unsanctioned strikes and lends strength to the baronial accusation that American labor unions are not responsible organizations able to control their membership. The union man's way of looking at it, of course, is that the gum manufacturers could free themselves of the threat of sit-downs by granting recognition to the unions and entering into signed agreements with them. Second of the major sit-downs hit Goodyear immediately on the heels of the Firestone settlement. It was in protest of a 10% cut in piece work rates affecting the pit and was abandoned after one night in favor of regular action through the union. In a few days there was one at Goodrich which closed down the plant, affecting 8,500 production and maintenance employees. It originated in a dispute over the time allowance given the Bedaux serfs for changing from one kind of machine to another. This dispute was settled in a lengthy conference between workers' committee and company officials, but the sit-down went on as the men demanded pay for the time lost in the original squat. It was called off after 24 hours with the initial protesting department receiving no compensation but part time payment being made in other departments where idleness had been enforced. The next passive resistance campaign struck Goodyear on Friday February 16th, and involved 140 tire builders protesting the laying off of 60 of their comrades. It soon tied up the whole of plant two, employing 1,800 men in four shifts. The company took the attitude that the original 140 had quit and announced that their jobs were gone. With the plant emptied for the weekend and scheduled to reopen Tuesday morning, Goodyear on Monday announced that it would rehire the 140 individually and would postpone putting the original layoffs into effect until further study had been given to production possibilities. A tire builder committee countered with a demand for three days' notice in cases of layoffs. Management refused and certainly without expecting that this meant it had a strike on its hands. It had one. Erupting as spontaneously as those workers who had set the General Tire conflict going, Goodyear local at its Monday night meeting voted strike and chased pickets out into the near zero weather to cover the plant's two gates. By Wednesday the three great Goodyear plants were tightly bottled with chain picketing going forward at all gates and too few men inside to start operations. Among the 1,000 executives, company policemen, maintenance workers and office workers inside was Paul W. Litchfield. He and the other barons had been more than fully prepared for that non-materializing March 1935, strike after the union had insisted on serving notice on them for many weeks. Wholly unprepared for this one was Litchfield, however, and never able to get the wheels turning, although he tried.

To cover all the gates and railroad tracks leading into Goodyear, the union had to scatter picket squadrons at intervals over an e1even mile front. It accomplished that. Fires blazed up in metal drums, shelters and wind-breaks were whacked together from two-by-fours and canvas and the men dug in to defy the zero blasts. Inside the gates Goodyear's movie cameraman turned his lens on the picket crowds, grinding out reel after reel of film to be flashed on a screen for company officials so that they might identify faces for future reference. For 11 to 12 days the 1,000 men stuck it out inside the buildings with successive mimeographed issues of the usually chaste and printed Wingfoot Clan house organ growing bawdier and bawdier as the confinement told. There was no bar to their leaving, but the idea that held them so long was that the plants were going to be opened up any minute.

Hardly was the strike two days old before the company sued to restrain the union from blocking the gates and "engaging in other unlawful acts." All six common pleas judges of Summit county ganged up to hear this one, no one nor three of them being willing to take the rap. Their mass decision handed down on February 22nd enjoined mass picketing and was effective on all union men as soon as they heard of it. The order never was enforced, the picket blockade never was broken and the judges never said anything about this contemptuous stuff. There were, of course, some attempts at enforcement but very fluttery and timid they were. With coppers in the ranks in sympathy with the strikers, with Mayor Lee Schroy and Police Chief Frank Boss deciding that it was wise to have no part in this business, Sheriff Flower and force marched out to Goodyear to make arrests. Strikers and sympathizers swarmed into the street to shout defiance at him and he marched right back again. Between March, 1935, when he was so busy preparing for action and this February of 1936, the military sheriff had had a taste of first hand strike work during the Columbia Chemical Co., battles at Barberton. So now he got on the telephone and vainly bellowed for the governor to send in troops. This within a few hours he would deny, admit, and deny again but the fact remains that he did just that. Also the anguished fellow appealed to all those judges, but they referred him to the county prosecutor and the prosecutor, having nervously observed the army before Goodyear, had no proposals on tap. And Goodyear remained closed.

There was, of course, the usual formation of a Citizens' Committee by civic leaders thinking along 1913 lines, but on reflection it found it advisable to confine its activities to a couple of full page newspaper advertisements bemoaning the collapse of law and order implicit in the continuing of picketing. Litchfield took to the radio and broadcast similar anguished cries about the "defiance of duly constituted government" from within the Goodyear plant. And one of the great scarlet violences of the strike occurred when a couple of deputies, unaccompanied by their sheriff, were jostled and wrestled to the ground as they made feeble attempt to clear the way for entrance of a freight car of crude rubber into the surrounded factory. The other was when strikers and coppers harmlessly pushed one another around as the flatfeet sought to remove picket shelter eye-sores from the street in none too strenuous compliance with orders from higher up.

Government mediators shuttled back and forth between the two camps, eminent citizen after eminent citizen was paraded to the microphone by Goodyear, and there were offers, counter-offers and refusals and the setting up of an organization of non-striking Goodyear employees who appealed to the sheriff and then to the prosecutor and then to the governor to take them to work. On March 2nd — a week after the strike had been made "official" by the AFL — Litchfield came out of hibernation, dug back in again at a downtown hotel and sent an attorney to demand that the dolorous sheriff set a date for re-opening the plant. The sheriff didn't think much of this idea, so the non-strikers' committee tried it on the mayor and city council — with equal success. Whereat loud threats of political extermination were hurled by Goodyear. And to the tune of the familiar music about "outside agitators" being responsible for everything, a Red-hunting ex-mayor launched a Law and Order League to force opening of the plant and was put on the air to yap appeals to "gang up for constituted law and order" and chase the "radical leaders" out of town. These bids died with a complete thud as did the duplicates provided by a selected evangelist. Goodyear now flatly announced a decision to force its plants open. Officialdom unanimously agreed that this would mean bloodshed and unanimously decided to remain neutral.

And still protesting the "breakdown of government," Goodyear finally got together with the union and a peace was concluded March 22nd, ending one of the most remarkable strikes in American labor history. Affecting some 14,000 to 16,000 Goodyearites, it has gone down into 'the records as perhaps the largest labor war ever carried on without violence. No heads were clubbed, no shots were fired, no tear bombs were hurled on the picketing scene and away from the battleground the sum total of carnage was a fist fight or two and the heaving of bricks through the windows of two "loyal" workers' homes. Yet if the company ever had prevailed upon officialdom to attempt forcing open its factories, one of the bloodiest clashes in industrial history undoubtedly would have resulted.

None of these officials, it is interesting to note, is a labor sympathizer. Each is or rather was as much a friend of the "better element" as were those Akron officeholders of 1913 who ostentatiously lined up with the factories and cracked down on the strikers with every nightstick, ordinance and proclamation at their command. Two things deterred 1936 officialdom from following suit: the temper of the strikers and the general attitude of the public at large. In part, the Akron public's inclination towards the workers was based on disgust with the often denounced price cutting wars of rubber and the fact that labor long had been sweated to pay for them. In greater part it was due to the superior caliber of the publicity with which the strikers appealed to them. In the past, labor, suspicious and secretive, invariably had lost out in this direction. In 1936, with the aid of Powers Hapgood, Adolph Germer, McAlister Coleman and other publicists and organizers sent in by John L. Lewis, labor steadily scored by presenting the basic issues of speedup and lack of security, while its opponents concentrated on shooting the outmoded Red scare stuff that no one took seriously.

Aides were sent, too, by Green, who long before had been cured of the sulks as he thought of what they might cost him. And both AFL and the Lewis Committee for Industrial Organization came through with coin in amounts of $2,000 and $3,000 at a time. As IWW and AFL had disputed for the soul of Everygummer in 1913, so now these successors contested for favor. Lewis, as we have said, stands on part of the IWW platform of the old days, and Green, it may be noted, has advanced to a point where he is at least up with the 1913 Lewis. The Lewis deputies stayed on in Akron long after the strike, incidentally, and left only as John L. began to lay his plans for trying to unionize steel. In that the rubber worker international is aligned with him, its executive committee acting in July 1936, to approve affiliation with the Committee for Industrial Organization. Which means that the rubber group is liable to any suspension or other "punishment" meted out to the United Mine Workers by the AFL.

Of the temper of the Goodyear strikers we have made mention. This is the one thing that provided the greatest contrast of all with the 1913 affair affecting approximately the same number of men. Not in 1936 was there any of the vagueness and uncertainty of the earlier protester who knew only that there had been a pay cut in a Firestone department but who had no clear idea of what he wanted. The new striker knew not only that there had been a layoff dispute in a Goodyear department touching things off, but knew exactly the things he wanted now that conflict was joined. And he was very earnest, very determined and very businesslike about it. Song and horseplay there was to cheer the long hours on the icy picket lines, but no trace of the rollicking attitude towards the strike as a whole that marked the 1913 jamboree.

Two decades of a constantly increasing speed-up have left none of that' carefree spirit. The iron has entered into the gummer soul.

That the Goodyear strike produced no sympathetic walkouts at other plants was due to President Dalrymple of the Rubber Workers International, who specifically asked that no such moves be made. And by concentrating on one factory the strikers were able to hold the whiphand from start to finish and achieve by far the best settlement ever obtained in a rubber walkout. It provided for giving notice of layoffs — the denial of which had produced the strike. It established the six-hour shift for the tire and tube division and provided that any change in this would have to be arranged for by vote of the employees in the departments affected. It provided for giving notice of changes in wage rates. And there were a number of other written concessions. Verbally the tire company's attorney agreed to a variety of provisions setting up collective bargaining machinery. These gave the union shop committees opportunity to deal with their foremen during working hours, gave the chairman of each division committee the right to enter into any department in his division to investigate complaints and effect settlements, gave the president of the local the right to accompany any committee conferring with personnel or factory heads on any matters in dispute. This was by all odds the farthest the union ever had got its foot into the Goodyear door. Within a week, Litchfield in a message to all Goodyear employees assured them that "special concessions" had been made to no group and pointedly stated that the complete terms on which the strike was settled were the written ones. This aroused no spirit of jolly good fellowship. Nor did an attempt to oust Akron mayor and police chief made through a non-striker group. Nor, later on, the Goodyear backing of county political candidates in an attempt to smack down Sheriff Flower and the other officials who had failed it (the Goodyear candidates were snowed under). Inside the plant there was plenty of bitterness without addition of any of this fuel. Strikers were refusing to work with "scabs," non-strikers were burning up the strikers with much sounding off over what their rewards would be, employees sworn in as special deputies during the strike still were swaggering about with guns on hips. Between March 24th and May 25th just 15 sit-downs ranging from a few minutes to 24 hours were recorded. And there were two mob scenes. On one occasion three of the ex-strikers allegedly worked out on the leader of the organization of non-strikers and a henchman after 25 tirebuilders brandishing the "tomahawks" of their trade had chased the two the length of plant two. On the other occasion, union employees in plant two allegedly "kidnapped" 20 non-union members, including three supervisors, and held them prisoners in an improvised pen while they took possession of the factory. Twenty-eight affidavits charging violation of the Ohio riot law were sworn out in connection with the second incident, three assault and battery affidavits in connection with the first one. One accused rioter has been tried at the date of writing. The jury disagreed and was dismissed.

A very interesting development of early May was a teamwork revival of the timeworn factory threat to move out of Akron. Most of this was horse feathers, some of it was straight stuff. No hokum was there about Goodyear's stepping up Gadsden production from 6,000 to 8,000 tires a day and starting the manufacture of inner tubes there, and none about Goodrich's starting the remodeling of its former reclaiming plant at Oak near Philadelphia which will go into production this fall at the rate of 5,000 tires a day. Akron lost by these moves, but immensely less of course than when Akron companies originally built at Gadsden and Los Angeles and transferred entire departments to Chicopee Falls — all in the days before labor trouble. Also likely to eventuate are Goodrich and General announced plans to build small mechanical goods factories outside Ohio. This is a logical business development aside from any labor angle. Which brings us to the pure hooey talk of complete decentralization, abandonment of Akron and so on. For one thing, it is nonsense to believe that any of these companies ever would abandon buildings which have cost them millions. For another thing, these debt-burdened corporations could not afford to erect such mammoth plants elsewhere even if they were foolish enough to want to. For another, decentralization by the picking up of existing bust or going bust plants around the country can be carried just so far without getting into inefficient operation and excessive losses as the classic case of US Rubber so completely demonstrates. Akron factories will continue to build tires where it is most economical to do so and for the most part this means at Akron despite higher wages there. An increase in small plants elsewhere to take advantage of freight rates there well may be, and a corporation in Goodyear's situation always can score on labor by keeping Cumberland, Gadsden and Los Angeles going at maximum while holding Akron at minimum during any given period of trouble in the Ohio city. Granted, that is, that these other cities aren't organized to the degree currently prevailing at Akron.

This is recognized, of course, by United Rubber Workers' leaders and they recently launched a campaign to raise $100,000 for organizing activities in every caoutchouc manufacturing city of the country. To what degree they ever can organize fascist Los Angeles is, of course, open to question. As to vigilante-owned Gadsden, the record so far is one of utter failure. Beating up of native union men and such visiting dignitaries as President Dalrymple of the Rubber Workers International, President John House of the Goodyear Akron local, organizer George W. Roberts of the Rubber Workers and numerous others is the current favorite pastime in the Alabama community.

These "malicious and brutal assaults" have been "encouraged and acquiesced in" by Goodyear according to a complaint issued on July 11th by the National Labor Relations Board. In addition to hanging these terrorism charges on the company, the document cites it for discharging Gadsden employees on the count of union affiliation, attempting to coerce its Akron employees by circulating decentralization threat "propaganda," dominating and financing the Industrial Assembly at Akron and encouraging the formation of the recently organized Stahl-Mate club of Goodyear Akron employees "which, upon information and belief, is militaristic in its tendencies and has indulged in terrorist activities." Modeled after Federal Trade Commission procedure are these labor board citations and the penalty is a "cease and desist" order backed by untried fine or imprisonment provisions in event of non-compliance. Hearing on the charges was set by the board for July 21st at Akron but the company attacked the constitutionality of the National Labor Relations Act in the District of Columbia Supreme Court and won an order temporarily enjoining the hearing. There the matter stood as this was written.

In Akron it is possible that the biggest showdown of all will come with the layoffs this fall as the heavy production season ends. Litchfield, commenting on termination of the Sears contract, already has intimated that this means a considerable employee reduction at Akron.

And aside from this development, town talk is that off-season production in all of the Rubber City's factories will be down as never before, with all of the union redhots being laid off and never rehired. This well may be sheer moonshine, but it offers interesting possibilities of a first-rate war. Popularity king in Akron at the moment, oddly, is none other than Harvey Firestone, long the most unpopular by reason of such broadcasts as his depth-of-the-depression classic to the effect that "anyone who wants a job bad enough could go out and get one." While Goodrich, Goodyear, General were doing all the heavy breathing about abandoning the Rubber City to the chipmunk and the hoot-owl, Harvey ostentatiously refrained from joining in. And somehow the word got around that he not only would like to see the other fellows decentralize so that he could whipsaw hell out of them through his centralized control, but that he actually is about to start throwing up a new mechanical rubber goods building in Akron. The factory hasn't materialized, but Firestone still is profiting by the tale. And by his May 5th action in accompanying an industry-wide 10% boost in tire prices with a voluntary 5% to 10% wage tilt for all factory employees, first time in casing history these things occurred simultaneously. All the other companies had to follow suit and with Goodrich very much incensed since the price hoist affected tires only and the extra pay had to be handed producers of all its vaunted 30,000 articles. And in addition to losing money on the deal where tire specialist Harvey was making it, Goodrich ran into the situation of the new factory hoist being the last straw in the case of the extremely underpaid office employees. Now the white collar boys are swarming to join, not clerical unions, but the actual rubber workers unions, and management is very much perturbed.

Firestone's one major sit-down of recent months came right after his wage increase and apparently was designed to show him that it takes more than a raise to call off the union. Goodyear's June ones protested the assault on Dalrymple at Gadsden and, of all things, the burning of three huge crosses in a field in view of fourth shift pitmen as they came back to work after the 24 hour Dalrymple protest. The men darkly charged that this stuff was the work of one of the anti-strike societies. Goodyear has had three more sizable sit-downs since then and Goodrich one. Additionally, there have been scores of minor ones never heard of outside the factories. At Firestone, currently 90% organized by the union, the sit-downs have been settled as fast as they come up by the skillful diplomacy of W.R. Murphy in charge of labor relations. Harvey, although lusting to break out and address the men personally, is wisely refraining, and the gifted Murphy is the only official in the whole organization who is permitted to handle his own work without any sort of interference by the ubiquitous board chairman. At Goodrich, the other union stronghold, peace has been preserved by the company's giving in to the dictatorial gummers in every slightest controversy. Here it was that the sit-down reached the ultimate ridiculous heights when one was staged by a softball team in protest against the company recreational department's assigning a nonunion umpire to officiate at one of its clashes. Seriously, Goodrich foremen virtually have been powerless for months and the corporation is unable even to make changes in any of its compounds. The workers, suspecting or pretending to suspect that any alterations in the batches mean extra labor for them, are insisting that there be no changes, and technicians have even quit recommending them. That this situation will prevail beyond the current season of maximum production demand is not anticipated.

Goodyear's most recent major tie-up — July 13th — sent that company's officials to city hall where, speaking for all of Akron's caoutchouc corporations, they put the ultimate pressure on the mayor. Promptly the next day, Hon. Lee D. Schroy issued an announcement that there would be no more sit-downs, that the Goodyear factories would be kept running at all costs and that police had orders to "go in and clean out the plants as soon as Goodyear officials see fit to call us." He has not yet had to make good on this inspired statement, the Goodyear union itself having adopted a policy of discouraging any and all unauthorized sit-downs in favor of united action as an organization whenever that becomes necessary. Prior to the descent on Schroy, Goodyear's course had differed from Goodrich's only in that it was making its concessions to end major sit-downs instead of to head them off. In this connection, Litchfield frankly stated, "In most instances resumption of production has been accomplished only by substantial concessions on the part of the management in the interest of peace and continuing production during the present peak period." In that there is more than an implication that things will be different as the dull season arrives. And the union worker, as we have seen, is at least equally determined that concessions shall continue to be made and in even more substantial form. Rubber's next historian is going to have a far longer labor story to tell.

SELECTED BIBLIOGRAPHY

BOOKS

C. F. Akers, *The Rubber Industry in Brazil and the Orient*, London, 1914.

H. Allen, *The House of Goodyear*, Akron, 1936.

Petro Martyre d'Anghiera, *Decades of the New World*, (Translated by M. Lok), London, 1582. [In Volume 5 of Hakluyt's *Divers Voyages*, etc.)

H.W. Bates, *The Naturalist on the River Amazons*, London, 1875.

N. Beasley, *Men Working, A Story of The Goodyear Tire & Rubber Co.*, New York,1931.

C.W. Bedford, and H.A. Winkelmann, *Systematic Survey of Rubber*

Chemistry, New York, 1923.

T. Belt, *The Naturalist in Nicaragua*, London, 1874.

D.C. Boulger, *The Reign of Leopold II*, Two volumes, London, 1925.

D.C. Boulger, *The Congo State*, London, 1898.

H. Brown, *Rubber, Its Sources, Cultivation and Preparation*, London, 1914.

J.M. Brown, *The Dutch East*, London, 1914.

R.L. Buell, *The Native Problem in Africa*, Two volumes, New York, 1928.

C. Christy, *The African Rubber Industry*, London, 1911.

F. Clouth, *Rubber, Gutta-Percha and Balata*, London, 1903.

F.S. Cocks, *E.D. Morel: The Man and His Work*, London, 1920.

A Commercial Handbook of the Netherland's East Indies, London, 1927.

C.M. de la Condamine, *A Succinct Abridgment of a Voyage made within the Inland Parts of South America; from the Coasts of the South-Sea to the Coasts of Brazil and Guiana, down the River of Amazons*, London, 1747.

L. Couperus, *Eastward*, New York, 1924.

C.L. Johnston in *One Hundred Years of American Commerce*, Volume 2, Ed., C.M. Depew, New York, 1895, p.498-504, *American Rubber Manufactures.*

E. Descamps, *New Africa*, London, 1903.

O. De Vries, *Estate Rubber, Its Preparation, Properties and Testing*, Batavia, 1920.

Centennial History of Summit County, Ohio and Representative Citizens, Ed., W.B. Doyle, Chicago, 1908.

A. Dubosc and A. Luttringer, *Rubber, its Production, Chemistry, and Synthesis in the Light of Recent Research*, London, 1918.

H.O. Duncan, *The World on Wheels*, Paris, 1926.

J.B. Dunlop, *The History of the Pneumatic Tyre*, Dublin, 1924.

L.E. Elliott, *Brazil Today and Tomorrow*, New York, 1917.

H. Fetzer, and others in *A Centennial History of Akron 1825-1925*, Akron, 1925.

H.S. Firestone and S. Crowther, *Men and Rubber*, New York, 1926.

H.A. Franck, *Vagabonding Down the Andes*, New York, 1917.

W.C. Geer, *The Reign of Rubber*, New York, 1922.

A. Gide, *Travels in the Congo*, Translated by Dorothy Bussy, New York, 1929.

A.C. Grimley in *The Development of American Industries, Their Economic Significance*, Eds., J.G. Glover, and W.B. Cornell, New York, 1932, p.227-244, The Rubber Industry.

C. Goodyear, *Gum-Elastic and Its Varieties*, New Haven, 1853.

K. Gottlob, *Technology of Rubber*, Translated by J.L. Rosenbaum, London, 1927.

K.G. Grubb, *Amazon and Andes*, London, 1930.

D. Gwynn, *Traitor Or Patriot: The Life and Death of Roger Casement*, New York, 1931.

T. Hancock, *Personal Narrative of the Origin and Progress of the Caoutchouc or India-Rubber Manufacture in England*, London, 1857.

P.W. Hansl, *Years of Plunder*, New York, 1935.

W.E. Hardenburg, *The Putumayo, The Devil's Paradise*, London, 1912.

E.A. Hauser, *The Colloid Chemistry of the Rubber Industry*, London, 1928.

E.A. Hauser, *Latex, Its Occurence, Collection, Properties and Technical Applications*, Translated by W.J. Kelly, New York, 1930.

A. de Herrera, *The General History of the Vast Continent and Islands of America*, Translated by J. Stevens, London, 1725.

G. Iles, *Leading American Inventors*, New York, 1912, p.176-217, Charles Goodyear.

W.H. Johnson, *Cultivation and Preparation of Para Rubber*, London, 1909.

H. Johnston, *Liberia*, Two volumes, New York, 1906.

C.F. Jones, *South America*, New York, 1930.

Akron and Summit County, Ohio, Three Volumes, Ed. S.D. Kenfield, Chicago and Akron, 1928.

D.P. Kidder and J.C. Fletcher, *Brazil and the Brazilians*, Philadelphia and Columbus, 1857.

G. Oenslager in *Twenty- Five Years of Chemical Engineering Progress*, Ed., S.D. Kirkpatrick, New York, 1933, p.174-194, Chemical and Engineering Advances in the Rubber Industry.

S.A. Lane, *Fifty Years and Over of Akron and Summit County*, Akron, 1982.

A. Lange, *The Lower Amazon*, New York, 1914.

J.C. Lawrence, The World's Struggle with Rubber, New York, 1931.

F.E. Lloyd, *Guayule, A Rubber Plant of the Chihuahuan Desert*, Washington, 1911.

R.H. Lock, *Rubber and Rubber Planting*, New York, 1913.

B.D.W. Luff, *Chemistry of Rubber*, London, 1923.

G. MacCreagh, *White Waters and Black*, New York, 1926.

F. Marchionna, *Latex and Its Industrial Applications*, New York, 1933.

R.O. Marsh, *White Indians of Darien*, New York, 1934.

The Science of Rubber, Ed., K. Memmler, Translated by R.F. Dunbrook, V. N. Morris, and others, New York, 1934.

E.D. Morel, *Red Rubber*, Revised edition, Manchester, 1919.

E.D. Morel, *Great Britain and the Congo*, London, 1909.

E.D. Morel, *Black Man's Burden*, London, 1920.

S. Morgan and H.P. Stevens, *The Preparation of Plantation Rubber*, London, 1927.

V. Olarte Camacho, *Las Crueldades de los Peruanos en el Putumayo y en el V Caqueta*, Bogota, 1932.

J. Parton, *Famous Americans of Recent Times*, Boston, 1881, p.309-346, Charles Goodyear.

G.S. Paternoster, *The Lords of the Devil's Paradise*, London, 1913.

R. Paterson, *Memorials on the Life of James Syme*, Edinburgh, 1874.

H.C. Pearson, *What I saw in the Tropics*, New York, 1906.

H.C. Pearson, *The Rubber Country of the Amazon*, New York, 1911. H.C. Pearson, *Crude Rubber and Compounding Ingredients*, New York, 1920.

H.C. Pearson, *Rubber Machinery; An Encyclopedia of Machines used in Rubber Manufacture*, New York, 1920.

B.K. Peirce, *Trials of an Inventor: Life and Discoveries of Charles Goodyear*, New York and Cincinnati, 1866.

C.J. Post, *Across the Andes*, New York, 1912.

H.E. Potts, *Chemistry of the Rubber Industry*, London, 1920.

E.E. Pratt, *International Trade in Staple Commodities*, New York, 1928, p.173-238.

J. Priestley, *A Familiar Introduction to the Theory and Practice of Perspective*, London, 1770.

M. Romero, *Coffee and India-Rubber Culture in Mexico*, New York, 1898.

T. Roosevelt, *Through the Brazilian Wilderness*, New York, 1914.

P. Schidrowitz, *Rubber*, London, 1911.

P. Schidrowitz, *Recent Progress in Rubber Chemistry and Technology*, London 1922.

S.P. Schotz, *Synthetic Rubber*, London, 1926.

T. Seeligmann, G.L. Torrilhon, G. L. and H. Falconnet, *India Rubber and Gutta Percha*, Translated by J.G. McIntosh, London, 1910.

H.E. Simmons, *Rubber Manufacture*, New York, 1921.

P. Stamberger, *The Colloid Chemistry of Rubber*, London, 1929. H.M. Stanley, Through the Dark Continent,Two Volumes, New York, 1898.

H.M. Stanley, *The Congo and the Founding of its Free State*, Two Volumes, New York, 1885.

H.P. Stevens and C. Beadle, *Rubber, Production and Utilization of the Raw Product*, London, 1915.

H.L. Terry, *India Rubber and Its Manufacture*, New York, 1907. J. de Torquemada, *De la Monarquia Indiana*, Volume 2, Madrid, 1615.

J.B. Tuttle, *The Analysis of Rubber*, New York, 1922.

F.W Up de Graff, *Head Hunters of the Amazon*, New York, 1923. H.W. Wack, *The Story of the Congo Free State*, New York, 1905. B.B. Wallace and L. Ramsay, *International Control of Raw Materials*, Washington, 1930 p.172-218, British Export Restrictions on Rubber.

P.W. Litchfield in *Representative Industries in the United States*, Ed., H.T. Warshow, New York, 1928, p.574-594, Rubber.

C.O. Weber, *Chemistry of India Rubber*, London, 1903.

L.E. Weber, *The Chemistry of Rubber Manufacture*, London, 1926. G.S. Whitby, *Plantation Rubber and the Testing of Rubber*, London, 1920.

W. Wicherley, *The Whole Art of Rubber-Growing*, London, 1911.

H.A. Wickham, *Rough Notes of a Journey Through the Wilderness*, London, 1872.

H.A. Wickham, *On the Plantation, Cultivation, and Curing of Para Indian Rubber*, London, 1908.

J.F. Woodroffe, *The Upper Reaches of the Amazon*, New York, 1914.

J.F. Woodroffe and H.H. Smith, The Rubber Industry of the Amazon, London, 1915.

Ed., A. Wright, *Twentieth Century Impressions of British Malaya*, London, 1980.

H. Wright, *Rubber Cultivation in the British Empire*, London, 1907.

H. Wright, *Hevea Brasiliensis, or Para Rubber: Its Botany, Cultivation, Chemistry and Diseases*, London, 1912.

J.C. Young, *Liberia Rediscovered*, New York, 1934.

PAMPHLETS

D.M. Beights, *Financing American Rubber Manufacturing Companies*, Urbana, 1932.

D.C. Boulger, *The Congo State is NOT a Slave State*, London, 1903.

R.L. Buell, *Forced Labor: Its International Regulation*, New York, 1930.

R.L. Buell, *The Reconstruction of Liberia*, New York, 1932.

A.J. de Souza Carneiro, *Rubber in Brazil*, Rio de Janeiro, 1913.

A Complete Congo Controversy, Illustrating the Controversial Methods of Mr. Morel, Edinburgh, 1905.

C.M. Cumming, *Rubber Planting in Malaya*, London, 1919.

C.F. Dawn, *The Future of Rubber: Whither and How?* London, 1934.

Decision in the Great India Rubber Case of Charles Goodyear versus Horace H. Day, New York, 1852.

H.S. Firestone, *America Should Produce its own Rubber*, Akron, 1923.

H.S. Firestone, *The World Rubber Crisis*, Akron, 1923.

H.S. Firestone, Jr., *The Romance and Drama of the Rubber Industry*, Akron, 1932.

The Florida Rubber Culture Company, *The Gold which Grows on the trees in California of South America, to be transplanted to South Florida*, Pittsburg.

H.M. Hall and T.H. Goodspeed, *A Rubber Plant Survey of Western North America*, Berkeley, 1919.

H.M. Hall and F.L. Long, *Rubber Content of North American Plants*, Washington, 1921.

F.R. Henderson, *Rubber, Its Production and Marketing*, New York, 1926.

W. Koren, Jr., *Liberia, The League and the United States*, New York, 1934.

J.C.G. Kunhardt, *The Future of Rubber*, London, UK, 1930.

J.A. McLaughlin, *Charles Goodyear of Woburn, Massachusetts*, Woburn.

J.A. Nieuwland, *The Story of Synthetic Rubber*, Proceedings of the Indiana Academy of Science, 1935.

E.W. Perry, Rubber: The Handmaid of Civilisation, Naugatuck, 1903.

The Petition of Charles Goodyear, Jr., Executor for the Extension of Letters Patent, granted to Charles Goodyear, deceased, New York, 1864.

Petition to US Senate and House of Representatives from Manufacturers, Dealers & Workmen engaged in the Business of Fabricating India Rubber, in Reference to the Present Defective Patent Laws and the Recent Unlawful Re-issue of a Patent to Charles Goodyear, Washington, 1850.

A. Phillipson, The Rubber Position and Government Control, London, 1924.

B.D. Porritt, Chemistry of Rubber, London, 1913.

B.D. Porritt, The Early History of the Rubber Industry, London, 1927.

W.A. Reid, *Rubber, Commodities of Commerce Series No.15,* Pan American Union, Washington, 1932.

Report of the Conference of Rubber, Automotive and Accessory Manufacturers, Washington, DC, February 27th 1923, Akron, 1923.

Rubber Uses and Their Development, London, 1923.

W.E. Simpson, *Rubber: Its Commercial and Financial Development from the tree to Payment of Dividends Upon Capital Invested*, New York, 1900.

A.M. Smith, The Ford Detroit Project, Detroit, 1928.

A Statement upon Questions Relating to the Salaries, Emoluments and the General Conditions and Terms of Service of Planters on Rubber Estates and other Plantations in British Malaya, The Incorporated Society of Planters,Kuala Lumpur, Federated Malay States, 1934.

H.P. Stevens and M.B. Donald, *Rubber in Chemical Engineering*, London, 1933.

H.P. Stevens and W.H. Stevens, *Rubber Latex*, London, 1933.

The Story of Goodyear, Akron, 1925.

The Story of the Tire, Akron, 1929.

D. Webster, *Speech of the Hon. Daniel Webster in the Great India Rubber Suit, Heard at Trenton, New Jersey, in March, 1852*, New York, 1852.

What a Business Man wants to know about the Goodyear-Sears Contract, Akron, 1934.

What is Happening in Akron, Akron, 1936.

A Wonder Book of Rubber, Akron, 1932.

H. Wright, *My Tour in Eastern Rubber Lands*, London, 1908.

GOVERNMENT DOCUMENTS AND PUBLICATIONS

Angola, Handbooks Prepared Under the Direction of the Historical Section of the Foreign Office, No.120, London, 1920.

P.W. Barker, *Rubber: Some Facts on Its History, Production and Manufacture*, United States Department of Commerce, Bureau of Foreign and Domestic Commerce, Washington, 1936.

W. Bowden, *Employment Hours, Earnings, and Production, January 1933 to January 1935*, United States Department of Labor, Bureau of Labor Statistics, Serial No. R.222, Washington, 1935.

British Colonial Reports on the Rubber Situation, United States Department of Commerce, Burean of and Domestic Commerce, Trade Information Bulletin No.603, Washington, 1929.

British Government Blue Books:

Correspondence Respecting the Treatment of British Colonial Subjects and Native Indians Employed in the Collection of Rubber in the Putumayo District, Miscellaneous No.8, 1912, Cd. 6266), London, 1912.

Report and Special Report from the Select Committee on Putumayo, Together with the Proceedings of the Committee, Minutes of Evidence and Appendices, London, 1913.·

Government White Papers:

Correspondence and Report from his Majesty's Consul at Boma Respecting the Administration of the Independent State of the Congo, Africa, No.1, 1904, Cd, 1933, London, 1904.

Further Correspondence Respecting the Administration of the Independent State of the Congo, Africa, No.7, 1904, Cd, 2097, London, 1904.

Further Correspondence Respecting the Taxation of Natives, and other Questions in the Congo State, Africa, No.1, 1909, Cd, 4466, London, 1904.

Report by his Majesty's Consul at Iquitos on his Tour in the Putumayo District, Miscellaneous No.6, Cd. 6678, London, 1913.

Census of Manufactures (1914, 1925, 1927, 1929, 1931: The Rubber Industries), United States Department of Commerce, Bureau of the Census, Washington, 1918, 1927, 1929, 1931, 1933.·

Ceylon: its Industries, Resources, Trades and Methods of doing Business, United States Department of Commerce, Trade Information Bulletin No.601, Washington, 1929.·

O.F. Cook, *Rubber Cultivation for Porto Rico*, United States Department of Agriculture, Division of Botany, Circ. 28, Washington, 1900.

O.F. Cook, *Culture of the Central American Rubber Tree*, United States Department of Agriculture, Washington, 1903.

O.F. Cook, *Rubber Possibilities of Many Kinds Exist in the United States*, United States Department of Agriculture, No.993, Washington, 1927.

Digest of Material on Technological Chnages, Productivity of Labor, and Labor Displacement, United States Department of Labor, ?Bureau of Labor Statistics, Washington, 1932.

E.A. Dunn, *A Commercial and Industrial Handbook*, United States Department of Commerce, Bureau of Foreign and Domestic Commerce, Trade Promotion Series No.25, Washington, 1925.

D.M. Figart, *The Plantation Rubber Industry in the Middle East*, United States Department of Commerce, Bureau of Foreign and Domestic Commerce, Trade Promotion Series No.2, Washington, 1925.

Foreign Combinations to Control Prices of Raw Materials, United States Department of Commerce, Bureau of Foreign and Domestic Commerce, Trade Information Bulletin No.385, Washington, 1926.

J.A. Fowler, *Plantation Rubber in the Netherlands East Indies and British Malaya*, (United States Department of Commerce, Bureau of Foreign and Domestic Commerce, Trade Information Bulletin No.27, Washington, 1922.

Netherlands East Indies and British Malaya, United States Department of Commerce, Special Agents Series No.218, Washington, 1923.

A. Hamilton, *Industrial Poisons Used in the Rubber Industry*, United States Department of Labor, Bureau of Labor Statistics, Bulletin No.179, Washington, 1915.

E.G. Holt, *Marketing of Crude Rubber*, United States Department of Commerce, Bureau of Foreign and Domestic Commerce, Trade Promotion Series No.55, Washington, 1927.

International Commission of Enquiry in Liberia, Communication by the Government of Liberia, Transmitting the Commission's Report, League of Nations, Official Report No.C.658, M.272, VI, Geneva, 1930.

C.D. La Rue, *The Hevea Rubber Tree in the Amazon Valley*, United States Department of Agriculture, Department Bulletin No.1422, Washington, 1926.

L. Harrison, *Rubber Regulation and the Malayan Plantation Industry*, United States Department of Commerce, Bureau of Foreign and Domestic Commerce, Trade Promotion Series No.159, Washington, 1935.

Liberia, Documents Relating to the Plan of Assistance Proposed by the League of Nations, United States Department of State, Publication No.535, Washington, 1933.·

I. Lubin, *Prices of Rubber and Rubber Products*, United States War Industries Board, WIB Price Bulletin No.30, Washington, 1919.

A.T. McPherson, *Reclaimed Rubber*, United States Bureau of Standards Circular No.393, Washington, 1931.

The President of the United States, *Slavery in Peru*, Message transmitting report of the Secretary of State, with accompanying papers concerning the alleged existence of slavery in Peru, Washington, 1913.

Record Book of Business Statistics, Part 3: Fuels, Automobiles and Rubber, United States Department of Commerce, Bureau of the Census, Washington, 1929.

Report of a Committee Appointed by the Secretary of State for the Colonies to investigate and Report upon the Present Rubber Situation in British Colonies and Protectorates, Cmd 1678, London, 1922.

Rubber and Its Use in the United States, United States Department of Commerce, Bureau of Foreign and Domestic Commerce, Special Circular No.1111, Washington, 1932.

Rubber Industry Letters Nos.1-23, United States Department of Commerce, Bureau of Foreign and Domestic Commerce, Washington, 1933-1935. Schur2, L.: Bolivia: A

W.L. Schurz, *Bolivia: A Commercial and Industrial Handbook*, United States Department of Commerce, Special Agents Series No.208, Washington, 1921.

W.L. Schurz, and others, *Rubber Production in the Amazon Valley*, United States Department of Commerce, Bureau of Foreign and Domestic Commerce, Trade Promotion Series No.23, Washington, 1925.

B. Stern. *Labor Productivity in the Automobile Tire Industry*, United States Department of Labor, Bureau of Labor Statistics, Bulletin No.585, Washington, 1933.

Sumatra: Economic and Commercial Survey, United States Department of Commerce, Washington, 1927.

Supplementary Report of the Committee Appointed by the Secretary of State for the Colonies, to Investigate and Report upon the Present Rubber Situation in British Colonies and Protectorates, Cmd 1756, London, 1922.

L.E. Talbert and B.B. Herman, *Trend of Employment*, United States Department of Labor, Bureau of Labor Statistics, Serial No. R221, Washington, 1935.·

J.C. Treadwell, *Possibilities for Para Rubber Production in Northern Tropical America*, United States Department of Commerce, Bureau of Foreign and Domestic Commerce, Trade Promotion Series No.40, Washington, 1926.

United States Senate Documents:

Memorial Concerning Conditions in the Independent State of the Kongo, 58th Congress, second session, Document No.282, Washington, 1904.

Conditions in the Kongo State, 58th Congress, third session, Document No.102, Washington, 1905.

Alleged Conditions in the Kongo Free State, 59th Congress, first session, Document No.316, Washington, 1906.

Report on Senate Resolution No.267 in Regard to the Kongo Free State, Committee on Foreign Relations, Washington, 1907.

C.F. Vance and others, *Possibilities for Para Rubber Production in the Philippine Islands*, United States Department of Commerce, Bureau of Foreign and Domestic Commerce, Trade Promotion Series No.17, Washington, 1925.

Wages and Hours of Labor in the Automotive Tire Industry, United States Department of Labor, Bureau of Labor Statistics, Bulletin No.358, Washington, 1924.

H.N. Whitford and A. Alfred, *Rubber Production in Africa*, United States Department of Commerce, Bureau of Foreign and Domestic Commerce, Trade Promotion Series No.34, Washington, 1926.

MAGAZINE ARTICLES

L. Adamic, *Will Rubber Snap?*, The Nation, March 20th 1935.

J. Anderson, The Bee or Literary Weekly Intelligencer, March 1791.

Development of the Rubber Industry, Economic Review of the Soviet Union, October 1931.

F.F. Duncan, *Cyrus S. Eaton, New Lord of Steel*, Review of Reviews, May 1930.

L. Durtain, *World-Sap from Indusrialised Forests*, Asia, February 1931.

M. Faraday, *On Pure Caoutchouc and the Substances by which it is Accompanied in the State of Sap or Juice*, Quarterly Journal of Science and the Arts, London, 1826.

H.L. Foster, *Ghost Cities of the Jungle*, New York Herald Tribune, Sunday Magazine Section, March 20th 1932.

E.J. Glave, *Conditions in Central Africa*, The Century Magazine, April 1897.

E.J. Glave, *Cruelty in the Congo Free State*, The Century Magazine, September 1897.

E.G. Holt and W.F. Lockwood, *How International Rubber Restriction Came About*, Asia, June, 1935.

E.G. Holt and W.F. Lockwood, *The Rubber Control Scheme at Work*, Asia, July, 1935.

J. Howison, *Some Account of the Elastic Gum Vine, etc.*, Asiatic Researches, 1798, 5, 157.

A. Kudner, *A Defense of Tire Profits*, Printers' Ink, May 5th 1932.

J.C. Lawrence, *Pioneers in the Commercial Development of Rubber*, Journal of Chemical Education, August 1930.

W.M. Morrison, *Personal Observations of Congo Misgovernment*, Review of Reviews, July 1903.

W. Roxburgh, *A Botanical Description of Urecola elastica, or Caoutchouc Vine of Sumatra and Pullo-pinang etc.*, Asiatic Researches, 1798, 5, 167.

Rubber and Mr. Hoover, The Nation, January 1926.

G.S. Schuyler, *Uncle Sam's Black Step-Child*, American Mercury, June 1933.

US Rubber, Fortune, February 1934.

H. Whitaker, *Barbarous Mexico: Rubber Slavery in the Mexican Tropics*, American Magazine, February 1910.

B.J. Widick, *Two Defeats for William Green*, The Nation, October 9th 1935.

F.J. Williams, *A New Leader in Finance: Clarence Dillon*, Review of Reviews, February 1926.

A. Winton, *Get A Horse*, Saturday Evening Post, February 8th 1930.

H. Wolf, *The Rubber Barons Fight to the Death*, American Mercury, June 1931.

H. Wolf, *Bankers at the Bedside*, Purchasing, August 1932.

A. Woollcott, *Aladdin on Broadway*, Collier's, February 18th 1928.

E.M. Woolley, *Akron: Standing Room Only!*, McClure's, July 1917.

Index

A

B

C

D

E

L

M

N

O

P

S

T

U

V

W

Y

Lightning Source UK Ltd.
Milton Keynes UK
01 March 2010

150784UK00001B/34/P

9 781847 353757